# VASCULAR PLANT TAXONOMY

■ **Fifth Edition**

Dirk R. Walters

David J. Keil

■ California Polytechnic State University

Zack E. Murrell

■ Appalachian State University

■ Illustrations by

Bonnie K. Walters

KENDALL/HUNT PUBLISHING COMPANY
4050  Westmark  Drive  Dubuque,  Iowa  52002

Chairman and Chief Executive Officer   Mark C. Falb
Director of National Book Program   Paul B. Carty
Editorial Development Manager   Georgia Botsford
Developmental Editor   Tina Bower
Vice President, Production and Manufacturing   Alfred C. Grisanti
Assistant Vice President, Production Services   Christine E. O'Brien
Prepress Editor   Kimberly Terry
Permissions Editor   Colleen Zelinsky
Designer   Jenifer Chapman
Senior Vice President, College Division   Thomas W. Gantz

Printed in the United States of America
10 9 8 7 6 5 4 3 2 1

# Contents

■ Chapter 4

## Phylogenetic Classification     59

■ Chapter 5

## Floras, Manuals, and Botanical Descriptions     77

■ Chapter 6

## Artificial and Phenetic Systems of Classification     89

## Chapter 11

### Introduction to the Flowering Plants 187

## Chapter 12

### Early Evolution of Flowering Plants: Basal Angiosperms, Magnoliids, and Basal Eudicots 221

## ■ Chapter 15
### Asterids 359

## ■ Chapter 16
## Monocots 421

## Chapter 17
## Experimental Plant Systematics     473

## Chapter 18
## Revisions and Monographs     487

# Preface

The preparation of a text for an introductory plant taxonomy course is a daunting experience. Our goal has been to prepare a text that can serve as an introduction to taxonomy and that is thorough enough to provide a solid background for students, but not so detailed that students feel overwhelmed by the subject matter. There is a fine line between oversimplification and burdensome detail. We have attempted to establish a balance in our presentation of the topics. The subject matter of systematic botany is broad and ever-expanding. There have been many new developments in taxonomy in the past few years, especially in the explosive growth of molecular phylogeny. Many relationships have been called into question, and much of the higher classification of flowering plants is in a state of flux. We faced many difficult decisions regarding what to include and at what level.

This text, like its subject, is a product of evolution. The first two editions of *Vascular Plant Taxonomy* were written in the form of a study guide/laboratory manual and much potentially useful information was only mentioned or was left out entirely. This format served only minimally and we realized that a more thorough text would be useful both to us and to our students. Our survey of taxonomy texts indicated that most covered either plant families or experimental methods in taxonomy but generally not both. We redesigned the third edition to present an introductory survey of both of these important topics. We attempted to include enough information to provide students with a basic understanding of the topics and to enable and encourage students to seek more information in the many popular and technical sources that are available in libraries. To this end we included a set of selected references at the end of each chapter that can serve as sources of additional information for students and instructors alike. The present edition continues this format with updated coverage of many topics. In addition, we have included an exercise at the end of each chapter to assist students in organizing their studies. Some of these are designed for individual or class activities in the lab, some as outside projects, and some as study questions for individual review.

This new edition includes some changes in the format of this book. It is organized into three parts. The first three chapters involve an introduction to the study of taxonomy, vegetative terminology and collecting techniques. This portion emphasizes plant structure as a way to engage students in the wonders of plant form and function. Many students come to a taxonomy/systematics class having very little experience with plant diversity. Hopefully, these chapters provide an opportunity for the teacher to create excitement with an initial introduction to plants. The next three chapters (Chapters 4–7) provide an exploration of the classification systems that have been used in the past and the currently accepted system that we use to name plants and recover their evolutionary history. Chapters 8–15 include an introduction to plant diversity among the "seed free" ferns and fern allies, the gymnosperms and the flowering plants. Finally, Chapters 16 and 17 describe ways that we can document plant diversity through experiments, revisions and monographs. Two appendices offer the student 1) an artificial key to families of vascular plants and 2) a glossary to the terms used in the text.

The field of plant taxonomy/systematics has undergone tremendous change over the past twenty years. We have developed new methods to explore evolutionary relationships of plants and new kinds of data (especially molecular data) that can be used to shed some light on our understanding of plant relationships. These new methods and data are generating changes in our classification system and we have attempted to reflect some of these changes in the layout of the diversity chapters (Chapters 8–15). The system presented here generally follows the phylogeny advocated by the Angiosperm Phylogeny

Group. This is a departure from the Cronquist system used in previous editions. Although this new system is not as clear-cut as Cronquist's subclass system of classification, it does present the up-to-date concepts of relationships. Family circumscription was maintained from the previous edition. This was done to maintain some continuity between editions.

We do not discuss each family equally. We have chosen to emphasize certain families that are accessible to botanists in temperate regions and that are commonly encountered or that are considered to be evolutionarily significant. These important families are illustrated and have at least a full page discussion of morphology, distribution and economic importance. Families of secondary importance are presented with their technical descriptions, size, and distribution. Those families deemed of least importance are listed by name with only size and distribution.

The family descriptions that are included in this book have been prepared with a standardized format and a limited technical vocabulary. The descriptions have been deliberately written without geographical limits and include variations that are not necessarily encountered in representatives of the families growing in growing in a particular area of North America. This inclusiveness was designed to avoid the problem of locally biased descriptions that are often a feature of regional floras. Cultivated plants, often from distant locations may not "fit" descriptions based solely on local wild representatives of a family.

The keys are designed for use in determination of family for plants regardless of their wild or cultivated status. We have attempted to include most of those families likely to be encountered by an individual residing in the continental United States but because of the diversity and discontinuous nature of the cultivated flora there have undoubtedly been omissions. We are quite aware of the likelihood that we have overlooked species that express combinations of features that do not fit the keys. We welcome comments on the structure and usefulness of the keys.

Most of the illustrations in this book are original line drawings prepared from living plants or photographic slides. Some are new with the fifth edition. Attempts have been made to illustrate features likely to be encountered among members of a family and some of the variations students can expect to see. For a large and diverse family, however, only a few of the representatives can be illustrated and students can expect to encounter variations that are not illustrated.

Zack E. Murrell
Dirk P. Walters
David J. Keil
Bonnie K. Walters

# Acknowledgments

A book of this type must of necessity be based on the work of others. We have tried to give credit to those whose knowledge and worked we have borrowed. Any errors of interpretation are ours and not those of the original authors. We would like to acknowledge the suggestions and encouragement made by our students and colleagues. Special thanks go to Kathy Keil for her encouragement, support, and patience during the preparation of the Fourth Edition of this text.

Taxonomy or systematics is a dynamic field that has undergone significant changes in the past three decades. The more recent revolution in molecular biology in the past ten years has brought a tremendous amount of new information to systematics. New methods and new data have changed some of our views of plant relationships and, in turn, our views of classification. This Fifth Edition represents an attempt to address those changes.

We can trace our understanding of plants back to folk taxonomies and the work of Theophrastus. The collectors and taxonomists in the 17th, 18th and 19th centuries created a tremendous increase in our knowledge of plants and classification. The more recent extraordinary efforts of Taktajan, Cronquist and others have given us a solid synthesis of our understanding of plant evolution. All of these efforts must be acknowledged as critical to our current system of classification.

The general classification structure presented here is based upon the recent collaborative efforts of the Angiosperm Phylogeny Group. This Fifth Edition represents an attempt to provide a bridge between the Cronquist system and our more recent phylogenetic analyses. There are many places where compromises were made that reflect the lack of consensus regarding circumscription of various families or orders. Hopefully, as more bright young students go into the field of plant systematics, we will learn enough about plants to converge on an accepted understanding of plant relationships. However, it is important for students to realize that science is a process of learning and this dynamic process should be reflected in our classification system.

The Fifth Edition would not have been possible without the help of S. Kristen Bright and Emily L. Gillespie. Their efforts in helping the third author generate some order in the chaos were crucial to making this edition a reality.

Most botanists can thank some mentor who opened their eyes to the world of plants. My Grandmothers, Louise C. Murrell and Helen S. Gresham, were two extraordinary women who did this for me. It was their gift of botanical awareness, a gift that I now see in the delight that my sons, Andrew and Gregory Murrell, take in showing plants to their Daddy that the third author must acknowledge as the encouragement that made this text possible.

Acknowledging the past

# About the Authors

**Dirk R. Walters** grew up in western Illinois where his interest in plants was kindled through Boy Scouting, first as a member and later as a leader. At Western Illinois University his general interest in botany was strengthened as he obtained a B.S. in Education (1965). Further work in Systematic Botany at Indiana University at Bloomington resulted in his M.A. (1966) and a Ph.D. (1969). He is Professor of Biology at California Polytechnic State University at San Luis Obispo, where he teaches courses in general biology, general botany, plant taxonomy, local flora, and plant ecology. His research interests include local flora and plant demographics.

**David J. Keil** began his lifelong interest and enthusiasm for botany as he grew up in Illinois. By the end of his sophomore year in high school he had decided on a career in botany. He received his B.S. (1968) and M.S. (1970) from Arizona State University and his Ph.D. (1973) from Ohio State University. He is Professor of Biology at California Polytechnic State University where he teaches courses in plant taxonomy, field botany, evolution, and biogeography. He had authored scientific papers, textbooks, and study guides, and was a major contributor to *The Jepson Manual* project. His research interests include Asteraceae systematics and floristics of western North America.

**Bonnie Kirby Walters** grew up in Northern Indiana where she has been drawing and chasing little animals ever since she can remember. In her high school days she was drawn to science and math, with art a welcome diversion. But, an art correspondence course during that period gave her untold hours of practice with drawing pens, a skill that would be useful later. College was a mixture of biology and art courses, until lack of time and money forced a decision that the major would be a B.S. (1966) in biology. Ecological and evolutionary study dominated her graduate work until she received her M.A. in Zoology (1972) from Indiana University. Current art work includes drawing covers for the local California Native Plant Society Newsletter and illustrating occasional research papers and lab manuals.

**Zack E. Murrell** grew up in western North Carolina and his grandmothers, Louise C. Murrell and Helen S. Gresham, instilled a love of plants at an early age. Hiking and backpacking in the Southern Appalachians led him toward a career in botany. His general interest was nurtured by Dr. Tom Daggy at Davidson College, where he obtained his B.S. (1980). He received his M. S. (1985) at the University of Tennessee under the mentorship of Dr. B. E. Wofford, and his Ph.D. (1992) at Duke University working with Dr. Brent Mishler. He is an Associate Professor at Appalachian State University where he teaches courses in general botany, evolution, plant systematics, agrostology, and anatomy and morphology. His research interests include systematics of the Cornales and Aristolochiales, species and speciation, biogeography, and rare plant conversation.

A young botanist should get his bearings well and early in life. There is no such thing as method in scientific discovery. Research cannot be done by rule. But if one master for himself a large and usable body of facts, then there is laid good foundations for the proper discipline of the scientific imagination. In the field of observation, says Louis Pasteur, success comes to him who is well prepared.

Willis Linn Jepson

On the outer rim of
Strawberry Cañon, at
11 Mosswood Road, Berkeley,
December 29, 1943.

# Introduction to Plant Taxonomy

## chapter 1

Plant taxonomy is the naming, describing and classifying of plants. The goal of plant taxonomy is to recognize and properly explain the variation of plants, and to organize this variation in a way that reflects what we know about their evolutionary relationships. The methods of plant taxonomy are as old as language itself, and as modern as the most advanced techniques in biochemistry, molecular biology, and computer science.

The field of taxonomy has undergone a revolution in the past 50 years. This revolution began in the 1960s and has been the subject of study by several philosophers of science, such as Thomas Kuhn. As methods changed over time, the people who utilized the newer methods often re-named themselves to indicate their use of the more recent methodology. Therefore, plant taxonomy is likely to be called plant systematics, systematic botany, or phylogenetics in textbooks and university course catalogs.

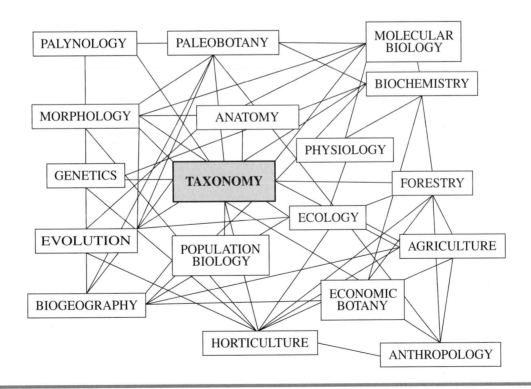

**Figure 1-1** Areas of science and their relationship to plant taxonomy.

# Relationship of Taxonomy to Other Areas of Biology

Plant taxonomy is the center of a web of inter-related areas of botany (Figure 1-1). Taxonomists use information about structure, chemical composition, physiology, ecology, genetics, evolution, geographical distribution and many other forms of data in interpreting relationships among plants. A very important function of taxonomy is the organization of diverse forms of information about plants. In turn taxonomists provide information of value to workers in other areas of biology and to non-specialists as well. One of the most common questions asked of a taxonomist is "What kind of plant is this?" This is an important question because the name of a plant is often the key to finding out additional information. If the name is known, previous research regarding that plant can be located and is potentially available for use. Equally important is the broad scope of information drawn together as a part of a taxonomic study. This information has application in many other areas of biology.

# Why Should You Study Taxonomy?

As humans have settled across the earth we have drastically changed our environment. We have fragmented the landscape with highways and cities and we are now realizing that our actions can even change the earth's climate. It is estimated that there are between 3 and 15 million species of plants on the earth today. We don't know how our actions will effect these species, but we do know that we have only begun to catalog what species are present.

Aldo Leopold was a conservation biologist who said that the key to "intelligent tinkering" is to save all the pieces. As we "tinker" with our environment, it is imperative that we determine 1) what the pieces (or species) are and 2) how they interact with one another, or how they fit together. This is the great challenge facing us today in biology.

The most exciting aspect of this daunting task in front of us is the amazing set of tools we now have to do this work. Recent advances in molecular biology allow us to understand the genetic basis for various plant traits. Computational tools are available to help us analyze large amounts of data. Recent field studies have given us new information about plant diversity of the past, as well as the present. This is an exciting time to be a plant taxonomist.

Taxonomists can find positions in research and teaching at universities, colleges, botanical gardens, herbaria, and museums. Because taxonomy is central to much of biology, taxonomists tend to be broadly trained and are often hired over other types of biologists as naturalists in parks and recreation areas, inspectors of agricultural products, in research positions for product development, plant breeding and pest control. There is a large and growing field of environmental monitoring and habitat restoration that needs well-trained taxonomists. But, it must be noted, an **excellent taxonomist** will have little trouble finding a position whereas a merely good taxonomist may go wanting.

Many students who take taxonomy classes will be users of taxonomy, not practitioners. They are individuals who will deal directly or indirectly with vascular plants as a part of their jobs. Workers in various areas of biology, agriculture, forestry and management need to know enough taxonomy to use the techniques and to understand the basic aspects of classification and the correct use of scientific names. Much of the information that has been recorded about a plant can be retrieved only if the name of the plant is known.

Some students take taxonomy classes because they are interested in the subject. Amateur naturalists often are curious about the plants around them. An individual may be employed in a position that has no direct connection with plants and still be an enthusiastic and knowledgeable amateur botanist. Many members of organizations dedicated to conservation and enjoyment of natural resources have benefited from training in taxonomy. You may find that studying taxonomy sparks an interest in plants that you did not previously have. Your experiences in plant taxonomy will undoubtedly change the way that you observe the plants around you.

# Basic Activities of Systematic Botany

Taxonomy involves some processes basic to all human cultures: cataloging, identifying and classifying (Figure 1-2). From the time that you as a child began to observe your surroundings and to organize

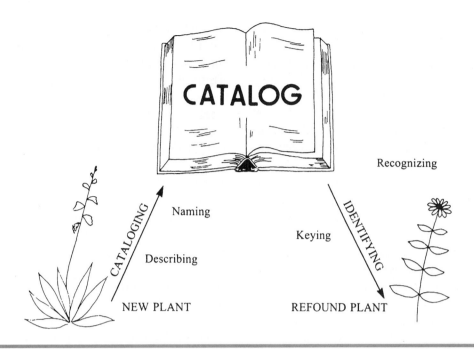

**Figure 1-2** Relationship among some of the tasks performed by taxonomists.

your accumulating knowledge and experiences, you have been carrying out these processes. We do these things automatically to make sense of the world around us.

Consider the objects illustrated in Figure 1-3. You had no difficulty in recognizing that objects A and B are two different kinds of spoons. A is a teaspoon and B is a tablespoon. Objects C and D are members of a second group; they are both knives. All six of the objects are kinds of silverware. Early in your life you cataloged these objects; you learned that certain objects with a particular shape and size are associated with the word spoon. Other objects are associated with other words. At the same time you learned a classification for these objects. Teaspoons, tablespoons and other similar objects (species) could be grouped into a more inclusive group (genus) for which you learned the name, spoon. Spoons, forks and knives together formed a still more inclusive group (family) for which you learned the name, silverware.

This organization of objects into groups and ranking of these groups is classification. Identification is the process of recognizing that a particular object is a member of a group and associating it with the name of that group. Although you had never before seen the particular knives, forks or spoons illustrated, you recognized them (identified them with something you had previously cataloged) and associated them with the appropriate code-words.

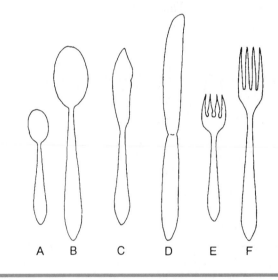

**Figure 1-3** A classification of the silverware family.

## Cataloging

**Cataloging** involves three steps—naming, describing and recording. **Naming** is essential to communication. We assign names to plants, to people, and to everything else around us. Names are the nouns of our language; they are codes that stand for objects or concepts. The coding process for assigning names to plants results in a **nomenclature** or system of names. Botanical nomenclature is discussed in Chapter 2.

**Describing** is our means of assembling and communicating information about the attributes of the objects or concepts to which we have assigned names. For many of the objects with which we interact we maintain mental concepts—descriptions of a sort that have accumulated through our experiences. If we wish to communicate these concepts to other people we generally use words that describe the attributes of the object. Most of the words we use are part of our everyday vocabularies. However, when describing complex structures (e.g., the parts of an automobile) we must have additional descriptive terms that enable us to communicate with precision. For plants too, the words of our everyday vocabulary are not always sufficient for the description of structures or phenomena that we encounter. One of the first steps for anyone studying plants is to learn some of the basic terminology used in botanical descriptions. Chapters 3 and 11 provide an introduction to botanical terms.

Information about objects that have been named and described must be **recorded** in some fashion if it is to be of use. Our memories serve as a repository for much information. Prior to the invention of writing, transmission of descriptive information from one individual to others was limited to oral descriptions. Each generation had to re-record descriptive information if it was to be passed along. Certain individuals served as living repositories of information. The written records of our society now far exceed the memories of any one individual and have a permanence that exceeds human lifespans.

Information about plants that have been named and described must also be properly recorded if the data are to be of use to laymen or other biologists. Names, descriptions and other data are published in books or journals that potentially can be read by anyone interested in the information. Over 300,000 different kinds of plants have been cataloged. Each of them has been described and named at least once and some several times. Each year, however, botanists discover additional plants that have never been cataloged before. The procedures for naming, describing and recording a previously unknown plant are described in Chapter 2. Other biologists may record additional information as more is learned about the plant. Updating the records is periodically necessary as additional information accumulates. The updated information is often recorded by specialists in the form of revisions and monographs (Chapter 18).

There is a constant need for information about all the plants that occur in a particular region. This information is recorded in the form of floras and manuals (Chapter 5). Much of the information that is recorded in floras, manuals and other botanical works comes from the study of individual specimens. Each individual plant is itself a record of much information. Herbaria (collections of preserved plants) are much like libraries in the large amounts of information that is recorded as individual specimens. In Chapter 7 methods for collecting and preserving plant specimens are discussed.

## Identification

**Identification** of plants can take place in several ways. You probably already are able to recognize some kinds of plants. You might identify these plants in the same way that you identify teaspoons, butter knives, and other familiar objects for which you already have a mental concept. But what if you have never encountered the plant before? Even the most highly trained botanists can recognize on sight only a small fraction of the species of the world. For determining the names of unknown plants, both you and the specialist need dependable techniques. One of these is to learn to recognize major groups of plants. Although there are hundreds of thousands of species of vascular plants there are only a few hundred families. Learning to recognize major families (by learning the features of the families and developing mental concepts for them) is a major step in plant identification. If you can recognize that a plant is a member of a particular family, you have eliminated all the other families (and all of the many species that they include). Discussions of families are presented in Chapters 8–16.

Another way is to use a taxonomic key. The process of **keying** is an artificial method of locating information by emphasizing similarities and differences among plants. A key is prepared by taking an array of plants and splitting them into two groups on the basis of different expressions of the same character (e.g., flower color red or white), and subsequently splitting each new group formed into two groups again until the group formed can no longer be logically split (i.e., the group consists of a single individual or variation among individuals is continuous such that a separation must be arbitrary). Using a key is to determine the name of a plant is a step by step process of elimination that involves comparing the features of an unknown plant with the choices in a key until all of the possibilities but one have been eliminated. Chapter 5 describes the construction and use of keys. Keying is a skill that taxonomy students are expected to learn and perfect. Determining the name for a plant is often a combination of recognition and keying. Appendix I is a key to plant families.

## Classification

The basic processes of classification are **grouping** and **ranking.** Grouping involves the sorting of objects into groups that have some features in common. Ranking is the organization of the groups into a nested hierarchy of small groups within progressively more inclusive groups. Both of these are processes we do all the time—usually without thinking about it. The result of this process, when applied to living organisms, is a formal classification hierarchy (Table 1-1). Each group that is recognized in the classification hierarchy is called a **taxon** (plural, **taxa**). In introductory plant taxonomy you will be taking your inherent abilities in classification and applying them to plants. You will not actually do much classifying yourself, but instead will be learning a system of classification that has resulted from the accumulated studies of many botanists (Chapters 10–12).

The classification of flowering plants outlined in Chapter 11 is devised by Dr. Arthur Cronquist, a botanist at the New York Botanical Garden. A system of classification for 300,000+ plants is necessarily more complicated than the classification of silverware or other items of our everyday surroundings. Dr. Cronquist's system of classification rests on the combined work of many botanists, past and present. In your studies you will learn the features of some of the major groups of plants and their position in the classification hierarchy of Cronquist's system. Major emphasis is placed on selected families of plants you are likely to encounter. These familiar plants will help you to understand the outline of the classification scheme.

## Data Gathering and Analysis

The classification of living organisms involves processes that are much more structured than those we use to classify the inanimate items of our everyday surroundings. It matters little if the classification that you unconsciously apply to your surroundings differs from that of your classmates. However, living organisms are connected through ancestor-descendant relationships. Modern taxonomists seek to determine the nature of these relationships and to construct classifications that accurately reflect the evolutionary history of the organisms. To do this taxonomists use various data-gathering techniques that can provide information useful in classification. The construction and testing of theories to explain the underlying relationships within natural groups is called **biosystematics.** Some of these methods are described in Chapters 3–4. Not all taxonomists are in agreement on the methods that are most useful in interpretation of the data. In Chapter 4 methods of constructing classifications are discussed.

Kingdom
  Division = Phylum
    Class
      Subclass
        Order
          Family
            Subfamily
              Tribe
                Genus
                  Subgenus
                    Section
                      Species
                        Subspecies
                          Variety
                            Form

Table 1-1    The classification hierarchy. In the classification of a particular kind of plant, the taxonomic ranks noted in boldface are always assigned. Additional ranks are included only as needed.

## What Must a Beginning Student in Taxonomy Accomplish?

A beginning course in plant taxonomy cannot present all there is to know about a field of study so broad, but it can introduce some of the problems and techniques used to solve them. There are two basic skills that a taxonomist is presumed to hold. These are (1) the recognition of a fairly large number of plants and (2) knowledge of techniques used to identify any that are not known. It is in these two areas that you as a student must become proficient. In order for you to do this you will have to become familiar with plant structures and some of the terms used to describe them. You will learn how to interpret the many variations of plant form that you will encounter as you are introduced to various groups of plants. You also will be exposed to at least the rudiments of nomenclature and classification.

## References

Crovello, T. J., and K. Adler. 1986. *Careers in Biological Systematics,* 2nd ed. revised by G. J. Anderson and J. A. Slater. American Society of Plant Taxonomists and Society of Systematic Zoologists.

Curtis, C. R., S. A. Graham, O. Tippo, and W. L. Stern. undated. *Careers in Botany,* 2nd ed. revised by R. H. Saigo and B. W. Saigo. Botanical Society of America.

Kuhn, T. 1970. *The Structure of Scientific Revolutions.* University of Chicago Press, Chicago.

Leopold, A. 1987. *A Sand County Almanac.* Oxford University Press, New York.

Stuessy, T. F., and K. S. Thomson (eds.). 1981. *Trends, Priorities and Needs in Systematic Biology,* 2nd ed. A Report to the Systematic Biology Program of the National Science Foundation. Association of Systematics Collections, Lawrence, Kansas.

Systematics Agenda 2000. 1994. *Systematics Agenda 2000: Charting the Biosphere.* A global iniative to discover, describe and classify the world's species. American Society of Plant Taxonomists, the Society of Systematic Biologists, and the Willi Hennig Society, in cooperation with the Association of Systematics Collections.

# Exercises

Some of the simpler concepts of classification can be readily demonstrated using a selection of dried fruits or seeds from various cultivated grains, legumes, and other large fruited/seeded cultivated plants. These structures are not only small and strongly different from each other, but their constituent individuals vary as well. In doing this exercise, you should base all decisions on the characteristics of the fruits and seeds before you and not on prior knowledge of what others have said the relations should be based on studies of the plant's total life cycle. Since all of us see and interpret the world around us slightly differently due to individual differences in background and training, this exercise is best done in a group or class situation using an identical set of seeds/fruits for each group. It is expected that different students will "see" different numbers of species in the selection and will maintain these difference throughout the exercise.

1. Arrange the selection of seeds given you by your instructor into "species." Base your decision on the attributes of the seeds and fruits themselves and not on any knowledge you have of the plants that produced the seeds.

    How many "species" do you have in the jar? _____

    If others in the class arrived at a different number of species than you did, explain how this could happen.

    Which group has the *correct* or *true* interpretation?

    What is the species concept (definition of species) used in this exercise?

    What other species definitions have you heard about?

    Which is the best definition?

2. Now arrange your species into only two groups of equal size. What feature(s) did you use to make your separation?

    After you have made your separation, check with the other groups in the room. Did they use the same features you did? _____ List the ways they separated the "species" of seeds.

Were there some "species" that did not seem to fit your groups? _____ Why didn't they fit?

3. Resolve any of the discordant elements in your two groups. That is, group your "species" into as many "genera" as makes sense to you. How many genera have you formed? _____ In the space below describe the genera you have formed.

Look at the "species" in each genus. Are the differences between the species of one genus the same as those in other genera?

Describe the differences among species within two genera. (Use the blank space at the bottom of the page if you need more room.)

Define the concept known as a **genus.**

4. Group your genera into "families." Do you have any "genera" or "species" that do not seem to fit with any other? Why do they not fit?

Define the concept known as a **family.**

Can a genus, family, or order be **monotypic,** consisting of a single species, genus or family respectively?

Name an example:

Can you describe the genera in one family with the same characters used in a different family?

Explain your answer.

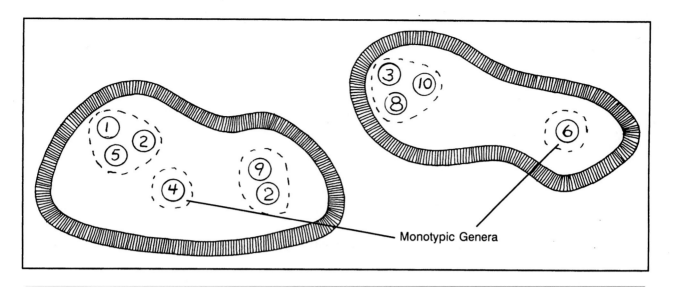

Monotypic Genera

**Figure 1-4** Hypothetical phenetic relationships among ten items.

5. On a separate sheet of paper, diagram your classification of the fruits and seeds for family, genus and species in the manner shown in Figure 1-4. Compare yours with the instructor's and/or your classmates'. Why are they not all the same? Which one is the correct one? Did you find evidence of differences in ability to perceive minor differences among the fruits and seeds among the classification systems you compared? Which ones would you classify as being made by taxonomic "**lumpers**" and and which by "**splitters**"? Can you think of other legitimate reasons for differences of opinion among "equally qualified" taxonomists?

# Botanical Nomenclature

## chapter 2

Nomenclature is the giving and using of names. Naming is a basic component in our language. Our speech requires it. Grammatically, names are the nouns we use to represent the objects and phenomena around us. Names allow us to communicate without resorting to lengthy descriptions. For example we can say "cow" to represent the large four-legged animated hairy object with a large sack underneath that gives us milk. Communication is aided by names that are readily understood by those hearing or reading the words.

## Common Names

Historically, there have been two more or less independent kinds of nomenclature devised for plants. These are **common names** and **scientific names.** Common names are the older. Common names are created by people doing the everyday living of life. They are words in the language of the layman and thus are easy to understand and use. There actually are many different systems of common names that have developed in different cultures and languages.

Although they are widely used, there are some disadvantages to common names. Common names develop in the language of a given people and may not be useful to people with a different language or dialect. Wide-ranging plants often have several different common names, some of them widely used and others unique to a given locality or language. The common garden pansy, for example, has at least two hundred known common names, about one fourth of which are Spanish, and another fourth

each in French, German and English. This multiplicity of common names can cause confusion. Would you suspect from the names that California bay-laurel, pepperwood, and Oregon-myrtle are the same plant?

Sometimes two or more kinds of plants share the same common name. Rattlesnake plant, for instance, is used for several different plants that have dry pods with loose seeds that rattle when the wind blows. On a local basis such common names may be very useful, but over a broader geographical area they may lead to problems in communication. Loosestrife is the widely used common name for two unrelated and dissimilar genera, *Lythrum* and *Lysimachia,* that have widely overlapping geographical ranges in eastern North America. The common name could be a barrier to communication between individuals using the name for different genera.

People develop words only for those objects they see or use. If a plant is inconspicuous or rare, it may not have a common name. Often the common or conspicuous members of a genus have their own common names. The less common or less conspicuous species of the same genus may be known only by the common name of the genus (if it has one). Sometimes common names represent what a botanist would consider to be a single species and sometimes they represent a genus or still larger grouping. Daisy, for instance, might be used to refer to a single species or to any of an indefinite number of more distantly related plants.

Common names have a place in everyday speech, but often they are not precise enough to be used in serious scientific studies or even in horticultural activities. Indeed they can be misleading. Since

11

there is no possible way to regulate the formation of common names or to legislate which ones gain acceptance, there is no way to be sure the names used are accurate. For example, there is a plant with bright red fruits that ripen around Christmas time. Its common name is Christmas-cherry. Unfortunately, it is not an edible cherry but a poisonous nightshade. The name Christmas-cherry suggests a set of relationships that is dangerously misleading.

Some botanists use hyphenation to indicate which common names show false relationships. Black oak (not hyphenated) is a true oak; poison-oak (hyphenated) is not an oak at all, but instead is a member of the cashew family. If everyone followed this practice, at least some of the confusion caused by common names would be eliminated. Unfortunately not all botanists use hyphenation in this fashion, and very few members of the general public are even aware of this convention. In this book we are using hyphenation of common names as described above. Kartesz and Thieret (1991) have published a lengthy discussion of recommendations for construction of common names.

Certain "common names" are not in common use by laymen. Many have been coined by botanists to satisfy a perceived need for non-technical names for communication with members of the general public. For example, the common name, "thread-leafed eriastrum" is merely a translation of the scientific name, *Eriastrum filifolium.* Laymen probably use this "common name" about as often as they use the scientific name. Some translations such as these are actually erroneous. One botanist who was attempting to standardize common names for California plants listed the common name of *Lupinus ludovicianus* as Louisiana lupine under the mistaken impression that the word, "*ludovicianus*", was derived from Louisiana. Actually, the plant is the county flower of San Luis Obispo County, California, the only place in the world where it grows. The word *ludovicianus* was derived from Luis instead of Louisiana, and the plant is usually called the San Luis Obispo lupine. Some translated common names have gained wide acceptance, though, particularly if a plant is conspicuous and there are no competing common names. For example, Jeffrey pine is the widely accepted common name for *Pinus jeffreyi.*

# Scientific Names

Scientific names evolved out of the study and reporting of plants in books during the period from the 13th to 18th centuries. This was a period when Europeans were finding out about the world beyond their continent and many unfamiliar plants and animals were being seen for the first time. There was a need to catalog these new finds, some of which had potential economic or medicinal value.

Because the language of learned men of this period was Latin, names of plants were written in Latin. This use of Latin has been passed on in the formal naming of plants. Even today scientific names are considered to be a part of the Latin language regardless of their original derivation.

Initially the process of naming plants varied from one botanist to another and there was little standardization of the procedures whereby names were created and assigned. A plant was often named differently by different individuals. The name of a species typically consisted of a generic name (a noun) followed by one or more Latin modifiers (usually adjectives or descriptive phrases). This system worked reasonably well as long as the number of kinds (species) of a given genus (those described with the same noun) was small. When the number of species became large, the names became quite long. For example, a species of buttercup ended up with the lengthy phrase-name, *Ranunculus calycibus retroflexis, pedunculis falcatis, caule erecto, foliis compositis,* which means "the buttercup with bent-back sepals, curved flower-stalks, erect stems and compound leaves." These long, cumbersome names are now known as **polynomials.** Such a name carried two roles. It was a name and it was also a **diagnosis** [a set of distinguishing features] of the plant.

Starting with the work of the great Swedish naturalist, **Carolus Linnaeus** (1707–1778) botanists gradually adopted a set of standard procedures for naming plants. Linnaeus began to formulate a set of underlying principles for botanical nomenclature in the 1730's and in 1751 he published a book called ***Philosophia Botanica*** *[Botanical Philosophy]* in which he set forth the principles of nomenclature he had adopted. These principles served as a sound foundation for later botanists. Today we have a set of formalized rules for naming plants called the **International *Code of Botanical Nomenclature*** (discussed below). Many of the rules of nomenclature found within this book had their beginnings with Linnaeus.

In 1753 Linnaeus published a very influential work called ***Species Plantarum [Species of Plants]***. In this book Linnaeus used **binomial nomenclature** instead of the unwieldy polynomials. Binomials are two-word names written in Latin. The first word is a noun, the name of the genus to which the plant is assigned. The second word is usually either an adjective or a possessive noun. For the buttercup described above, Linnaeus coined the name, *Ranunculus bulbosus*. Linnaeus did not originate the idea of two-word names. These had been used occasionally by other workers in the 16th and 17th centuries, but these early taxonomists had not used binomials consistently, and use of these names had not become widely accepted. Linnaeus was the first to use binomial nomenclature consistently for all the plants known in his day. His system was simple to use and very logical and was accepted readily by most of his contemporaries. He is credited with being the father of the modern scientific nomenclature.

One of the major benefits of binomial nomenclature was almost immediately apparent. The name no longer had to serve as a diagnosis of the plant. The name of a plant could remain stable even though new discoveries might require a modification of the list of characteristics that distinguished the plant from others. Today the use of the binomial system of scientific nomenclature is universally accepted. It has changed little from the time it was first applied and perfected by Linnaeus. The system of naming plants works because it has been agreed upon by botanists throughout the world.

## Names of Taxa above the Genus Level

The *International Code of Botanical Nomenclature* provides for the naming of plants at various taxonomic ranks. The names of taxa above the genus level have standardized suffixes that indicate the rank of the taxon (Table 2-1). The taxonomic rank can automatically be determined from the name. Except at the kingdom level, the names are formed by appending the appropriate suffix to the name of a genus. Thus Magnoliopsida and Magnoliophyta are based on the genus *Magnolia*. Asteridae, Asterales, and Asteraceae are based on *Aster*. Lactucoideae and Lactuceae are based on *Lactuca*. The names of

| Taxon | Scientific name (with suffix emphasized) |
|---|---|
| Kingdom | Meta**phyta** |
| Division | Magnolio**phyta** |
| Class | Magnoli**opsida** |
| Subclass | Aster**idae** |
| Order | Aster**ales** |
| Family | Aster**aceae** |
| Subfamily | Lactuc**oideae** |
| Tribe | Lactuc**eae** |
| Genus | *Taraxacum* (no standardized suffix) |
| Species | *Taraxacum officinale* (no standardized suffix) |

**Table 2-1** The classification hierarchy for the common dandelion (in the system of Cronquist, 1981).

taxa above the genus level are *not* underlined or italicized.

The *International Code of Botanical Nomenclature* provides for an exception to the use of standardized suffixes in the case of eight families. Each of these families has two different equally correct names (Table 2-2). These distinctive families were recognized by early taxonomists and given names that came to be widely accepted. This took place long before the formation of names by appending a suffix to the end of a genus name was standardized. These traditional names were so well established and widely used that there was resistance to their

| Standardized name | Traditional name |
|---|---|
| Apiaceae | Umbelliferae |
| Arecaceae | Palmae |
| Asteraceae | Compositae |
| Brassicaceae | Cruciferae |
| Clusiaceae | Guttiferae |
| Fabaceae | Leguminosae |
| Lamiaceae | Labiatae |
| Poaceae | Gramineae |

**Table 2-2** Families with alternate names.

replacement by less familiar names based on genus names. In recognition of the preferences of some botanists for standardization and of others for tradition, the *Code* allows the use of either name. Each of the eight families thus has both a traditional name and a name with the standardized *-aceae* suffix. Both names are equally acceptable. It is recommended, however, that botanists be consistent within the context of a particular work (e.g., the flora of a region). If the traditional name is used for one family, traditional names should be used for the others as well.

In the survey of plant families (Chapters 8–16) in this book, we have listed families by their standardized names, but both names are presented. As you, as a taxonomy student, learn the features of these families you should learn both names. Some manuals and floras use the traditional names and others use the standardized names.

## Generic Names

A **generic name** (the name of a genus) is treated as a Latin noun. It should be capitalized and either underlined or written in italics. Generic names come from various sources (Table 2-3). Some are ancient Latin plant names. Many others have been coined by botanists. Many generic names are derived from Greek or Latin word roots. Some are formed from the names of mythological beings or historic personages. Taxonomists have often named genera in honor of a contemporary, such as an explorer or fellow botanist.[1] A generic name may even be a meaningless [but pronounceable] combination of letters.

Latin nouns have a grammatical gender. This usually has nothing to do with the actual gender, if any, of the object being named. A generic name may be masculine, feminine, or neuter. The ending of a name often indicates its grammatical gender.[2] Names ending in *-a* usually have feminine gender (e.g., *Potentilla*). Those ending in *-us* are commonly masculine (e.g., *Lupinus*), though a significant minority are feminine (especially genera of woody plants such as *Pinus* and *Quercus*). Names ending in *-is* may be masculine or feminine though the major-

| Generic name | Derivation |
|---|---|
| *Acer* | ancient Latin for maple |
| *Conium* | Latinized form of ancient Greek name |
| *Tsuga* | Japanese name for hemlock |
| *Pseudotsuga* | Greek and Japanese meaning false hemlock |
| *Ribes* | Arabic name for gooseberry |
| *Artemisia* | Greek mythology—Artemis, the goddess of the hunt |
| *Dimorphotheca* | Greek, meaning two forms of fruits |
| *Trifolium* | Latin, meaning three leaves |
| *Campanula* | Latin, meaning little bell |
| *Kalmia* | Named for Peter Kalm, a student of Linnaeus and botanical explorer in eastern North America |
| *Brandegea* | Named for T. S. Brandegee, California botanist |
| *Allium* | Ancient Latin for garlic |
| *Muilla* | Anagram of *Allium* |

**Table 2-3** Examples of derivation of generic names.

ity are feminine (e.g., *Physalis*). Names ending in *-um* (e.g., *Cirsium*) are neuter. Names ending in *-e* may be feminine *(Chorizanthe)* or neuter *(Secale)*.

## Species Names

A **species name** is a *two-word name* that consists of a **generic name** and a **specific epithet.** For example, *Trillium grandiflorum* is the scientific name of a common spring wildflower in eastern North America. *Trillium* is the generic name, and *grandiflorum* is the specific epithet. The specific epithet by itself is *not* a species name. A specific epithet is usually written in lower case and underlined or italicized. [Certain specific epithets may be capitalized (e.g., those derived from a person's name), but the *International Code of Botanical Nomenclature* recommends that specific epithets always be written in lower case].

A specific epithet is usually either a Latin adjective, or a noun in the genitive (possessive) case (Table 2-4). If the specific epithet is an adjective it must agree with the noun it modifies in gender and

---

[1]When a genus is named in honor of an individual, the name is usually formed by adding -ia to the individual's surname, except in those cases in which the surname ends in a vowel.

[2]Names may end with letters other than those discussed here. You may wish to consult Botanical Latin (Stearn 1993) for additional information about Latin grammar. It is often possible to determine the gender of a name from the adjectives that modify it (see footnote 3).

| Epithet | Derivation |
|---|---|
| *albiflora* | Latin compound meaning white-flowered |
| *chrysantha* | Greek compound meaning yellow-flowered |
| *serrata* | Latin for saw-toothed |
| *arvensis* | Latin meaning of fields |
| *borealis* | Latin for northern |
| *americana* | from America |
| *pensylvanica* | from Pennsylvania |
| *neomexicana* | from New Mexico |
| *engelmannii* | for George Engelmann |
| *greenei* | for Edward Lee Greene |
| *eastwoodiae* | for Alice Eastwood |
| *eastwoodiana* | for Alice Eastwood |

**Table 2-4** Examples of derivation of some specific epithets (adjectival epithets all with feminine ending).

in number (singular) and thus often has the same ending as the noun).[3] Such a specific epithet may be descriptive of the plant, it may refer to its geographical range or it may be descriptive of habitat or some other important information.

Specific epithets are often coined to honor a person [generally someone who has some connection with botany]. When a specific epithet is derived from a man's surname, it is usually formed by adding the suffix *-ii* to the man's name (e.g., *wrightii*), except in those cases where the surname ends in the letter "r" or a vowel (including "y"). The suffix *-ii,* which is the possessive case for the Latinized version of a man's name, is the equivalent of *-'s* in English. *Eriogonum wrightii,* for instance, means Wright's buckwheat. If the surname ends in r or a vowel, the suffix is usually *-i* (e.g., *porteri, parryi, greenei*). For a woman's name the suffixes are *-iae* and *-ae* respectively. A specific epithet may also be formed as an adjectival form of a name. For instance, *Eriogonum butterworthianum* means the Butterworth buckwheat.

---

[3]The ending of a Latin adjective depends on the gender of the noun it modifies (e.g., *Scrophularia californica* [f], *Caulanthus californicus* [m], and *Taraxacum californicum* [n]). The endings of a specific epithet can often be used to determine the gender of the generic name it modifies (e.g., *Rhamnus californica*). From the feminine ending of the specific epithet it can be determined that *Rhamnus* has feminine grammatical gender.

There are exceptional cases in which the specific epithet is a second noun that is not possessive. Examples are discussed in Nicolson (1986).

Scientific names are often written with their **author** or **authors,** the individual or individuals who are responsible for having given the plants their names. Consider the following examples of species names.

*Lotus corniculatus* L.

*Lotus heermannii* (Dur. & Hilg.) Greene

In both cases the genetic name is *Lotus,* a genus in the pea family. The specific epithet of the first species is an adjective that in Latin means "bearing a horn-like projection". The second species was named in honor of A. L. Heermann, a nineteenth century plant collector. The latter name means Heermann's lotus.

The name or names of the authors follow the binomials. The surnames of these individuals are often abbreviated. The author for the first species is Linnaeus, whose name as an author is customarily abbreviated as L. The history of the second name is a bit more complicated. The species was originally named by two naturalists from Philadelphia, E.M. Durand and T.C. Hilgard, as *Hosackia heermannii.* Several years later, E.L. Greene, a California botanist, concluded that the genus *Hosackia* should be merged with *Lotus,* and transferred the specific epithet, *heermannii* from *Hosackia* to *Lotus.* Durand and Hilgard **(the parenthetical authors)** get credit for having published the epithet, *heermannii.* Greene **(the combining author)** gets credit for transferring the epithet to *Lotus* and publishing the combination, *Lotus heermannii.*

Sometimes two or more authors are listed after a species name. These names may be separated by an ampersand (&) or by the prepositions, "ex" or "in". The ampersand indicates that two authors worked together on the description, whereas "ex" means that the second author published a name proposed by the first. The word "in" is used to designate the situation where the first author wrote a section of a book or article edited by the second.

In formal taxonomic publications such as floras, identification manuals, revisions, and monographs, the names are generally written with their authors. This helps to trace the history of a name and to avoid confusion, since a scientific name may have been independently applied to different plants [accidentally] by different taxonomists. For example, *Erigeron canescens* Torr. & A. Gray applies to an entirely different plant than does the *Erigeron canescens* Hook. & Arn. Some manuals include a list of the abbreviations of authors' names with the name spelled out and a bit of biographical information provided for each.

## Names of Infraspecific Taxa

Taxonomists sometimes encounter species that are variable. If these variations represent geographical races, a botanist may recognize them as **subspecies** or **varieties.** Subspecies are more inclusive than varieties. In a highly variable species both subspecies and varieties may be recognized. [This does not happen very often]. If a variant is sporadic in its occurrence and does not have a geographical range of its own, most taxonomists will not formally name it. Some botanists, however, recognize such plants as **forms.** The name of an infraspecific taxon is an epithet similar in construction to a specific epithet and preceded by a word or abbreviation that indicates its rank (e.g., *Penstemon heterophyllus* ssp. *australis; Lupinus densiflorus* var. *aureus*). It should be emphasized here that the horticultural "cultivar" is *not* a part of the formal system of botanical nomenclature.

When a species is divided into infraspecific taxa or when a taxonomist recognizes that two or more taxa that previously had been recognized as separate species are actually geographical races or forms of a single species, names must be adjusted accordingly. One of the infraspecific taxa that results includes the type specimen of the species.[4] Its subspecific (varietal, formal) epithet will be exactly the same as the specific epithet, and the name will be written as in the following example:

*Hemizonia parryi* Greene ssp. *parryi*

Notice that the epithet of the subspecies that contains the type specimen is written without an author. This is often referred to as the "typical" subspecies because it contains the type specimen of the species, but other subspecies may be more common or widespread. We prefer to call it the **type subspecies.** All other subspecies are written together with their author(s):

*Hemizonia parryi* Greene ssp. *australis* Keck

*Hemizonia parryi* Greene ssp. *congdonii* (Robinson & Greenman) Keck

*Hemizonia parryi* Greene ssp. *rudis* (E. Greene) Keck

---

[4]See discussion of General Principles of Botanical Nomenclature below.

# Pronouncing Scientific Names

Latin is now a seldom-spoken language, and we do not know precisely how it was spoken in the Roman world. Many scientific names are words that were not a part of ancient Latin and would sound as foreign to the Romans as Latin does to us. Many English-speaking botanists pronounce Latin names as if the words were written in English. This is known as the Traditional English system. There are many variations and these are often passed on from teacher to student. Individual botanists are not always consistent in pronunciation, often pronouncing names as they first learned them, even if words of similar construction end up with differing pronunciations.

On the other hand, most classicists and many European botanists prefer Reformed Academic Latin in which strict rules govern the pronunciation of particular letters or combinations of letters. Phonetically the latter undoubtedly comes closer to the Roman pronunciation than does the English system. As an example, the family name, Rosaceae, is usually pronounced Ro-záy-see-ee by most English-speaking taxonomists and Ro-sáh-seh-ah by continental botanists. Differing pronunciations can hinder communication in some cases. In our increasingly internationalized world it is increasingly likely that American botanists will come in contact with scientists from other regions. A good source of information on pronunciation in the Reformed System is *Botanical Latin* by W. T. Stearn (1993). Weber (1986) suggested a set of pronunciation guidelines for American botanists who wish to communicate with botanists educated in other countries.

Although there is no consensus among botanists of the world regarding the pronunciation of vowel sounds, there are some general guidelines. Look at the word carefully and pronounce the word phonetically. All vowels in Latin are sounded. In some cases two vowels are pronounced together as a diphthong, making a single syllable. The most common of these are *ae* and *oe*. In most other two-letter combinations, both are pronounced. In an English word such as *advance* that ends in *e,* this terminal letter is usually silent, but in Latin a terminal *e* is always pronounced, as in the word *arvense* (ar-vén-se *not* ar-véns).

The authors of some floras provide accent marks to indicate the syllable of a word to be stressed. In the absence of these marks you may have to guess or ask how to pronounce the name. There are various guidelines for accenting Latin words but these are too detailed for presentation here. Consult a Latin dictionary or *Botanical Latin* if you are interested in learning more about the language and its pronunciation. The editors of *The Jepson Manual* (Hickman, 1993) presented the following recommendations regarding pronunciation:

1. Divide the word carefully into syllables (it is safest to assume that every vowel belongs to a different syllable).

2. Pronounce each syllable (e.g., "co-to-ne-as-ter", not "cot-on-east-er).

3. Listen to others and practice what sounds good to your ear; conviction is important.

4. Attempt to accent all syllables equally; this is likely to show you where accents fall naturally (some manuals, but not this one, specify accents with stress marks).

5. Develop your own standards for pronouncing common endings like "-aceae", "-iae", "-ensis", etc.

6. Retain pronunciation of proper names used in scientific names ("jones-eeee", not "jo-nes-ee-eye").

7. When someone presumes to correct your pronunciation, a knowing smile is an appropriate response.

# The International Code of Botanical Nomenclature

The rules for use of scientific names are maintained and updated periodically at meetings of botanists called **International Botanical Congresses.** The updated rules and other guidelines are published after each Botanical Congress in a new edition of the ***International Code of Botanical Nomenclature.*** It must be emphasized that neither the congress nor the rules have any executive or judicial power. The rules must be accepted or rejected on

their own merits. There is no way, outside of peer pressure, to force a recalcitrant taxonomist into line.

A second problem with applying the rules concerns the way in which an experienced taxonomist interprets the variation of plants. The rules say nothing about this. One taxonomist may see three distinct groups of plants that he recognizes as species, whereas another equally qualified taxonomist sees only one. In one case there are three names for a set of plants while in the other there is only one. The "correct" interpretation is the one that becomes used.

With these limitations in mind it is time to make a quick survey of the rules. You should use the current edition of the *International Code of Botanical Nomenclature* as the final word on problems arising from the use of scientific names. The selected reading section has a number of references that discuss the rules section by section. In this exercise only the broad principles upon which the rules are based are discussed.

# General Principles of Botanical Nomenclature

1. **Botanical nomenclature is independent of zoological nomenclature.**

The *Code* applies equally to names of all taxonomic groups treated as plants, including fungi and algae (the code does not include bacteria other than blue-green algae as plants) whether or not these groups were originally so treated. This is a problem only for some small organisms that are treated as protozoans by zoologists and as algae or fungi by botanists (and that may have separate names under the two different codes). Although both botanists and zoologists adhere to the same general principles of naming, there are many differences in detail. For example, the earliest date for which priority applies for most plant names is 1753, whereas it is 1758 for most animals. There are also differences in citation of authors and in criteria for valid publication. It is possible for a plant and animal to have exactly the same generic or even species names.

2. **The application of names of taxonomic groups is determined by means of nomenclatural types.**

When a botanist names a new species or infra-specific taxon he or she designates a particular specimen, the **holotype,** to serve as a permanent reference point for the name. This specimen is deposited in a particular herbarium (plant museum) where it may be consulted by other botanists. The holotype is the tangible expression of the description and thus becomes the basis of comparison. Duplicate specimens of the holotype (specimens of the same plant collected at the same time in the same place) are called **isotypes.** Type specimens are often specially curated in herbaria and are particularly important to a taxonomist who is attempting to determine the correct application of a name.

If the taxon being named is a genus, family, or order, the nomenclatural type is a species, genus, or family, respectively. Each genus has a type species, each family has a type genus, etc. Ultimately the type of a genus or higher category is the type specimen of the type species.

Occasionally a taxonomist discovers that no holotype exists for a particular taxon. In some cases the holotype has been lost or destroyed. During World War II thousands of type specimens burned up when the Berlin Herbarium was destroyed during an Allied bombing raid. In other cases a holotype was never designated (the requirement of designating a type is a comparatively recent rule). A botanist may have listed several specimens as types or may have merely listed the specimens he had examined; these are called **syntypes.** If no holotype exists, a knowledgeable taxonomist (generally someone who is studying the plants in question) must choose a specimen to serve as if it were the holotype. The Rules specify from what group of specimens a type can be chosen. For example, an isotype or syntype would have the highest priority. Second in priority would be any other specimen seen and cited by the original author as being included in the taxon. These are known as **paratypes.** A type specimen chosen from these sources is called a **lectotype.** If none of the original specimens are extant, then a new type or **neotype** must be chosen.

### 3. The nomenclature of taxonomic groups is based on priority of publication.

Said in its simplest terms, this means that the first correctly published name for a taxon is the one to use. However, this principle has been found to require limitation in practice. Because the publication date of Linnaeus' **Species Plantarum** effectively marks the end of polynomial nomenclature and the beginning point for binomial nomenclature, the date of its publication (May 1, 1753), is considered the earliest listing to which priority of publication applies for all vascular plants, some bryophytes, some algae, and some fungi. Any name published before this date (even if it was a binomial) has no standing as far as priority is concerned. Any binomial correctly published after that date must be considered for priority purposes. For certain plants (and fossils) a later date is the starting point.

Priority is very important when two or more names are discovered to apply to the same taxon. It is not hard to see how a plant could be named more than once. As early taxonomists attempted to sort out and name the many plants that were being sent to them or that they were collecting, it was inevitable that sometimes two or more taxonomists would independently name the same plant. It is not at all unusual for a species to have been discovered and named by several different botanists. Additionally some species are sufficiently variable that their extremes look quite different. Early taxonomists often worked from very limited samples and were not aware of the natural variation encountered in the wild. The importance of priority is that it allows us to decide which of the competing names to use—the earliest one published.

Priority has been limited in certain cases by the concept of **nomina conservanda** (conserved names). In some cases a botanist discovers that a very widely used name is actually predated by an obscure, largely unknown name. The more commonly used name may be conserved or retained as the valid name, but this takes a special action of a Botanical Congress. The name with the earlier publication date becomes a rejected name. The *International Code of Botanical Nomenclature* contains a list of conserved and rejected names. Until recently only generic and family names could be conserved. Under the provisions of the most recent editions of the code, conservation of species names is possible, but only under very limited conditions.

### 4. Each taxonomic group with a particular circumscription, position, and rank can bear only one correct name, the earliest that is in accordance with the Rules, except in specified cases.

Sometimes botanists disagree as to the placement of a particular taxon. One taxonomist may consider a plant to be a member of one genus whereas a second botanist thinks it belongs in a different genus. Taxonomists may disagree as to the rank of a taxon, one treating a plant as a variety and another treating the same plant as a distinct species. In these cases there is a correct name for the plant in each of the alternative taxonomic placements. In the example discussed above, *Hosackia heermannii* Dur. & Hilg. is the correct name for the species if the genus *Hosackia* is considered to be distinct from *Lotus*. If *Hosackia* is merged with *Lotus,* the correct name is *Lotus heermannii* (Dur. & Hilg.) Greene. Each name is correct in a given context. There is still another interpretation for this taxon. It has been considered to be a variety of another species: *Lotus eriophorus* Greene var. *heermannii* (Dur. & Hilg.) Ottley.

Alternate names for a plant are **synonyms**. There are two types of synonyms. **Taxonomic synonyms** are synonyms in the *opinion* of a taxonomist. They are names based on different type specimens and are considered to be synonyms because a taxonomist who has studied the plants has concluded that they apply to the same taxon. One taxonomist may consider them to be synonyms whereas another does not. **Nomenclatural synonyms** are based on the same type specimen (and almost always have the same specific or infraspecific epithet). Nomenclatural synonyms are always linked because they share the same type specimen. Synonyms of either type can differ in rank.

Certain names cannot be used because they do not conform to the rules of nomenclature. These are called **illegitimate names.** For instance it is against the rules to use a name that has already been validly published for a different kind of plant. Such **later homonyms** must be rejected. Also against the rules are **tautonyms** (binomial names in which the same word is used for the generic name and for the specific epithet). Certain procedures must be followed when a name is first published (described below); if these procedures have not been followed the name must be rejected (or in some cases it can be published correctly at a later date).

5. **Scientific names are Latin or treated as Latin regardless of their derivation.**

The origin of the word that is used to make the name can be Latin or any other language if it is written in the Roman alphabet. It is treated as if it were a part of Latin and is subject to the rules of grammar of that language. The words chosen to be Latinized for use as names usually have some relationship to botany. [See the discussion above under Generic Names and Species Names].

6. **The Rules of nomenclature are retroactive unless expressly limited.**

Early botanists did not have a code of nomenclature. The rules we follow gradually developed through the practices of early taxonomists and were later codified. Not all botanists, of course, did things the same way and certain early practices are now considered to be against the rules. In order to consistently apply the code it has been necessary to retroactively outlaw certain early procedures. Other practices are considered acceptable from 1753 to some later date and after that date they are disallowed. For instance, the requirement that a holotype be designated dates only from January 1, 1958. Prior to that date a species could be named without designation of a type.

# Naming a Newly Discovered Plant

Although taxonomists have been naming plants for over two hundred years, new species are still being discovered. Occasionally taxonomists even discover previously unrecognized genera and families. Most of the new taxa now come from tropical areas that have been poorly explored by trained botanists. On occasion, however, new taxa are discovered even in well-botanized temperate regions.

There are several steps that a taxonomist must follow to **validly publish** the name of a new plant (i.e., publish it in an acceptable format). (1) The name must be properly constructed and it must not be a name anyone has ever used before. (2) The taxonomist must clearly indicate the rank of the taxon being described. (3) A specimen must be designated as the holotype. (4) The taxonomist must publish a description or diagnosis written in Latin. A diagnosis is a brief statement that indicates the ways in which the newly described taxon differs from other plants. A description of the plant in the language of the taxonomist and an illustration are often prepared but these are not requirements. (5) The name and the accompanying information must be **effectively published** (i.e., printed in a publication that would be generally available to other botanists). New taxa are usually described in botanical journals or books such as manuals or floras. Publishing the name of a new plant in a seed catalog, a newspaper, or some other ephemeral publication is not acceptable (although in earlier times such publications were allowed).

The requirement that a description or diagnosis written in Latin be part of the publication of a new taxon may seem archaic, but it actually is a perpetuation of the use of Latin as an international scientific language. Although the remainder of the publication may be in English, German, Russian, Japanese, or some other language, the essential description or diagnosis is in Latin. A taxonomist does not have to learn all the languages of the world to be able to understand the publication of a new species. Because so much of the early taxonomic work was written in Latin, a professional taxonomist must be able to work with this language. Some botanists who are more skilled than others in the use of Latin often are asked by their colleagues to prepare the required Latin descriptions or diagnoses. Stearn's *Botanical Latin* is a very useful resource for writing and reading botanical Latin.

Proposals have been made to drop the requirement for Latin descriptions or diagnoses and to use English instead. English is more and more becoming the dominant international scientific language. Although there are some good arguments in favor of this proposal it has not received broad international support to date.

The recent changes in the field of systematics, including the new methods of phylogenetics, have led many systematists to believe that we need a new method of naming life. The Phylocode has been developed as a formal set of rules to govern phylogenetic nomenclature. This system is designed to be used concurrently with the existing codes based on Linnaean nomenclature (i.e., *International Code of Botanical Nomenclature* (ICBN), *International Code of Zoological Nomenclature* (ICZN), *International Code of Nomenclature of Bacteria: Bacteriological Code* (BC), *International Code of Virus Classification and Nomenclature* (ICVCN)). The basic idea behind this code is to have a system of naming that matches how we think organisms are related to one another. For example, every organism would be placed in at least one group (life, as a clade), and then would be placed in other nested clades within that group. As this code develops, you will be able to see how a system of naming is developed (see *http://www.ohiou.edu/phylocode/*).

One of the interesting issues involved in this new naming system is the naming of species. The group of people developing the code is not in agreement, so they have postponed development of this part of the code. A second issue of interest is how to implement this code, and when it will be used by the scientific community. These issues will need to be addressed over the next few years as this new code develops.

# Why Do Botanists Change Plant Names?

Scientific names are often considered to be more stable than common names (see discussion of common names above). This is often, but not always, the case. Now and then a non-taxonomist is dismayed to learn that the scientific name of a familiar plant has been changed. When this affects an economically important species, there are howls of protest.

For many years the standard reference for California plants was *A California Flora* by Philip A. Munz and David D. Keck. When it was published in 1959 it was the most up to date compilation of information about the state's flora. Within a few years, however, so much new information had accumulated that Munz published a lengthy supplement. Over the years since the publication of Munz's supplement, an accumulation of new information and corrections of old information made a new flora more and more necessary. When a new flora, *The Jepson Manual* was finally published in 1993, many professional and amateur botanists pored over the pages to find out what names had been changed. Many were dismayed to learn that the long-familiar names of some plants had been replaced by different, unfamiliar names. Why were there so many name changes?

The answer to the question is not simple. Some changes are a result of the application of the rules of nomenclature. Earlier California botanists in some cases had made errors in the names they accepted and the rules require other names to be used. Some of the changes resulted from an application of the principle of priority of publication. Even today we are still discovering instances where the name that has long been in use for a plant is predated by another name. As early taxonomists attempted to sort out and name the many plants that were being sent to them from western North America or that they were collecting, sometimes two or more taxonomists independently named the same plant. Early botanists who visited California included Russians,

Spanish, English, and Americans. It is not at all unusual for a species to have been discovered and named by several different botanists. Some early names that were published in obscure books or serials have only recently come to the attention of California botanists.

In other cases it was the advance of our knowledge of California's plants that brought about reinterpretations of the previous taxonomic treatments for particular genera and species. New methods of gathering or interpreting data sometimes resulted in evidence that relationships were not what earlier botanists had thought. Frequently these realignments result in name changes.

In some cases modern studies indicate that two or more kinds of plants that had previously been treated as members of different species are actually so closely related that they belong in the same species. Some species are sufficiently variable that their extremes look quite different. Early taxonomists often worked from very limited samples and were not aware of the natural variation encountered in the wild. When species are merged, sometimes it turns out that the name that has been applied to a widespread, conspicuous species is not the oldest available name and the combined species end up with the name that had once been applied to a much less widespread plant.

To professional botanists many of these changes were old news. During the 34 years between the publication of *A California Flora* and *The Jepson Manual* many studies of California plants had been published in botanical journals. It was only when the changes that resulted from these studies were gathered together in *The Jepson Manual* that they seemed so numerous.

The nomenclature of a group of plants is tightly tied to its classification. Sometimes botanists disagree as to the placement of a particular taxon or group of taxa. An early taxonomist may have considered a plant to be a member of one genus whereas a later botanist has evidence that it belongs in a different genus. When this happens, the later botanist may transfer the species from one genus to another. Some taxonomists (splitters) tend to see differences as more important than similarities and divide large groups into smaller taxa. Others (lumpers) may see similarities as more important than differences and merge small taxa into larger groups. As a result of differing taxonomic treatments a species may have correct names in several genera. A taxonomist authoring a particular treat-

ment has to decide which classification is best supported and therefore which name to use.

Taxonomists may disagree as to the rank of a taxon as well, one treating a plant as a subspecies or variety and another treating the same plant as a distinct species. Once again in these cases there is a correct name for the plant in each of the alternative taxonomic placements.

# References

Bailey, L. H. 1933. *How Plants Get Their Names.* MacMillan Co., New York (reprinted in 1963 by Dover Publication, New York).

Benson, L. 1962. *Plant Taxonomy. Methods and Principles.* Ronald Press, New York, pp. 341–387.

Berlin, B. 1973. Folk systematics in relation to biological classification and nomenclature. *Annual Review of Ecology and Systematics* 4:259–271.

Brown, R. W. 1956. *Composition of Scientific Words.* Smithsonian Institution Press, Washington, D.C. (1991 reprint).

Brummitt, R. K., and C. F. Powell (eds.). *Authors of Plant Names.* Royal Botanic Gardens, Kew, England.

Greuter, W., J. McNeill, et al. (eds.). 1994. International Code of Botanical Nomenclature as adopted by the Fifteenth International Botanical Congress, Yokohama, August–September 1993. *Regnum Vegetabile* 131.

Hickman, J. C., ed. 1993. *The Jepson Manual. Higher plants of California.* University of California Press, Berkeley, Los Angeles, and London.

Jeffrey, C. 1977. *Biological Nomenclature.* The Systematics Association, London.

Kartesz, J. T., and J. W. Thieret. 1991. Common names for vascular plants. Guidelines for use and application. *Sida* 14:421–434.

Nicolson, D. H. 1974. Orthography of names and epithets: latinization of personal names. *Taxon* 23:549–561.

———. 1986. Species epithets and gender information. *Taxon* 35:323–328.

Savory, T. 1962. *Naming the Living World.* John Wiley & Sons, New York.

Stearn, W. T. 1993. *Botanical Latin. History, Grammar, Syntax, Terminology and Vocabulary,* 4th ed. David & Charles, Newton Abbot, England.

Weber, W. A. 1986. Pronunciation of scientific names. *Madroño* 33:234–235.

# Exercises

1.  Arrange this list of taxa from the largest (most inclusive) to the smallest (least inclusive) and indicate the rank of each taxon: Fagaceae, *Quercus,* Magnoliopsida, Fagales, Hamamelidae, *Quercus marilandica,* Magnoliophyta.

| Taxon | Rank |
|-------|------|
|       |      |
|       |      |
|       |      |
|       |      |
|       |      |
|       |      |

2.  Explain the parts of the following name:

    *Brassica nigra* (L.) Koch

    a.  *Brassica*

    b.  *nigra*

    c.  (L.)

    d.  Koch

3.  The following names have been discovered to apply to the same taxon. They were published independently by different botanists working in different places at different times.

    *Erythronium obtusatum* Goodd. (1902)

    *Erythronium nuttallianum* Regel (1871)

    *Erythronium utahense* Rydb. (1917)

    *Erythronium grandiflorum* Pursh (1814)

    a.  Which of these names should be used for the plant in question?

    b.  Why?

4. One of the combinations above had already been published for a different plant by two other botanists.

    *Erythronium nuttallianum* Roem. & Schult. (1820)

    The name *Erythronium nuttallianum* Regel is illegitimate because it is a(an):

5. Suppose you determine that *Herba foetida* Benth. is really a member of the genus *Planta*. Write the name as it would appear in the genus *Planta* with yourself as the combining author.

6. Explain the difference between a holotype and an isotype.

7. The following names have all been considered to represent the same taxon:

    *Centaurium curvistamineum* (Wittr.) Abrams (1951)

    *Centaurium minimum* (Howell) Piper (1915)

    *Centaurium muhlenbergii* (Griseb.) Wight ex Piper (1906)

    *Centaurium muhlenbergii* (Griseb.) Wight ex Piper forma *albiflorum* (Suksd.) St. John (1937)

    *Centaurium muhlenbergii* (Griseb.) Wight ex Piper var. *albiflorum* Suksd. (1927)

    *Centaurodes muhlenbergii* (Griseb.) Kuntze (1891)

    *Erythraea curvistaminea* Wittr. (1886)

    *Erythraea minima* Howell (1901)

    *Erythraea muhlenbergii* Griseb. (1839)

    What kind of synonyms are *Erythraea minima* and *Centaurium minimum?*

    What kind of synonyms are *Centaurium curvistamineum* and *Centaurium muhlenbergii?*

    What is the difference between the two kinds of synonyms?

    What did Piper do?

    Why is there a different ending for the specific epithets of *Erythraea minima* and *Centaurium minimum?*

    Why is the epithet *muhlenbergii* spelled the same in each genus?

    Which specific epithet has nomenclatural priority?

    If this species is considered to be a species of *Centaurium,* what is the correct name?

8. Suppose you were going to name a species of the genus *Planta* in honor of your lab instructor. Write out the name in full including the author.

9. Describe three steps that you would have to follow to name this species of *Planta*.

   a.

   b.

   c.

10. Lists of authors are often included in regional or state floras. Find out who the following people were. Determine the information requested.

| Abbreviation | Full name | Country | Years lived |
|---|---|---|---|
| L. | | | |
| Pursh | | | |
| Nutt. | | | |
| Torr. & A. Gray | | | |
| DC. | | | |
| Benth. & Hook. | | | |
| Engelm. | | | |

11. How many occupations can you list that would need to use scientific names for plants? How would these groups of people be impacted by a change in names for plants?

# Taxonomic Evidence

chapter 3

There are many kinds of evidence taxonomists can use to identify species, genera, and families of plants. The traditional evidence used to classify plants has been the morphology, or the shape and size of the various parts of a plant. During the 20th century, taxonomists began to use other evidence, such as anatomy, chromosomes, secondary plant compounds, and proteins. In the past two decades there has been an explosion in the use of nucleic acids (DNA and RNA) to aid in plant classification.

This chapter will explore some of the forms of evidence used in classification today. Taxonomists use evidence such as leaf shape, petal color, or DNA sequence of two different plants to determine if they are the same species, or to determine how closely related they are to each other. A critical aspect of this process is the issue of **homology.**

## Characters and Character States

An aspect of a plant, such as petal color, is called a **character.** Each character can have different **character states,** such as red or white petals. In the case of petal color, it would seem easy to determine that the character states of red and white are homologous, or derived from a common ancestry. However, it is often difficult to be certain of homology of character states. In general, similarity of position, developmental pathways, or intermediate forms of character states can provide some evidence of homology.

This chapter will focus on vegetative characters, the terms used to describe the roots, stems, and leaves of vascular plants. Reproductive terminology will be introduced with each of the groups of vascular plants. In the first half of the chapter we will provide definitions of selected terms. The second half of the chapter will explore ways to examine character variation and to determine the presence of character states.

Character states that can be counted or that can be measured against a standard (e.g., a ruler) are **quantitative characters.** Others that cannot be represented as a numerical value (e.g., ovary position, fruit type, or presence or absence of a structure or chemical compound) are **qualitative characters.** Measurements of quantitative characters yield **continuous data** if any value within a range is possible (e.g., plant height, leaf length and petal width) or **discontinuous data** if only discrete, integer, or count values can be obtained (e.g., numbers of leaves, petals or seeds). Many qualitative characters have only two character states (e.g., leaves punctate or not punctate). Multistate characters have three or more alternative expressions (e.g., parietal, axile, basal, apical, and free-central placentation). Some characters that are often treated as qualitative (e.g., shapes) can also be expressed as a series of measurements. Characters such as color may be either qualitative or quantitative depending on the amount of effort expended. Color quantification, for instance, can be accomplished by comparison to a continuously varying standard such as a color atlas.

Taxonomists employ many different kinds of characters in their studies of plant variation. Some are macroscopic features, readily visible to the naked eye or requiring only slight magnification to be observed. You have undoubtedly become familiar with many such features and the terminology that describes them. Microscopic features can be equally useful. These include the fine structures of seeds and other small plant parts and the even finer structures of pollen grains, the internal anatomy of stems, leaves, and flower parts, the patterns of embryogenesis, etc. The microscope has made it possible for taxonomists to examine the number, structure, and behavior of chromosomes.

In recent years molecular characters have become important in taxonomic studies. To a certain extent taxonomists have always been interested in molecular features. Such characters as odor and color have an underlying molecular basis. However, modern methods of analysis of the chemical constituents of plants have made these data more accessible to taxonomists and more quantifiable. Taxonomists have used comparisons of secondary plant products such as phenolics, alkaloids, and terpenoids as well as the associated biosynthetic pathways in their analyses. The great strides in biotechnology that have made genetic engineering a household phrase have also brought the analysis of DNA and related macromolecules into the labs of many taxonomists. Characters of interest and importance in taxonomic studies include single base pair changes in the structure of genes. This topic is discussed in greater detail in Chapter 17.

# Vegetative Terminology

Botanists long ago recognized the need for words that accurately describe the variable features of plants. Latin was the language used by scholars in various European countries during the early history of plant taxonomy, and training in classical languages was a standard part of a person's formal education. It was during this time that most botanical terms were coined. Some terms employed to describe plant variations are words used as in classical Latin. Other classical words were redefined with new meanings for use as descriptive terms. Still others were derived from Latin or Greek roots. Most of the terms we use today can be traced to these classical languages. Many of the terms also share roots in Latin or Greek with words that are a part of your everyday vocabulary. If you can see the connection between the terms and words that you already use, you will probably find it easier to learn the terms.

The following is an illustrated list of *selected terms* that describe some of these variations. These terms are chosen because of their widespread use. Many additional terms have been coined that describe conditions occasionally encountered in plants. No attempt has been made to list here all the possible terms and synonyms. To do so would be confusing and overwhelming to beginning students. The selected terms serve merely as a beginning point. From time to time you will encounter words that are not included in this list. The glossary (p. 495) in this book is much more comprehensive. Most floras and manuals that you are likely to use also have glossaries. Occasionally you may encounter slight differences from book to book in the definition or use of a particular term. In this situation use the term as defined and applied by the author of the book you are using.

**A WORD OF CAUTION**—Terms were coined to represent clearly distinguishable differences. However, plants often vary continuously, and the structures that you see on a particular plant may not precisely match the illustrations or the definitions below. As you use a term you gradually will develop an understanding of the range of variability that it encompasses. You may encounter a feature that seems to be intermediate between two terms. In this situation, terms are often used in combination (e.g., leaves linear-lanceolate) or with modifying words (e.g., leaves broadly ovate). Organs on a plant may vary to the extent that it is necessary to describe a range of conditions (e.g., leaves linear to lanceolate). Sometimes a structure will look almost like the condition described by a particular term, but not quite, and a modifying prefix may be used (e.g., a subsessile leaf is almost sessile; it has a very short petiole). Some terms are diminutives of other terms and are recognizable by the presence of the letters, -ul-, intercolated into the word (e.g., a serrulate leaf has tiny teeth on its margin whereas a serrate leaf has larger teeth).

# DEFINITIONS OF SELECTED TERMS _____

## Life Span

**annual**—A plant that completes its life cycle in one growing season. It begins as a seedling, grows to reproductive maturity, flowers, sets seed and dies.

**biennial**—A plant that completes its life cycle in two growing seasons. Vegetative growth takes place in the first year. In the second year the plant grows to reproductive maturity, flowers, sets seed and dies.

**perennial**—Any plant that lives for more than two growing seasons. It may be herbaceous and die back to underground perennating structures such as rhizomes, tubers, corms or bulbs or it may be woody. All trees and shrubs are perennials.

## Plant Habit

**herb**—A plant with little or no above-ground perennial woody tissue. In temperate regions most herbs die back to the ground at the end of each growing season. However, many herbaceous plants of warm regions do not die back. An herb may be annual (underground parts die), biennial or perennial (underground roots, rhizomes, corms, bulbs, tubers survive to next growing season).

**subshrub**—A plant having the stature of a shrub but not completely woody. The lower stems of the plant have some stiff perennial woody tissue but the upper portions are essentially herbaceous. The term **suffrutescent** is often used to describe such plants.

**shrub**—A woody perennial plant of comparatively low stature with one to many relatively slender trunks from near its base.

**tree**—A large woody plant with one to several relatively massive trunks and an elevated crown. The boundary between trees and shrubs is not sharp, and growing conditions may determine whether a particular woody plant matures as a tree or as a shrub.

**succulent**—A plant with thick, usually soft, watery leaves and/or stems (e.g., ice plants or cacti).

**vine**—A woody or herbaceous plant that trails on the ground or climbs over other plants or obstacles by various means. The stem of a vine is often so weak that it cannot support the plant by itself.

## Plant Organs

**roots**—Usually cylindrical plant axes that lack external nodes and buds. Branch roots arise from the interior of other roots. Roots are usually the portion of a plant that absorb water and minerals. Most commonly roots are underground structures, but some plants produce above-ground roots. Roots are positively geotropic.

**stems**—The usually cylindrical leaf-bearing main axes of a plant that are externally divided into nodes and internodes. Branches and flowers arise from external buds. Stems may be aerial or subterranean. Stems have negative or neutral geotropism.

**leaves**—The lateral appendages on a stem that usually serve as the primary photosynthetic surface of a plant. Leaves are attached at nodes and buds arise in their axils (sometimes not evident in the axils of young leaves).

## Root Types (Fig. 3-1)

**tap root**—A central main root that descends vertically; it is larger than any branch roots.

**fibrous root**—A thin root arising from another root or from stem tissue. The total root system can be fibrous.

**adventitious root**—A root that originates from stem or leaf tissue rather than from the interior of another root.

## Stem Types (Fig. 3-1)

**aerial stem**—Prostrate to erect, above-ground stems. These are the most commonly encountered types of stems.

**rhizome**—An underground, horizontal stem that spreads and perennates a plant. The leaves of a rhizome are often reduced to scales. Rhizomes usually bear adventitious roots.

**stolon**—A horizontal stem at or just above the surface of the ground that gives rise to a new plant at its tip or from axillary branches. It may bend to the ground and surface.

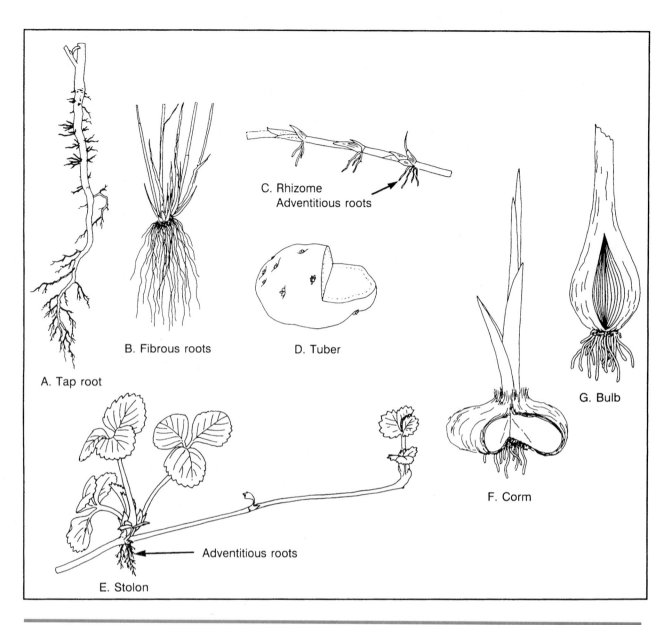

**Figure 3-1** Major types of underground root and stem structures.

**bulb**—A thickened, short vertical stem axis with large, fleshy storage leaves attached; usually borne below ground.

**corm**—A solid, erect, enlarged underground stem with leaves absent or dry and scalelike; usually borne below ground.

**tuber**—A solid enlarged, horizontal, shortened, stem, usually borne below ground.

## Stem Features (Fig. 3-2)

**node**—The position on a stem where a leaf is or was attached.

**internode**—The part of a stem axis between two successive nodes.

**axil**—The upper angle formed by a leaf and the twig to which the leaf is attached. Typically an axillary bud (lateral bud) forms in each leaf axil.

**bud**—An external meristem (either naked or protected by bud scales) found on stems. A bud may

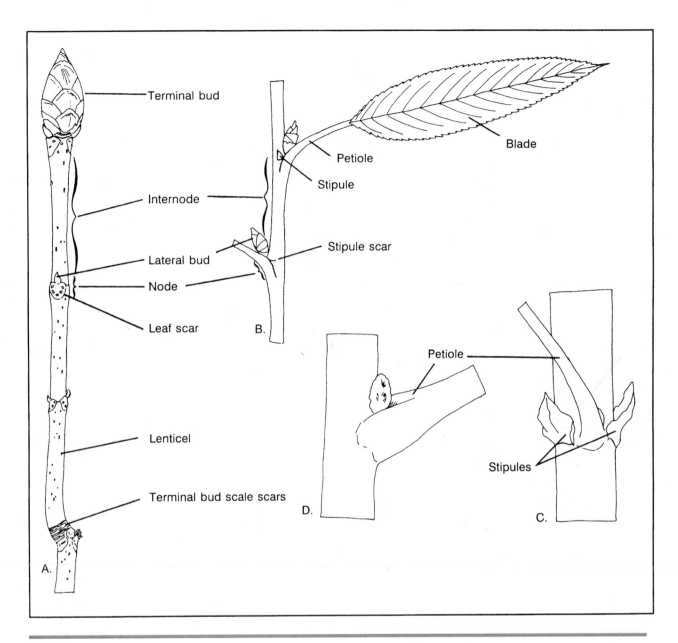

**Figure 3-2** Structures associated with twigs. A. Twig in dormant condition. B. Segment of leafy twig. C. Base of stipulate leaf. D. Base of estipulate leaf.

give rise to a leafy stem, one or more flowers, or a combination of vegetative and reproductive structures.

**axillary bud**—A bud borne in the axil of a leaf; a lateral bud.

**terminal bud**—A bud borne at the end of a stem.

**bud scales**—Scale leaves that cover and protect terminal and axillary buds.

**bud scale scars**—The rough places or scars left on a stem when bud scales fall off.

**stipule scars**—A pair of rough places or scars (or sometimes a single ringlike scar) left when stipules fall from a twig; they may be found on either side of the leaf attachment or leaf scar and may be the only indication that stipules were present on the plant.

**leaf scar**—The rough place or scar left when a leaf falls from a twig. Leaf scars contain one or more dot-like scars called vascular bundle scars. The pattern of bundle scars along with the shape of the leaf scar can sometimes be used to determine the genus or species of a plant.

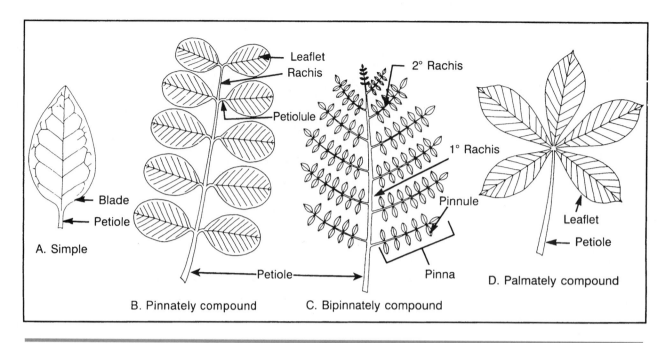

**Figure 3-3**  Features of simple and compound leaves. A. Simple leaf. B. Pinnately compound leaf. C. Bipinnately compound leaf. D. Palmately compound leaf.

**lenticel**—Specialized corky outgrowths from the bark of twigs that serve as passage-ways for gas exchange.

## Leaf Parts (Fig. 3-2)

**leaf blade**—The portion of a leaf that is thin and flattened. In most plants leaf blades are the principal photosynthetic surfaces. Some leaves (e.g., pine needles) lack flattened blades.

**petiole**—The stalk of a leaf, sometimes absent.

**stipules**—A pair of appendages located at the base of a leaf. They are sometimes attached to the petiole and sometimes to the stem; occasionally they are fused into a ring around the stem. A leaf may be **stipulate** (with stipules) or **estipulate** (without stipules). Stipules can take on many forms, from leaflike to thin and papery to hard and woody (see Fig. 3-11). In some plants stipules fall off soon after a leaf develops, leaving behind stipule scars on the stem or petiole.

## Leaf Complexity (Fig. 3-3)

**simple leaf**—A leaf with a single blade.

**compound leaf**—A leaf with more than one blade per petiole or leaf base.

**leaflet**—One of the segments of a compound leaf. A leaflet has its own blade; it may be sessile or it may have its own stalk, a **petiolule.**

**rachis**—The axis of a pinnately compound leaf; the rachis is the continuation of the petiole above the attachment of the lowermost leaflet.

**pinnately compound**—A compound leaf with leaflets attached like the vanes of a feather to an elongated rachis.

**bipinnately compound**—A compound leaf divided twice with leaflets attached to secondary rachises.

**pinna** (pl., **pinnae**)—The primary division(s) of a pinnately compound leaf.

**pinnule**—The ultimate division of a two or more times pinnately compound leaf blade.

**petiolule**—The stalk of a leaflet.

**palmately compound**—A compound leaf with leaflets radiating from a common point of attachment like the fingers of a hand; no rachis is present.

## Leaf Attachment (Fig. 3-4)

**petiolate**—A leaf possessing a stalk or petiole; opposite of sessile.

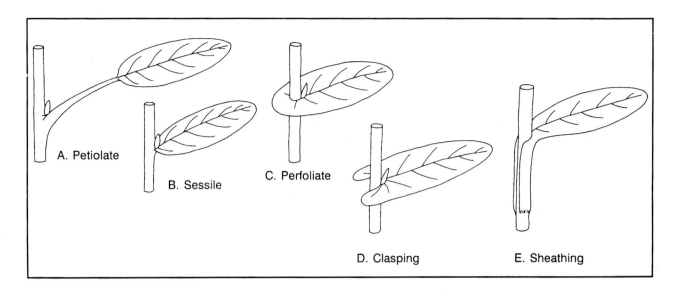

**Figure 3-4** Attachment of leaves to stems.

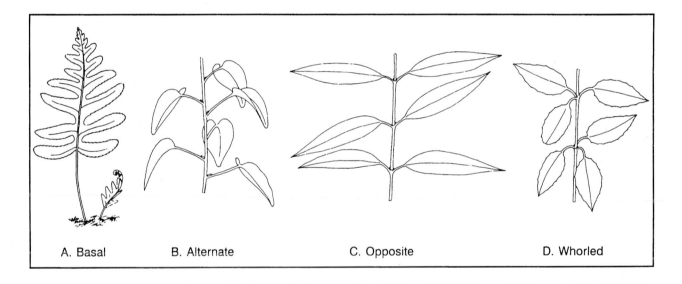

**Figure 3-5** Leaf arrangement (phyllotaxy).

**sessile**—A leaf blade attached directly to a twig, lacking a petiole. Sessile also is used to describe any other structure that lacks a stalk (leaflet, flower, fruit, etc.)

**subsessile**—A leaf with a very short petiole.

**perfoliate**—A sessile leaf with the stem passing through the blade. A pair of opposite sessile leaves fused to each other around the stem are **connate-perfoliate.**

**clasping**—A sessile leaf with lobes of blade tissue projecting backwards on either side of the stem.

**sheathing**—A leaf blade or petiole that is prolonged into a tube that partially or completely surrounds the twig above the node to which the leaf is attached. Grasses and most other monocots have sheathing leaves.

## Leaf Arrangement (Phyllotaxy) (Fig. 3-5)

**basal**—A pattern of leaf distribution where leaves are attached to an underground stem or rhizome. Such leaves appear to arise directly from the ground.

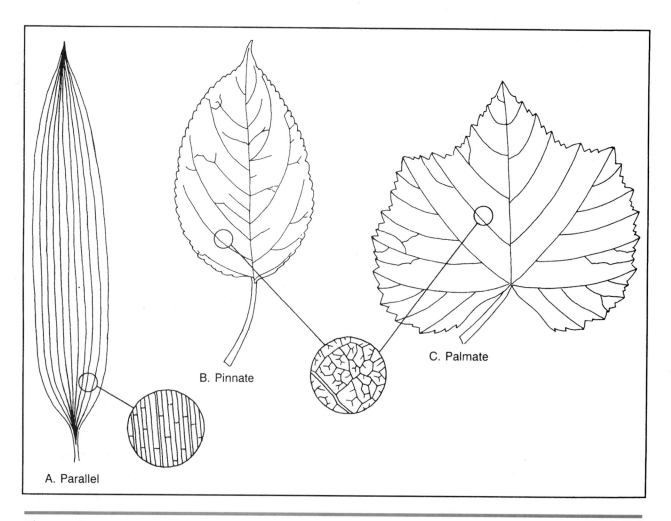

**Figure 3-6** Common types of leaf venation. Both pinnate and palmate venation usually are associated with netted venation (see inset).

**cauline**—A pattern of leaf distribution where leaves are attached to an above-ground stem.

**alternate**—Only one leaf attached at each node; the commonest condition.

**opposite**—Two leaves attached at each node, borne on opposite sides of the stem.

**whorled**—Three or more leaves symmetrically distributed around the stem at the same node. (Note: Leaves attached to an axillary bud or to a short lateral branch should *not* be confused with whorled leaves.

## Leaf Venation (Fig. 3-6)

**pinnate venation**—Pattern of veins in a leaf where there is a single large midvein and secondary veins diverging from it.

**palmate venation**—Pattern of veins in a leaf in which three or more primary veins diverge from a common point at the base of a leaf blade; secondary veins diverge from each primary vein.

**parallel venation**—Pattern of veins in a leaf where several primary veins extend side by side throughout the full length of a leaf; secondary veins are obscure or pass ladder-like between the primary veins; characteristic of most monocots.

## Leaf Blade Shape (Fig. 3-7)

**needle-like**—Leaves very long and narrow, relatively thick; in the shape of a needle with little or no differentiation of a blade.

**scale-like**—Leaves in the shape of small, flat or thickened triangles with little or no differentiation of a blade.

**linear**—Long and narrow, the sides parallel or nearly so, leaf blade more than 4× longer than broad.

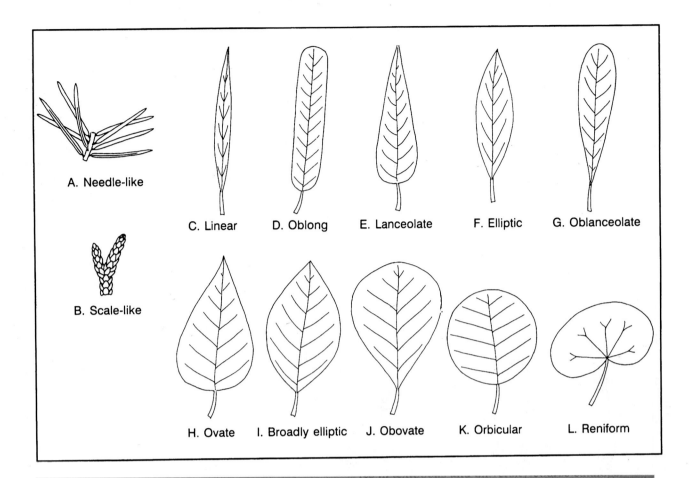

**Figure**
**3-7** Common shapes of leaf blades.

**oblong**—Nearly rectangular, 2× to 4× longer than broad with sides nearly or quite parallel most of their length.

**lanceolate**—Shaped like a spearhead; 3× to 4× longer than broad; widest near base and tapering to the apex.

**oblanceolate**—The inverse of lanceolate, widest at tip and tapering to the base.

**ovate**—Egg shaped; leaf blade less than 3× longer than broad and widest near base.

**obovate**—The inverse of ovate; widest near the apex.

**elliptic**—Shaped like an ellipse; widest near middle and tapering equally toward both ends.

**orbicular**—Circular, about as long as broad.

**reniform**—Kidney shaped, wider than broad and notched at base.

## Leaf Apices (Fig. 3-8)

**acute**—Sharp-pointed with an angle of less than 90 degrees, often broadly used for any type of pointed leaf.

**acuminate**—Long-tapering to a point, often involves a change of angle of margin with respect to midrib.

**obtuse**—Blunt with an angle of greater than 90 degrees, often broadly applied also to leaves that are rounded or truncate.

**emarginate**—Blade with a notch at apex.

**truncate**—Appearing as if cut off at end; tip nearly straight across.

**rounded**—Forming a smooth continuous curve.

**mucronate**—Blade with a small abrupt point at apex.

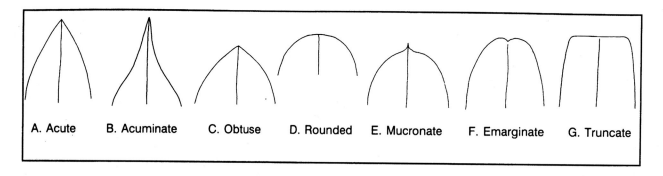

**Figure 3-8** Common leaf blade apices.

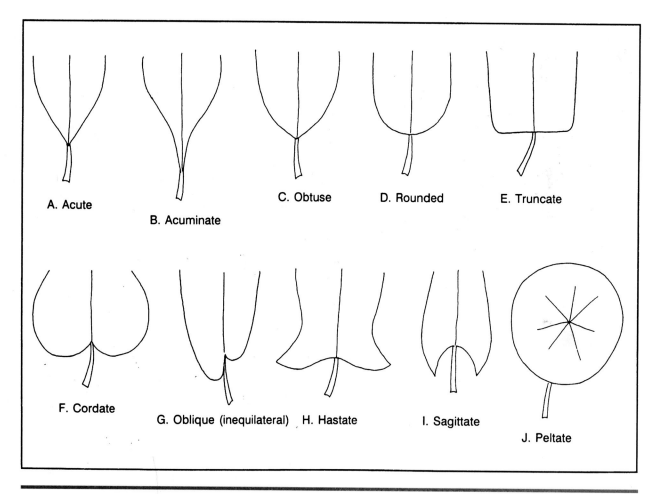

**Figure 3-9** Common leaf blade bases.

# Leaf Bases (Fig. 3-9)

**acute**—Sharp pointed with an angle of less than 90 degrees, often broadly used for any type of pointed leaf base.

**acuminate**—Long tapering to the base, with a change of angle of margin with respect to midrib.

**obtuse**—Blunt with an angle of greater than 90 degrees, often broadly applied to leaf bases that are rounded or truncate.

**rounded**—Forming a smooth continuous curve.

**truncate**—Appearing as if cut off at the base, nearly straight across.

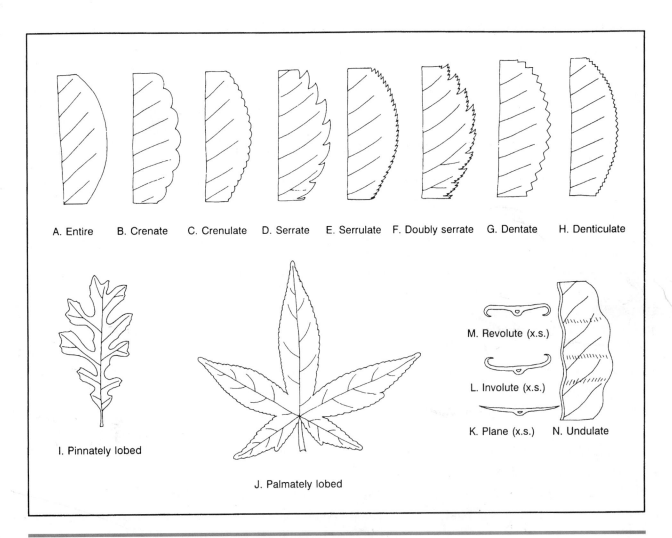

**Figure 3-10** Common leaf blade margins.

cordate—Shallowly notched at the base.

oblique—Slanting unequally or unequally sided at base; inequilateral; asymmetrical.

hastate—Of the shape of a spearhead with sharp basal lobes sharply divergent; used for a shape or base.

sagittate—Blade with two basal lobes, lobes parallel or converging to petiole; arrowhead shaped with tails long and narrow and pointing downward or inward.

peltate—Umbrella-like; the petiole attached to blade inside the margin. Peltate leaves are often orbicular.

## Leaf Margins (Fig. 3-10)
## –as viewed from top–

entire—Margin that is smooth or of unbroken outline, without teeth or lobes.

crenate—Margin with regular rounded teeth.

crenulate—Diminutive of crenate; margin with very small rounded teeth.

serrate—Saw-toothed margin with sharp teeth bent toward apex of the leaf.

serrulate—Diminutive of serrate, very small teeth bent toward apex of the leaf.

doubly serrate—Margin with teeth of two sizes (small teeth on the big teeth); teeth bent toward apex.

dentate—Margin with teeth pointed perpendicular to margin.

denticulate—Diminutive of dentate, teeth small and pointed perpendicular to margin.

lobed—Margin incised or dissected with very large teeth (lobes).

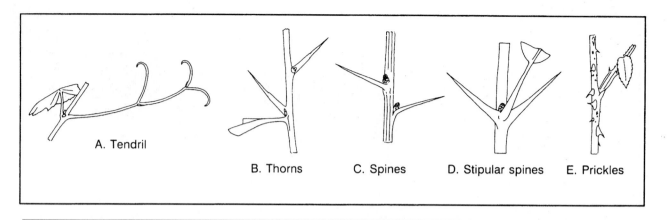

**Figure 3-11** Special modifications of plant parts.

A. Tendril    B. Thorns    C. Spines    D. Stipular spines    E. Prickles

**pinnately lobed**—lobes along the sides of an elongate leaf; there is a single clear midvein.

**palmately lobed**—lobes radiating from the petiole or point of attachment of the blade. Usually several major veins radiate from a common origin.

## —as viewed from the side—

**plane**—Flat; a standard leaf with midrib and margin all in one plane or nearly so.

**involute**—A leaf that rolls upward; making the top concave.

**revolute**—A leaf with margins rolled or bent downward, making the top convex.

**undulate**—Wavy-margined; a leaf margin that bends gently above and below the plane of the midrib.

## Special Features (Fig. 3-11)

**tendril**—A spirally coiled or clawlike structure by which a plant grasps an object and clings to it. A tendril may originate by modification of a leaf or a stem.

**thorn**—A woody, sharp-pointed modified stem. It may be terminal or borne in the axil of a leaf or leaf scar.

**spine**—A woody, sharp-pointed modified leaf or leaf-part.

**foliar spine**—modification of a whole leaf into a spine; buds or branches are often borne in the spine's axil.

**petiolar spine**—persistence of a petiole as a spine after the leaf's blade has fallen.

**stipular spines**—spines borne in pairs and lateral to leaf or leaf scar.

**prickle**—A small, sharp, non-woody structure developed from outgrowth of the surface of bark or epidermis.

## Surface Features (Fig. 3-12)

**glabrous**—Surface smooth, lacking trichomes (plant hairs).

**pubescent**—A surface with trichomes. Sometimes used to designate pubescence composed of short, straight soft hairs.

**pilose**—Pubescent with straight, soft hairs.

**villous**—Pubescent with more or less curly, soft hairs that are not tangled or matted.

**strigose**—Pubescent with short hairs that lie flat against surface.

**hirsute**—Pubescent with rather rough, coarse hairs.

**hispid**—Pubescent with stiff bristle-like hairs.

**scabrous**—Pubescent with short, stout hairs; sandpapery to the touch.

**tomentose**—Pubescent with long, curly, matted hairs; woolly.

**puberulent**—Pubescent with very short hairs.

**stellate**—Pubescent of hairs that branch at or near the base.

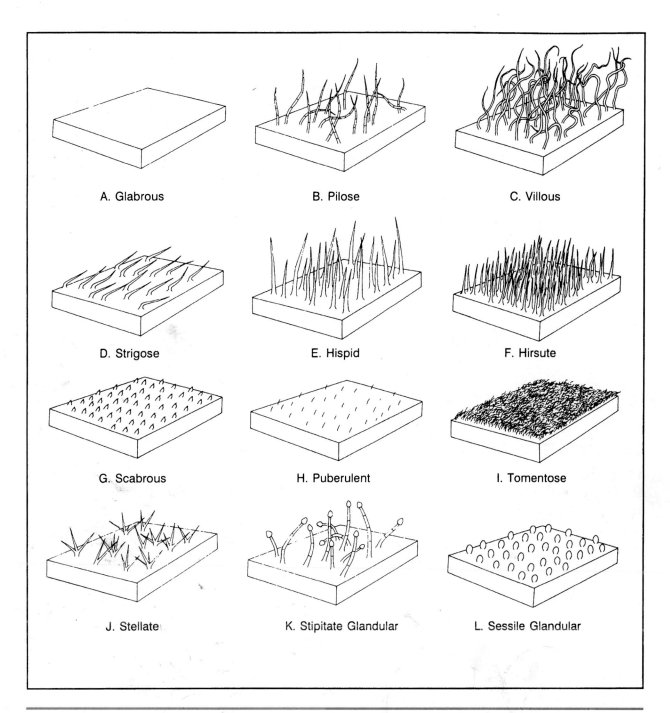

**Figure 3-12** Surface features.

**glandular**—Pubescent with glandular hairs. The glands often break down into sticky beads of fluid. The glands may be stalked (**stipitate glands**) or sessile.

**glandular punctate**—Surface with glands sessile or embedded. In some leaves embedded glands are visible as translucent dots when light is transmitted through blade.

**glaucous**—A surface with a waxy covering so thick that it gives a white or bluish color to leaf or other plant part. In some cases the waxy coating can be rubbed off, revealing a darker underlying color.

# Character Variation

Anyone who has observed plants carefully is aware of the great variability of plant structures. The recognition of patterns in this variability is one of the fundamental observations that makes possible the classification of plants into logical groupings. A second and sometimes disquieting observation is that some plant variations are much more useful than others in determining relationships and establishing classifications. Indeed, some variations can be very misleading. It is not always apparent which variable features are useful and which are not. One of the important functions of systematic botany is to determine the nature of the variation observed in plants, to describe and interpret the variable features correctly, and to use them at the appropriate taxonomic level.

## Distinguishing Inherited Variation Patterns from Environmentally Induced Variation

The two principal sources of variation are inheritance and environment. All measurable or observable aspects of a plant's form, chemical constitution and physiological responses comprise its **phenotype.** The phenotype represents the interaction of a plant's heredity or **genotype** and its environment. A particular characteristic of a plant's phenotype may be controlled primarily by its heredity, induced by the environment in which the plant is living, or determined by a combination of both.

Some characters are more useful in taxonomic studies than others. Taxonomists are particularly interested in features that are useful in determining relationships and in organizing stable classifications. Since genetically controlled characters are more stable than environmentally induced features and are capable of being passed from parent to offspring, inherited characteristics are considered to be of greater informational value in taxonomic studies. Unfortunately, the underlying genotype can only be estimated by thorough study of a plant's phenotype in conjunction with genetic experiments.

One problem that taxonomists often face is to determine which observed phenotypic variations are environmentally induced **(plastic responses)** and which are inherited. Variation that is induced by the environment can be very striking but is easily changed and is not passed on to the next generation. The water buttercup, *Ranunculus aquatilis,* for instance, has dissected leaves under water and crenate-cordate ones above. The ability to make both kinds of leaves is inherited, but the form of a particular leaf is determined by whether it develops above or below the water surface. Shade leaves of trees and shrubs tend to be larger than ones growing in the sun. Variation from individual to individual may represent the conditions under which the plants are growing rather than genetic differences among the plants.

How does a taxonomist studying a series of specimens separate environmentally induced variation from inherited genetic differences? If the only plants available for study are dried specimens, it may be very difficult to separate the two. This is one of the principal sources of error in taxonomic studies. Early botanists working with very limited samples often had no way of knowing which characters were inherited and which represented environmentally induced variation. Many species were named that later were determined to be mere variants. Even today this problem still exists and the names published by early workers must be sorted out.

Botanists who have observed the plants growing in the wild often can recognize the effects of environmentally induced variation. If live plants are available for study it is often possible to separate genetic and environmental responses under experimental conditions. One method of doing so is to use transplantation studies (see Chapter 17). Plants that express phenotypic variation may be grown together in a **uniform experimental garden** where environmental variation is minimized. Variations that persist when plants are grown under uniform conditions are probably determined by the plants' genetic makeup. On the other hand if character differences noted in wild-grown plants disappear under uniform conditions, environmental factors are probably responsible for the observed variation.

## Determining the Usefulness of Particular Variables

The patterns of variability are important in determining the level at which a particular character is meaningful. Although a feature may be genetically determined, it may not be useful in a particular taxonomic study. A character that does not vary from one taxon to another within the group under study is not useful in constructing classifications within that group. On the other hand a character may be variable but lacks meaningful correlations with other characters and thus has little predictive value. The usefulness of a character may depend on the level in the taxonomic hierarchy or the taxonomic group with which one is working. Some characters vary from one species to another within a genus or family, whereas others are constant within a particular family but vary among families. A character may vary among individuals in one taxonomic group and among species in another. The usefulness of each character must be determined on the basis of thorough knowledge of plants being studied. A character that is useful in the study of one taxonomic group cannot be *assumed* to be useful in another. It must be tested first.

Individual plant characters do not exist alone, but are associated with all other characters. Many individual characters are correlated with one another because they allow each other to exist. For example, if a plant has very large leaf blades, then there must be stout petioles to support them. Characters such as zygomorphy can be split into more basic characters **(unit characters)** such as stamen number or petal shape or arrangement. The emphasis of both complex characters and their dependent constituent unit characters can bias a taxonomic study.

Botanists have to deal with a kind of variation that zoologists seldom encounter: variation among the parts of a single individual. Because almost all animals are determinate in their growth, among zoologists it is generally conceded that an individual animal does not vary and that variation is a characteristic of a population of organisms. However, because an individual plant grows by serially active meristems, each of which is active under slightly different environmental conditions than those that precede or follow it, organs derived from these different meristems may have different phenotypes. Therefore an individual plant can be thought to represent a population of meristems which show variation in their expression at different times during the life of the plant. For example, in spring when there is adequate water, the internodes formed by trees and shrubs of temperate zones are often elongated, but in late summer when water is limited growth slows down and the internodes are often much shorter. The basal leaves of an herbaceous plant may be quite different in shape, size, and other features than those produced higher up on the stem. In California, poison-oak is a shrub on dry open sites and a vine in shady canyon bottoms. To minimize differences caused by environment, taxonomists comparing individuals for statistical studies should choose plant organs to measure that are produced under similar growing conditions at the same time in the growing season. Determining the growing conditions from dry specimens is not always possible.

A taxonomist also needs to be careful to not add a bias to either character selection or to the plants chosen to measure. Collectors were unconsciously (or deliberately) choosing smaller plants so they would fit on the limited size of the herbarium sheets. Other biases include the collection of individuals with pretty flowers, the sampling of only the lower branches from trees (easiest to reach), and the gathering of specimens from roadside habitats that often differ from less-disturbed environments.

Another problem is that many plants are capable of vegetative reproduction, and often this results in the production of many apparently separate individuals of exactly the same genotype. If all these individuals were to be measured, a bias from many duplicate measurements would be created. Clones of some species may be quite large. Quaking aspen *(Populus tremuloides),* for instance, has clonal populations that sometimes cover several hectares. Collecting individuals no closer than a set distance (determined from study of the plants' growth characteristics) will generally minimize this problem. It is important to remember that if the sample of plants chosen does not represent the population in nature that we think it does, or if it does and the characters are poorly chosen or incorrectly measured then the analyses based upon these observations will be wrong. If the measurements are wrong, all of the procedures that follow will simply magnify the errors.

## Convergence and Parallelism

An important part of the analysis of characters is to distinguish character states resulting from convergent or parallel evolution from those inherited from a common ancestor. Distantly related plants living in similar environments often have remarkably similar features. Thus members of the Cactaceae and certain African species of *Euphorbia* have similar ribbed succulent stems that are armed with sharp spiny structures. From a distance these plants are remarkably similar. Their similarities are not indicative of a close relationship, however, but rather of convergent evolution under similar environmental conditions. Their very different reproductive features betray their lack of relationship as does a close inspection of their superficially similar vegetative structure [Cacti have true spines; succulent Euphorbias have thorns or stipular spines].

Convergent evolution has also occurred in various other groups of plants. Thus many species of rainforest trees of diverse families have similar entire-margined leaves with elongated "drip tips." Unrelated shrubs in widely separated areas with a "Mediterranean" climate commonly have sclerophyllous leaves (small, leathery, and thick). Desert shrubs of various families often have small, grayish leaves. Character similarities resulting from convergence must be discounted in determining relationships.

The effects of parallel evolution are more subtle. Related plants often respond to similar environments in the same way. The character states that they share may result from the same genetic background. However these features may be misleading if they evolved independently in different lineages within a group of related plants. Character states that evolved in parallel may suggest different relationships than those that were inherited from common ancestors. Such distinctions become very important in cladistic analyses (Chapter 4).

Vegetative parts of plants are often considered to be of lesser value in determining relationships than are reproductive features. This is in large part a concession to both the plasticity of vegetative structures and the effects of convergent and parallel evolution. Reproductive structures are not free from convergence and misinterpretation though. Catkins or catkin-like structures have apparently evolved independently in several unrelated lineages. In the early phylogenetic classification proposed by Engler and Prantl all of the catkin-bearing dicots were grouped together as a single taxon, the Amentiferae. Selection by both pollinators and seed dispersal agents can result in parallel or convergent character states in reproductive features.

## Description of Variation in Quantitative Characters

When a plant taxonomist measures a plant, he or she is usually less interested in that individual plant, than what that individual plant can tell about the rest of the plants similar to the one measured. The taxonomist is attempting to characterize a group of items (plants or plant parts) as they exist in nature. This group is the **population** (in the statistical sense). A population is usually so large or inaccessi-

ble that it cannot be measured in its entirety. Even though it cannot be measured directly, this population is *assumed* to exist and therefore we can *assume* it can be described. Although we cannot measure even selected characters of every individual of a population, we can measure selected characters of a few plants isolated from the population in nature. This smaller subset of individuals is called a **sample.** All taxonomic work is done with samples. From these samples we infer the characteristics of the population from which the sample was drawn.

As a relatively simple example, we might want to know if average leaves on the east and west side of the same California live oak tree are the same length. [As discussed above, an individual tree can be considered to be a group of many meristems]. The populations being compared comprise all of the leaves on the two sides of the tree. To measure all of them would be overwhelmingly time-consuming. Fortunately by statistical sampling methods we can avoid this tedious process and still arrive at meaningful results. We would first chose a random sample of leaves from each side of the tree, measure each leaf and record the lengths in a table such as Table 3-1. Note that the values within a column vary and that many of the values overlap those in the second column. Probably you will also note that very

| Individual leaf | East side | West side |
|---|---|---|
| 1 | 4.2 | 3.5 |
| 2 | 4.2 | 3.4 |
| 3 | 4.9 | 3.6 |
| 4 | 4.5 | 4.1 |
| 5 | 4.6 | 5.0 |
| 6 | 4.6 | 4.6 |
| 7 | 5.4 | 4.0 |
| 8 | 4.1 | 2.0 |
| 9 | 5.0 | 3.3 |
| 10 | 5.1 | 4.1 |
| 11 | 5.2 | 3.3 |
| 12 | 4.5 | 3.6 |
| 13 | 3.6 | 4.8 |
| 14 | 4.3 | 4.7 |
| 15 | 4.6 | 4.2 |

**Table 3-1**  Table 3-1 compares the length (in cm) of leaves collected from the east and west sides of the same California live oak tree (*Quercus agrifolia*).

| | SPECIES A | | SPECIES B | |
|---|---|---|---|---|
| Individual plant | Corolla tube length | Calyx lobe length | Corolla length tube length | Calyx lobe length |
| 1 | 2.5 | 5.0 | 3.0 | 3.8 |
| 2 | 2.5 | 6.5 | 5.5 | 5.5 |
| 3 | 3.0 | 9.0 | 4.0 | 3.5 |
| 4 | 2.5 | 4.0 | 3.3 | 3.5 |
| 5 | 2.3 | 3.0 | 2.5 | 3.5 |
| 6 | 3.0 | 5.5 | 3.0 | 4.0 |
| 7 | 2.5 | 4.0 | 5.0 | 4.0 |
| 8 | 4.0 | 5.0 | 3.7 | 2.5 |
| 9 | 2.5 | 5.5 | 2.8 | 4.5 |
| 10 | 2.5 | 5.7 | 4.0 | 3.8 |
| 11 | 4.0 | 5.6 | 3.0 | 2.8 |
| 12 | 4.0 | 6.0 | 2.5 | 5.5 |
| 13 | 3.0 | 4.5 | 3.0 | 3.8 |
| 14 | 3.0 | 5.0 | 3.0 | 3.5 |
| 15 | 4.0 | 4.5 | 3.5 | 4.0 |

**Table 3-2** Table 3-2 lists measurements (in mm) of two characters taken from herbarium specimens of two species of *Schizanthus*

little meaning can be deduced from the raw numbers in the two columns. How can we make these numbers more meaningful so that we can interpret them? How do we know if the variation described in one column is different from that in another column? These questions can be answered by using either graphical or statistical methods. First we will look at some simple graphical methods and later some statistical procedures that accomplish the same results.

## Graphical Depiction of Character Variation

Two methods of graphically displaying the data in Tables 3-1 and 3-2 are commonly used—histograms and scatter plots. The histogram (bar-graph) is used when only one variable is to be displayed. The **histogram** is constructed by counting the number of observations in each of a set of previously defined intervals and making a bar equal in length to the count of observations in the interval. Figure 3-13 is a histogram depicting the data in Table 3-1. From this simple histogram we can see that leaves on the east side of the tree tend to be larger than those on the west side. Advantages of a histogram are its ease of construction and interpretation. Computer graph-

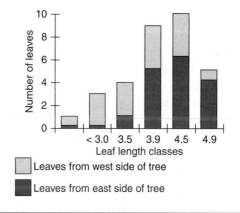

**Figure 3-13** Histogram comparing samples of coast live oak leaves collected on the east and west sides of the same tree. (Data from Table 3-1).

ics programs are widely available that can easily convert a data matrix such as that in Table 3-1 into histogram. A disadvantage of a histogram is that it can only display one character at a time.

A second type of graphic display that is often used for taxonomic data is the scatter plot (scatter diagram). The scatter plot is used to display or compare data representing two variables. The horizontal (x) axis is marked off in units appropriate for one of the variables (dimensions) and the vertical (y) axis the other. Then each pair of values is plotted to

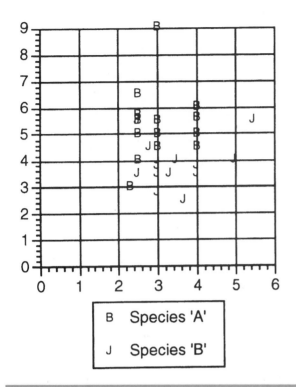

B Species 'A'

J Species 'B'

**Figure 3-14** A scatter plot comparing two species of *Schizanthus* for a pair of characters sometimes used to distinguish them (columns 1–4 in Table 3-2).

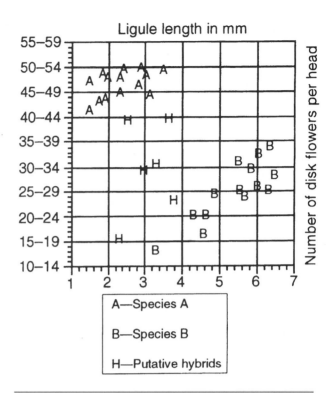

Ligule length in mm

Number of disk flowers per head

A—Species A

B—Species B

H—Putative hybrids

**Figure 3-15** A scatter diagram of two species and their putative hybrids.

produce a scatter of points. Two or more data sets can be plotted on the same scatter plot by using different symbols for the points. Figure 3-14 is an example of a scatter plot using the data from Table 3-2. From the distribution of points in Figure 3-2, it is evident that although the characters, corolla tube length and calyx lobe length, show a trend, there is much overlap between individuals of the putative species.

Scatter plots are sometimes used to graphically display evidence of hybridization (Fig. 3-15). In this example the two species are distinctly different in the characters, ligule length and number of disk flowers. The individuals suspected of being hybrids show intermediacy in both characteristics. Note that some of the suspected hybrids trend toward one or the other of the two species. This suggests the possibility of back crossing or hybridization beyond the first generation. Note that a graphic display such as this is not proof that hybridization has taken place at all.

A scatter diagram has its spatial limitations. Only two characters can be plotted on a two-dimensional display. A 3-dimensional scatter plot can be visualized though it is difficult to display without distortion. Computer programs are available that can plot multiple characters in N dimensional space, but these lose the value of a visual display. Any attempt to visually display multidimensional plots results in much distortion and loss of information.

There is an alternative way to display more than two characters on a scatter plot. The scatter plot can be elaborated so that additional characters are coded by the symbols that are used to note the data points of the two-dimensional plot. Figure 3-16 is an example of a **pictorialized scatter plot or diagram** that compares 4 varieties of *Lupinus nanus* in San Luis Obispo County, California. Note that in this scatter diagram, shading and different shapes of dots have been used to indicate different character states. Small tick marks have been added to the tops and sides of the dots to note the variations of additional characters. In this example, five different characters have been plotted. In a pictorialized scatter diagram it is possible to plot qualitative characters as well as the quantitative. There are drawbacks to the use of pictorialized scatter diagrams. As the symbology becomes more complex, it becomes progressively more difficult to see and interpret the patterns, especially in examples such as depicted in Figure 3-16.

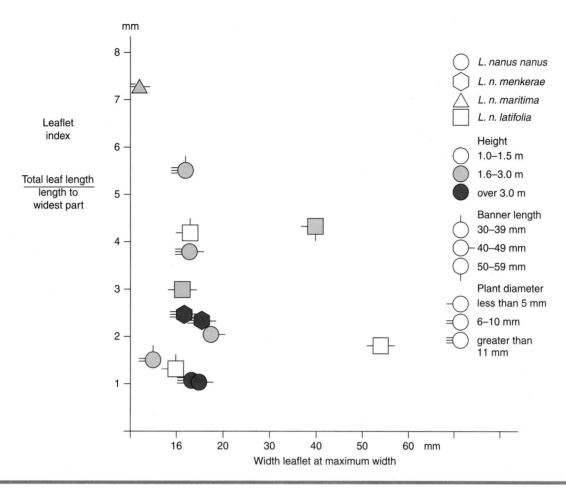

**Figure 3-16** Pictorialized scatter diagram showing the relationship among 5 characters and 14 populations of *Lupinus nanus* from San Luis Obispo County, California. Does the diagram indicate clearly the four varieties said to range into this county?

## Statistical Analysis of Character Variation

Character data are often subjected to statistical analysis. Qualitative data have the advantage that they are relatively easy to obtain, but must be modified (coded) before they can be handled by numerical means. Quantitative characters can be handled readily by numerical methods and are therefore easier to use in comparisons requiring statistical procedures. These distinctions are important when choosing the appropriate type of statistical analysis but have no significance in determining the importance of a character in classification or other taxonomic applications.

Graphical methods work well when the samples compared are (or are said to be) quite distinct from one another. If the range of values in one sample is duplicated to a considerable extent by the values in the other sample(s), statistical methods must be used. Statistical methods can determine whether there are statistically significant differences between populations. It is important to realize that each step must be done in the proper order and that a number of criteria (often assumptions) must be true for statistical analysis to yield proper results. The following is a short discussion of some of the fundamental considerations that must be kept in mind before statistical analyses are attempted. There are many sources with more complete discussions available (see Selected References at end of the chapter).

Statistical procedures are dependent on a predictable relationship between the sample and the sample's population. Just as we are not really interested in the individual plant we measured to get the sample, our interest in the sample is only as an estimate of the population. Statistical analyses are designed to compare parent populations, not the samples.

Before we can discuss statistical procedures we must introduce the concept of the **null hypothesis.** When we compare two or more populations by their samples we are really asking, "What is the probability that the samples represent the same population?" Our assumption, the null hypothesis, is that the samples do represent the same population. With statistical tests we can determine the likelihood that this is a true statement. If the null hypothesis is rejected as a result of statistical tests, an alternative hypothesis must be accepted.

The columns of numbers in Tables 3-1 and 3-2 represent variation in the lengths of three different characters. We can derive two characteristics from each of the columns. We can easily figure out a "center" value and a measure of "variability" among the numbers within the column. There are three commonly used measures of **central tendency** or "center" of the sample. The **mode** is the most frequently encountered observation. The **median** is the center observation after all the values in the column have been ranked from smallest to largest. The **mean** (arithmetic average) is the most important and widely used of the three and is defined by the following formula:

$$\bar{x} = \frac{\Sigma X}{N}$$

Where $\bar{x}$ = the arithmetic mean (average),

X = any count, measurement, etc.,

N = the number of observations (sample size), and

$\Sigma$ = the symbol for summation or addition of all the values of 'X' for the sample or

$\Sigma X = X_1, + X_2, + X_3, + \ldots X_n$

The other characteristic of the column (sample) of numbers is the measure of variability. Again, a number of them have been used. In taxonomic work the most commonly used are the range (the largest and smallest value in the sample) and the variance. The **range** is often used in plant descriptions because it is easily obtained, but since it is strongly dependent on sample size and may be dependent on a single extreme measurement, it cannot be used in statistical work. The **variance** ($s^2$) is the average difference between each observation and the mean and can be calculated by the following formulas:

$$s^2 = \frac{\Sigma X^2 - \dfrac{\Sigma(X)^2}{N}}{N-1}$$

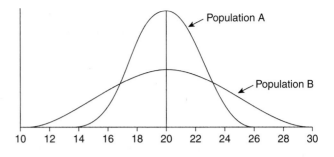

**Figure 3-17** The frequency distribution of two populations, both with a mean of 20, but with different variances. Population B has the greater variation and thus would need a larger sample number of individuals in its sample.

or more simply

$$s^2 = \frac{\Sigma(\bar{x} - X)^2}{N-1}$$

Where $\Sigma X$ = the sum of each observation,

X = the value of an observation,

N = the number of observations (items in a column), and

$\bar{x}$ = the mean or average.

Any column of numbers (sample) can be reduced to two characteristics **(parameters)**—the mean and the variance. Both parameters are required. Given that you have two samples both with an average of 20, can it be said with certainty that the samples are identical? The answer is no. Figure 3-17 shows the distribution of two samples both with the same mean $\bar{x}$ = 20). The difference between these two samples is the variance (distribution of the values about the mean). Sample A has a smaller variance than sample B. A similar situation could be constructed showing the same variance and different means.

We have said that the sample is a small group of plants or plant parts chosen to represent the larger population of similar plants or plant parts that exist in nature. Although we cannot actually measure all individuals in the population, we can still assume that they exist. If the population is real, then it, like the sample chosen to represent it can be assumed to have an arithmetic mean or average (m) and a variance ($s^2$). The population mean and standard deviation must be *estimated* by selecting the proper sample. This sample has a measurable mean [called O(x,$^-$) to distinguish it from the population mean ($\mu$)] and variance [called $s^2$ to distinguish it from the

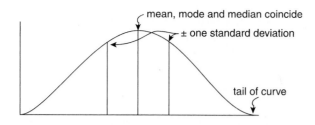

**Figure 3-18** Normal distribution and the relationship among parameters used in its description.

population's variance ($s^2$)]. Our task is to determine how closely our sample represents the population in nature that we are trying to characterize. In order to do this we must make a few more assumptions.

Although there are several types of data distribution patterns, for the purposes of this discussion, we will assume that we are dealing with a population that is **statistically normal.** A normal population is one in which the mean, mode (most commonly encountered observation) and the median (the middle value of a group of observations) are equal and the dispersion around the mean is such that the percentage of observations on either side of the mean are related to the standard deviation as follows:

a. one standard deviation on either side of the mean includes 68% of the observations.

b. two standard deviations on either side of the mean includes 95% of the observations.

c. three standard deviations on either side of the mean include 99% of the observations.

A picture of the distribution of a "normal" population would correspond to what is known as a normal or bell-shaped curve. (Fig. 3-18). This assumption is usually a fairly safe one for a taxonomist to make since most measures and counts on biological material conform to a normal distribution when sample sizes are adequately large (20 or more).

The standard deviation (square root of the variance) can be used to obtain an estimate of how close the sample mean is to the population mean. The confidence one has in the means of the samples depends on the amount of variation in the population being sampled. For example Population B in Figure 3-17 has much more variation than population A. The more variation in the population, the less confidence there will be that the mean is correctly describing or summarizing it. The statistic used to determine the confidence in a sample mean is the **standard error.**

The standard error ($s_{\bar{x}}$) is defined by the following formula:

$$s_{\bar{x}} = \frac{s^2}{N}$$

Using the standard error as a base it is possible to put numerical limits or an interval of confidence about the sample mean which indicates how often we would expect chance to produce a sample mean different from the true mean. This interval is called the confidence interval and is represented by:

$$x \pm t(s_{\bar{x}})$$

The value (t) is based on a distribution devised by W. S. Gossett in 1907 and is called Student-t. A table of t-values will be provided by your instructor or is available in any book on statistics. In order to use the table you will first have to decide how sure you want to be that you have a true estimate of the population mean. Do you want to be correct 19 times out of 20, or 99 times out of 100? This would mean that you would expect to come up with a wrong estimate 1 time out of 20, or 1 time out of 100, respectively. Here then is our test of the null hypothesis. Let us say you are willing to accept being wrong 1 time out of 20 or 5%, then under the value 95% at the top of the table, find the value that corresponds to the degrees of freedom you have in your sample. Degrees of freedom are calculated by the formula:

$$d.f. = N - 1$$

Where N = the number of sample measurements

If the null hypothesis is rejected, the taxonomist will be faced with deciding whether or not two different means and standard deviations are different due to sampling error or because they represent different parent populations (alternate hypotheses). This question can be answered two ways. The investigator can make histograms or scatter plots and decide visually whether two groups seem to exist or can resort to various inferential statistical methods. Graphical methods are easier but are more subject to personal bias or error. Inferential statistical methods are more complicated but are less subject to personal bias if used properly. There are two types of inferential statistics. **Parametric statistics** have four stringent requirements in order for them to be used. These are:

1. Random or at least an unbiased samples.

2. Statistically equal variances.

3. Normal distributions.

4. Data continuous or of count type.

**Non-parametric** statistics have only the requirement for unbiased or random samples. Parametric statistics are able to discriminate finer differences than non-parametric ones. It must be re-emphasized that both require a random, or at least an unbiased, sample to make any valid inferences about the populations sampled.

Both parameters, the mean and the standard deviation, have their own special statistical tests of inference. The specific test used depends on the number of groups being compared, the level of measurement, and the number of variables being compared. The selection and application of these is called univariate statistics when comparisons are made between a parameter and its population or between two populations based on their sample parameters and multivariate statistics when comparisons are made among more than two population parameters. Taxonomists must be familiar with the basic tenets of sampling, but will rarely be called upon to use univariate inferential statistical tests per se. This is because most taxonomic problems deal with large numbers of sample means and/or variances for each of many characters and taxa. The characters (description) of one taxon cannot be compared to the characters of one or more other taxon with univariate methods.

Multivariate analyses are statistical tools that enable an investigator to examine and statistically compare patterns of variation based upon variation in several to many characters and/or several to many populations. The details of multivariate statistical methods and the specialized procedures known as classification theory are beyond the scope of this text. Detailed information about these sophisticated techniques is included in some of the references listed below.

The availability of computers and the great strides that have taken place in computer technology in the past quarter century have made it possible for taxonomists to carry out statistical analyses without laborious calculations. The ease of use places a responsibility on the investigator to design investigations that are meaningful. Poorly designed experiments can yield questionable data that remain questionable even when subjected to statistical analysis. Properly designed studies, on the other hand, enable a taxonomist to make meaningful analyses of complicated patterns of variation.

## Coding Quantitative Morphological Data for Phylogenetic Analyses

There is considerable debate in the literature concerning the use of quantitative morphological data in phylogenetic analyses (see Gift and Stevens 1997, Thiele 1994). In general, in seeking separate character states we look for gaps in the variation, and this is called "gap coding." Statistical tests outlined above can help evaluate the presence of gaps in the variation. However, there may be times when rare variants or hybrids blur the gaps.

The recommendations by Peter Stevens (1991) offer a reasonable way to approach the problem. Stevens asserts that we should, as a general rule, use character states that do not have overlap. However, if the variation is appropriately analyzed, there may be cases where discontinuities (not only gaps) can be used. The critical issue for Stevens is that the justification of character states be well documented. This entails adequate sampling of the variation, appropriate analysis of the variation, and careful description of the methods used in your systematic analyses.

## References

Abbott, L. A., F. A. Bisby and D. J. Rogers. 1985. *Taxonomic Analysis in Biology. Computers, Models, and Databases.* Columbia Univ. Press, New York.

Bailey N. T. J. 1984. *Statistical Methods in Biology,* 2nd ed. Hodder and Stoughton, London.

Blackith, R. E., R. A. Reyment and N. A. Campbell. 1987. *Multivariate Morphometrics,* 2nd ed. Academic Press, N.Y.

Bookstein, F. L. 1982. Foundations of morphometrics. *Annual Review of Ecology and Systematics* 13:451–470.

Briggs, D., and S. M. Walters. 1984. *Plant Variation and Evolution,* 2nd ed. Cambridge Univ. Press, Cambridge.

Campbell, R. C. 1989. *Statistics for Biologists.* Cambridge University Press, New York and Cambridge.

Crovello, T. J. 1970. Analysis of character variation in ecology and systematics. *Annual Review of Ecology and Systematics* 1:55–98.

Davis, P. H., and V. H. Heywood. 1963. *Principles of Angiosperm Taxonomy.* D. Van Nostrand Co., Princeton, New Jersey.

Felsenstein, J. (ed.). 1983. *Numerical Taxonomy.* Springer-Verlag, New York.

Finney, D. J. 1980. *Statistics for Biologists.* Chapman and Hall, New York.

Gift, N. and P. F. Stevens. 1997. Vagaries in the delimitation of character states in quantitative variation—an experimental study. *Systematic Botany* 46(1): 112–125.

Grant, V. 1963. *The Origin of Adaptations.* Columbia University Press, New York.

Judd, W. S., C. S. Campbell, E. A. Kellogg, P. F. Stevens, and M. J. Donoghue. 2002. Plant Systematics: A Phylogenetic Approach, 2nd Edition. Sinauer Associates: Sunderland, MA.

Lewis, A. E. 1984. *Biostatistics.* 2nd ed. Van Nostrand, Reinhold, Co., New York.

Oxnard, C. E. 1978. One biologist's view of morphometrics. *Annual Review of Ecology and Systematics* 9:219–242.

Pankhurst, R. J. 1992. *Practical Taxonomic Computing.* Cambridge University Press, New York and Cambridge.

Parker, R. E. 1973. *Introductory Statistics for Biology.* Edward Arnold Publ., London.

Pimentel, R. A. 1979. *Morphometrics.* Kendall-Hunt, Dubuque, Iowa.

———, and J. D. Smith. 1986. *BIOΣTAT I. A Univariate Statistical Toolbox.* Sigma Soft, Placentia California.

———. 1986. *BIOΣTAT II. A Multivariate Statistical Toolbox,* 2nd ed. Sigma Soft, Placentia, California.

Schlichting, C. D. 1986. The evolution of phenotypic plasticity in plants. *Annual Review of Ecology and Systematics* 17:667–693.

Sivarajan, V. V. *Introduction to the Principles of Plant Taxonomy.* 2nd ed. edited by N. K. B. Robson, Cambridge University Press, Cambridge and New York.

Sokal, R. R., and F. J. Rohlf. 1969. *Biometry.* W. H. Freeman & Co., San Francisco.

Stevens, P. F. 1991. Character states, morphological variation, and phylogenetic analysis: a review. Systematic Botany 16(3): 553–583.

Stuessy, T. F. 1990. *Plant Taxonomy. The Systematic Evaluation of Comparative Data.* Columbia University Press, New York.

Thiele, K. 1994. The holy grail of the perfect character: The cladistic treatment of morphometric data. Cladistics 9: 275–304.

Zar, J. H. 1984. *Biostatistical Analysis.* Prentice-Hall, Englewood Cliffs, New Jersey.

# Exercises

## Part I

1. Collect leafy twigs 1 inch or more long with at least two leaf-bearing nodes from 10 different trees or shrubs (see Fig. 3-19). These samples should represent as much of the variation discussed in this chapter as possible. (**NOTE:** Look at the labels in worksheet 3-1 before you begin to collect specimens! The first three items should be written down at the time of collection.)

2. Press and dry these leafy twigs in several folds of newspaper.

3. After the leafy twigs are more or less dry (1–2 days), mount them on a sheet of heavy paper (biological paper is best). Use either transparent tape or white glue.

4. In the lower right-hand corner of each sheet paste one of the labels from worksheet 3-1 and fill in the information indicated.

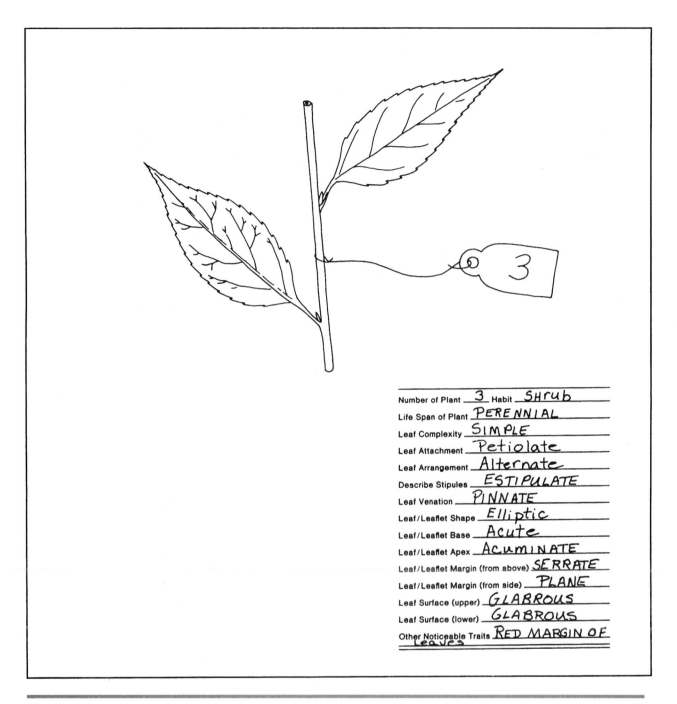

Number of Plant __3__ Habit __SHRUB__
Life Span of Plant __PERENNIAL__
Leaf Complexity __SIMPLE__
Leaf Attachment __Petiolate__
Leaf Arrangement __Alternate__
Describe Stipules __ESTIPULATE__
Leaf Venation __PINNATE__
Leaf/Leaflet Shape __Elliptic__
Leaf/Leaflet Base __Acute__
Leaf/Leaflet Apex __ACUMINATE__
Leaf/Leaflet Margin (from above) __SERRATE__
Leaf/Leaflet Margin (from side) __PLANE__
Leaf Surface (upper) __GLABROUS__
Leaf Surface (lower) __GLABROUS__
Other Noticeable Traits __RED MARGIN OF Leaves__

**Figure 3-19** An example of leafy twig specimen and descriptive label required for the Exercise 3. Note: one leaf blade should have the lower surface showing and the other the upper surface. Two full pages of labels, ready to be cut apart, can be found in Worksheet 3-1.

**Worksheet 3-1.** Labels to be cut apart and glued to the lower right corner of the sheets holding the leafy twigs collected for Exercise 3.

---

Number of Plant _____ Habit _____
Life Span of Plant _____
Leaf Complexity _____
Leaf Attachment _____
Leaf Arrangement _____
Describe Stipules _____
Leaf Venation _____
Leaf/Leaflet Shape _____
Leaf/Leaflet Base _____
Leaf/Leaflet Apex _____
Leaf/Leaflet Margin (from above) _____
Leaf/Leaflet Margin (from side) _____
Leaf Surface (upper) _____
Leaf Surface (lower) _____
Other Noticeable Traits _____

---

Number of Plant _____ Habit _____
Life Span of Plant _____
Leaf Complexity _____
Leaf Attachment _____
Leaf Arrangement _____
Describe Stipules _____
Leaf Venation _____
Leaf/Leaflet Shape _____
Leaf/Leaflet Base _____
Leaf/Leaflet Apex _____
Leaf/Leaflet Margin (from above) _____
Leaf/Leaflet Margin (from side) _____
Leaf Surface (upper) _____
Leaf Surface (lower) _____
Other Noticeable Traits _____

---

Number of Plant _____ Habit _____
Life Span of Plant _____
Leaf Complexity _____
Leaf Attachment _____
Leaf Arrangement _____
Describe Stipules _____
Leaf Venation _____
Leaf/Leaflet Shape _____
Leaf/Leaflet Base _____
Leaf/Leaflet Apex _____
Leaf/Leaflet Margin (from above) _____
Leaf/Leaflet Margin (from side) _____
Leaf Surface (upper) _____
Leaf Surface (lower) _____
Other Noticeable Traits _____

---

Number of Plant _____ Habit _____
Life Span of Plant _____
Leaf Complexity _____
Leaf Attachment _____
Leaf Arrangement _____
Describe Stipules _____
Leaf Venation _____
Leaf/Leaflet Shape _____
Leaf/Leaflet Base _____
Leaf/Leaflet Apex _____
Leaf/Leaflet Margin (from above) _____
Leaf/Leaflet Margin (from side) _____
Leaf Surface (upper) _____
Leaf Surface (lower) _____
Other Noticeable Traits _____

---

Number of Plant _____ Habit _____
Life Span of Plant _____
Leaf Complexity _____
Leaf Attachment _____
Leaf Arrangement _____
Describe Stipules _____
Leaf Venation _____
Leaf/Leaflet Shape _____
Leaf/Leaflet Base _____
Leaf/Leaflet Apex _____
Leaf/Leaflet Margin (from above) _____
Leaf/Leaflet Margin (from side) _____
Leaf Surface (upper) _____
Leaf Surface (lower) _____
Other Noticeable Traits _____

---

Number of Plant _____ Habit _____
Life Span of Plant _____
Leaf Complexity _____
Leaf Attachment _____
Leaf Arrangement _____
Describe Stipules _____
Leaf Venation _____
Leaf/Leaflet Shape _____
Leaf/Leaflet Base _____
Leaf/Leaflet Apex _____
Leaf/Leaflet Margin (from above) _____
Leaf/Leaflet Margin (from side) _____
Leaf Surface (upper) _____
Leaf Surface (lower) _____
Other Noticeable Traits _____

**Worksheet 3-1.** Labels to be cut apart and glued to the lower right corner of the sheets holding the leafy twigs collected for Exercise 3.

Number of Plant _____ Habit _____
Life Span of Plant _____
Leaf Complexity _____
Leaf Attachment _____
Leaf Arrangement _____
Describe Stipules _____
Leaf Venation _____
Leaf/Leaflet Shape _____
Leaf/Leaflet Base _____
Leaf/Leaflet Apex _____
Leaf/Leaflet Margin (from above) _____
Leaf/Leaflet Margin (from side) _____
Leaf Surface (upper) _____
Leaf Surface (lower) _____
Other Noticeable Traits _____

Number of Plant _____ Habit _____
Life Span of Plant _____
Leaf Complexity _____
Leaf Attachment _____
Leaf Arrangement _____
Describe Stipules _____
Leaf Venation _____
Leaf/Leaflet Shape _____
Leaf/Leaflet Base _____
Leaf/Leaflet Apex _____
Leaf/Leaflet Margin (from above) _____
Leaf/Leaflet Margin (from side) _____
Leaf Surface (upper) _____
Leaf Surface (lower) _____
Other Noticeable Traits _____

Number of Plant _____ Habit _____
Life Span of Plant _____
Leaf Complexity _____
Leaf Attachment _____
Leaf Arrangement _____
Describe Stipules _____
Leaf Venation _____
Leaf/Leaflet Shape _____
Leaf/Leaflet Base _____
Leaf/Leaflet Apex _____
Leaf/Leaflet Margin (from above) _____
Leaf/Leaflet Margin (from side) _____
Leaf Surface (upper) _____
Leaf Surface (lower) _____
Other Noticeable Traits _____

Number of Plant _____ Habit _____
Life Span of Plant _____
Leaf Complexity _____
Leaf Attachment _____
Leaf Arrangement _____
Describe Stipules _____
Leaf Venation _____
Leaf/Leaflet Shape _____
Leaf/Leaflet Base _____
Leaf/Leaflet Apex _____
Leaf/Leaflet Margin (from above) _____
Leaf/Leaflet Margin (from side) _____
Leaf Surface (upper) _____
Leaf Surface (lower) _____
Other Noticeable Traits _____

Number of Plant _____ Habit _____
Life Span of Plant _____
Leaf Complexity _____
Leaf Attachment _____
Leaf Arrangement _____
Describe Stipules _____
Leaf Venation _____
Leaf/Leaflet Shape _____
Leaf/Leaflet Base _____
Leaf/Leaflet Apex _____
Leaf/Leaflet Margin (from above) _____
Leaf/Leaflet Margin (from side) _____
Leaf Surface (upper) _____
Leaf Surface (lower) _____
Other Noticeable Traits _____

Number of Plant _____ Habit _____
Life Span of Plant _____
Leaf Complexity _____
Leaf Attachment _____
Leaf Arrangement _____
Describe Stipules _____
Leaf Venation _____
Leaf/Leaflet Shape _____
Leaf/Leaflet Base _____
Leaf/Leaflet Apex _____
Leaf/Leaflet Margin (from above) _____
Leaf/Leaflet Margin (from side) _____
Leaf Surface (upper) _____
Leaf Surface (lower) _____
Other Noticeable Traits _____

# Part II

1. Two populations of plants or plant parts will be provided to you. They may all be from one species or they may be from two species. In consultation with your instructor, chose two or three characters to measure, measure them, and record the results in Table 3-3.

   After completing Table 3-3, compare the samples by producing a histogram or scatter diagram to compare your measurements with those of your neighbors.

   What is the null hypothesis you are testing visually?

   Using the computer or computer terminal available, compare the means from your sample with those of your neighbors using the parametric "t-test" or the non-parametric "Mann-Whitney-U test" using an alpha value of .05. The alpha value refers to the probability of rejecting a true null hypothesis.

| POPULATION 1 | | | POPULATION 2 | | |
|---|---|---|---|---|---|
| Specimen | Character A | Character B | Specimen | Character A | Character B |
| 1 | | | 1 | | |
| 2 | | | 2 | | |
| 3 | | | 3 | | |
| 4 | | | 4 | | |
| 5 | | | 5 | | |
| 6 | | | 6 | | |
| 7 | | | 7 | | |
| 8 | | | 8 | | |
| 9 | | | 9 | | |
| 10 | | | 10 | | |
| 11 | | | 11 | | |
| 12 | | | 12 | | |
| 13 | | | 13 | | |
| 14 | | | 14 | | |
| 15 | | | 15 | | |
| 16 | | | 16 | | |
| 17 | | | 17 | | |
| 18 | | | 18 | | |
| 19 | | | 19 | | |
| 20 | | | 20 | | |
| mean | | | mean | | |
| Standard deviation | | | Standard deviation | | |

**Table 3-3**   Blank table for recording data for analysis for question 1.

2. On your own collect twenty samples each from two separate populations of the same species. In consultation with your instructor, choose two or three characters to measure, measure them, and record the results as in Table 3-4. At least two of the characters chosen should be quantitative.

After completing Table 3-4, compare the samples by producing a histogram or scatter diagram. What is the null hypothesis you are testing visually?

Using the computer or computer terminal available, analyze the data to determine if the two populations are statistically different.

If they are, can you suggest a basis for the differences?

How might you test your hypotheses?

| | POPULATION 1 | | | | | POPULATION 2 | | | |
|---|---|---|---|---|---|---|---|---|---|
| Specimen | Character A | Character B | Character C | Character D | Specimen | Character A | Character B | Character C | Character D |
| 1 | | | | | | | | | |
| 2 | | | | | | | | | |
| 3 | | | | | | | | | |
| 4 | | | | | | | | | |
| 5 | | | | | | | | | |
| 6 | | | | | | | | | |
| 7 | | | | | | | | | |
| 8 | | | | | | | | | |
| 9 | | | | | | | | | |
| 10 | | | | | | | | | |
| 11 | | | | | | | | | |
| 12 | | | | | | | | | |
| 13 | | | | | | | | | |
| 14 | | | | | | | | | |
| 15 | | | | | | | | | |
| 16 | | | | | | | | | |
| 17 | | | | | | | | | |
| 18 | | | | | | | | | |
| 19 | | | | | | | | | |
| 20 | | | | | | | | | |
| mean | | | | | | | | | |
| Standard deviation | | | | | | | | | |

**Table 3-4**  Blank table for recording character data for comparison of two plant populations.

# Phylogenetic Classification

chapter 4

In 1950 a German entomologist, **Willi Hennig** (1913–1976), proposed a new method of phylogenetic classification. His approach was radically different from previous techniques in that it was based upon the adherence to a strict and repeatable method of evaluating relationships. He argued that classification should reflect the branching pattern of evolution (genealogy) rather than the degree of advancement and divergence. He stated that only groups that are strictly **monophyletic,** composed of an ancestral taxon and *all* of its descendants should be recognized. He stated that monophyletic groups should be defined by **shared derived characters,** features that changed states when the ancestor to the monophyletic group evolved. He further noted that groups defined by shared primitive features are not monophyletic.

Hennig's ideas received little attention at first. His work was available only in difficult-to-read technical German and was unknown to many systematists. Additionally Hennig coined a formidable set of new terms for use with his methodology. It was not until the mid 1960s that Hennig's ideas became widely available in English and began to gain widespread attention. At first Hennig's ideas had a much greater impact among zoologists than among botanists. However, since the late 1970s many plant systematists have adopted Hennigian methodology for their own research. These methods are now a part of the standard systematic toolbox for many taxonomists.

In the 1960s and 1970s the field of taxonomy received considerable criticism for being more "art" than science. Taxonomists were attacked for not using a specific methodology in their research. The classification systems that were developed by tradi-

tional or evolutionary taxonomists were criticized for being subjective. The field of population genetics was exploring variation at the level of genes and proteins using electrophoresis, and funding and university faculty positions began to shift away from the "low tech" field of taxonomy. Many taxonomists began to call themselves systematists, partly to shed the "old fashioned" label, and began to use "modern" techniques of chromosome analysis (1950–1960), chemosystematics (1970–1980), and ultimately molecular (DNA) data (1990–present).

These newer methods of analysis had a major impact on a field of study that had relied on morphology and anatomy for centuries to classify plants. Although these new methods did change systematics, the major shift occurred during the 1970s and 1980s as followers of Hennig began to argue that their method of evaluating plant relationships was better than traditional or phenetic taxonomy. Specifically, they argued that Hennigian methods were repeatable (unlike traditional methods) and reflected evolutionary history (unlike phenetic methods).

One of the products of an analysis using Hennig's methods is a type of branching diagram called a **cladogram.** Hennig wanted classifications to reflect the relationships displayed in the cladograms. An opponent of the Hennigian methods coined the name **cladistics** as a derogatory term for the new approach and it was widely used through the 1990s. However, much like in battle, when the victor writes the history, it appears that the term **phylogenetic systematics** or simply **phylogenetics** has replaced cladistics as the name used to refer to this method. Today we see Hennigian methods of phylogenetic systematics applied to ecology, medicine, and even linguistics, and it is the method of choice currently used to reconstruct evolutionary history and to classify life.

Although it took a while for Hennig's ideas to gain attention in the botanical community, a botanist played a major independent role in the development of cladistic concepts. **Warren Wagner** (1920–), an American fern specialist, began to develop ideas similar to those of Hennig in the late 1950s and early 1960s. Wagner developed what he called the **groundplan/divergence** method for constructing phylogenetic trees. This approach embodied attempts to display not only the branching sequence of evolutionary diversification but also the amount and direction of change. Like Hennig, Wagner placed primary emphasis on determining the genealogy of taxonomic groups. Wagner used different terms and a different type of display than did Hennig, and at first few systematists were aware of the similarity of methods. By the late 1960s, however, the connection had been made and Wagner's ideas have become the basis for some of the computer algorithms used to analyze relationships.

# Phylogenetic Methodology

Let's examine phylogenetic methodology. Imagine for the moment a group of three species, A, B, and C. These three species are related to each other, but we are not sure how. Perhaps B and C are more closely related to each other than either is to A. Or perhaps A and B are most closely related; or maybe A and C are most closely related. Maybe all three are equally related. The point is that any of these relationships is possible. Let's look at a method of presenting these possible relationships. Examine Figure 4-1. The four diagrams summarize the possible relationships we have been discussing. These are cladograms. They represent the four hypotheses of relationship described previously.

Before we go on, let's look at what we mean by relationship. Cladogram 1 above suggests that B is more closely related to C than either is to A because B and C share a common ancestor (at the point where the lines split) not shared with A. Cladograms 2 and 3 suggest alternate relationships with a similar interpretation. In cladogram 4 no two of the three taxa share a common ancestor not shared with the third. The nodes where two or more lines join are considered to represent "speciation events" (i.e., evolutionary changes that marked the splitting of one ancestral lineage into two or more descendant lineages). Time is represented on the vertical axis.

Suppose we wanted to decide which of the four cladograms accurately depicts the relationships of these three species. Only one of them can actually indicate the real relationship. How can we decide which one? What we need is evidence that supports one of these hypotheses and none of the others. Without a complete fossil record [and even if we had one, how would we know it was complete?], we must rely on the evidence provided by the organisms themselves. This evidence comes in the form of measurable attributes (characters) of the plants. The problem is how to interpret these characters.

As discussed in Chapter 6, since the 1800s botanists have attempted to group taxa based largely on the assumption that the groups that were more closely related evolutionarily would share more characters in common. The actual implementation of this idea was often less than perfect. Botanists tended to choose characteristics of the plants that "worked" to make the groups, not necessarily all the features. They gave more weight to some features than to others.

Why is it that some features work better than others? If we examine the more reliable characters we

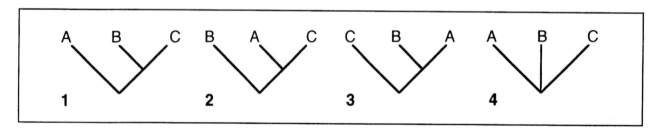

Figure  Possible relationships among three taxa as expressed by cladograms.
4-1

| Taxon | Characters | | | | | | | |
|-------|---|---|---|---|---|---|---|---|
| | 1 | 2 | 3 | 4 | 5 | 6 | 7 | 8 |
| A | + | − | − | − | + | + | − | − |
| B | − | + | − | + | + | + | + | + |
| C | − | − | + | + | + | + | + | + |

Table 4-1   Characteristics of hypothetical taxa A, B, and C in the form of a taxon-character table.

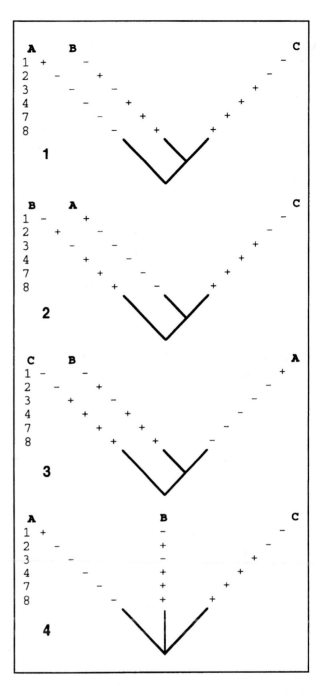

Figure 4-2   Distribution of character states of taxa A, B, and C on cladograms 1–4. Character numbers are listed on the left.

find that they tend to be the ones specific to the group we are studying or to a part of it. Those features that are shared with other groups of plants are often much less helpful. Those which can help us to understand relationships within the group of plants we are studying are those features that evolved among the members of our group. They can help us to decide which of the hypotheses of relationship is most likely to be correct. Those features that evolved before our study group came into existence are not likely to be of much help to us in evaluating the relationships within the taxa we are studying. They may help us to see relationships of our study group to other groups, however.

Let's take a look at some evidence regarding Taxa A, B, and C. In Table 4-1 several features of plants A, B and C are presented. In this example each of these characters exists in two versions (**character states**). Several of these characters appear to show relationships. Characters 1(−) and 4(+), 7(+), and 8(+) suggest a grouping of taxa B and C. Character 2(−) suggests a grouping of A and C, and character 3(−) suggests a grouping of A and B. Features 5(+) and 6(+) are uninformative because all three taxa have them. They might define our study group or they might occur in other groups as well. The six apparently informative characters are displayed on the four cladograms in Figure 4-2. There definitely appear to be some conflicts among the characters. Some seem to suggest one sort of relationship and others suggest a different sort. If we were to base our decision on overall similarities and differences, it would not be a very clear-cut decision.

The discussion above suggested that those characters that have evolved along with the organisms being studied (**apomorphies**) are the most useful features in an analysis of relationships. The most informative features are shared derived features (**synapomorphies**). Such features are displayed by those organisms derived from a common ancestor.

[The character evolved in the ancestor and was passed on to its descendants]. Those characters that were already present when the ancestor evolved (**plesiomorphies**) and have merely been passed along to the descendants (**symplesiomorphies**) may actually cause confusion by suggesting false relationships. Is there some way of discovering which of the characters are the useful synapomorphies and which are the not-so-useful symplesiomorphies?

What we need to know is which character states are ancestral and which are derived. Did 1(+) give rise to 1(−) or vice versa? How can we tell? Fossils might be helpful, but they are seldom available, and the fossil record is so spotty that it is very easy to misinterpret. We might look to see if some eminent botanist has made a pronouncement on the issue, but eminent botanists might be mistaken [there are certainly some very contradictory opinions expressed in the literature (see Chapter 6)]. We could guess, but that's not very good science. What we need is an unbiased objective method.

Let's look outside of our study group to see what characters the most closely related taxon has. We will call this close relative the **outgroup** and our study group the **ingroup** (see Phylogenetics Glossary on p. 000). Features that are shared by the members of the outgroup and by one or more of the members of the ingroup very likely were present in the common ancestor of ingroup and outgroup. Now we have some additional evidence (Table 4-2). Character state 1(+) is present in taxon A and in the outgroup. This suggests that it was present in the ancestor of both the outgroup and the ingroup and therefore within the ingroup it is a plesiomorphy. By the same reasoning we can tentatively conclude that character state 1(−) is an apomorphy shared by B and C. Character state 2(+) is apparently an apomorphy unique to taxon B. That would make character state 2(−) a symplesiomorphy. Character state 3(+) is an apomorphy of taxon A, and character state 3(−) is another symplesiomorphy. Character state 4(−) is a plesiomorphic feature, and 4(+) is an

apomorphy shared by B and C. Character state 5(+) was apparently present in the ancestor of both the ingroup and the outgroup and is still uninformative to us. Character state 6(+) *may* be a synapomorphy that defines the ingroup. The features of the outgroup have suggested the direction that the characters have evolved in our study group.

On the basis of on this evidence, which of the cladograms best reflects the evolution of the ingroup? Before we can decide we need some criterion for deciding what "best" means. The criterion we are going to use to evaluate cladograms is that the simplest explanation that is in accord with the data is most likely the correct one. This is sometimes called the **parsimony criterion.** It should be noted here that the parsimony criterion does not mean that evolution occurs in the most parsimonious fashion, but that the shortest cladogram is the best for the data at hand. In Figure 4-3 the various cladograms are presented with the derived features (apomorphies) indicated as black squares and the plesiomorphies as hollow squares.

Notice that in cladogram 1 taxa B and C, which share apomorphic states of characters 1, 4, and 8, occur together. In cladograms 2, 3, and 4, on the other hand, these taxa are separated. In cladogram 1 we can assume that character state 1(−) evolved once in the ancestor that gave rise to B and C. In the same fashion 4(+) evolved from 4(−) only once, and 8(+) evolved once from 8(−) once. In cladograms 2–4, however, character state 1(−) must have evolved twice, once in the line leading to C and once in the line leading to B. Character states 4(+) and 8(+) must also have evolved twice. We would have to assume that the features shared by B and C are the result of parallel evolution. This hypothesis is not supported by any evidence. The simplest explanation for our data is that B and C share a common ancestor not shared by A. This explanation requires only three mutations for characters 1, 4, and 8. The other three hypotheses all require six steps. We therefore accept cladogram 1 as the explanation that best fits the data.

Phylogenetic analysis is a rather complicated and rigorous procedure. It is designed to be fully repeatable and to remove the personal opinion that so often clouds taxonomic decision-making. Phylogenetic studies require a thorough knowledge of the organisms being studied. The larger the group being analyzed and the greater the number of characters being evaluated, the more complicated it gets. Fortunately computers are now available to us. Several

| Taxon | Characters | | | | | | | |
|---|---|---|---|---|---|---|---|---|
| | 1 | 2 | 3 | 4 | 5 | 6 | 7 | 8 |
| A | + | − | − | − | + | + | − | − |
| B | − | + | − | + | + | + | + | + |
| C | − | − | + | + | + | + | + | + |
| Outgroup | + | − | − | − | + | − | + | − |

Table 4-2  Characteristics of Hypothetical Taxa A, B, C, and their Outgroup. Plesiomorphic character states (those shared by the outgroup and at least part of the ingroup) are shaded. Only the apomorphic character states (not shaded) are useful in the phylogenetic analysis of the ingroup.

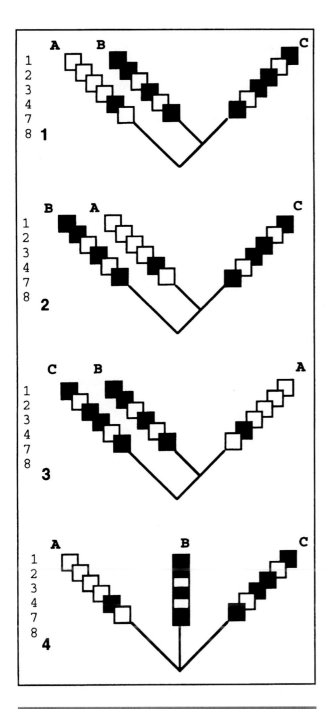

**Figure 4-3** Distribution of apomorphic and plesiomorphic character states for taxa A, B, and C on cladograms 1–4. Apomorphic character states are marked by dark boxes. Plesiomorphic states are marked by unfilled boxes.

computer programs are now in wide use that can compare and analyze many cladograms very rapidly. Felsenstein maintains a large listing of these computer program at http://evolution.genetics.washington.edu/phylip/software.html.

As with other approaches there are some problems with cladistics. Parallel evolution in different lineages does take place sometimes and can create false interpretations of character polarity. Reversals in character evolution can also take place. Within the course of evolution, a character can revert from the apomorphic state to what appears to be the ancestral, plesiomorphic condition. Characters that are assumed to be of taxonomic value may actually be environmentally induced variations. All of these will complicate the analysis of character state change that is the foundation of cladistic methodology. On the other hand, it is these conflicts between characters that may enable a systematist to detect problem characters and to investigate the reasons for the conflicts. The character conflicts that result from parallel evolution, reversals, and misinterpretations are collectively described as **homoplasy.** The greater the amount of homoplasy present in a data matrix, the more difficult cladistic analysis of that matrix is likely to be. An analysis using the parsimony criterion may yield several different cladograms with the same or nearly the same number of steps (assumed mutations required by the structure of the cladogram). The taxonomist must then decide which of the cladograms to accept or must look for additional sources of data. Often a **consensus tree** is constructed that shows the parts of the alternative cladograms that are shared. When only small parts of the alternative equally parsimonious cladograms change, the information contained in a consensus tree can be very informative about evolution within the group.

There are some problems that must be addressed in the use of this methodology to generate a classification system. Some of these problems are challenges that impact botanists to a greater extent than zoologists. For instance, the "open" growth system of plants allow them to respond phenotypically to environmental variation to a greater extent than can animals with their "closed" growth system. Reticulate evolution through polyploidy is seldom encountered in animals but is common in plants. Phylogenetics is designed to sort out the sequence of evolutionary changes as a branching pattern of divergences. When these branches come back together via hybridization, the method has difficulty discerning this change.

The interface between molecular biology and cladistics has opened a new and exciting field called **molecular phylogenetics.** The new techniques for extracting, amplifying, and sequencing DNA and other macromolecules yield individual base or amino acid substitutions as character state changes that can be analyzed cladistically. The large data sets that result from these studies require much computer time for cladistic analysis. In some cases processing a large data set may take weeks or even months for a high speed computer. The results of molecular phylogenetic studies have suggested some major realignments in the systematics of various plant groups. The hypotheses of relationship generated in such studies can then be investigated using other types of data.

Botanists are now using phylogenetics as an effective tool for analyzing relationships. The Angiosperm Phylogeny Group (APG) is a group of systematists that are using phylogenetic methods and various forms of data (molecular, morphological, etc.) to understand the relationships of flowering plants. This project is ongoing and has resulted in some changes from the traditional classification systems produced by Cronquist and Takhtajan (Stevens 2001).

# PHYLOGENETICS GLOSSARY

**Apomorphy** (apomorphic character state)—An evolutionarily derived character state, important for determining relationships in cladistic analyses. In a character transformation series with more than two states, a particular character state may be apomorphic relative to some states and plesiomorphic (q.v.) relative to others.

**Autapomorphy**—An apomorphic character unique to a single taxon in a cladogram and thought to have originated with that taxon.

**Bootstrap analysis**—A technique sometimes used in cladistic analyses to place confidence intervals on the resulting cladograms. Substitutions are made for individual characters in a data set creating a series of new data sets equal in size to the original. Each of these data sets is then analyzed cladistically. A majority rule consensus tree is created that includes all the monophyletic taxa that show up in at least 95 percent of the resulting cladograms.

**Character**—Any attribute of an organism that can be consistently measured, counted, or otherwise defined.

**Character state**—One of the expressions of a character; a character may have as few as two states (e.g., presence-absence) or it may have several to many states (e.g., color).

**Character transformation series**—A hypothesized sequence of changes of states in a character from the most plesiomorphic (primitive) to the most apomorphic (derived). The sequence may be linear or branched.

**Cladistics**—A type of analysis of phylogenetic relationships in which monophyletic taxa are grouped on the basis of shared derived features (synapomorphies).

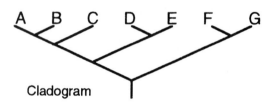
Cladogram

**Cladogram**—A graphic display of a hypothesis of relationship among a group of taxonomic entities in the form of a branching diagram in which the branching sequence is based on inferred historical connections between the entities. A cladogram can be rotated at any node (q.v.) and still have the same information content.

**Consensus tree**—A branching diagram that shows the branching patterns shared among a group of alternative cladograms. Areas of disagreement result in polychotomies (unresolved areas in the cladogram where multiple branches arise together). A **strict consensus tree** includes only those parts of the cladogram that are shared by *all* of the alternative cladograms. A **majority rule consensus tree** includes those parts of the cladogram that are shared by most of the alternative cladograms.

**Convergent evolution**—The situation in which two or more kinds of unrelated or distantly related organisms independently acquire what appears to be the same character state from different preexisting ancestral character states. The difference between convergent and parallel evolution cannot always be determined. Convergent evolution is a source of homoplasy in cladistic analyses.

**Data matrix**—A taxon by character table displaying the character states present in each taxon (Table 4-1).

**Homoplasy**—The result of parallel or convergent evolution in which two or more kinds of organisms independently acquire the same (or apparently the same) character state. Homoplasy sometimes also results from reversals when in a particular lineage a character reverts from an apomorphic to a plesiomorphic state. In phylogenetic analyses homoplasy can cause character conflicts or misinterpretation of relationships.

**Ingroup**—In cladistics a monophyletic group under investigation for character polarity by comparison with an outgroup (q.v.).

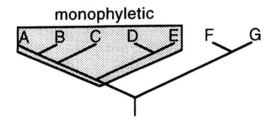

**Monophyletic**—In cladistic analyses a natural taxonomic group composed of an ancestral taxon and *all* of its descendants. Many taxonomists also use the term monophyletic in the sense of paraphyletic (q.v.) to include taxa derived from a common ancestor but not necessarily including all the descendants.

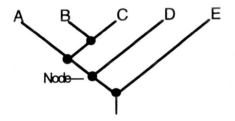

**Node**—A branching point in a cladogram, representing a speciation event.

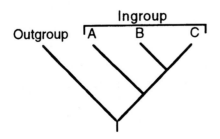

**Outgroup**—A monophyletic taxon that is used in a cladistic analysis to determine which of two or more states of a character in the group being investigated (the ingroup) is apomorphic. Ideally the sister group of the ingroup is used in this analysis; sometimes more than one outgroup must be used.

**Parallel evolution**—The situation in which two or more kinds of related organisms independently acquire the same character state from the same preexisting ancestral character states. The difference between convergent and parallel evolution cannot always be determined. Parallel evolution is a source of homoplasy in cladistic analyses.

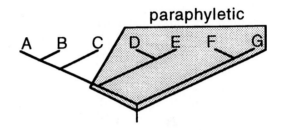

**Paraphyletic**—A taxonomic group that includes an ancestral species and some, but not all of its descendants. In cladistic analyses such a group is rejected in favor of monophyletic taxa. Many currently recognized taxa are probably paraphyletic.

**Parsimony criterion**—A way of choosing between conflicting hypothesis, often used in cladistic analyses. Given a choice of two or more cladograms (hypotheses of relationship) the one involving the fewest assumptions is most likely to be correct. Also called Ockham's Razor (it was first proposed by William of Ockham in 1347) or the principal of simplicity.

**Phylogenetic tree**—A postulated set of relationships among a group of organisms. A phylogenetic tree may or may not be based on cladistic analyses. Often a phylogenetic tree includes postulated ancestor-descendant relationships and a time scale.

**Phylogeny**—The evolutionary history of a group of organisms.

**Plesiomorphy** (plesiomorphic character state)—An evolutionarily primitive (ancestral) character state. In a character transformation series the original state is plesiomorphic to all others. In a character transformation series with more than two states, a particular character state may be apomorphic (q.v.) relative to some states and plesiomorphic relative to others.

**Polarization**—Determining which of a pair of character states is plesiomorphic (ancestral) and which is apomorphic (derived).

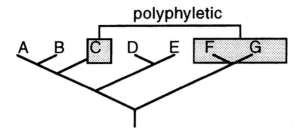

**Polyphyletic**—A non-natural taxonomic group composed of organisms without a common history, hypothesized to have origins in two or more different evolutionary lines; the most recent common ancestor of the two lineages is excluded from the group. Such a taxonomic grouping is generally rejected when it is demonstrated to be polyphyletic and the component taxa are reorganized into monophyletic groups. Some taxa that are currently recognized may actually prove to be polyphyletic when subjected to thorough analysis.

**Reversal**—The situation within a monophyletic lineage in which an apomorphic character state reverts to the plesiomorphic condition. Reversals represent a source of homoplasy in cladistic analyses.

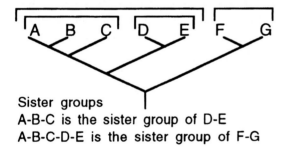

Sister groups
A-B-C is the sister group of D-E
A-B-C-D-E is the sister group of F-G

**Sister group**—A monophyletic taxon that is hypothesized to be the closest relative of a given taxon. Sister taxa are hypothesized to share a common ancestor not shared with any other organisms.

**Step**—Any hypothesized character state change within the structure of a particular hypothesis of relationship (cladogram)

**Symplesiomorphy** (symplesiomorphic character state)—A plesiomorphic character state shared by two or more taxa (shared ancestral character state). In cladistic studies, symplesiomorphies are not considered to be evidence for linking taxa in a cladogram.

**Synapomorphy** (synapomorphic character state)—An apomorphic (derived) character state shared by two or more taxa, considered in cladistic studies to be evidence of relationship.

**Tree length**—The total number of steps (hypothesized character state changes) required by the structure of a particular cladogram based upon a given set of data. Cladograms of different tree length are judged by the parsimony criterion.

# PHYLOGENETIC VS. TRADITIONAL CLASSIFICATION

Not all taxonomists have welcomed cladistics. It has been one of the most controversial developments in systematics. The rationale and methodology have been challenged by taxonomists who (1) prefer phenetic method, (2) who adhere to traditional, non-rigorous approaches to evaluating relationships, (3) who do not understand phylogenetics or (4) who understand phylogenetics and do not like its effects on nomenclature and classification. During the 1970's and 1980's caustic exchanges have taken place between pheneticists and cladists and between traditional systematists and cladists, particularly in the journal *Systematic Zoology*.

Many of the taxa recognized in traditional classifications are not monophyletic in a cladistic sense. "Grade" taxa have often been recognized that are defined by ancestral characters (symplesiomorphies) or by a mixture of ancestral and derived features. These taxa are often **paraphyletic** or **polyphyletic** (see Phylogenetic Glossary above). Only monophyletic taxa can be defined by synapomorphies. Neither paraphyletic nor polyphyletic taxa include all of the descendants of their most recent common ancestor. A paraphyletic taxon is what is left when the more specialized members of a lineage have been pared off and given taxonomic recognition. Often a paraphyletic taxon is defined by what it does not have [the features of its more advanced relatives]. It includes some but not all the descendants of its most recent ancestor. A polyphyletic taxon is composed of selected members of two or more lineages.

Many traditional systematists have been unwilling to accept the rigor of cladistic methodology and prefer a less formal approach to classification. For instance, traditionalists often use a different definition of monophyletic, one that includes both monophyletic and paraphyletic taxa. Arthur Cronquist (1987) wrote that

> If the monophyletic requirement is interpreted loosely rather than strictly, most of the conflict between phylogeny and taxonomy disappears. Monophylesis and polyphylesis are not such utterly distinct things as the terms would suggest. There is a continuous gradation from the strictest monophylesis to the most utter polyphylesis in proposed taxonomic groups. . . . In order to be natural and acceptable, a taxonomic group must fall somewhere toward the monophyletic end of this scale. Simpson has proposed . . . the useful rule of thumb that if a taxonomic group of a particular rank is derived wholly from another group of lesser rank, that is a sufficient degree of monophylesis for taxonomic purposes.

A second objection has been the insistence by some phylogeneticists (including Hennig) that **sister groups** (monotypic groups sharing the same common ancestor) should be accorded the same taxonomic rank, no matter how different they are in appearance or in number of subtaxa. Hennigian methods require that a classification reflect the cladogram on which it was based and vice versa. Given either the cladogram or the classification it should be possible to determine the structure of the other.

Figure 4-4 illustrates the problem. In the traditional classification two families exist, "a" and "b." Cladistic analysis reveals that although family "a" is monophyletic, family "b" is paraphyletic. If a cladist wishes to retain "a" as a family, then "b" must be broken up into several families. But only "b" is a sister group of "a." What rank do we use for the group comprising "a" and "b" (the sister group of "c")? Each pair of sister groups requires a new taxonomic rank and each of the resulting taxa needs a name. This grossly inflates the nomenclature. An alternative is to lump together all the taxa formerly included in both "a" and "b" as a single monophyletic family. However, this monophyletic taxon may be rather heterogeneous. If the phylogeneticists wishes to recognize "a" at the subfamilial level, then he must be willing to shoehorn in a set of extra ranks and names between the family and subfamily level. In published phylogeneticists studies involving a large number of taxa the number of taxonomic ranks and names is sometimes quite intimidating.

A possible way around this, called **phyletic sequencing** was suggested by Wiley (1981). This approach eliminates the requirement of granting each pair of sister groups its own taxonomic rank by requiring that they be listed in the sequence that they appear in the cladogram. In this system taxa a, b, c, d and e in Figure 4-4 would be treated at the same rank (family) and listed in that sequence. As long as the sequence is preserved the structure of the cladogram can be recovered. However, this suggestion has not been accepted within the systematic community.

Existing classifications are full of paraphyletic taxa. Systematists traditionally have recognized some taxa on the basis of their advanced features, others because of primitive characteristics, and still others as somewhere in between. Often very distinctive members of a group of plants or animals have been accorded taxonomic recognition because of some conspicuous feature not present in related taxa. Some groups are in effect defined by the features that they do not have. The strict application of phylogenetics to classification of the angiosperms is resulting in some major reinterpretations at the family and generic levels.

Some of these changes have a actually been advanced by non-phylogeneticists. Robert Thorne, for instance, in 1973 argued for the union of the Apiaceae and Araliaceae on the grounds that the Apiaceae, as traditionally defined, is polyphyletic. According to Thorne the Apiaceae consists of more than one lineage independently derived from different parts of the Araliaceae. In cladistic terminology the Apiaceae is polyphyletic because it includes some of the descendants of two different lineages while excluding their common ancestors. The characters that define the Apiaceae evolved in parallel in two or more different ancestors. The Araliaceae, as traditionally recognized, is a paraphyletic taxon

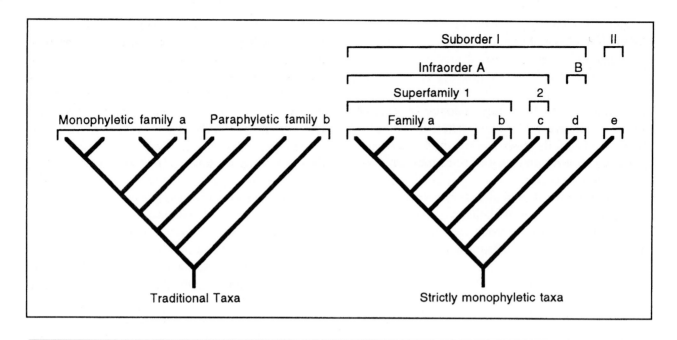

**Figure 4-4** Implications of classifying sister groups at the same taxonomic rank.

defined by symplesiomorphies. Cantino (1992) presented evidence that the Lamiaceae is similarly polyphyletic.

There are several other family pairs that *may* represent a narrowly defined monophyletic taxon classified together with a more generalized paraphyletic taxon. Examples include Asclepiadaceae (almost certainly monophyletic) and Apocynaceae, Brassicaceae and Capparaceae, Orobanchaceae and Scrophulariaceae and various others. The redefinition of widely accepted taxa and the nomenclatural changes that are likely to result may not be received with much enthusiasm by the botanical public, regardless of their justification.

# Phylogenetic vs. Numerical Phenetics

Traditional systematists are not the only systematists with whom cladists have clashed. Numerical pheneticists have taken part in vigorous exchanges with the cladists. Claims and counterclaims have been made about the virtues and faults of the two approaches.

A brief comparison of cladistics and numerical phenetics will show some of the major differences between these methods of evaluating relationships. Both are methods of evaluating characters of organisms and both yield **dendrograms**. [Compare the phenogram illustrated in Figure 6-2 with the cladograms illustrated above]. Cladistic analyses separate the shared derived characters from the ancestral features. Numerical phenetic analyses mix derived and ancestral characters indiscriminately. The grouping of organisms in phylogenetic analyses is based on shared derived characters (synapomorphies) and the best cladogram is determined by the parsimony criterion. The grouping of organisms in numerical phenetics is based on some mathematical method of measuring overall similarity (e.g., a coefficient of similarity) and the resulting phenogram groups taxa or OTU's on the basis of these numerical values.

The purpose of phylogenetics is to reconstruct the evolutionary history of a group of organisms. The purpose of phenetics is to group together the most similar organisms, perhaps with the hope that the clustering will represent relationships. With the proper analyses a data matrix can be recovered in its entirety from the structure of a cladogram, but there is no way to recover the data that were used to construct a phenogram from the structure of the phenogram or from the numerical measures of similarity.

# Other Approaches to Phylogenetics

The methods of generating cladograms described above are those used by the "parsimony school" of cladistics. Several alternate methods have been proposed. The most widely used of these are Wagner's **groundplan/divergence** method, **maximum likelihood analysis,** and **character compatibility.** These methods, like the parsimony method, begin with a data matrix of taxa and characters. The methods differ in the way in which the characters are used.

## The Groundplan/Divergence Method

Warren Wagner, the originator of groundplan/divergence method, hypothesized the existence of a generalized plesiomorphic or ancestral condition for each taxonomic group. This is the groundplan for that group of organisms. Evolutionary changes from that groundplan can be measured in terms of (1) patristic distance (the number of changes from the generalized plesiomorphic condition), (2) divergence (the specific characters that change in one lineage or another) and (3) cladistic branching sequences.

Phylogenetic relationships are presented on a diagram resembling half of a bullseye with a series of concentric semicircles (Fig. 4-5). Each semicircle represents a level of advancement (patristic distance) from the groundplan which is located at the zero point the center of the diagram. Dots representing species are plotted on the diagram on the semicircle representing the degree of advancement. The divergence and sequence of character state changes are plotted as lines connecting the dots. Dots may be connected in a linear sequence indicating progressive advancement within a particular lineage in which some species have advanced more than others. Dots may be connected in a branching sequence where some members of a lineage have advanced by changes in one character and other members have changed in other characters. If a taxon exists that combines the pre-branching character states, it is plotted at the point of branching. If no extant taxon has the character combination hypothesized to exist at the branching point, the space is left empty. Taxa that are suspected or known to be of hybrid origin can be plotted on the diagram in a position suggested by the combination of characters that they possess and connected by dashed lines to the putative parents.

As in the other cladistic systems there is the problem of translating the diagram into a classification. The groundplan/divergence cladograms

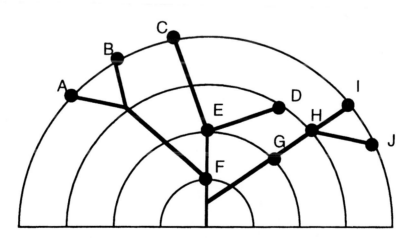

**Figure 4-5**    Example of a Wagner Groundplan/Divergence cladogram plotted on a bullseye diagram. The distance of a taxon from the origin indicates the number of apomorphic features of that taxon relative to the groundplan. Divergence indicates that taxa have advanced independently with apomorphic states of different characters. Taxa A, B, C, I, and J are equally advanced relative to the groundplan but each has its own unique set of apomorphic features. Taxon H is more advanced than G, but less advanced than I and J. G and H do not have any unique derived features not shared with I and J. No extant taxon combines the features of the presumed ancestor of A and B.

contain information about both branching sequence and amount of advancement. In Hennigian cladograms the branching sequence is paramount and degree of advancement plays no part in determining the resulting classification. A Wagner groundplan/divergence cladogram can be interpreted in that fashion. Other approaches are possible, however. Wagner (1969) discussed various alternatives in the interpretation of such a cladogram of relationships of fern families. Wagner and some other users of his approach have been willing to accept paraphyletic in addition to monophyletic taxa.

The principal criticism that has been directed toward users of the groundplan/divergence approach has been the methods that some investigators use to determine the sequence of changes in character states that is fundamental to determining the groundplan for a particular group of organisms. Methods of character polarization other than outgroup comparison are mostly subjective and suspect. If the groundplan is incorrect then the analyses that follow are certain to yield incorrect results. Churchill et al. (1984) determined that groundplan/divergence studies yielded the same results as Hennigian methods if outgroup comparison was used to polarize the characters.

## Maximum Likelihood, Bayesian Analysis, and Distance Methods

Maximum likelihood and genetic distance methods are statistical approaches that have been applied primarily to nucleotide sequence data in which there is a high probability that parallel mutations have occurred repeatedly. In these situations direction of change is difficult to determine and parallelisms and reversals may be expected to occur. Rates of mutation may vary as well. The algorithms involved in using these methods are beyond the scope of this text. Felsenstein (1981), Goldman (1990), Yang (1994), and Huelsenbeck and Ronquist (2001) discuss the strengths and weaknesses of these methodologies. Some workers prefer one or another of these methods over parsimony analysis for nucleic acid sequence data.

## Character Compatibility Methods

In character compatibility studies (see papers below by Duncan, Meacham and Estabrook) each character is examined to determine the logical sequence of character state changes that took place as evolution occurred within the group of organisms being studied. For two-state characters the transformation is merely from one state to the other or vice versa. For multistate characters a character state tree is prepared. The directionality of evolutionary transformation of the characters may be unknown at the time the character state tree is prepared. Each character state transformation supports a particular phylogenetic hypothesis for the group of organisms being investigated.

Next the data are analyzed to determine the largest **clique** of mutually compatible character states. Compatible character states are those that support the same phylogenetic hypothesis. If two characters are incompatible, they support different phylogenies. They cannot both be true. The largest clique is then used as the starting point for construction of a cladogram. Characters that are not a part of the largest clique are initially disregarded. After the largest clique of mutually compatible characters has been defined, a cladogram that displays the resulting hypothesis of relationship is prepared. Because of character conflicts, some of the branches in the cladogram may not be wholly resolved.

All characters that were incompatible with the largest clique are evidence of homoplasy. Each presumably represents an example of parallel evolution or reversal, or a misinterpretation of the character. If it can be shown that a character state transformation that was incompatible with the largest clique really represents parallel evolution or reversal, the incompatible character can be treated as having two or more similar but separate character states. When reinterpreted in this way the incompatible character may become compatible with the members of the largest clique. If particularly firm evidence exists that one or more of the characters excluded from the largest clique represents a true phylogeny, then the largest clique may actually represent homoplasy, and a different cladogram may be a more accurate representation of the data.

A cladogram derived by character compatibility may be rooted or unrooted. In an unrooted cladogram determinations of ancestral and derived conditions are not required although logical changes of character states must be postulated. The resulting diagram does not indicate the sequence of speciation events but is designed to illustrate relationships based upon the clique of compatible characters. Rooting of a cladogram occurs when a particular character combination is determined to be ancestral (by outgroup comparison or other criteria).

Several aspects of character compatibility analysis are controversial. For example, the cladogram that results from a character compatibility analysis is not necessarily the most parsimonious reconstruction of phylogenetic relationships. This has resulted in a heated exchange between the advocates of character compatibility and those of parsimony methods. This exchange has been exacerbated by the willingness of advocates of character compatibility methods to accept paraphyletic as well as monophyletic taxa. This represents a deep philosophical difference between members of the two camps regarding the purposes and methods of determining evolutionary relationships.

# References

Angiosperm Phylogeny Group [APG]. 1998. An ordinal classification for the families of flowering plants. *Ann. Missouri Bot. Gard.* 85: 531–553.

———. 2003. An update of the Angiosperm Phylogeny Group classification for the orders and families of flowering plants: APG II. *Bot. J. Linnean Soc.* 141:399–436.

Ashlock, P. D. 1974. The uses of cladistics. *Annual Review of Ecology and Systematics* 5:81–100.

Bremer, K., and H.-E. Wanntorp. 1978. Phylogenetic systematics in botany. *Taxon* 27:317–329.

Brooks, D. R., J. N. Caira, T. R. Platt, and M. R. Pritchard. 1984. *Principles and Methods of Phylogenetic Systematics.* A Cladistics Workbook. University of Kansas Museum of Natural History Special Publication 12. Lawrence Kansas.

Cantino, P. D. 1992. Evidence for a polyphyletic origin of the Labiatae. *Annals of the Missouri Botanical Garden* 79:361–379.

Churchill, S. P., E. O. Wiley and L. A. Hauser. 1984. A critique of Wagner ground-plan divergence studies and a comparison with other methods of phylogenetic analysis. *Taxon* 33:212–232.

Cronquist, A. 1987. A botanical critique of cladism. *Botanical Review* 53:1–52.

Crowson, R. A. 1970. *Classification and Biology.* Atherton Press, New York.

Dahlgren, R., and K. Bremer. 1985. Major clades of the angiosperms. *Cladistics* 1:349–368.

———, and F. N. Rasmussen. 1983. Monocotyledon evolution. Characters and phylogenetic estimation, Pp. 255–395 *in* M. K. Hecht, B. Wallace and G. T. Prance (eds.), *Evolutionary Biology,* Vol. 16. Plenum Publ. Co.

de Queiroz, K., and S. Poe. 2001. Philosophy and phylogenetic inference: a comparison of likelihood and parsimony methods in the context of Karl Popper's writings on corroboration. *Systematic Biology* 50:305–321.

Duncan, T. 1980. Cladistics for the practicing taxonomist—an eclectic view. *Systematic Botany* 5:136–148.

———, and T. F. Stuessy (eds.). 1984. *Cladistics: Perspectives on the Reconstruction of Evolutionary History.* Columbia Univ. Press, New York.

———, (eds.). 1985. *Cladistic Theory and Methodology.* Benchmark Papers in Systematic and Evolutionary Biology Series. Van Nostrand Reinhold Co., New York.

Dupuis, C. 1984. Willi Hennig's impact on taxonomic thought. *Annual Review of Ecology and Systematics* 15:1–24.

Eldredge, N., and J. Cracraft. 1980. *Phylogenetic Patterns and the Evolutionary Process.* Columbia Univ. Pr., N.Y.

Estabrook, G. F. 1972. Cladistic methodology: a discussion of the theoretical basis for the induction of evolutionary history. *Annual Review of Ecology and Systematics* 3:427–456.

Farris, J. S. 1970. Methods for computing Wagner trees. *Systematic Zoology* 19:83–92.

——— 1988. *Hennig86,* version 1.5. Published by the author, Port Jefferson Station, New York.

Felsenstein, J. 1981. Evolutionary trees from DNA sequences: a maximum likelihood approach. *Journal of Molecular Evolution* 17:368–376.

———. 1982. Numerical methods for inferring evolutionary trees. *Quarterly Review of Biology* 57:379–404.

———. 1983. Parsimony in systematics: biological and statistical issues. *Annual Review of Ecology and Systematics* 14:313–333.

——— (ed.). 1983. *Numerical Taxonomy.* Springer-Verlag, New York.

———. 1985. Confidence limits on phylogenies: an approach using the bootstrap. *Evolution* 39:783–791.

Funk, V. A. 1985. Phylogenetic patterns and hybridization. *Annals of the Missouri Botanical Garden* 72:681–715.

Gaffney, E. S. 1979. An introduction to the logic of phylogeny reconstruction, pp. 79–111, in Cracraft, J. and N. Eldredge (eds.) *Phylogenetic Analysis and Paleontology,* Columbia University Press, New York.

Goldman, N. 1990. Maximum likelihood inference of phylogenetic trees, with special reference to a poisson process model of DNA substitution and parsimony analysis. *Systematic Zoology* 39:345–361.

Hennig, W. 1966. *Phylogenetic systematics*. University of Illinois Press, Urbana.

Hill, C. R., and P. R. Crane. 1982. Evolutionary cladistics and the origin of angiosperms. Pp. 269–361 *in* K. A. Joysey and A. E. Friday (eds.), *Problems of Phylogenetic Reconstruction*. Systematics Association Special Volume no. 21. Academic Press, London.

Hillis, D. M. 1987. Molecular versus morphological approaches to systematics. *Annual Review of Ecology and Systematics* 18:23–42.

———, and C. Moritz. 1991. *Molecular Systematics*. Sinauer Associates, Sunderland, Massachusetts.

Huelsenbeck J. P., and F. Ronquist. 2001. MrBayes: Bayesian inference of phylogeny. Version 2.01. Distributed by the author. Department of Biology. University of Rochester.

Hull, D. L. 1970. Contemporary systematic philosophies. *Annual Review of Ecology and Systematics* 1:19–54.

———. 1979. The limits of cladism. *Systematic Zoology* 28:416–440.

Humphries, C. J, and V. A. Funk. 1984. Cladistic methodology, pp. 323–362 *in* V. H. Heywood and D. M. Moore (eds.), *Current Concepts in Plant Taxonomy*. Systematics Association Special Volume no. 25. Academic Press, London.

Joysey, K. A., and A. E. Friday. 1982. *Problems of Phylogenetic Reconstruction*. Systematics Association Special Volume no. 21. Academic Press, London.

Kitching, I. J., P. L. Forey, C. J. Humpheries, and D. M. Williams. 1998. *Cladistics: The Theory and Practice of Parsimony Analysis*. Second Edition. The Systematics Association Publication no. 11. Oxford University Press, Oxford.

Maddison, W. P., M. J. Donoghue, and D. M. Maddison. 1984. Outgroup analysis and parsimony. *Systematic Zoology* 33:83–103.

———, and D. R. Maddison. 1992. *MacClade, version 3.0*. Sinauer Associates, Sunderland, Massachusetts.

Mayr, E. 1981. Biological classification: toward a synthesis of opposing methodologies. *Science* 214:510–516.

Meacham, C. A. 1980. Phylogeny of the Berberidaceae with an evaluation of classifications. *Systematic Botany* 5:149–172.

———. 1981. A Manual Method for character compatibility analysis. *Taxon* 30:591–600.

———. 1984. The role of hypothesized direction of characters in the estimation of evolutionary history. *Taxon* 33:26–38.

———, and G. F. Estabrook. 1985. Compatibility methods in systematics. *Annual Review of Ecology and Systematics* 16:431–446.

Miyamoto, M. M., and J. Cracraft (eds.). 1991. *Phylogenetic Analysis of DNA Sequences*. Oxford University Press, N.Y.

Pankhurst, R. J. 1992. *Practical Taxonomic Computing*. Cambridge University Press, New York and Cambridge.

Patterson, C. 1982. Morphological characters and homology. Pp. 21–74 *in Problems of Phylogenetic Reconstruction*. Systematics Association Special Volume no. 21. Academic Press, London.

———. 1982. Cladistics and classification. *New Scientist* 94:303–306.

Platnick, N. I., and V. A. Funk. 1983. *Advances in Cladistics,* Volume 2. Proceedings of the Second Meeting of the Willi Hennig Society. Columbia University Press, New York.

Ross, H. 1974. *Biological Systematics*. Addison Wesley, Reading, Massachusetts.

Sneath, P. H. A., and R. R. Sokal. 1973. *Numerical Taxonomy*. W. H. Freeman, San Francisco.

Soltis, P. E., D. R. Soltis, and J. J. Doyle (eds.). 1992. *Molecular Systematics of Plants*. Chapman and Hall, New York.

Stevens, P. F. Evolutionary Polarity of Character States. *Annual Review of Ecology and Systematics* 11:333–358.

Stevens, P. F. (2001 onwards). Angiosperm Phylogeny Website. Version 5, May 2004 [and more or less continuously updated since]. http://www.mobot.org/MOBOT/research/APweb/.

Stuessy, T. F. 1983. Phylogenetic trees in plant systematics. *Sida* 10:1–13.

———, and D. J. Crawford. 1983. Flavonoids and phylogenetic reconstruction. *Plant Systematics and Evolution* 143:83–107.

———. 1990. *Plant Taxonomy. The Systematic Evaluation of Comparative Data*. Columbia University Press, New York.

Swofford, D. L. 1990. *PAUP: Phylogenetic analysis using parsimony*. Illinois Natural History Survey, Champaign.

Wagner, W. H., Jr. 1969. The construction of a classification. Pp. 67–103 *in* C. G. Sibley (chairman), *Systematic Biology.* Publication 1692, National Academy of Sciences, Washington, D.C.

———. 1980. Origin and philosophy of the ground-plan-divergence method of cladistics. *Systematic Botany* 5:173–193.

Wiley, E. O. 1981. *Phylogenetics: The Theory and Practice of Phylogenetic Systematics.* John Wiley & Sons, N.Y.

———, D. Siegel-Causey, D. R. Brooks, and V. A. Funk. 1991. *The Compleat Cladist. A Primer of Phylogenetic Procedures.* University of Kansas Museum of Natural History Special Publication 19. Lawrence, Kansas.

Yang, Z. 1994. Statistical properties of the maximum likelihood method of phylogenetic estimation and comparison with distance matrix methods. *Systematic Biology* 43:329–342.

# Exercises

## ■ Discussion Questions

1. What aspects of phylogenetic systems do cladistic approaches correct?

2. What are the underlying assumptions of cladistic analysis?

3. What is meant by parsimony?

4. Explain the reasons for outgroup comparison.

5. Why are polyphyletic taxa undesirable?

6. What is the difference between monophyletic and paraphyletic?

7. Why do some taxonomists prefer to recognize paraphyletic systems?

## Problem Solving

8. Consider the data displayed in the table below for a hypothetical group of plants. Assume that Taxa 1–3 comprise a monophyletic group.

| | (A) Leaves | (B) Petals | (C) Filaments | (D) Flower color | (E) Ovary chambers | (F) Fruit | (G) Seed coat | (H) Seed number |
|---|---|---|---|---|---|---|---|---|
| **CHARACTERS AND CHARACTER STATES** | | | | | | | | |
| Taxon 1 | opposite | acute | hairy | yellow | one | fleshy | smooth | many |
| Taxon 2 | alternate | acute | hairy | purple | two | fleshy | smooth | many |
| Taxon 3 | alternate | obtuse | smooth | red | one | dry | rough | few |
| Outgroup | alternate | obtuse | smooth | red | two | dry | rough | many |

In the space below construct cladograms representing the *possible* relationships among these four taxa.

Which of these is the most parsimonious cladogram consistent with the data in the table? Why?

Which of the eight characters support this conclusion? List them by letter.

Which of the eight character states represent autapomorphies? List them by letter and taxon.

Which of the eight characters represent homoplasy? List them by letter.

9. With the aid of your instructor, find a published data matrix. Using a cladistics program on a microcomputer (PAUsP, MacClade, etc.) carry out a cladistic analysis of this data set.

What is the source of the data set you analyzed?

What is the taxonomic group under investigation?

How many taxa are included in the ingroup?

Did the author of the study include more than one outgroup?

How many characters were included in the data set?

Did the analysis yield one cladogram that was more parsimonious than the rest?

If so, how many steps did it require?

Attach a printout of this cladogram.

If there is more than one equally most parsimonious cladogram, how many different trees are there?

How many steps did they require?

How is it possible for an analysis to give more than one equally most parsimonious tree?

Are alternate cladograms similar in most respects or are there major differences?

What characters in the data matrix are examples of homoplasy in the most parsimonious cladogram(s)?

# Floras, Manuals, and Botanical Descriptions

chapter 5

One of the most valuable skills that you as a taxonomy student can gain is the ability to correctly determine the names for plants on your own. Occasionally almost everyone needs to know the name of a plant. Anyone whose business, research, hobbies or interests includes plants can appreciate the skills of a taxonomist who not only knows the names of various plants but also can determine the names of unknown plants as well. This chapter examines some of the methods used by taxonomists in plant identification.

The potential scope of the problem of determining the correct name for an unknown plant might seem overwhelmingly large. It certainly would be if you had to use trial and error methods. Potentially you would have to compare the unknown plant with all of the many thousands of kinds of plants in the world. Such an approach would be prohibitively difficult and time-consuming. Fortunately there are various shortcuts that you can learn to use. *ask Dr. Keil*

The easiest method of identification is to ask an "expert" who has already done the comparisons. "Experts" vary in competence, with the most reliable usually found in localized centers of botany such as colleges and universities, museums and herbaria, botanical gardens, and governmental agencies involved with the interpretation of nature for the public. "Experts" specializing in cultivated plants often work in nurseries. There is no question that such a person can be very helpful. There are some problems though with this approach. It may be difficult to transport identifiable material to the "expert" in a timely fashion if the "expert" is not close at hand. Then too, "experts" cannot always make the time available for everyone who needs a plant identified. There is also the very human prob-

lem of judging the competency of your "expert." Some "experts" will answer any question with confidence even though they do not really know the answer. It may be difficult to tell when this is happening.

Often you will have to determine the name of the plant on your own. One approach, unfortunately very subject to error, is to compare the unidentified plant with photographs or drawings in popular or scientific books. Plant species are often distinguished on the basis of technical characters that may not be evident in a picture. Printing costs often prevent the publication of pictures of all the plants known from a region. Often only the most conspicuous plants or those for which photographs were available are illustrated. You may well have a plant that is not illustrated. Because no two individual plants are exactly alike, your unknown plant will undoubtedly differ in one feature or another from all of the illustrations. Some picture books have descriptions of the distinguishing features in addition to illustrations and these may help in identification. The best use of illustrations, however, is as a backup to other methods of identification.

Comparison of an unknown plant to previously determined specimens can also be very helpful. However, you have to have access to an herbarium (a collection of dry plant specimens), a botanical garden or some other institution with previously identified specimens, and you have to know where to look. If you do not have any idea what the plant is, it is very unlikely that you will ever find it in an herbarium. Moreover, most institutional herbaria are not willing to allow just anyone to rummage through the specimens. Another problem is that no two specimens look exactly alike.

Unless you know the features that can be used to distinguish one species from another, you can easily be misled by the similarities and differences among the specimens.

Many amateurs and students have their own small herbaria that serve as a repository for specimens that they have previously determined. Some of the major herbaria of the world started as the personal collection of some early botanist. You might want to establish a collection of your own. Procedures for preparing plant specimens are discussed in Chapter 7.

Comparison of the unknown plant with written descriptions is potentially a very reliable method of making your decision. However, some of the problems mentioned above apply if you are looking up descriptions. If you do not already have a good idea what the plant might be, you could search for a long time. Descriptions vary greatly in their quality. The best are composed by taxonomists who have studied enough specimens to document the range of variability present in each taxon. If you have a good idea of what the unknown plant might be, well-written descriptions can greatly help you in making your final decisions. The features of botanical descriptions are discussed later in this chapter.

One of the best approaches is to become an "expert" yourself. In studying plant taxonomy you will gain some expertise by learning to recognize the features of some important groups of plants (particularly some of the major plant families). You will undoubtedly learn to recognize some genera and species as well. The more you know, the easier it will become for you to add to your knowledge. All experts started out knowing no more than you, as an introductory student, know now. If you know that a plant is a member of a particular family or of a particular genus, you are already well along on your way to determining the name of the plant. You have eliminated the members of all the other families in the world.

As a taxonomy student, you will often be asked to determine the correct name for an unknown plant [and it may not be a member of a group that you know]. One of the fastest and most reliable methods of doing this is to use a taxonomic key. A key is designed to help you to eliminate wrong answers and to arrive at the correct determination without laboriously searching at random. After arriving at an answer, you then can check to find out whether you are correct by checking with an expert, an illustration, a previously determined specimen, or a description. Keys, descriptions and sometimes illustrations are found in books called floras and manuals. Proficiency in the use of taxonomic keys is one of the most important skills of an expert taxonomist.

# Construction of Keys

The keys found in most floras and manuals are dichotomous keys. They consist of a series of two-way choices along the route from the unknown to the known. At each fork in the road the user is asked a pair of questions. If the user answers these questions correctly each time he or she will end up in the right place. The first dichotomous key was published in 1805 by a French botanist, Jean de Lamarck, in his **Flora Francois** as a shortcut method of comparing plant descriptions. Other taxonomists soon adopted the use of keys and they have become a standard taxonomic tool in biology. The following paragraphs examine the construction of taxonomic keys and also their use.

The structure of a key is very simple. It is a series of paired, mutually exclusive statements that divides a set of objects or classes into progressively smaller subsets until each subset consists of only one object or kind of object. Let's examine this definition in detail. If a key is a series of paired statements, then good grammar would dictate that we start the statements with capital letters and end them with periods. A properly. constructed statement generally comprises a subject, a verb, and a predicate. The subject is a noun, the object the key writer wants the reader to look at. Our key statement differs from most statements in that the verb is implied and not written. It would always be a form of the verb "to be", so we omit it. The predicate consists of one or more adjectives or other modifiers that describe the subject. Examine the pair of statements below:

**Apex of leaf [is] acute to acuminate.**
subject - [verb] - predicate
**Apex of leaf [is] obtuse.**

Notice that both statements begin with the same subject. The predicates describe alternate, non-overlapping conditions of that subject. If you had a set of leaves, some with acute tips and others obtuse, this pair of statements could be used to divide the set into two subsets. Other pairs of statements could then be employed to further divide each of the subsets.

We can write a key to the five common fasteners illustrated in Figure 5-1. When we divide a set of

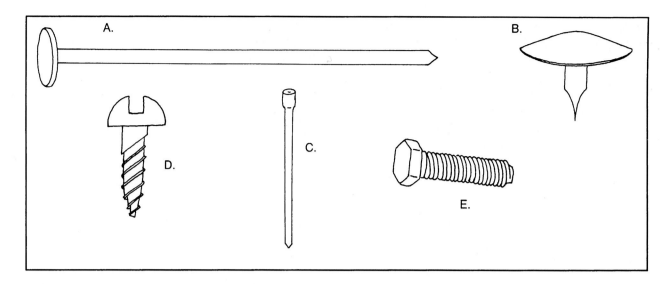

**Figure 5-1** Five common fasteners used in the discussion of keys.

objects into subsets, we must account for all the objects. We cannot leave any out. We can divide our set of five fasteners into a subset of four and a subset of one, or we can divide it into subsets of three and two. In this case let's separate the objects on the basis of whether the shaft is grooved (D and E) or not (A, B, and C). Our first pair of statements is written as follows:

1. Shaft of fastener with spiral grooves.

1' Shaft of fastener smooth.

Statement 1 defines a subset of two objects and statement 1' (one prime) defines a subset of three. Let's consider the first subset. We now have a set of two objects and we need a pair of statements that will distinguish them.

1. Shaft of fastener with spiral grooves.
    2. Tip of shaft pointed . . . . . D-screw
    2' Tip of shaft blunt . . . . . . . E-bolt
1' Shaft of fastener smooth.

Notice that we indented statements 2 and 2' to indicate that they are subsets of the group defined by statement 1. When a key statement defines a subset with only one object in it we supply the name of the object. Now let's look at the fasteners with smooth shafts. We can easily divide them into two groups.

1. Shaft of fastener with spiral grooves.
    2. Tip of shaft pointed . . . . . . D-screw
    2' Tip of shaft blunt . . . . . . . E-bolt

1' Shaft of fastener smooth.
    3. Head cylindric, nearly as tall as wide . . . . . . . . . . . . . . . . . C-finishing nail
    3' Head flat, much wider than thick.

Statements 3 and 3' are subsets of 1'. Notice that they are indented the same distance from the left margin as 2 and 2'. Now all we have to do is separate the nail from the tack.

1. Shaft of fastener with spiral grooves.
    2. Tip of shaft pointed . . . . . . D-screw
    2' Tip of shaft blunt . . . . . . . . E-bolt

1' Shaft of fastener smooth.
    3. Head cylindric, nearly as tall as wide . . . . . . . . . . . . . . C-finishing nail
    3' Head flat, much wider than thick.
        4. Shaft one inch or more long . . . . . . . . . . . . . . A-nail
        4' Shaft less than 1/2 inch long . . . . . . . . . . . . . . B-tack

Our key is now complete. Notice that we started with five objects and we used four pairs of statements in the key. If n = number of objects in a set, a properly constructed key to those objects will have n – 1 pairs of key statements. If you end up with some other number, you have made a mistake.

The key we have written is an "indented key." **We will use the indented form of key exclusively in this text.** However you may encounter another form of key in some manuals. Let's write our key in the format of a "bracket key."

1.  Shaft of fastener with spiral grooves . . . . . . . . 2

1'  Shaft of fastener smooth. . . . . . . . . . . . . . . . 3

   2 (1).  Tip of shaft pointed . . . . . . . . . D-screw

   2'  Tip of shaft blunt . . . . . . . . . . . . . . . E-bolt

3 (1').  Head cylindric, nearly as tall as
      wide . . . . . . . . . . . . . . . . . C-finishing nail

3'  Head flat, much wider than thick . . . . . . . . . . 4

   4 (3').  Shaft 1 inch or more long . . . . . . . A-nail

   4'  Shaft less than 1/2 inch long . . . . . . B-tack

Notice that in a bracket key you need "road signs" to tell you where to go. Without the numbers on the right, there would be no way for you to know which key statements to read next. Without the numbers in parentheses it is very difficult to retrace your steps. It is easy to get lost in a bracket key.

When you are writing a key, remember that it is being written for someone else to read. Be as neat and clear as possible. Below is a list of guidelines for key writing.

1.  Start each key statement with the subject. The subject is what you want the reader to look at. This should be a noun. Remember—subject first, then the modifiers, not the other way around. Whenever possible use the same first word for both statements.

    **[Right]**

    1.  Leaves opposite.

    1'  Leaves alternate.

    **[Wrong]**

    1.  Opposite leaves.

    1'  Alternate leaves.

2.  Avoid unnecessary words. They get in the way and obscure the meaning of the key.

    **[Right]**

    1.  Leaves opposite.

    1'  Leaves alternate.

    **[Wrong]**

    1.  Plants with opposite leaves.

    1'  Plants with alternate leaves.

    Note that in this example, the leaves are what the key writer wants the reader to look at. But

instead of drawing attention to the leaves, the writer has buried them at the end of the key statement.

3.  Whenever possible use different words to start successive pairs of key statements. It makes the key easier to use and presents fewer opportunities to make mistakes.

4.  Avoid broad, all-inclusive negatives. In general avoid using the word "not." It is much preferable to write positive statements.

    **[Right]**

    1.  Flowers purple.

    1'  Flowers white or pink.

    **[Wrong]**

    1.  Flowers purple.

    1'  Flowers not purple.

    What color is not purple? It could include green, yellow, brown, blue, etc. *Never* use "not as above." It is the broadest and least informative of all the negatives.

5.  Use "absent" instead of "not present," "without" in place of "not with," "glabrous" instead of "not pubescent," etc. Reserve the use of "not" for absolute negatives that cannot be easily expressed otherwise.

    1.  Petioles prickly.

    1'  Petioles not prickly.

6.  Use measurements rather than "large," "small," "tall," "short," etc. These words require a frame of reference. A sunflower may be tall when compared to a dandelion, but it is very short when compared to a redwood. Your reader may have a different frame of reference than you have, but if you use centimeters, meters, etc., you are setting the frame of reference to an absolute scale.

7.  Whenever it is at all practical to do so, divide the set of objects into equal or subequal subsets. Avoid the practice of picking off the items in the set one at a time. However, if the set includes a truly exceptional plant or object, it is often best to eliminate it early to get it out of the way.

8.  Use features of the plant that are constant in preference to variable features, particularly if the members of the resulting subsets overlap.

**[good]**

1. Flowers purple.

1′ Flowers white.

**[bad]**

1. Leaves 8–12 cm long.

1′ Leaves 6–10 cm long.

What is the user of the key to do if the plant being keyed has leaves 9 cm long?

9. Three-way choices are definitely against the rules. What do you do with a group of three items? Divide them into subsets of 2 and 1 and then divide the subset of 2 in a second pair of statements. Remember the n – 1 rule. It will always take two pairs of statements to separate three objects.

**[right]**

1. Flowers white . . . . . . . . . . . Plant A

1′ Flowers red or yellow.

2. Petals red . . . . . . . . . . . . . . Plant B

2′ Petals yellow . . . . . . . . . . . Plant C

**[wrong]**

1. Flowers white . . . . . . . . . . . Plant A

1′ Flowers red . . . . . . . . . . . . . Plant B

1′ Flowers yellow . . . . . . . . . . Plant C

# Use of Keys

Most of the time you will be using a key written by someone else. Remember that the key is a reconstruction of another person's interpretation of how a set of plants can be broken up into subsets. The ease of use and clarity of a key are dependent upon several factors. Some botanists are more skilled than others at seeing and using meaningful characters. Even the best key writers sometimes face the task of writing a key to a taxonomically difficult group in which characters vary or overlap. The key writer may not have known about a variation present in the plant that you are attempting to key. Before you criticize a key too severely, think about the difficulty of writing a key to several hundreds or thousands of plants and getting all the details right.

Do not expect to be able to pick up a key and immediately be able to use it rapidly and without error. Keying is a skill, and like any other skill it can be improved with practice. Your skill at using a key

is dependent on several factors. One of the most important is vocabulary. If you have to look up every other term you encounter, it will take a long time for you to make your way through a key. You also have to correctly interpret the features of the plant. How carefully you work will directly affect your ability to correctly use a key. Keys are very unforgiving. If you make a mistake in using a key and do not correct your error, you almost certainly will arrive in the wrong place. If you were to guess rather than accurately determine the answer at three places in keying out a plant, you would have only one chance in eight of arriving at the correct answer. Making four guesses reduces your chance to one in sixteen. **It does not pay to guess when keying.**

There are several basic guidelines to follow when using a key. (The following are from **Taxonomy of Flowering Plants,** Second Edition, by C. L. Porter. W. H. Freeman and Company, Copyright © 1967.)

1. **Always read both choices**—even if the first seems to be the logical one to take. The second may be even better.

2. Be sure you understand the meaning of the terms involved. **Do not guess.**

3. When measurements are given, use a calibrated scale. **Do not guess.**

4. When minute objects are concerned, use a lens of sufficient magnifying power to show clearly the feature you need to see.

5. Since living things are always somewhat variable, do not base your conclusion on a single observation, but arrive at an average by studying several parts or specimens. It is surprising how often students will find the one unusual or aberrant sample in a large assortment of normal things!

6. As in travelling a forking road, if the choice of division is not clear, or if you have no way of making a choice because you do not have sufficient information, try both divisions: arrive at two possible answers by doing so, and then read descriptions of each in order to make a choice. A key is only a shortcut to identification; it is not essential if descriptions are available.

7. Finally, having arrived at an answer in a key, do not accept this as absolutely reliable, but check a description of the plant to see if it agrees with the unknown specimen. If not, an error has been made somewhere, either in the key or in its use.

The ultimate check on identification is a comparison of the unknown with an authentically named specimen in an herbarium.

# Computers and Plant Identification

The explosive growth in computer technology of the past quarter century has had an impact upon both the generation and use of keys. It is possible to construct a detailed data base of taxa and their characteristics and to sort this information into the structure of a taxonomic key. A computer can make many comparisons within the structure of the data base and determine a logical sequence of steps for constructing a key. This technology was developed in the 1960's and 1970's for use with main-frame computers. Similar programs are now available in several different formats for various microcomputer systems.

Computers have made possible various types of interactive keys for identification. One of the drawbacks of using a conventional dichotomous key is the limitation of having to progress through the key in the sequence of steps determined by the author of the key. This at times can be extremely frustrating, particularly when the key is not clearly written or does not initially request the features that are most obvious to the user. Additionally, keys sometimes require characters that are absent from the specimen being examined. With a computer it is possible to construct a multiple-entry key that allows the user to access the key with the features that are readily available. As these programs are perfected, their use is likely to increase.

## List of Online Keys

http://www.hua.huh.harvard.edu/FNA/

http://www.eFloras.org

http://www.Ucjeps.berkeley.edu

http://www.MountainNature.com

http://www.herbarium.lsa.umich.edu

## Online Volumes

Volume 1 Introduction

Volume 2 Pteridophytes and Gymnosperms
31 families

Volume 3 Magnoliophyta: Magnollidae and Hamamelidae
32 families

Volume 4 Magnoliophyta: Caryophyllidae, Part 1
10 families

Volume 5 Magnoliophyta: Caryophyllidae, Part 2
3 families

Volume 22 Magnoliophyta: Alismatidae Arecidae, Commelinidae (in Part), and Zingiberidae
30 families

Volume 23 Magnoliophyta: Commelinidae (in part): Cyperaceae
1 family

Volume 26 Magnoliophyta: Liliidae: Lillales and Orchidales
11 families

# Floras and Manuals

**Floras** and **manuals** are among the most widely used sources of taxonomic information about plants. As a student in an introductory taxonomy class you will undoubtedly be required to learn how to use such a publication. In these references professional taxonomists have recorded the results of their studies and observations on the plants growing in a specified geographical region or a particular ecological situation. Some floras and manuals cover a wide geographic range, and others are restricted to a small area. Most include all of the vascular plants of the region being considered, but some deal only with certain types of plants (e.g., aquatic plants, ferns, shrubs, etc.). This chapter will serve as a guide to the kinds of information found in floras and manuals.

The words manual and flora are often used interchangeably, but they are not necessarily the same. A flora may be merely a list of the plants of an area or it may provide the reader with keys, information about the plants' distribution, habitats, etc. Manuals are generally more complete sources of information. In addition to keys, a manual includes a description for each family, genus and species, and some manuals provide descriptions for taxa of other ranks as well. **Illustrated manuals** supplement the descriptions with line drawings that illustrate features of the plants. Line drawings can be very useful, but the user should be aware that only one individual of a species is generally illustrated. The description may present variations not indicated by the drawing.

Observations and measurements of individual plant specimens are the ultimate sources of information for the preparation of the keys, botanical descriptions and statements of flowering time, habitat and geographical range that are included in a flora or a manual. These measurements and observations may have been made by the author(s) of the flora or by specialists who have studied particular taxa in detail. Each individual specimen is a record of the appearance of an individual plant and also documents the plant's occurrence at a particular location and time. Ideally the author(s) of a flora consults a large number of specimens before publishing the work. If the area covered by the flora has been poorly sampled in the past, the author(s) may have to rely on a less than ideal number of specimens.

The publication of a flora is often the stimulus to specialized studies. Because a flora generally deals with all the species of an area, taxonomic problems are often uncovered during the preparation of the work. Species that are poorly distinguished or that exhibit unexplained patterns of variation have to be dealt with along with those that are well understood. Often the author(s) of a flora cannot take the time necessary to resolve all of these taxonomic problems if the flora is to be published. Decisions may have to be made on the basis of incomplete data. The author may include a statement that a group of taxa are in need of further study. Other taxonomists can investigate the problem taxa later in their own specialized research projects.

The relationship between floras and monographic research is reciprocal. Just as a flora can serve to point out taxa in need of further study, revisionary and monographic studies (Chapter 18) provide a foundation for the treatment of many of the plants that occur in the area covered by the flora. The author of a flora cannot be a specialist in all groups of plants. Published research by various members of the taxonomic community assist the author(s) of a

flora in the interpretation of many of the taxa in the study area.

The preparation of a flora or manual in a timely and thorough fashion is becoming increasingly difficult for an individual taxonomist. Some floras have been essentially the life work of their authors. More and more in recent times the preparation of a major flora has come to be a collaborative work involving several to many contributors. This has the advantage of bringing together the expertise of various specialists who prepare the treatments for the plants they know well. It has the disadvantage of making the resulting work somewhat uneven in style and vocabulary. The problems of the editor who attempts to coordinate such a study are many, and the expense of both time and money can be great.

Any flora is incomplete to a certain extent. Specimens or live plants not seen by the author(s) or by specialists often express extremes of variation that exceed those presented in the descriptions. Plants sometimes occur in habitats or localities outside the range given in the flora. Additional taxa may be discovered in the area covered by the flora after the work has been published. The publication of revisions and monographs by specialists may modify the interpretations of taxonomic limits and the nomenclature used in the flora. Because of ongoing taxonomic research, floras periodically have to be updated or are superseded by later works (see discussion at end of Chapter 2).

## Botanical Descriptions

The plant descriptions included in a manual are a very important source of information for anyone attempting to determine the correct name for a plant. To find the description of a taxon in a manual you need to have the plant's name. If you already know or suspect the name of a plant, you can look it up in the index and go directly to the description. In many cases, though, you will arrive at the name of a plant through use of a taxonomic key. If you are attempting to determine the name of an unknown plant, it is very important for you to compare the features of the plant you are examining with the description of the taxon you have arrived at through use of the key. In most manuals, every time a key ends in a name, there is a description available with which to check yourself. Ideally, the determination of a plant with a key should not be considered complete until you have ascertained that the features of the plant are in accordance with the taxon's description. This is where the difference between a manual and a flora becomes very important. If you are using a flora that lacks descriptions you will not be able to check your determination unless you have available a description of the taxon in another reference.

As you read the description and observe the plant you should not find any 'major' discrepancies between the printed description and the characters visible on the plant. However, in descriptions of genera, families, or higher level taxa, do not expect that the description will always fit exactly. Organisms are variable and exceptions occur. Words such as 'usually,' 'mostly,' 'rarely,' etc., mean what they say. You will occasionally encounter plants which display the unusual condition. However, if you do meet a *major* conflict between a plant and the description you are reading, the error is usually in your keying and not in the description or the plant.

Some characteristics are invariant for a particular taxon whereas others may be highly variable. Variation in quantitative characters (counts, measurements) is expressed as a range of numbers. Descriptions, unlike statistical analyses, generally do not include mean values, standard deviations, etc. Variation in qualitative characters is expressed as a series or range of descriptive terms. Variations that are seldom encountered may be listed in parentheses (as in the descriptions of families in Chapters 8–16 of this book).

## Sample Treatments

In most manuals the descriptions correspond to the end points in a sequence of keys. There usually is a key to families and a taxonomic description for each family. For each family that contains more than one genus there is a key to genera and a description of each genus. For each genus that has more than one species there is a key to species and a description of each species. Keys and descriptions for varieties or subspecies may be included as well. Descriptions vary greatly from one manual to another in their completeness. **A California Flora** (Munz and Keck, 1959) [actually a manual] contains some of the more complete descriptions found in regional floras. The sample descriptions that follow are slightly modified from those published by Munz and Keck (pp. 196–197). Used by permission of University of California Press.

## Eschscholzia Cham. in Nees.
## California Poppy

Annual or perennial herbs with colorless juice. Leaves alternate, mostly glabrous, ternately finely dissected. Flowers yellow to red-orange, or in cultivated forms with other colors, peduncled. Torus dilated to form a funnel-shaped base for the pistil. Sepals 2, completely united into a cap (calyptra) pushed off by the expanding petals. Petals 4 rarely 6 or 8. Stamens many to as few as 16; filaments short; anthers linear. Ovary cylindric, 1-celled, with 2 placentae; styles short; stigma with 4–6 linear divergent lobes. Capsule elongated, 10-nerved, 2-valved from base toward apex. Seeds subglobose to slightly elongate, reticulate or rough-tubercled or pitted. Of much horticultural interest, many color-forms having been developed. Ca. 8 or 10 species from Columbia River to northern Mexico. (Named for Dr. J. F. *Eschscholtz,* 1793–1831, surgeon and naturalist with Russian Expeditions to Pacific Coast in 1816 and 1824.)

**Eschscholzia caespitosa** Benth. [*E. tenuifolia* Benth., *E. rhombipetala* Greene, *E. caespitosa* var. *rhombipetala* Jeps., *E. hypecoides* Benth., *E. caespitosa* var. *hypecoides* Gray.] Annual, usually with several stems from a tuft of basal leaves, glabrous or with patches of short stiff hairs, somewhat glaucous, 1–3(–4) dm. high; stems scapose to leafy; leaves dissected into many narrow divisions which are mostly under 1 cm. long; buds erect; torus turbinate, without a spreading outer rim; calyptra ovoid-elliptic, apiculate, 10–18 mm long; petals bright yellow, 1–2.5 cm. long; capsule 5–8 cm long, 1.5–2 mm. thick; seeds subglobose to somewhat longer than thick, 1–1.3 mm. long, gray-brown, with a network of thin low roughened ridges forming rows of ca. 5–6 irregular meshes; $n=6$ (Smith, 1937).—Dry flats and brushy slopes, below 3500 feet; Valley Grassland, Foothill Woodland, Chaparral; about the Great Central Valley south at altitudes up to 5000 ft., to San Bernardino and Orange Counties. March–June.

Let's examine the information that is included here. One of the important roles of a flora is to stabilize the names being used for plants that occur in the study area. The accepted name for this genus is *Eschscholzia*. It was first recognized as being distinct and correctly published by a German botanist, A. L. von Chamisso in a publication by C. G. D. Nees von Esenbeck, another German botanist. [Many manuals provide a list of the author names where you can look up the abbreviation and find out the individual's name and a bit of biographical information.] The whole genus has the common name California Poppy.

The correct name for the species of California Poppy described above is *Eschscholzia caespitosa*. However there has been some confusion over its correct name and as a result there are several synonyms, names that in the opinion of Dr. Munz all apply to the same taxon. In *A California Flora* synonyms are always listed within square brackets following the accepted species name. Other manuals may list synonyms differently. A list of synonyms can be extremely useful. It is sometimes necessary to compare treatments in different floras. Other authors may have treated the taxa differently. For instance, Dr. Robert Hoover recognized both *Eschscholzia rhombipetala* and *E. hypecoides* as distinct from *E. caespitosa* in *The Vascular Plants of San Luis Obispo County, California.* (To find out why Dr. Munz used *E. caespitosa* instead of any of the other names, see Chapter 2).

The descriptions generally follow a standard sequence. Features applicable to the whole plant generally come first. Parts of the plant are generally described from the base to the apex, and flowers are described from the outside to the inside. Inflorescences are described before flowers, and flowers before fruits. Additional information may include remarks about the plants' economic value (if any), the size of the taxon (if it is a family or a genus), its geographic range, habitat, elevation, flowering period, chromosome numbers, derivation of the name, etc.

Descriptions vary in their technical vocabulary. The writers of some manuals (including Munz's *A California Flora*) used the full range of technical terminology. This is a formidable barrier to introductory students who are just beginning to learn the botanical vocabulary. Other manuals (such as *The Jepson Manual*) use a minimum of technical terms. This enables individuals unfamiliar with the broad scope of terminology to use the book more readily. We encourage serious students to learn more than the minimum of terminology so as to not limit themselves to only a few references. The descriptions and keys in this book are written at a middle level and the glossary includes a much wider scope of terminology.

## Selected List of North American Floras and Manuals

Abrams, L. 1923–1960. *Illustrated Flora of the Pacific States.* Washington, Oregon and California. Stanford Univ. Press, Stanford. 4 vols. (vol. 4 by Abrams and R. S. Ferris).

Correll, D. S., and M. C. Johnston. 1970. *Manual of the Vascular Plants of Texas.* Texas Research Foundation, Renner, Texas.

Cronquist, A., A. H. Holmgren, N. H. Holmgren, J. L. Reveal and P. K. Holmgren. 1972–1984. *Intermountain Flora.* Volumes 1 (1972; Introduction, ferns, fern allies and gymnosperms), 3, part B, (1989; Fabales); 4 (1984; Asteridae except Asteraceae), 5 (1994; Asterales), and 6 (1977; monocots) of a projected 6 volumes have been published to date.

Crow, G. E., and C. B. Hellquist. 2000. Aquatic and wetland plants of northeastern North America. The University of Wisconsin Press, Madison.

Dorn, R. D. 1992. *Vascular Plants of Wyoming,* 2nd ed. Mountain West Publishing, Cheyenne, Wyoming.

Fernald, M. L. 1950. *Gray's Manual of Botany,* 8th ed. American Book Co., New York.

Flora of North America Editorial Committee. 1993. *Flora of North America, North of Mexico.* Oxford University Press, New York and Oxford. Vols. 1 (Introduction) and 2 (Pteridophytes and Gymnosperms) of a projected 14 volumes have been published to date.

Gleason, H. A. 1952. *The New Britton and Brown Illustrated Flora of the Northeastern United States and Adjacent Canada.* New York Botanical Garden, New York.

Gleason, H. A., and A. Cronquist. 1991. *Manual of Vascular Plants of Northeastern United States and Adjacent Canada,* 2nd ed. New York Botanical Garden, Bronx.

Great Plains Flora Association. 1986. *Flora of the Great Plains.* University Press of Kansas, Lawrence.

Harrington, H. D. 1954. *Manual of the Plants of Colorado.* Sage Books, Denver.

Hickman, J. C. (ed.). 1993. *The Jepson Manual: Higher Plants of California.* University of California Press, Berkeley, California.

Hitchcock, C. L., A. Cronquist, M. Ownbey and J. W. Thompson. 1955–1969. *Vascular Plants of the Pacific Northwest.* University of Washington Press, Seattle. 5 volumes.

———. 1973. *Flora of the Pacific Northwest.* University of Washington Press, Seattle.

Holmgren, N. H. et al., eds. 1998. Illustrated companion to Gleason and Cronquist's manual. New York Botanical Garden, New York, NY.

Jepson, W. L. 1923–1925. *A Manual of the Flowering Plants of California.* Associated Students Store, University of California, Berkeley.

Jones, R. L. 2005. Plant Life of Kentucky: An Illustrated Guide to the Vascular Flora. The University Press of Kentucky, Lexington.

Kartesz, J. T. 1994. A synonymized checklist of the vascular flora of the United States, Canada, and Greenland, 2nd ed. Timber Press, Portland, OR. 2 vols.

Kearney, T. H., and R. H. Peebles. 1960. *Arizona Flora,* 2nd ed. with supplement by J. T. Howell and E. McClintock. University of California Press, Berkeley.

Long, R. W., and O. Lakela. 1976. *A Flora of Tropical Florida. A Manual of the Seed Plants of Southern Peninsular Florida.* Banyan Books, Miami.

Martin, W. C., and C. R. Hutchins. 1980–81. *A Flora of New Mexico.* 2 vols. J. Cramer, Vaduz, Liechtenstein.

Munz, P. A., and D. D. Keck. 1959. *A California Flora.* University of California Press, Berkeley.

———. 1968. *Supplement to A California Flora.* University of California Press, Berkeley.

Radford, A. E., H. E. Ahles and C. R. Bell. 1964. *Manual of the Vascular Flora of the Carolinas.* University of North Carolina Press, Chapel Hill.

Scoggan, H. J. 1978. *The Flora of Canada.* 4 parts. National Museum of Natural Sciences, National Museums of Canada, Ottawa.

Shreve, F., and I. L. Wiggins. 1964. *Vegetation and Flora of the Sonoran Desert.* Stanford Univ. Press, Stanford.

Small, J. K. 1933. *Manual of the Southeastern Flora.* Published by the Author, New York. Facsimile reprint in 2 vols., 1972, Hafner Publ. Co., New York.

Steyermark, J. A. 1963. *Flora of Missouri.* Iowa State Univ. Press, Ames.

Van Bruggen, T. 1976. *The Vascular Plants of South Dakota.* Iowa State University Press, Ames.

Voss, E. G. 1972, 1985. *Michigan Flora: a Guide to the Identification and Occurrence of the Native and Naturalized Seed Plants of the State.* Volumes 1 (1972; gymnosperms and monocots) and 2 (1985; dicots, Saururaceae–Cornaceae) of a projected 3 volumes have been published to date.

Weakley, A. S. In preparation. Flora of the Carolinas, Virginia, and Georgia. Copies are available from the author at the University of North Carolina Herbarium (herbarium.unc.edu).

Weber, W. A. 1987. *Colorado Flora. Western Slope.* Colorado Associated University Press, Boulder.

———. 1990. *Colorado Flora. Eastern Slope.* University Press of Niwot.

Welsh, S. L. 1974. *Anderson's Flora of Alaska and Adjacent Parts of Canada.* Brigham Young Univ. Press, Provo, Utah.

———, N. D. Atwood, S. Goodrich, and L. C. Higgins, 1993. *A Utah Flora,* 2nd ed. Print Services, Brigham Young University, Provo, Utah.

Wiggins, I. L. 1980. *Flora of Baja California.* Stanford University Press, Stanford.

Wofford, B. E. 1989. Guide to the Vascular Plants of the Blue Ridge. The University of Georgia Press, Athens, GA.

Wood, C. E., et al. 1958–. *Generic Flora of the Southeastern United States.* Published one or a few families at a time in the Journal of the Arnold Arboretum.

## References

Abbott, L. A., F. A. Bisby and D. J. Rogers. 1985. *Taxonomic Analysis in Biology. Computers, Models and Databases.* Columbia University Press, New York.

Bailey, L. H. 1949. *Manual of Cultivated Plants Most Commonly Grown in the Continental United States and Canada,* revised edition. Macmillan Publishing Co., New York.

———, and E. Z. Bailey. 1976. *Hortus Third. A Consise Dictionary of Plants Cultivated in the United States and Canada.* Revised and expanded by the staff of the Liberty Hyde Bailey Hortorium. Macmillan Publishing Co., New York.

Benson, L. 1979. *Plant Classification,* 2nd ed. D. C. Heath & Co., Lexington, Massachusetts. pp. 5–6.

Duncan, T., and C. A. Meacham. 1986. Multiple-entry keys for the identification of angiosperm families using a microcomputer. *Taxon* 35:492–494.

Frodin, D. G. 1983. *Guide to Standard Floras of the World.* Cambridge University Press, Cambridge.

Harrington, H. D. and L. W. Durrell. 1957. *How to Identify Plants.* Sparrow Press, Chicago.

Heywood, V. H. 1984. *Designing Floras for the Future.* pp. 397–410 *in* V. H. Heywood and D. M. Moore (eds.), Current Concepts in Plant Taxonomy. Systematics Association Special Publication no. 25. Academic Press, London.

Huxley, A., M. Griffiths, and M. Levy (eds.). 1992. The New Royal Horticultural Society Dictionary of Gardening. 4 vols. The MacMillan Press, Ltd, London and The Stockton Press, New York.

Krüssmann, G. 1976–1978. Manual of Cultivated Broad-leaved Trees & Shrubs. Timber Press, Beaverton, OR. 3 vols.

Lawrence, G. H. M. 1951. *Taxonomy of Vascular Plants.* MacMillan, New York. pp. 225–228.

Metcalf, Z. P. 1954. The construction of keys. *Systematic Zoology* 3:38–45.

Pankhurst, R. J. (ed.). 1975. *Biological Identification with Computers.* Systematics Association Special Volume No. 7. Academic Press, London.

———. 1978. *Biological Identification.* University Park Press, Baltimore.

Porter, C. L. 1967. *Taxonomy of Flowering Plants,* 2nd ed. W. H. Freeman & Co., San Francisco. pp. 82–85.

Radford, A. E., W. C. Dickison, J. R. Massey and C. R. Bell. 1974. *Vascular Plant Systematics.* Harper & Row, New York, pp. 522–536.

Sneath, P. H. A., and A. O. Chater. 1978. Information content of keys for identification, pp. 79–95 *in* H. E. Street (ed.), *Essays in Plant Taxonomy.* Academic Press, London.

# Exercises

1. On the basis of the descriptions made of the leaves and twigs collected for the exercise at the end of Chapter 3, construct an indented dichotomous key to at least ten of the specimens in your collection using terms from your descriptions.

   Hand in:

   a.  The leaf specimens and description labels that you prepared for Exercise 3-3.

   b.  The key that you have written to at least ten of the specimens.

2. Your instructor will provide you with a key to use in identifying some unknown samples. Practice using this key so that you become familiar with both the key and the terminology used in the key.

3. Your instructor will have available in the laboratory several different books that might be used for plant identification, ranging from wildflower guides to floras and manuals. Compare these as to the quality and nature of the information available, completeness, types of keys [if any], and ease of use.

# Artificial and Phenetic Systems of Classification

chapter 6

C lassification is a process by which we organize the world around us. It involves two primary operations, grouping and ranking. Grouping results in a horizontal or coordinate arrangement of the objects or organisms being classified on the basis of some perceived similarities. Ranking results in a vertical organization of small groups into a nested hierarchy of progressively larger groups. There are many different ways of carrying out the processes of grouping and ranking. We might ask such questions as: "How do we recognize groups?" "What is the basis for ranking?" "Is one method of classifying better than another?" "How many ranks should we recognize?" In this chapter we examine some of these questions.

People have named and classified plants for thousands of years. Folk taxonomy is the system of naming that has occurred in every culture, as people learned to use plants for food and medicine. Many of those folk taxonomies recognized groups that we still recognize today, such as oaks and dogwoods. These early systems of naming were well suited for people to be able to communicate about the plants in their own regions of the world. However, as humans began to travel widely and bring specimens home for study in the 17th and 18th centuries, the number of kinds of plants known in the Western culture jumped from a few hundred to several thousand. At this point, scientists and explorers needed a more formal system to name plants.

Early formal classification systems were based upon what is called an "artificial" system, where the groupings of different kinds of plants were not necessarily those that were most similar to each other. In the 17th and 18th centuries botanists began to construct classification systems that recognized that plants (and animals) could be grouped into nested clusters. Early groupings were generally "phenetic" systems that used overall similarity to group plants.

## Artificial Classification

Artificial classification is based on *a priori* delimitation of the groups to be recognized. The classifier decides beforehand how the objects or organisms to be classified will be grouped and then fits them into the system. Consider the following [rather extreme] example. Suppose that we want to devise a classification for all objects with legs. One way of grouping them would be to organize them according to the number of legs. All one-legged objects would fit into one group, all two-legged objects into a second, all three-legged objects into a third, and so on (Figure 6-1). You might notice that humans, birds and extension ladders are all bipeds in this system and that homes, folding chairs and stepladders are all quadrupeds. The advantage of this system is that it is stable and that anyone who can count can use it.

**Figure 6-1** Some objects with legs used to make an artificial classification.

The disadvantage is that the groups do not seem "real." There is very little predictive value to such a system. Knowing the number of legs characteristic of a group tells you very little about the other features of the objects comprising the group.

The first systems of plant classification were quite simple and highly artificial. The earliest classification that we are aware of was devised by the ancient Greek philosopher, **Theophrastus** (ca. 370–285 B.C.). The major groups in his classification scheme were based on the plants' habit: trees, shrubs, subshrubs, and herbs. Subsidiary groupings were based on several other features. Habit remained a feature of many classification systems until well into the Renaissance. Classifications based on habit separate members of groups of plants that we recognize today as having more than one habit class. They also group together plants that are otherwise very dissimilar.

The purpose of most early compilations of plants was the recording of information about plants of some economic or purported medicinal value. The authors of these works are known as **herbalists.** The quality of their work varied greatly. Most used habit or the purported medicinal value to group plants together. Some later systems were devised solely for the purpose of recalling plant names and descriptions. Although these systems are still of limited use today, they cannot retain enough information to be emphasized.

**Carolus Linnaeus** (1707–1778) produced an excellent artificial system that organized the scattered array of taxonomic descriptive observations available in the early 1700s. In his "Sexual System" system the plant kingdom was divided into 24 classes, based almost exclusively on the number and nature of the stamens. Subsidiary groupings were based on the number of styles. Some of the classes created by this approach were composed of plants that did not seem to "fit" together on the basis of other characteristics. Some plants that otherwise resembled each other had different numbers of stamens and were placed into separate classes. Linnaeus was well aware that his system was artificial. It was primarily an aid to identification. His classification was easy to understand, though, and anyone could use it. At the time that it was proposed it did much to popularize botany.

## Phenetic Classification

The artificiality of Linnaeus's classification system was widely acknowledged. Some of his contemporaries and later workers attempted to group plants in a very different fashion. Plants were allowed to "classify themselves." One approach was to group plants in a botanical garden until all the individuals of a group appeared to "fit" with one another. This was an approach used by a French botanist, **Bernard de Jussieu** (1699–1776) to display the features of a classification system he had devised. His nephew, **Antoine Laurent de Jussieu** (1748–1836), used this as a basis for an improved classification that he published in 1789 in a book entitled *Genera Plantarum* (Genera of Plants). Groups were recognized only after observations of their features had been made. This is *a posteriori* reasoning. Groupings of plants done in this way produce **Phenetic** systems. One important observation that was made by early pheneticists is that the set of features that can be used to classify the members of one group of plants may not work for another. Attempts to apply any set of features on a universal basis will yield an artificial classification. This was demonstrated most convincingly by **Michel Adanson** (1727–1806), a French botanist who compared the features of plant families in a tabular format.

The phenetic systems were considered be more "natural" than the artificial systems and were often called **natural systems.** There are several meanings of the term "natural." To some this word implied the existence of a "plan of nature" or a created order. The system of classification was an attempt to organize the plants according to the divine plan. Others considered "natural" to mean merely that the groups recognized had a "reality" not present in artificial groupings because they were based on their own shared characteristics. A natural system had greater predictivity than an artificial system; a person could predict the features of a plant if he knew the features of other plants classified together with it. One presupposition inherent in all of the natural systems was that species were fixed, unchanging entities. These systems were essentially static relative to time.

The earliest attempts to devise a natural system of classification actually predated Linnaeus' artificial system. Some groups of plants so obviously "fit" together that they were recognized very early in the history of botany. Most of the early systems were actually a mixture of artificial and natural groups. Those post-Linnaean botanists who attempted to devise natural systems of classification were thus building upon a long tradition and history of ideas.

The most important of the natural systems were published in the 18th and 19th centuries. Many workers contributed toward the accumulating knowledge but three groups of botanists stand out because of the impact of the systems that they proposed. Several members of a French family, the de Jussieu's, contributed to the development of a natural system of classification. Their classification was outlined in the book entitled *Genera Plantarum* published in 1789 by Antoine de Jussieu. The de Jussieu system served as the basis for much subsequent work by other botanists.

**Augustin Pyrame de Candolle** (1778–1841), a Swiss botanist, outlined his own system in a work entitled *Systema,* published in 1818. De Candolle and later his son, Alphonse, and other coworkers attempted to catalog all the plants of the world in a multivolume encyclopedic work called the *Prodromus Systematis Naturalis Regni Vegetabilis* (Forerunner to a Natural System of the Plant Kingdom). The publication of this massive work spanned over 50 years but because of the continued discovery of new species of plants, it never came close to achieving its goal.

The last of the great natural systems was published by two Englishmen, **George Bentham** (1800–1884) and **J. D. Hooker** (1817–1911). Their system was presented in a three-volume work, entitled *Genera Plantarum,* published from 1862 to 1883. This work was based on much careful original observation and attention to detail and included descriptions of all of the genera known at the time. Many other workers contributed to the expansion of knowledge that served as the base for constructing these classification systems.

# Charles Darwin, Evolution, and Systematics

The publication of Charles Darwin's *On the Origin of Species* (1859) gave botanists a new way to group plants, based upon evolutionary relationships. Botanists began to speculate about which plant characteristics were ancestral and which were derived (for example, wind pollination versus insect pollination). Sometimes fossils were used to support these determinations, but often logic or personal opinions were used to evaluate the direction of evolution. Over the next 100 years, three main systems of classification developed, each using a different system to group or classify. The field of taxonomy began to be called "systematics" and the three main types were

1. numerical phenetic systematics,

2. traditional or evolutionary systematics, and

3. cladistic or phylogenetic systematics.

By the 1960s these three methods of classification were well developed and the proponents of each method waged fierce battles in the literature and at scientific meetings to champion their own method, and to discredit the opposing methods. This debate had the effect of rejuvenating the field of systematics or taxonomy that had been overshadowed by the population genetics work of the "modern synthesis" from 1900–1980.

Although philosophers of science debate whether this change in the field of systematics was a "paradigm shift" or a gradual change in ideas (Hull 1988, Kuhn 1970, Popper 1968), today the field of systematics is dominated by cladistic or phylogenetic systematics. However, systematics students should be careful to recognize that the competing schools of thought left significant legacies that are still of value today. For example, although the phenetic method is not used at higher taxonomic levels, it is still a useful tool to make an initial ("alpha taxonomy") grouping of species. Phenetic methods are also used extensively at the species level to understand species structure. Although evolutionary systematics has been supplanted by phylogenetic systematics, much of the classification scheme generated by Cronquist, Taktajan and others has generally been supported by the newer method of classification, and also supported by new characters that have been developed for analysis, such as DNA sequences.

The remainder of this chapter will explore the numerical phenetic method and the evolutionary or traditional classification schemes that were developed in the 20th century. Chapter 7 will provide a more detailed examination of the cladistic or phylogenetic systematic methodology.

# Numerical Phenetic Systematics

The numerical phenetic approach developed from the early work of Antoine Laurent de Jussieu and Michel Adanson. Phenetic classifications were in large part dependent upon the ability of the individual taxonomist to perceive and correctly interpret complex patterns of variation. These classifications represent an attempt to base classification on patterns of overall similarity and dissimilarity. Each individual plant can be thought of as an assemblage of many characters. However, the ability of the human mind to understand simultaneous variation in more than a few characters is limited. Although many taxonomists have a marked ability to perceive complex patterns they are not necessarily able to properly analyze the independent variation of the component characters. Although the goal of phenetic methodology is to classify on the basis of *overall* patterns of similarity and differences, taxonomists often have actually based their classifications on a few "reliable" characters.

Overtly phenetic classification of the vascular plants as a whole was largely abandoned in favor of phylogenetic schemes after Bentham and Hooker published their system. A revival of the phenetic approach, generally applied only to lower and thus smaller taxa, began in the 1960's. This approach was designed to take advantage of the ability of computers to process large amounts of data in a short period of time. "Numerical Phenetics" or "Numerical Taxonomy" was the result. At the time that these methods were first proposed they were touted as providing a new and unbiased way of making comparisons based on the analysis of as many characters as possible.

A numerical taxonomist builds a phenetic system by mathematically comparing descriptions. He or she compares the states of the characters of one plant or taxon with the states of characters found in other plants or taxa, and those that share the most character states are interpreted as being most closely related in a phenetic or numerical sense. The steps in the process are described below:

1. Individual plants are described by observing and measuring **as many** characters as possible. Character states may be assigned numerical values or coded as present/absent.

2. The characters and organisms are arrayed in an OTU × character table similar to Table 6-1. OTU is an abbreviation for **O**perational **T**axonomic **U**nit. An OTU can be an individual plant, a field population, a species, or any other recognizable taxonomic category or entity.

3. After all the measurements are made and the OTU × character table is completed, some measure of similarity among the OTUs must be calculated for **every** pair of taxa in the table. This measure of similarity may be as simple as a percent similarity such as Simpson's index:

$$\text{Simpson's index} = \frac{\text{no. of shared characters}}{\text{total no. of characters}} \times 100\%$$

| Characters (attributes observed or measured) | | | | | | | | | | | | | | | | |
| OTUs | 1 | 2 | 3 | 4 | 5 | 6 | 7 | 8 | 9 | 10 | 11 | 12 | 13 | 14 | 15 | 16 | . . . n |
|---|---|---|---|---|---|---|---|---|---|---|---|---|---|---|---|---|---|
| A | + | – | + | + | – | – | + | + | + | + | + | – | – | – | + | – | |
| B | + | – | – | – | + | + | – | – | + | + | – | + | + | – | – | – | |
| C | + | – | – | – | + | + | – | + | – | + | – | + | + | – | – | – | |
| D | – | + | + | + | + | – | – | – | – | + | + | – | – | – | – | + | |
| E | – | + | + | + | – | – | + | + | + | – | + | + | – | – | + | + | |

*Present = +    Absent = – (Numerical values also may be entered.)

**Table 6-1** OTU × character table for five hypothetical OTUs. Depending on the study an OTU may be an individual or a taxon

More commonly, however, the indices of similarity used today are much more complicated. See Sheath and Sokal (1973) for a more complete discussion of the various types of similarity indices and their advantages and disadvantages. Today, the many required repetitive and time consuming calculations would be done by one of the several computer software packages available for personal, mini- and mainframe computers.

4. The calculated similarity values for every pair of taxa are placed in a second table known as an OTU × OTU TABLE. An example of is Table 6-2.

5. Although it is evident that some of the values in Table 6-2 are high (indicating that OTUs are very similar) and some are low (indicating low similarity) there is little obvious information in this table. Further the larger the number of taxa being compared, the more difficult it is to interpret the OTU × OTU table. The taxonomist can form a picture of the relationships found hidden within the table by resorting to some type of graphical summary. **Cluster analysis** can be used to generate a **dendrogram,** one of the most commonly used graphical summaries. Figure 6-2 is an example of a dendrogram.

6. The last step in the process is to interpret the OTU × OTU table or the derived dendrogram. Taxa with high similarity values would be interpreted as similar and those with low values interpreted as very dissimilar or unrelated. The exact level of similarity required for two taxa to be 'alike' or to be 'different' in practice must be determined subjectively by the taxonomist. The dendrogram is **not** a phylogenetic tree. The dendrogram can only be interpreted as indicating phenetic relationship. That is, taxa 'F' and 'A'

| OTU/OTU | A | B | C | D | E |
|---|---|---|---|---|---|
| A | 100 | 78 | 65 | 22 | 41 |
| B | 78 | 100 | 71 | 56 | 42 |
| C | 65 | 71 | 100 | 91 | 32 |
| D | 22 | 56 | 91 | 100 | 28 |
| E | 41 | 42 | 32 | 28 | 100 |

**Table 6-2** An example of an OTU × OTU table used to compare taxa. Note that the upper and bottom halves of the table are identical.

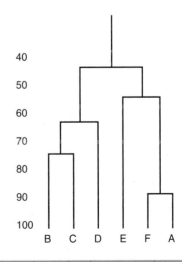

**Figure 6-2** An example of a dendrogram. The cross lines indicate the level of similarity. Taxa 'A' and 'F' share over 90% of their characters. Taxon 'E' shares only 60% of the measured characters with cluster formed by taxa 'A' and 'F'.

resemble each other and 'E' resembles this pair more than it does taxa 'B', 'C', or 'D'. It is not proper to conclude that taxa 'E', 'A', and 'F' are historically related based on evidence analyzed using this technique. Some taxonomists do give phylogenetic interpretations to the relationships displayed in a dendrogram based on their experience and the superposition of rules or 'dicta' discussed in the next chapter. Other taxonomists consider this to be a totally inappropriate thing to do.

Notice that in this analysis all characters are generally held to be of equal weight or importance. [It may be necessary to avoid obviously linked features to minimize correlations among characters]. No decision was required concerning the character's role in evolution and whether it is primitive or advanced in the group under study. In a phenetic study primitive and derived characters are not treated separately. This has been a major criticism of phenetic classification. This is a separate topic and it will be taken up in Chapter 7. Characters of demonstrated "importance" or "value" may be weighted equally with those known or suspected to be of lesser reliability. In response to these criticisms, methods of character weighting have been devised by some numerical pheneticists.

Once the clustering processes have taken place, the pheneticist can group the OTUs on the basis of their similarities into higher categories. Species can be grouped into genera; genera can then be grouped into families; families into orders, and so on. This process of classification yields a scheme that displays overlapping taxa. That is, genera are defined by the species included, and families are defined by the summation of the shared characters of the genera and thus the species that they contain. It is important to remember that each taxon is formed independently by coalescing its component taxa and therefore must be defined independently of other taxa at its same level.

# Traditional or Evolutionary Systematics

The theory of evolution had a very significant impact on botanical classification. The various artificial systems and the natural systems that followed were based upon the premise that the plants of the earth were unchanging. The popularization of the theory of evolution that followed the publication in 1859 of *On the Origin of Species* by **Charles Darwin** made botanists aware that classifications had a historical context. Plants had not always been the same as they are today. Instead they had evolved over long periods of time. Evolution implies change, with ancestors and their descendants differing in various ways. But how could a taxonomist determine which features were ancestral and which were derived? Taxonomists had been measuring and comparing the features of plants for many years. Now they had to interpret them in an evolutionary context.

Tantalizing evidence came from fossils. The history of the earth was long and many of the plants of the past differed from those of the present. Some major groups apparently predated others. The flowering plants seemed to be among the latecomers. However the fossil record seemed very sketchy and was difficult to interpret. It certainly could not provide answers to all of the questions taxonomists were asking. The fossils that are available indicate that certain groups of gymnosperms evolved before the angiosperms came into existence. Botanists have hypothesized that the angiosperms evolved from an ancestor that, if present today, would be classified as a gymnosperm. But what did this ancestor look like? What group of gymnosperms evolved into the angiosperms? The fossil record is silent on these questions. Several different groups of gymnosperms have been hypothesized as ancestors to the angiosperms and each hypothesis has led to a different classification of the flowering plants.

Without a dependable fossil record, taxonomists had to resort to the materials that they had available to them, the modern plants that they were already studying. They had to make decisions about the past based upon the features of the plants of the present. Much can be learned from such comparative studies. However, many of the methods taxonomists used were rather subjective. With their own combinations of logical deduction, educated opinion, and guesswork different taxonomists arrived at some very different classification schemes. One approach was to attempt to determine overall evolutionary trends. Physicists and chemists had discovered grand "laws" that seemed to govern the operation of the universe. Perhaps there were similar principals that applied to the evolution of plants. Various "laws of nature" were proposed, disproved, and discarded. Early phylogeneticists put together lists of their assumptions or left them unstated and then proceeded to attempt to classify the plant kingdom. One of the earliest attempts to utilize the concept of evolution as a guiding principle in classification was published in 1875 by a German taxonomist, **W. Eichler** (1839–1887). Eichler drew heavily on the phenetic classification systems of Jussieu and other early workers. Eichler's system was largely phenetic, but he acknowledged the role and importance of evolution. He considered plants with complex structures to be more advanced than those with simple structures. Eichler's classification was widely accepted in most of Europe.

## Engler and Prantl System

In 1892 another German botanist, **Adolf Engler** (1844–1936), published a revision of Eichler's system that became the most widely used classification scheme in the world. The system was elaborated over a twenty-four year period (1887–1911) in a twenty-volume work entitled *Die Naturlichen Pflanzenfamilien;* [The Natural Families of Plants]; edited by Engler and his colleague, **Karl Prantl** (1849–1893) and including the work of various other contributors. This massive work was essentially a world flora at the genus level. It provided descriptions, keys, illustrations and other information regarding all of the known plant genera and families. The publication of this monumental work took place at about the same time that most of the major herbaria of the world were being established. So influential was the work of Engler and Prantl that the collections in many of these institutions are arranged according to the Engler and Prantl system. Many local and regional floras are arranged according to the Engler and Prantl system as well. A brief outline is provided below for the Englerian treatment of the vascular plants; the system also included the nonvascular plants but these are omitted here. Knowledge of the order of placement of families can speed up the location of descriptions in these floras and specimens in herbaria.

Division **Pteridophyta**—The vascular plants that reproduce by spores and never produce seeds; divided into 5 classes.

Division **Embryophyta Siphonogama**—All seed-bearing plants.

Subdivision **Gymnospermae**—Plants with naked ovules that are directly pollinated.

Subdivision **Angiospermae**—The flowering plants; ovules enclosed in carpels.

Class I. **Monocotyledonae**

Class II. **Dicotyledonae**

Subclass I. **Archichlamydae**—Plants with flowers lacking petals (apetalae) or with flowers that have separate petals (Choripetalae).

Subclass II. **Metachlamydae**—Plants with flowers that have fused petals (Sympetalae).

Within each of the major groups in the Englerian system the plants are arranged with the presumably most primitive first and the presumably more advanced following. In the dicots, for instance, catkin-bearing plants such as willows and oaks with apetalous unisexual flowers were considered to be closest to the presumed gymnospermous ancestors [catkins resemble cones to some extent] and these plants are listed first among the dicots. Despite its wide acceptance, the Englerian classification was badly flawed. One of the main faults was that structurally simple flowers (apetalous flowers) were automatically considered to be more primitive than those with petals. Evidence supporting the alternative hypothesis of secondary reduction was largely disregarded. Several other features of the Englerian classification scheme are also subject to criticism. Imperfect flowers were usually considered more primitive than perfect flowers despite evidence suggesting alternate interpretations. Some of Engler's hypotheses regarding evolution of placentation types are now considered to be incorrect or overly simplified.

Various revisions of the Engler and Prantl classification were proposed in the years following the publication of *Die Naturlichen Pflanzenfamilien.* Engler and his coworkers modified the classification in various ways but retained most of the basic underlying assumptions. These revisions were outlined in successive editions of a work entitled *Syllabus der Pflanzenfamilien.* Edition 11 (1936), edited by Engler and **Friederich Diels** was the last in which Engler had a direct hand. The 12th (and most recent) edition, published in 1954 (vol. 1) and 1964 (vol. 2), was edited by **Hans Melchior** (vol. 1 with Erich Werdermann). Major independent revisions were published by **Richard von Wettstein**

(1862–1931), an Austrian botanist and by **Alfred Rendel** (1865–1938), an English worker.

## Bessey's System

A fundamentally different system of classification was proposed by an American botanist, **Charles Bessey** (1845–1915). Bessey's system can be considered a reinterpretation of the phenetic system of Bentham and Hooker in phylogenetic form. Bessey first proposed his system in 1893, revised it in 1897, and published the final version in 1915. The underlying assumptions of Bessey's classification were stated as a series of "dicta" that indicated his interpretation of evolutionary trends in the angiosperms. He based these dicta on a combination of evidence from the fossil record, studies of developmental morphology, and interpretations of comparative morphology.

Bessey chose as primitive a very different set of plants than did Engler and Prantl. In both classifications gymnosperms were considered ancestral to the angiosperms. The putative ancestors chosen by Engler and Prantl were hypothetical gymnosperms with unisexual catkinlike cones. They considered the monocots and dicots to have originated independently from different ancestral gymnosperms. Bessey, on the other hand, proposed that the ultimate angiosperm ancestors were members of a long-extinct group of gymnosperms now known as the Bennittitales. Bessey considered flowers to be congested stem axes with leaves modified as perianth parts, stamens and carpels. The reproductive parts of Bennitittalean plants had been reconstructed as bisexual flower-like cones that apparently represented congested stem axes with dense clusters of pollen-bearing and ovule-bearing leaves surrounded by sterile leaves. Bessey considered plants such as *Magnolia* and *Ranunculus* that have numerous separate perianth parts, stamens and carpels to be the most primitive angiosperms. He called the group of plants that included these putatively primitive angiosperms the Ranales. Bessey regarded the monocots to be an early offshoot from the dicots.

In Bessey's system the monocots and dicots were considered to comprise two classes within a single phylum. The modern monocots and dicots were all considered to have arisen from an ancestry in the Ranales. Bessey's system is outlined here:

Class **Oppositiflorae** (Dicots)
    Subclass **Strobiloideae**
        Superorder **Apopetalae-Polycarpellatae**
        Superorder **Sympetalae-Polycarpellatae**
        Superorder **Sympetalae-Bicarpellatae**
    Subclass **Apopetalae**
        Superorder **Apopetalae**
        Superorder **Sympetalae**

Class **Alterniflorae** (Monocots)
    Subclass **Strobiloideae**
    Subclass **Cotyloideae**

The taxa recognized by Bessey represent grades of evolutionary specialization [as determined by application of his "dicta"]. Bessey used a diagram often referred to as "Bessey's Cactus" to illustrate relationships among the various angiosperm taxa (Figure 6-3).

Bessey's system, like that of Engler and Prantl, had various weaknesses. The arrangement of taxa into evolutionary grades in some cases grouped plants that are probably not closely related. This is similar to the criticisms of artificial classifications that attempt to apply a predetermined set of criteria in classification. Some of Bessey's "dicta", though far more sophisticated than the characters applied in early artificial classifications, still represented attempts to find universally acceptable criteria for classification. A serious criticism of Bessey's reconstruction of angiosperm phylogeny was that he oversimplified relationships by indicating that various modern taxa are the evolutionary ancestors of other modern taxa. The diagrammatic presentation of his system connected the orders of angiosperms in a series of apparent ancestor-descendent relationships.

As evidence accumulated during the twentieth century, it became evident that Bessey's system, however flawed, was built upon a more solid foundation than was the widely used system of Engler and Prantl. In many cases the treatment of individual groups by Engler and Prantl was superior to that of Bessey, but the basic starting point of Bessey's system appears to be more sound. As a consequence, most of the classifications developed during the past seventy years (Table 6-3), resemble Bessey's system to a considerable extent.

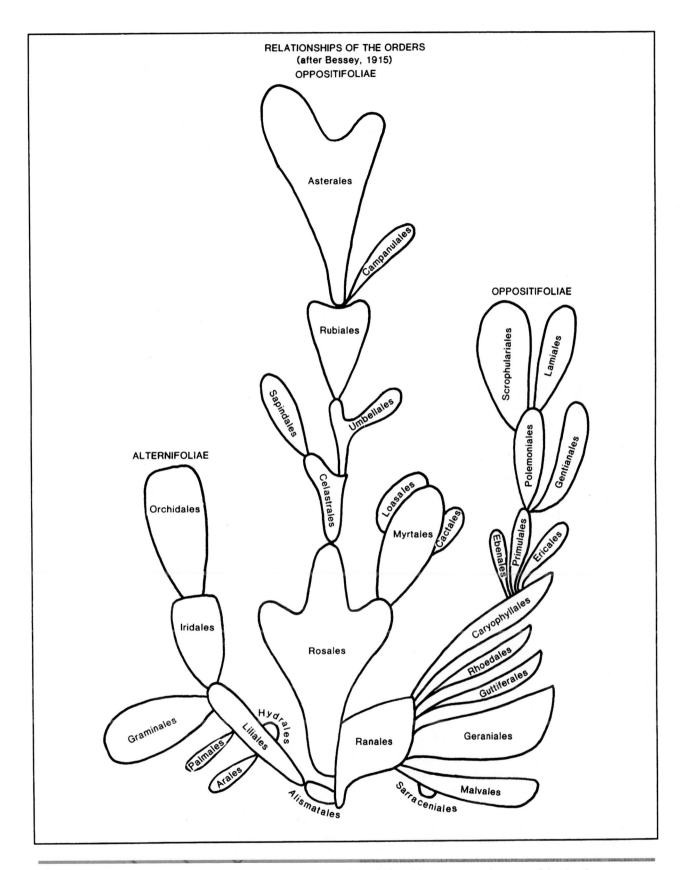

**Figure 6-3** Diagram used by Bessey (1915) to illustrate putative relationships among major taxa of the Angiosperms. Used by permission of the Missouri Botanical Garden.

| Botanist | System proposed | Revisions |
|---|---|---|
| John Hutchinson | 1926 | 1934 and 1973 |
| Armen Takhtajan | 1954 | 1954, 1959, 1966, 1980, 1987 |
| Arthur Cronquist | 1957 | 1968, 1981, 1988 |
| Robert Thorne | 1968 | 1976, 1981, 1983, 1992 |
| Rolf Dahlgren | 1975 | 1980, 1981, 1982 (monocots), 1983, 1985 (monocots) |
| Ernest Roleau | 1981 | |
| David Young | 1982 | |

**Table 6-3**   A Summary of Modern Classifications of the Flowering Plants.

## Hutchinson's System

**John Hutchinson** (1884–1972), an English botanist, proposed a classification similar in many ways to that of Bessey. Hutchinson considered the primitive angiosperms to have had a Ranales-like ancestor that gave rise to both monocots and dicots. He divided the dicots into two lines of evolution, one primarily woody and the other mostly herbaceous. Although Hutchinson's treatments at the family level were often very insightful, his separation of woody and herbaceous dicots is considered to be a basic flaw that rendered his overall system unacceptable.

Beginning in the early 1950's several new systems of classification were proposed. The workers who have developed these classifications have had available to them a continuously expanding base of knowledge of plant structure, physiology, and chemical composition. Numerous botanists contributed to this expansion of knowledge. Many discoveries in paleobotany have aided in interpretation of phylogeny. Botanists are still accumulating evidence that supports or refutes particular proposed relationships. However, there is still a wide latitude for interpretation of all the accumulated data. It should not be surprising that a number of phylogenetic systems have been proposed. The body of evidence that we have available to us today is much greater than that available to the early workers, but there still is much that remains unknown.

The four most widely used of the modern phylogenetic classifications are outlined and discussed below. For a detailed comparison at the family level see Brummitt (1992). Although the systems of Roleau and Young are listed above they have not been as widely used as have the systems of Takhta-jan, Cronquist, Thorne, and Dahlgren and are not further discussed here. Detailed tabular comparisons of modern and historical phylogenetic classification schemes are presented by Becker (1973), by Bedell and Reveal (1982), and by Brummitt (1992).

## Takhtajan's System

**Armen Takhtajan** (1910–), a Russian botanist, began publishing on the evolution of flowering plants during the 1940's. His major theories on angiosperm evolution and classification first became available to western botanists (in English translations) in 1950. Much of his early work was devoted to determining patterns of angiosperm evolution, especially in regard to the origin of the angiosperms. The 1980 version of his system emphasizes the ordinal and higher levels of classification. The most recent version of Takhtajan's system (1987) is not yet available in an English translation.

## Cronquist's System

**Arthur Cronquist** (1919–1992) was an American botanist who spent his professional career at the New York Botanical Garden where he was Senior Scientist. Cronquist's early work on angiosperm phylogeny took place in parallel with that of Takhtajan. Cronquist published an outline of his system (dicots only) in 1957 and soon afterward began an extensive correspondence with Takhtajan that helped to shape the ideas of both workers. In a book entitled *The Evolution and Classification of Flowering Plants* published in 1968 Cronquist set forth his views on the origin of angiosperms and the evolution of characters and presented his classification

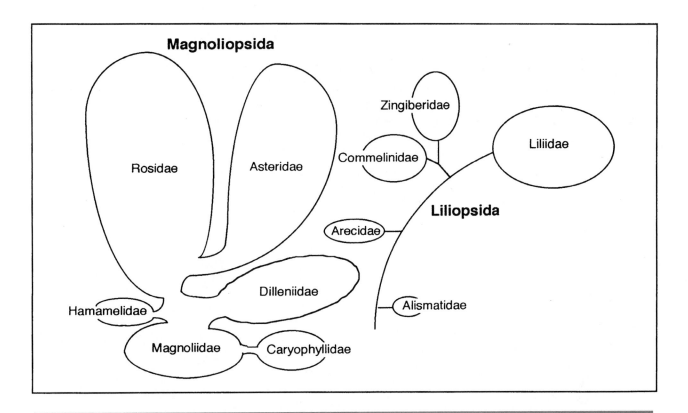

**Figure 6-4** Diagram used by Cronquist to illustrate relationships among subclasses of dicots (left) and monocots (right). From *Evolution and Classification of Flowering Plants,* 2nd ed., p. 502, by A. Cronquist, copyright © 1988, The New York Botanical Garden. Used by permission.

with emphasis at the ordinal level. In his 1981 book *An Integrated System of Classification of Flowering Plants* Cronquist made numerous modifications in his system and expanded the coverage to include detailed accounts of each family. In 1988 Cronquist published a much-expanded second edition of *The Evolution and Classification of Flowering Plants.* Cronquist's work was based on an encyclopedic knowledge of plants accumulated over a long and productive life and on his familiarity with the vast scope of published information.

One criticism of Cronquist's system has been directed toward the diagrammatic illustrations that he used to illustrate relationships. In his 1968 book Cronquist illustrated relationships with diagrams that suggest that taxa displaying advanced features had arisen from taxa of equal rank lacking those features (Fig. 6-4). This suggests that modern orders have given rise to other modern orders and that modern subclasses have given rise to other modern subclasses.

**G. Ledyard Stebbins** (1906– ) published a revision of the 1968 version of Cronquist's system as an

appendix to his book *Flowering Plants. Evolution above the Species Level* (1974). Accompanying Stebbins' revision was a two-dimensional diagrammatic "balloon diagram" of the classification (Fig. 6-5). This depiction of putative evolutionary relationships is an attempt to display present-day affinities at the subclass and ordinal levels without indicating ancestor-descendent relationships among modern plants. It displays relationships as a cross-section through the branches of a phylogenetic tree in the "now" time dimension. This type of display was pioneered by **Lyman Benson** (1909–1993) in a 1957 textbook entitled *Plant Classification* (revised 2nd edition in 1979). Benson's system is largely Besseyan and is organized according to degree of advancement from a putative Ranalian ancestral form. Stebbins' illustration and others of similar form also display relationships between taxa and their divergence from common ancestors by placement of the taxa on the diagram. Size of taxa is indicated by the size of the "balloons."

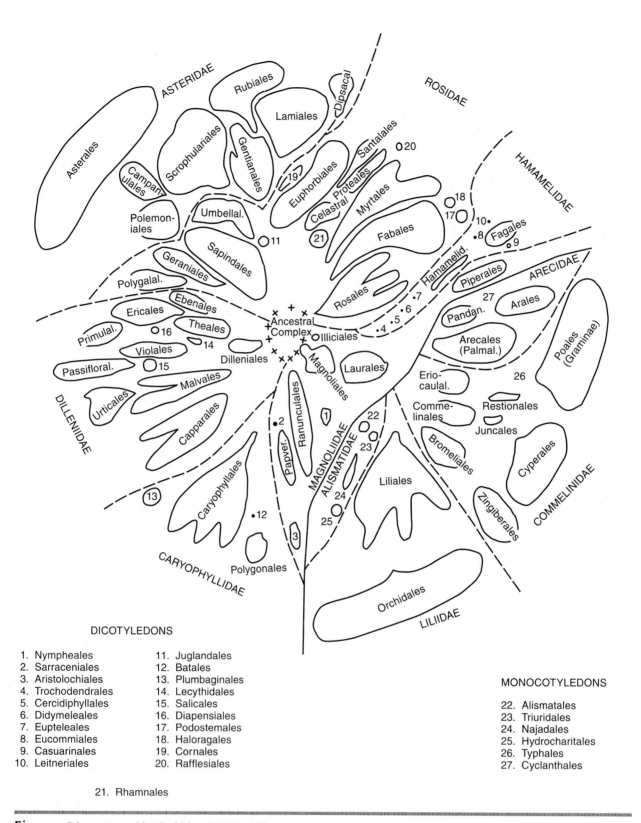

**Figure 6-5** Diagram used by Stebbins (1974) to illustrate putative relationships among subclasses and orders of the angiosperms. Used by permission of Belnap Press of Harvard University.

DICOTYLEDONS

1. Nympheales
2. Sarraceniales
3. Aristolochiales
4. Trochodendrales
5. Cercidiphyllales
6. Didymeleales
7. Eupteleales
8. Eucommiales
9. Casuarinales
10. Leitneriales

11. Juglandales
12. Batales
13. Plumbaginales
14. Lecythidales
15. Salicales
16. Diapensiales
17. Podostemales
18. Haloragales
19. Cornales
20. Rafflesiales

21. Rhamnales

MONOCOTYLEDONS

22. Alismatales
23. Triuridales
24. Najadales
25. Hydrocharitales
26. Typhales
27. Cyclanthales

## Thorne's System

**Robert Thorne** (1920–) is Curator emeritus of Botany at the Rancho Santa Anna in Claremont, California, and emeritus Professor of Botany at the Claremont Graduate College. He began his studies of angiosperm phylogeny in the 1950's. Thorne published an outline of his classification system in 1968. He has carried out extensive research on primitive angiosperms and patterns of evolutionary specialization. In 1976 he published an expanded outline of his classification together with discussions of primitive angiosperms, interpretations of evolutionary trends, and construction of phylogenetic classifications. As additional evidence has accumulated, Thorne has revised his classification system several times. Thorne has used a diagram similar to those constructed by Benson and Stebbins to illustrate putative relationships in his system. Figure 6-6 is a diagram of Thorne's 1992 classification scheme.

## Dahlgren's System

**Rolf M. T. Dahlgren** (1932–1987) was Professor of Botany at the University of Copenhagen. Dahlgren published the original version of his system in 1975 and revised it several times. One of Dahlgren's major interests was the evolution of characters within the angiosperms. He was particularly aware of the problems of interpretation of character conditions in an evolutionary context. Using a two-dimensional graphic display of his phylogenetic system (Fig. 6-7) as a template, Dahlgren displayed the distribution of many different character states within the angiosperms. These included chemical, developmental and morphological characters. In the last revision of his classification of the monocots Dahlgren et al. (1985) employed cladistic methodology (see Chapter 4) to interpret evolutionary character transformations. Dahlgren and his associates made especially valuable contributions regarding the evolution and classification of the monocots. Dahlgren's widow, Gertrud Dahlgren published revisions of Dahlgren's classification and the accompanying diagrams in 1989.

The authors of the various systems differ in their methods of evaluating character evolution and in using this information in erecting classification schemes. One method that is often used is to propose general trends in character evolution and to use these trends as guidelines in the interpretation of particular relationships (Table 6-4). Bessey's dicta are an example of this approach. This practice runs the risk of over-generalization. Wherever possible, botanists attempt to determine the pattern of character evolution individually for each group of taxa under investigation. Where this is based on careful evaluation of adequate data, the various authors are substantially in agreement. Where data are scanty, incomplete, or seemingly contradictory, the methods employed by the individual taxonomists and the weight placed on characters can greatly affect the resulting classifications. Learned opinion has sometimes substituted for solid data. The cladistic approach used by Dahlgren and his colleagues is the most rigorous method of character evaluation (see Chapter 4).

The systems of Cronquist (Table 6-5) and Takhtajan (Table 6-6) resembled each other in many respects prior to Takhtajan's 1987 revision. Both authors treated the angiosperms as a division with two classes, each composed of several subclasses. Within each subclass Takhtajan recognized one or more superorders, a category not employed by Cronquist.

Most of the subclasses recognized by Cronquist and by Takhtajan (1980) were similar in name, though not wholly congruent in composition. For example, Cronquist placed the Euphorbiales into the subclass Rosidae, whereas Takhtajan put the Euphorbiales into the Dilleniidae. Cronquist considered the Euphorbiales to comprise four families (Buxaceae, Simmondsiaceae, Pandaceae, and Euphorbiaceae). On the other hand Takhtajan circumscribed the Euphorbiales differently (Euphorbiaceae, Pandaceae, Dichapetalaceae, and Aextoxicaceae) and places the Buxaceae and Simmondsiaceae into the orders Buxales and Simmondsiales in subclass Hamamelidae. There are numerous other differences in placement of orders and families in the two systems. Takhtajan tended to split more liberally at the family level (410 families) than did Cronquist (384).

Takhtajan's newest (1987) version of his classification (Table 6-4) is markedly different. The system has undergone a remarkable taxonomic inflation with numerous new superorders and orders, many containing but a single family. The number of superorders has increased from 27 to 51, the number of orders from 91 to 165, and the number of families from 410 to 532. In contrast, Cronquist's system changed little between 1981 and 1988.

The systems of Thorne (Table 6-7) and Dahlgren (Table 6-8) have several superficial similarities. In

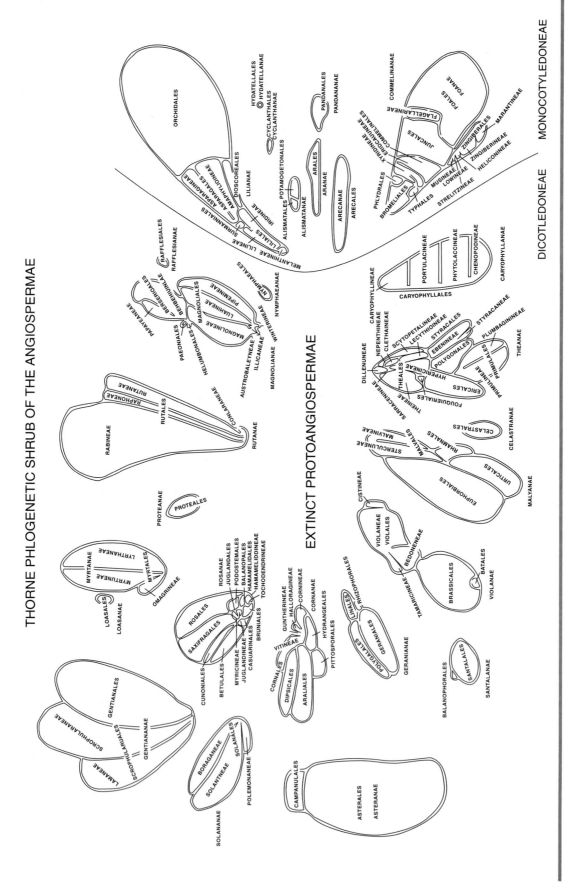

THORNE PHLOGENETIC SHRUB OF THE ANGIOSPERMAE

**Figure 6-6** Diagram illustrating putative relationships among superorders and orders of angiosperms in 1992 version of Thorne's classification system. Copy of diagram courtesy of Dr. Thorne.

**Figure 6-7** Diagram used by Dahlgren (1980) to illustrate putative relationships among superorders and orders of the angiosperms. Used by permission of Academic Press, London.

both classifications the angiosperms are treated as a subclass with numerous superorders. This sets them apart from Takhtajan and Cronquist, in whose systems the angiosperms are treated as a division with two classes, each with several subclasses.

The total number of superorders recognized by Thorne (28) and Dahlgren (35) far exceed the number of subclasses recognized by Cronquist (11) and Takhtajan (11), but are far fewer than the number of superorders recognized by Takhtajan in the 1987 revision of his system (51). There are numerous differences in familial composition of the orders and superorders defined in these systems. Dahlgren split much more liberally at the ordinal and family levels (85 orders, 463 families) than did Thorne in his 1983 (53 orders, 350 families). In the 1992 version of his classification, Thorne increased the number of these taxa recognized with a total of 71 orders and 437 families.

Both Thorne and Dahlgren employed the ending "-iflorae" to signify superordinal rank in earlier versions of their classifications whereas Takhtajan uses the suffix "-anae" for the same level of classification. Thorne adopted the "-anae" suffix in the 1992 version of his classification (presented in Table 6-7) whereas Dahlgren continued to use the "-iflorae" ending in all his revisions.

Although these different systems of classification can be confusing to the beginning systematics student, they do provide an interesting perspective on how perceptions of scientists can change over time. The previous edition (Walters and Keil 4th ed.) used Cronquist's system as the basis for their classification, this new edition uses the Angiosperm Plant Group (APG) system (Stevens 2001). In advanced studies you may wish to compare Cronquist's system with the APG system, as well as others. In such a study you will begin to learn the reasons employed by different botanists for the decision they have made and the many forms of evidence they used in arriving at their conclusions.

## VEGETATIVE TRENDS

| Primitive condition | | | Advanced condition |
|---|---|---|---|
| tap root system | | | fibrous root system |
| vessel-less wood | | | vessels in wood |
| trees | shrubs | perennial herbs | annual herbs |
| much cambial activity | | | little or no cambial activity |
| evergreen | | | deciduous |
| simple entire leaves | | toothed and lobed leaves | compound leaves |
| stipules persistent | | stipules ephemeral | stipules absent |
| leaves alternate | | opposite | whorled |

## REPRODUCTIVE TRENDS

### SPECIALIZATION

| | |
|---|---|
| beetle pollination | pollination by other insects |
| wind pollination | pollination by vertebrates |
| spirally attached parts | whorled parts |
| undifferentiated perianth | biseriate perianth |
| uniseriate perianth | no perianth |
| laminar filaments | thread-like filaments |

### FUSION

| | |
|---|---|
| free parts | adnate parts |
| no hypanthium | hypanthium |
| superior ovary | inferior ovary |
| distinct parts | connate parts |
| calyx | synsepalous calyx |
| apopetalous corolla | sympetalous corolla |
| androecium | androecium |
| apocarpous gynoecium | syncarpous gynoecium |
| marginal placentation | axile or parietal placentation |

### CHANGE OF SYMMETRY

| | |
|---|---|
| radial symmetry | bilateral symmetry |

### REDUCTION

| | |
|---|---|
| large parts | small parts |
| many part (e.g., stamens) | few parts |
| perfect flowers | imperfect flowers |
| apocarpous gynoecium | monocarpous gynoecium |
| axile placentation | free-central placentation |
| marginal, axile, parietal, or free-central | apical or basal placentation |
| bitegmic ovules | unitegmic ovules |
| crassinucellate ovules | tenuinucellate ovules |

### SECONDARY INCREASE

| | |
|---|---|
| few parts | many parts |
| small parts | large parts |

**Table 6-4**  Some commonly observed evolutionary trends in flowering plants.

Division **Magnoliophyta** (Angiosperms)
Class **Magnoliopsida** (Dicots)
Subclass **Magnoliidae**
Orders **Magnoliales, Laurales, Piperales, Aristolochiales, Illiciales, Nymphaeales, Ranunculales, Papaverales**
Subclass **Hamamelidae**
Orders **Trochodendrales, Hamamelidales, Daphniphyllales, Didymelales, Eucommiales, Urticales, Leitneriales, Juglandales, Myricales, Fagales, Casuarinales**
Subclass **Caryophyllidae**
Orders **Caryophyllales, Polygonales, Plumbaginales**
Subclass **Dilleniidae**
Orders **Dilleniales, Theales, Malvales, Lecythidales, Nepenthales, Violales, Salicales, Capparales, Batales, Ericales, Diapensiales, Ebenales, Primulales**
Subclass **Rosidae**
Orders **Rosales, Fabales, Proteales, Podosternales, Haloragales, Myrtales, Rhizophorales, Cornales, Santalales,**

**Rafflesiales, Celastrales, Euphorbiales, Rhamnales, Linales, Polygalales, Sapindales, Geraniales, Apiales**
Subclass **Asteridae**
Orders **Gentianales, Solanales, Lamiales, Callitrichales, Plantaginales, Scrophulariales, Campanulales, Rubiales, Dipsacales, Calycerales, Asterales**
Class **Liliopsida**
Subclass **Alismatidae**
Orders **Alismatales, Hydrocharitales, Najadales, Triuridales**
Subclass **Arecidae**
Orders **Arecales, Cyclanthales, Pandanales, Arales**
Subclass **Commelinidae**
Orders **Commelinales, Eriocaulales, Restionales, Juncales, Cyperales, Hydatellales, Typhales**
Subclass **Zingiberidae**
Orders **Bromeliales, Zingiberales**
Subclass **Liliidae**
Orders **Liliales, Orchidales**

**Table 6-5** Classification of the angiosperms according to Cronquist (1981).

---

Division **Magnoliophyta** (Angiosperms)
Class **Magnoliopsida** (Dicots)
Subclass **Magnoliidae**
Superorder **Magnolianae**
Orders **Magnoliales, Eupomatiales, Annonales, Winterales, Illiciales, Austrobaileyales, Laurales, Lactoridales, Chloranthales, Piperales, Aristolochiales**
Superorder **Rafflesianae**
Order **Hydnorales, Rafflesiales, Balanophorales**
Superorder **Nepenthanae**
Order **Nepenthales**
Superorder **Nymphaeanae**
Orders **Nymphaeales, Ceratophyllales**
Superorder **Nelumbonanae**
Order **Nelumbonales**
Subclass **Ranunculidae**

Superorder **Ranunculanae**
Orders **Ranunculales, Glaucidiales, Paconiales, Papaverales**
Subclass **Caryophyllidae**
Superorder **Caryophyllanae**
Order **Caryophyllales**
Superorder **Polygonanae**
Order **Polygonales**
Superorder **Plumbaginanae**
Order **Plumbaginales**
Subclass **Hamamelidae**
Superorder **Trochodendranae**
Orders **Trochodendrales, Cercidiphyllales, Eupteales**
Superorder **Eucommianae**
Order **Eucommiales**
Superorder **Hamamelanae**
Orders **Hamamelidales, Daphniphyllales, Balanopales, Didymeales,** *(continued)*

**Table 6-6** Classification of the angiosperms according to Takhtajan (1987). Takhtajan treats the angiosperms as a division with 2 classes, 11 subclasses, and 165 orders.

Myriothamnales, Buxales,
Simmondsiales, Casuarinales, Fagales,
Betulales
Superorder **Juglandanae**
Orders **Myricales, Rhoipteleales,
Juglandales**
Subclass **Dilleniidae**
Superorder **Dillenianae**
Order **Dilleniales**
Superorder **Theanae**
Orders **Actinidiales, Paracryphiales,
Theales, Medusagynales, Ochnales,
Ancistrocladales, Elatinales**
Superorder **Lecythidanae**
Order **Lectyhidales**
Superorder **Sarracenianae**
Order **Sarraceniales**
Superorder **Ericanae**
Orders **Ericales, Doapensiales, Ebenales,
Sapotales, Primulales**
Superorder **Violanae**
Orders **Violales, Tamaricales,
Fouquieriales, Salicales,
Dioncophyllales, Cucurbitales,
Begoniales, Capparales, Moringales,
Batales**
Superorder **Malvanae**
Orders **Bixales, Malvales**
Superorder **Urticanae**
Order **Urticales**
Superorder **Euphorbianae**
Orders **Euphorbiales, Thymelaeales**
Subclass **Rosidae**
Superorder **Rosanae**
Orders **Cunoniales, Bruniales,
Geissolomatales, Saxifragales,
Droserales, Gunnerales, Rosales,
Crossosomatales, Podostemales**
Superorder **Myrtanae**
Orders **Rhizophorales, Myrtales,
Haloragales**
Superorder **Fabanae**
Order **Fabales**
Superorder **Rutanae**
Orders **Connarales, Sapindales, Rutales,
Leitneriales, Coriales, Linales,
Geraniales, Balsaminales,
Tropaeolales, Limnanthales,
Polygalales**
Superorder **Celastranae**
Orders **Celastrales, Santalales**

Superorder **Rhamnanae**
Orders **Rhamnales, Elaeagnales**
Superorder **Proteanae**
Order **Proteales**
Superorder **Vitanae**
Order **Vitales**
Superorder **Cornanae**
Orders **Hydrangeales, Cornales,
Aralidiales, Torricelliales, Apiales,
Pittosporales, Biblydiales, Dipsacales,
Cynomoriales**
Subclass **Lamiidae**
Superorder **Gentianae**
Orders **Gentianales, Oleales**
Superorder **Loasanae**
Order **Loasales**
Superorder **Solananae**
Orders **Solanales, Convolvulales,
Polemoniales, Boraginales**
Superorder **Lamianae**
Orders **Scrophulariales, Hippuridales,
Lamiales, Hydrostacheales**
Subclass **Asteridae**
Superorder **Campanulanae**
Orders **Campanulales, Stylidiales,
Goodeniales, Calycerales**
Superorder **Asteranae**
Order **Asterales**
Class **Liliopsida** (Monocots)
Subclass **Alismatidae**
Superorder **Butomanae**
Orders **Butomales, Hydrocharitales,
Alismatales**
Superorder **Najadanae**
Orders **Aponogetonales, Scheuchzeriales,
Juncaginales, Potamogetonales,
Posidoniales, Zosterales,
Cymodoceales, Najadales**
Subclass **Triurididae**
Superorder **Triuridanae**
Order **Triuridales**
Subclass **Liliidae**
Superorder **Lilianae**
Orders **Liliales, Burmanniales,
Amaryllidales, Asparagales,
Smilacales, Dioscoreales, Taccales,
Alstroemeriales, Haemodorales,
Orchidales**
Superorder **Pontederianae**
Orders **Pontederiales, Phylidrales**
Superorder **Bromelianae**

**Table**
**6-6**  Continued

Orders **Bromeliales, Velzsiales**
Superorder **Zingiberanae**
Order **Zingiberales**
Superorder **Juncanae**
Orders **Juncales, Cyperales**
Superorder **Hydatellanae**
Order **Hydatellales**
Superorder **Commelinanae**
Orders **Commelinales, Restionales,
Poales**

Subclass **Arecidae**
Superorder **Arecanae**
Orders **Arecales, Cyclanthales**
Superorder **Pandananae**
Order **Pandanales**
Superorder **Aranae**
Order **Arales**
Superorder **Typhananae**
Order **Typhales**

**Table
6-6** Continued

Class **Angiospermae (Magnoliopsida)**
Subclass **Dicotyledonae (Magnoliidae)**
Superorder **Magnolianae**
Orders **Magnoliales, Ceratophyllales,
Nelumbonales, Paeoniales, Berberidales**
Superorder **Nymphaeanae**
Order **Nymphaeales**
Superorder **Rafflesianae**
Order **Rafflesiales**
Superorder **Caryophyllanae**
Order **Caryophyllales**
Superorder **Theanae**
Orders **Theales, Ericales, Fouquieriales,
Styracales, Primulales, Polygonales**
Superorder **Celastranae**
Order **Celastrales**
Superorder **Malvanae**
Orders **Malvales, Urticales, Rhamnales,
Euphorbiales**
Superorder **Violanae**
Orders **Violales, Brassicales, Batales**
Superorder **Santalanae**
Orders **Santalales, Balanophorales**
Superorder **Geranianae**
Orders **Linales, Rhizophorales, Geraniales,
Polygalales**
Superorder **Rutanae**
Order **Rutales**
Superorder **Proteanae**
Order **Proteales**
Superorder **Rosanae**
Orders **Hamamelidales, Casuarinales,
Balanopales, Bruniales, Juglandales,
Betulales, Rosales, Saxifragales,
Podostemales, Cunoniales**

Superorder **Cornanae**
Orders **Hydrangeales, Cornales,
Pittosporales, Araliales, Dipsacales**
Superorder **Asteranae**
Orders **Asterales, Campanulales**
Superorder **Solananae**
Orders **Solanales**
Superorder **Loasanae**
Order **Loasales**
Superorder **Myrtanae**
Order **Myrtales**
Superorder **Gentiananae**
Orders **Gentianales, Scrophulariales**
Subclass **Monocotyledonae (Liliidae)**
Superorder **Lilianae**
Orders **Liliales, Burmanniales, Asparagales,
Dioscoreales, Orchidales**
Superorder **Hydatellanae**
Order **Hydatellaceae**
Superorder **Triuridanae**
Order **Triuridales**
Superorder **Alismatanae**
Orders **Alismatales, Potamogetonales**
Superorder **Aranae**
Order **Arales**
Superorder **Cyclanthanae**
Order **Cyclanthales**
Superorder **Pandananae**
Order **Pandanales**
Superorder **Arecanae**
Order **Arecales**
Superorder **Commelinanae**
Orders **Bromeliales, Philydrales, Typhales,
Zingiberales, Commelinales, Juncales,
Poales**

**Table
6-7** Classification of the angiosperms according to Thorne (1992). Thorne treats the angiosperms as a class with 2 subclasses, 28 superorders, and 71 orders.

Class **Angiospermae**
  Subclass **Dicotyledonae**
    Superorder **Magnoliiflorae**
      Orders **Annonales, Aristolochiales,**
        **Rafflesiales, Magnoliales, Lactoridales,**
        **Chloranthales, Illiciales, Laurales,**
        **Nelumbonales**
    Superorder **Nymphaeiflurae**
      Orders **Piperales, Nymphaeales**
    Superorder **Ranunculiflorae**
      Orders **Ranunculales, Papaverales**
    Superorder **Caryophylliflorae**
      Order **Caryophyllales**
    Superorder **Polygoniflorae**
      Order **Polygonales**
    Superorder **Plumbaginiflorae**
      Order **Plumbaginales**
    Superorder **Malviflorae**
      Orders **Malvales, Urticales, Euphorbiales,**
        **Thymelaeales, Rhamnales, Elaeagnales**
    Superorder **Violiflorae**
      Orders **Violales, Cucurbitales. Salicales,**
        **Tamaricales, Capparales, Salvadorales**
    Superorder **Theiflorae**
      Orders **Dilleniales, Paeoniales, Theales**
    Superorder **Primuliflorae**
      Orders **Primulales, Ebenales**
    Superorder **Rosiflorae**
      Orders **Trochodendrales, Cercidiphyllales,**
        **Hamamelidales, Geissolomatales,**
        **Balanopales, Fagales, Juglandales,**
        **Myricales, Casuarinales, Buxales,**
        **Cunoniales, Saxifragales, Droserales,**
        **Gunnerales, Rosales**
    Superorder **Podostemiflorae**
      Order **Podostemales**
    Superorder **Proteiflorae**
      Order **Proteales**
    Superorder **Myrtiflorae**
      Orders **Haloragales, Rhizophorales,**
        **Myrtales, Chrysobalanales**
    Superorder **Fabiflorae**
      Order **Fabales**
    Superorder **Rutiflorae**
      Orders **Sapindales, Rutales, Polygalales,**
        **Geraniales, Balsaminales, Tropaeolales**

    Superorder **Santaliflorae**
      Orders **Celastrales, Vitales, Santalales**
    Superorder **Balanophoriflorae**
      Order **Balanophorales**
    Superorder **Araliiflorae**
      Orders **Pittosporales, Araliales**
    Superorder **Asteriflorae**
      Orders **Campanulales, Asterales**
    Superorder **Solaniflorae**
      Orders **Solanales, Boraginales**
    Superorder **Corniflorae**
      Orders **Fouquieriales, Ericales,**
        **Eucommiales, Sarraceniales, Cornales,**
        **Dipsacales**
    Superorder **Loasiflorae**
      Order **Loasales**
    Superorder **Gentianiflorae**
      Orders **Goodeniales, Oleales, Gentianales**
    Superorder **Lamiiflorae**
      Orders **Scrophulariales, Hippuridales,**
        **Lamiales, Hydrostachyales**
  Subclass **Monocotyledonae**
    Superorder **Liliiflorae**
      Orders **Dioscoreales, Asparagales, Liliales,**
        **Burmanniales, Melanthiales**
    Superorder **Ariflorae**
      Order **Arales**
    Superorder **Triuridiflorae**
      Order **Triuridales**
    Superorder **Alismatiflorae**
      Orders **Alismatales, Najadales**
    Superorder **Bromeliiflorae**
      Orders **Velioziales, Bromeliales, Philydrales,**
        **Haemadorales, Pontederiales, Typhales**
    Superorder **Zingiberiflorae**
      Order **Zingiberales**
    Superorder **Commeliniflorae**
      Orders **Commelinales, Hydatellales,**
        **Cyperales, Poales**
    Superorder **Cyclanthiflorae**
      Order **Cyclanthales**
    Superorder **Areciflorae**
      Order **Arecales**
    Superorder **Pandaniflorae**
      Order **Pandanales**

**Table 6-8**   Classification of the angiosperms according to Dahlgren (Dahlgren, 1983; Dahlgren et al., 1985). Dahlgren treats the angiosperms as a class with 2 subclasses, 35 superorders, and 108 orders.

# References

Abbot, L. A., F. A. Bisby, and D. J. Rogers. 1985. *Taxonomic Analysis in Biology. Computers, Models and Databases.* Columbia Univ. Press, New York.

Baum, B. R., T. Duncan, and R. B. Phillips. 1984. A bibliography of numerical phenetic studies in systematic botany. *Annals of the Missouri Botanical Garden* 71:1044–1060.

Becker, K. M. 1973. A comparison of angiosperm classification systems. *Taxon* 11:19–50.

Bedell, H. G., and J. L. Reveal. 1982. Amended outlines and indices for six recently published systems of angiosperm classification. *Phytologia* 51:65–156.

Benson, L. 1957. *Plant Classification.* D. C. Heath and Co., Lexington, Mass. (2nd Ed., 1979).

Benson, L. 1979. *Plant Classification,* 2nd ed. D. C. Heath & Co., Lexington, Massachusetts, pp. 507–532.

Bessey, C. 1915. Phylogenetic taxonomy of flowering plants. *Annals of the Missouri Botanical Garden* 2:1–155.

Brummitt, R. K. 1992. *Vascular Plant Families and Genera.* Royal Botanic Gardens, Kew.

Cronquist, A. 1957. Outline of a new system of families and orders of dicotyledons. *Bull. Jard. Bot. Etat.* 27:13.

———. 1968. *The Evolution and Classification of Flowering Plants.* Houghton Mifflin Company, Boston.

———. 1981. *An Integrated System of Classification of Flowering Plants.* Columbia University Pr., N.Y.

———. 1983. Some Realignments in the Dicotyledons. Nordic Journal of Botany 3:75–83.

———. 1988. *The Evolution and Classification of Flowering Plants,* 2nd ed. New York Botanical Garden, Bronx, N.Y.

Crowson, R. A. 1970. *Classification and Biology.* Atherton Press, New York.

Dahlgren, G. 1989a. The last Dahlgrenogram: system of classification of the dicotyledons, pp. 249–260 *in* K. Tan, R. R. Mill, and T. S. Elias (eds.), *Plant Taxonomy, Phytogeography, and Related Subjects.* Edinburgh University Press, Edinburgh.

———. 1989b. An updated angiosperm classification. Botanical Journal of the Linnaean Society 100:197–203.

Dahlgren, R. M. T. 1975. A system of classification of the angiosperms to be used to demonstrate the distribution of characters. *Botaniska Notiser* 128:119–147.

———. 1980. A revised system of classification of the angiosperms. *Botanical Journal of the Linnaean Society* 80:91–124.

———. 1983. General aspects of angiosperm evolution and macrosystematics. *Nordic Journal of Botany* 3:119–149.

———, S. Rosendal-Jensen, and B. J. Nielsen. 1981. A revised classification of the angiosperms with comments on correlations between chemical and other characters, p. 149; *in* D. A. Young and D. Seigler (eds.), *Phytochemistry and Angiosperm Phylogeny.* Praeger Publ., N.Y.

———, and H. T. Clifford, 1982. *The Monocotyledons: a Comparative Study.* Academic Press, London.

———, and P. F. Yeo. 1985. *The Families of the Monocotyledons.* Springer-Verlag, Berlin.

Duncan, T., and B. R. Baum. 1981. Numerical Phenetics: Its Uses In Botanical Systematics. *Annual Review of Ecology and Systematics* 12:387–404.

Engler, A., and L. Diels. 1936. *Syllabus der Pflanzenfamilien.* Ed. 11. Berlin.

Engler, A., and K. Prantl. 1897–1915. *Die Naturlichen Pflanzenfamilien.* 20 vol. Leipzig.

Felsenstein, J. (ed.). 1983. *Numerical Taxonomy.* Springer-Verlag, New York.

Heywood, V. H., and J. McNeill (eds.), 1964. *Phenetic and Phylogenetic Classification.* The Systematics Association, London.

Hull, D. L. 1988. Science as a process: an evolutionary account of the social and conceptual development of science. University of Chicago Press, Chicago.

Hutchinson, J. 1973. *The Families of Flowering Plants Arranged According to a New System Based on Their Probable Phylogeny.* 3rd ed. Clarendon Pr., Oxford.

Jones, S. B. Jr., and A. E. Luchsinger. 1979. *Plant Systematics.* McGraw-Hill Book Co., New York, pp. 2–21.

Kubitzki, K. (ed.). 1977. *Flowering Plants. Evolution and Classificatiom of Higher Categories.* Plant Systematics and Evolution Suppl. 1. Springer-Verlag, New York.

Kuhn, T. S. 1970. The structure of scientific revolutions, 2nd ed. University of Chicago Press: Chicago.

Lawrence, G. H. M. 1951. *Taxonomy of Vascular Plants.* MacMillan, New York, pp. 42–91.

Pankhurst, R. J. 1992. *Practical Taxonomic Computing.* Cambridge University Press, New York and Cambridge.

Popper, K. R. 1968. The logic of scientific discovery. Harper Torchbooks: New York.

Porter, C. L. 1967. *Taxonomy of Flowering Plants,* 2nd ed. W. H. Freeman & Co., San Francisco, pp. 7–29, 65–75.

Rouleau, E. 1981. *Guide to the generic names appearing in the Index Kewensis and its fifteen supplements.* Jules Chatelain Inc., Cowanville, Canada.

Sneath, P. H. A., and R. R. Sokal. 1973. *Numerical Taxonomy.* W. H. Freeman, San Francisco.

Sokal, R. R. 1986. Phenetic taxonomy: theory and methods. *Annual Review of Ecology and Systematics* 17:423–442.

———, and P. H. A. Sneath. 1963. *Principles of Numerical Taxonomy.* W. H. Freeman, San Francisco.

———. Phenetic taxonomy. *Annual Review of Ecology and Systematics* 17:423–442.

Sporne, K. R. 1956. The phylogenetic classification of the angiosperms. *Biological Review of the Cambridge Philosophical Society* 31:1–29.

———. 1959. On the phylogenetic classification of plants. *American Journal of Botany* 46:385–394.

———. 1976. Character correlations among angiosperms and the importance of fossil evidence in assessing their significance. Pp. 312–329 *in* C. B. Beck (ed.), *Origin and Early Evolution of Angiosperms.* Columbia University Press, New York.

Stebbins, G. L. 1974. *Flowering Plants. Evolution above the Species Level.* Belknap Press of Harvard University Press, Cambridge.

Stevens, P. F. Evolutionary polarity of character states. *Annual Review of Ecology and Systematics* 11:333–358.

Stevens, P. F. 1984. Metaphors and typology in the development of botanical systematics, 1690–1960, or the art of putting new wine in old bottles. *Taxon* 33:169–211.

Stevens, P. F. (2001 onwards). Angiosperm Phylogeny Website. Version 5, May 2004 [and more or less continuously updated since]. *http://www.mobot.org/MOBOT/research/APweb.*

Stevens, P. F. 2000. Botanical systematics 1950–2000: Change, progress or both? Taxon 49:635–659.

Stuessy, T. F. 1990. *Plant Taxonomy. The Systematic Evaluation of Comparative Data.* Columbia University Press, New York.

Takhtajan, A. L. 1954. *Origins of Angiospermous Plants.* Soviet Science Press. (in Russian). English translation: American Institute of Biological Sciences, Washington, D.C. 1958.

———. 1959. *Die Evolution der Angiospermen.* Fischer, Jena, East Germany.

———. 1969. *Flowering Plants. Origin and Dispersal.* Oliver and Boyd, Edinburgh.

———. 1980. Outline of the classification of the flowering plants (Magnoliophyta). *Botanical Review* 46:225–359.

———. 1987. *Systema Magnoliophytorum.* Nauka, Leningrad. [in Russian.]

———. 1991. *Evolutionary Trends in Flowering Plants.* Columbia University Press, New York.

Thorne, R. F. 1968. Synopsis of a putative phylogenetic classification of the flowering plants. *Aliso* 6:57–66.

———. 1976. A phylogenetic classification of the Angiospermae. *Evolutionary Biology* 9:35–106.

———. 1981. Phytochemistry and angiosperm phylogeny: a summary statement. Pp. 233–295 *in* D. A. Young and D. S. Seigler, l.c.

———. 1983. Proposed new realignments in the angiosperms. *Nordic Journal of Botany* 3:85–117.

———. 1992a. An updated phylogenetic classification of the flowering plants. *Aliso* 13:365–389.

———. 1992b. Classification and geography of the flowering plants. *Botanical Review* 58:225–348.

Young, D A., and D. S. Seigler. 1981. General classification and characteristics of vascular seed plants. Pp. 569–589 *in* A. Mitsui and C. C. Black (eds.), *Basic Principles,* Vol. 1, part 1, *CRC Handbook of Biosolar Resources.* CRC Publ. Co., West Palm Beach, Florida.

# Exercises

1. Determine the classification of the following families in the systems of Thorne, Dahlgren, Takhtajan, and Cronquist:

   Solanaceae, Ericaceae, Asteraceae, Ranunculaceae, Magnoliaceae, Caryophyllaceae, Polygonaceae, Poaceae, Arecacaeae, Araceae, Juncaceae

2. Find a plant with flowers big enough for you to dissect at home with the aid of a hand lens.

   a. On a separate sheet of paper, prepare a complete and accurate description of the plant using the same sequence of characters as in the descriptions above. Write the description in your own words. Do not copy it from another source. You do not have to supply supplementary information on range, habitat, etc.

   b. Using the information you have discovered from your dissections and the description you have written, key the plant to family. Use the family key in the back of this book (Appendix 2). Check your plant against the description of that family. Do they match? Are there any discrepancies? [If there are major differences, you'd better check the accuracy of your observations or the route you took through the key].

   c. Press a specimen of this plant in a fashion similar to that described in Chapter 7 for vegetative samples. Your specimen should include both vegetative and reproductive parts. On the sheet with your specimen provide a standard herbarium label with the following information (see Chapter 7).

---

**Family:**

**Collection site:**



**Field observations on the plant and its habitat:**



**Collector:**                                    **Date:**

---

# Discussion Questions

1. How does artificial classification differ from phenetic classification? What are the underlying philosophies of each?

2. What are the advantages of an artificial system? What are the disadvantages?

3. What are the advantages of a phenetic system? What are the disadvantages?

4. Were the 19th century natural systems really based on overall resemblance? Are overall resemblance and closeness of relationship the same thing? How practical is classification by overall resemblance?

5. Is it correct to interpret a dendrogram created by cluster analysis as a phylogenetic tree?

6. What aspect of phylogenetic systems is it that numerical phenetic approaches were desired to correct? How successful are numerical approaches in achieving these goals?

7. What are the differences among the higher taxonomic levels? What is the difference between a subclass and a superorder? How do the orders of Takhtajan differ from those of Thorne?

8. Can you explain how such expert botanists as Thorne, Dahlgren, Cronquist, and Takhtajan could prepare such different classifications?

9. What is a classification? Is there a "right" system?

10. How do phylogenetic systems differ from phenetic systems?

chapter 7

# Collecting and Preserving Plants for Study

No matter what type of work a taxonomist is doing, it sooner or later will involve study of preserved plants. Since the plants themselves are the best permanent record of their appearance in nature, it is very important to prepare specimens that retain the plant's features with a minimum of distortion. This chapter examines some of the methods and considerations involved in preparing plants for study.

Plant specimens serve several functions. They permanently record the appearance of individual plants. Because no two individuals of a species are identical, a series of specimens can more accurately represent the variation in a species than can any one individual. Many plants are seasonal in their growth and reproduction and are available for study in living condition only at certain times of the year. Preserved specimens, on the other hand, are available for study at any time of the year. Another function of specimens is to document where, when and under what conditions plants occur in nature. Properly prepared specimens bear labels documenting where and when they were collected and who did the collecting. A specimen without collection data generally has little scientific value. A set of dried specimens of a species often represents many more locations than an investigator would be able to visit.

Another function of plant specimens is to serve as permanent records of the identity of taxonomic names cited in a study. **Type specimens,** discussed in Chapter 2, are examples of one sort of documentation. Other specimens that serve as permanent records are **voucher specimens.** A voucher may record the plant from which a chromosome count was obtained, from which a chemical compound was extracted or on which an experiment was performed. It may also serve as documentation of the presence of a species in a particular locality. This may be a part of a floristic survey or an environmental impact report. The voucher serves as the permanent record of the identity of the plant studied. Occasionally the author of a study accidentally misidentifies a species. If there is no voucher specimen the data of the study may permanently be tied to the wrong plant name. On the other hand, if the author preserved a voucher, it can be reidentified and the data can be applied to the correct species.

Except for particularly bulky structures, plant specimens are usually pressed and dried. This results in a two-dimensional representation of an originally three-dimensional plant. These specimens are usually glued or otherwise affixed to sheets of paper, labeled, and filed in some arrangement that allows for later retrieval. A collection of dried plants is an **herbarium.** An herbarium may consist of only a few specimens mounted in a notebook or thousands to millions of specimens filed in large wooden or metal cabinets. Herbaria are museums of enormous value to plant taxonomists. The specimens they contain are a major source of the data used in taxonomic research. Large herbaria represent the cumulative efforts of many individual botanists.

# The Ethics of Plant Collecting

Many areas of the world are poorly known botanically. There are many areas that need to be documented floristically, for scientific and educational purposes. In general, any ecological or floristic study should include voucher specimens that are appropriately collected and deposited in a herbarium. On the other hand, humans have fragmented the environment with urbanization, suburbanization, farming and highways. This increasing impact on the biota of the earth has generally led to loss of species and the loss of suitable habitat for many imperiled species. Given this conflict between our need to learn how to be scientists and to do good science, and our need to conserve and protect the diversity that we do have, it is important to learn how to collect plants in an ethical fashion.

Much work has been done to understand the biology of imperiled plants in North America and Europe. There are government agency sites on the Internet that list imperiled species, and often provide information about distributions and abundance. **You should make yourself aware of the imperiled plants and animals in your area.**

There are state, federal, and global rankings of the level of endangerment of imperiled species. The International Union for Conservation of Nature and Natural Resources (now known as IUCN—The World Conservation Union) lists global rankings of rarity at *www.iucn.org* in "The Red Book." The U.S. Fish and Wildlife Service provides federal status designations and lists Endangered and Threatened species that are protected under the provisions of the Endangered Species Act of 1973, as amended through the 100th Congress. Within the United States, each state has its own Natural Heritage Program that is a member of the Natural Heritage Network of NatureServe *(www.natureserve.org)*. Each state designates a status for imperiled plants in that state. Ranking can range from plants that are very rare to those that are on "watch lists" to be studied to determine their rarity.

**Scientific studies or collecting should only be conducted with the permission of the landowner or appropriate government agency.** Respect the rights of property owners. If the land is posted as no trespassing, don't trespass. Many people are quite willing to grant permission to collect if you ask. If you are on public land (local, state, or federal park or forest) you should obtain a permit before gathering specimens. In some cases, an agency will not issue a permit for general collecting. Often collecting in these areas is restricted to specialists who are conducting specific scientific inquiries. Remember that the flowers you remove from a public place cannot be enjoyed by others.

**Collect specimens only when removal will not seriously harm the local populations of the species.** The guidelines below are particularly applicable to herbaceous plants. Look around you before collecting a specimen. If the plants are common, generally you can collect specimens without any significant long-term impact on the population. If there are fewer than 20 to 30 individuals in an area, leave it there. Photographs of specimens are often effective vouchers for a species, or can be used to help identify a specimen to determine if it is imperiled. If it is essential to take a specimen as a voucher, uncommon species can be vouchered with a fragment (leaves and flowers/fruit). Make sure to leave enough of a perennial plant to allow it to recover. If you are collecting from a woody plant, take no more than 5 percent of the plant and make "clean" cuttings with clippers to reduce the risk of infection to the plant.

The bottom line for collectors is to have a reason for killing or damaging a plant. Make a good specimen with all the parts and appropriate data for your purpose. **When in doubt, leave it there.** If it is an important find, you can come back later to appropriately document the occurrence.

# Collecting Plants

**Materials required:** Field notebook, pen or pencil, plastic bags, knife or clippers, trowel.

The preparation of a plant specimen involves several steps. The two initial steps are (1) recording of field data and (2) collection of the plant. Let's examine the field data first. Most taxonomists maintain a series of field notebooks that document the botanist's itinerary, observations and collections.

A **field notebook** can be any pocket-size or larger notebook that has the pages securely bound. Ring notebooks are not recommended because the pages can be easily detached and lost. Examine the sample page from a field notebook in Figure 7-1.

Field notebooks contain several types of information. A typical entry would include

1. the date of the observation or collecting activity,

2. the location,

3. a general summary of the ecological situation,

4. the elevation [very important in mountainous regions], and

5. specific information about each of the plants collected.

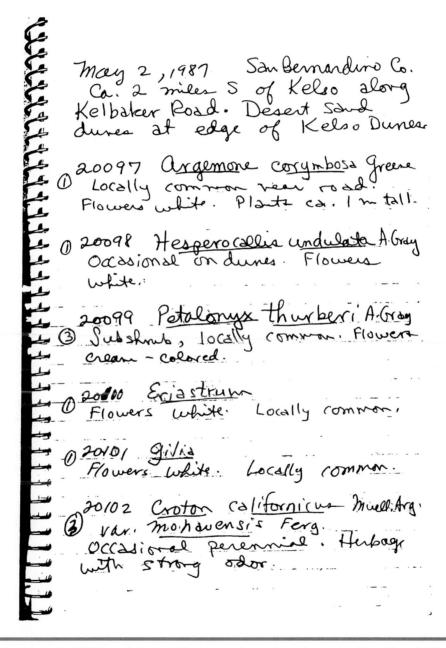

**Figure 7-1**  A page from a typical field notebook. The larger number is the collection number (Dr. Keil's collection number series began at 1 in 1966). The smaller number (circled) indicates the number of labels to be prepared. Where a species name has not yet been determined a space is left to write it in later.

The collector usually separates the observations about one plant from those about another by assigning each kind of plant collected at a site a number. Most taxonomists use a consecutive number system that began at 1 at some time in the past and that increases throughout the career of the worker. A collection number should never be reused, even if the specimen is later discarded.

There are several items of specific information that should be recorded for each plant collected. These data are:

1. relative abundance [i.e., how common was the plant?],

2. specific habitat,

3. flower color,

4. plant size [if only part of the plant is collected or if size is variable in the population from which your specimen was collected],

5. odor [if any], and other information that would not be evident on the dried specimen.

When choosing a plant to collect, remember that you are trying to collect as complete a picture of the plant as you can. This means that you may have to collect several parts of the plant showing, for example, roots[1], stems, leaves, flowers and fruits. For some plants, identification is impossible unless specific parts are present. For herbaceous plants roots and other underground structures are often very important. You might not be able to determine whether the plant is annual or perennial without these parts. On herbs the lower leaves are often markedly different from the upper leaves. Specimens should be in reproductive condition. Without flowers or fruits it is often impossible to identify a plant.

Herbarium sheets are typically 16 1/2 inches long and 11 1/2 inches wide. Individual plants of many species fit comfortably onto paper of this size. Try to avoid the practice of collecting *only* plants to fit the paper (see reasons for this in Chapter 3). For small plants you may want to collect several individuals. However, a representative fragment of a large plant may be all that will fit. This is true of all trees and most shrubs. Plants show much variation among individuals and often among parts of the same individual. Be conscious of this variation and try to represent it in your collection.

After a specimen has been collected you should keep it as fresh as possible until you are ready to press it. You may also wish to hold back a sample for dissection if you have not already identified the plant. [Professional taxonomists work mostly with dried specimens, but this is initially sometimes difficult for students.] A sealed plastic bag is usually sufficient to keep plants fresh. You may sprinkle a few drops of water on the plants to raise the humidity in the bag and reduce the wilting of the plant. Do not leave the bag in the open sun or in the warm interior of an automobile. If there will be a considerable time lag between collection and pressing, put the bag in a refrigerator.

# Pressing Plants

**Materials required:** 2 plant press frames [sheets of 1/4 inch plywood 12 × 18 inches or a lattice frame the same size], sheets of corrugated cardboard approximately 12 × 18 inches, sheets of blotter paper about the same size, sheets of newspaper, 2 lengths of strap or rope, and pen or pencil.

Preservation of specimens involves pressing the plants flat and drying them under pressure. The **plant press** (Figure 7-2) consists of the following parts: the press frames give strength to the press, allowing pressure to be spread uniformly over the plants being dried. The cardboards **(ventilators)** provide channels through which air can diffuse. This permits exchange of moist air from the interior of the press and dry air from the outside. The corrugations should run across the cardboard rather than along its length. Folds and wrinkles in the cardboard impede air movement and slow the drying process. The blotters perform several functions. By molding themselves around the stems and roots of the plant they allow pressure to be spread over the surface of the plant. They absorb moisture and allow it to slowly diffuse into the cardboards. They also cushion the cardboards against stiff stems, roots, and other structures that otherwise might crush the corrugations in the ventilators. The straps [much preferable to ropes] bind the press tightly together. A good strap has a buckle that can be cinched tight.

Some botanists use metal corrugates instead of cardboards and some use foam plastic instead of

---

[1]In some situations roots should *not* be collected. See section on ethics of plant collecting.

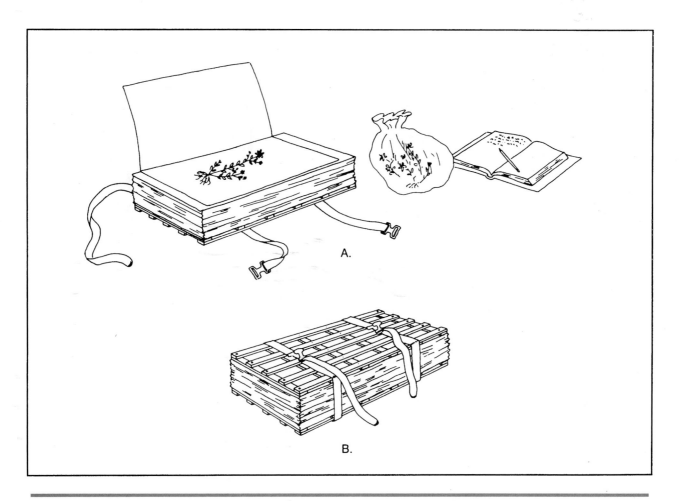

**Figure 7-2** Pressing plant specimens. A. Assembling a plant press and pressing plants. B. A fully assembled plant press.

blotters. Metal corrugates are less easily deformed by woody plant parts and retain the important water-transmitting air passages. They do not become damp and susceptible to the growth of mildew when used to dry plants with a lot of moisture. Metal corrugates have the disadvantage of being heavier and much more expensive than cardboard. Foam plastic has the advantage over blotter paper of molding itself more readily around plant parts and this allows pressure to be applied more uniformly to stems and leaves. Foam plastic has the disadvantage of being much more bulky than blotters when not under pressure in a plant press and presents a storage problem when not in use.

Plant specimens are placed into sheets of newspaper to be pressed. The newspaper should be either tabloid sized sheets (about 18 × 24 inches; 18 × 12 inches when folded) or *half* of a full sized newspaper sheet (about 16 × 24 inches; 16 × 12 inches when folded). When folded around the plant the newspaper should not extend outside the plant press. Only a single sheet of newsprint is needed for each specimen. Don't use a whole section from the newspaper. Place the specimen to be pressed into the sheet of newspaper and carefully arrange the leaves, flowers, etc., to minimize overlap. It is acceptable to bend stems, petioles, etc. if the bend allows the specimen to fit. Some of the leaves should be pressed with the upper surface exposed and others with the lower surface exposed. You may wish to spread out the petals of a flower to display the parts. Make the specimen look as attractive as possible. [This is easier for some plants than for others]. Remember that pressing will not improve the appearance of a poorly displayed specimen. The collection number (from your field notes) is written on the outside margin of the newspaper. This permits you to determine where and when you collected the plant, even if you forget. If you anticipate changing the papers before the plants are dry, you may wish to write the number on a small string tag and attach it directly to the specimen.

The plant press is assembled in the following sequence. First comes a press frame. Then a cardboard and a blotter. You are now ready for the first plant in its newsprint folder. Then comes a blotter, a cardboard and another blotter. You are now ready for another plant in its sheet of newspaper. Then comes another blotter-cardboard-blotter sequence. This continues until you run out of plants or plant press parts. A second press frame completes the sequence. When the press is assembled, wrap the straps around it and cinch it tight. You may want to sit or stand on the press to flatten out the plants or to retighten the press after the plants have had a chance to "relax" for a few hours. Make sure that the straps remain tight. Pressure is necessary for good specimens.

The final step involves drying the press and the plants it contains. This may be done in several ways. In hot dry areas the press can simply be placed into a warm, well-ventilated area. In a few days, the specimens will be dry. Outdoors, you can place the press a few inches above a hot asphalt surface. In cool areas or humid weather, however, plants may get moldy before they get dry. It may be necessary to change the papers one or more times. Drying can be speeded up by placing the plant press over a radiator, a heater or in a special plant press drier. **Do not** put the press in an oven—this could cause a fire. A convenient home-made press drier can be constructed by mounting a well-insulated light bulb inside a box and placing the plant press on top of the box with the corrugations of the ventilators vertical. The light bulb warms the air inside the box. It rises by convection through the ventilators and draws out the moisture of the plants. **Do not** allow the press, the plants or the sides of the box to come into contact with the hot light bulb.

After the specimens are dry they can be removed from the press. Keep them in the newspaper in which you pressed them. It bears your collection number and is much more convenient to handle than the brittle plant specimens. Until the specimens are mounted they should remain in their newspaper wrappers. It is important to keep your specimens in an insect-free environment. If an herbarium cabinet is available, it should be used. If not, sealing the specimens inside a heavy plastic bag when you are not working with the plants can keep out insect pests. If you keep the bag inside a box you can protect the specimens inside from damage.

Most plants can be pressed easily. Others are more challenging. Below are some problems often encountered in specimen preparation and some suggested solutions:

**Succulents:** The problem with succulents is the large quantity of water they contain and the thick epidermis that prevents water loss. Some succulents can live for months inside a plant press under conditions that would rapidly dry out most other plants. By the time the plant is dry it may have shriveled into a nearly unrecognizable remnant. On the other hand your specimen may get moldy. Techniques used to dry succulents involve (1) killing the plant and (2) removing the water. The plant may be killed by treating it with a chemical such as alcohol, by freezing it, or by placing it into a microwave oven for a couple of minutes. Removing the moisture may involve sectioning the plant and slicing away fleshy tissue and/or frequently removing damp blotters and cardboards and substituting dry replacements.

**Thick woody stems and roots:** These structures may be split lengthwise so that the plants will lie flat in the press.

**Very large leaves or other plant parts:** Palm trees, bamboos, tree ferns and other large plants present major challenges to the collector. Large parts can be pressed in pieces in several sheets of newspaper with labels indicating the part pressed. Use the same collection number for all the sheets of the series.

**Very small plants or plant parts:** Place the small plants into a paper envelope. Write the collection number on the envelope.

**Large woody or fleshy fruiting structures:** Pine cones, large dry fruits and other similar structures can be dried whole and filed separately from the pressed specimens. Some kinds of large fleshy fruits can also be dried. Fruits also can be sliced into sections and dried in a plant press. You may have to change cardboards, blotters and papers several times to remove excess moisture. Alternatively, fleshy fruits can be placed into jars of preservative fluids. Although this retains structure well, jars of fluid present a storage problem. Small fruits are usually pressed and dried with the vegetative and floral parts of the plant.

**Very spiny or thorny plants:** These can be very troublesome to press. Side branches can be clipped off or the specimen can be forced into the press (generally with a lot of crunching and popping). This sometimes mangles the ventilators rather badly. If the plant has delicate flowers or fruits, you may want to press them separately from the coarse vegetative structures. They can be recombined after the specimen is dry.

**Submersed aquatic plants:** The stems and leaves of aquatic plants are often so weak that they cannot support the plant outside of the water. When removed from water the plant is limp and difficult to arrange in a plant press. Aquatic plants can be "floated" onto a half sheet of paper in a shallow basin or tray. The stems and leaves can be arranged while still supported by water. The paper is gently removed from the water with the plant clinging to it and is transferred into a second piece of paper in a plant press.

# Identification of Plant Specimens

**Materials required:** Razor blade or scalpel, forceps, dissecting needles, hand lens or dissecting microscope, flora or manual with taxonomic keys and descriptions.

**Identification** is the process of determining the name of a plant. The procedures for determining the correct name of a plant are discussed in Chapters 4 and 5.

# Labeling Specimens

After a specimen has been collected, dried and identified, it is ready to be labeled. Actually, some specimens get labeled first and identified later. An herbarium label is a permanent record of the collection data. Most of these data are drawn from the field notebook of the collector. A sample label is illustrated below in Figure 7-3.

The important aspects of the label are summarized as follows. The label itself should be made of white acid-free bond paper with a high rag content. A properly prepared herbarium specimen may last for hundreds of years. An acid-containing paper such as newsprint or notebook paper yellows and becomes very brittle after only a few years. The size

---

**FLORA OF CALIFORNIA**
**SAN LUIS OBISPO COUNTY**
ROBERT F. HOOVER HERBARIUM
CALIFORNIA POLYTECHNIC STATE UNIVERSITY

*Verbena lasiostachys* Link.
   (Verbenaceae)

Oak Shores Development on northern shore of Nacimiento Lake. Foothill woodland dominated by *Quercus douglasii, Q. lobata, Q. agrifolia,* and *Pinus sabiniana.* Understory mostly grass-dominated. Small patches of chaparral. Flowers purple. Common.

David Keil no. 19575                26 May 1986
with V. L. Holland

**Figure** Sample herbarium label.
**7-3**

of the label varies but most are not more than about 4 × 5 inches, and they are often smaller.

The information on the label can be broken into three categories:

1. Information that remains the same on all labels. This may be preprinted on the label form. It often includes the heading and sometimes the collector and other data.

2. Information that remains the same for all plants collected together at the same location and the same time. This includes the collector, date, locality, elevation and general summary of the habitat.

3. Information that changes from plant to plant. This includes the scientific name, the family, the relative abundance and specific habitat, flower color and collection number.

Preparing labels requires the organization of your data into the proper format. Notice that the scientific name is italicized (it may be underlined if your typewriter or printer cannot print in italics) and that it is accompanied by the author. The collector's name is written out. Do not put just your surname or only your initials. The date is written out. Do not write dates in the form 4/6/84. In the United States this means April 6; in some countries it means June 4. If your specimen is permanently deposited in an herbarium it might be around for along time. Some future worker might even not know in what century you collected the plant! Label typing can be a slow, tedious process, but recent advances in computer technology and software make it possible to generate high quality labels quickly and easily. Some label-typing programs are designed for the student and some for the professional who must process numerous specimens efficiently. Labels can also be prepared easily with standard word-processing programs.

# Final Stages of Processing Specimens

If you plan to donate your specimens to an herbarium, they would now be ready. That is, your specimens would be unmounted, placed inside a folder made from newsprint, and accompanied by the completed label. The final steps of processing involve mounting the plant, stamping it with the herbarium's official stamp(s), and filing it away in the collection.

Plants are mounted on sheets of thickish, high rag-content paper (preferably acid-free) especially designed for herbarium specimens (Figure 7-4). Standard paper size in the United States is 16 1/2 × 11 1/2 inches. The label is usually glued in the lower right corner of the paper. An envelope made of rag-bond paper is affixed to the sheet to hold any fragments, seeds, loose flowers or other parts that are (or might become) detached from the plant. The plant is attached to the paper by glue spread on the back of the leaves, stems, etc. White glue (Elmers, etc.) is inexpensive, readily available, and suitable for specimens. It may be diluted 50:50 with water for ease of application. Weights are usually placed on the specimens until the glue is dry. Waxed paper can be placed between the weights and the specimens to keep them from sticking together. Straps of cloth tape or thread are often used to make sure that stiff stems and roots are firmly attached to the paper. Mounted specimens are usually stamped with an institutional stamp and often with a consecutive number stamp as well.

After the specimens have been stamped they are ready to be filed in the herbarium. The arrangement of individual specimens is almost always by family. The sequence of families varies from herbarium to herbarium. Many are arranged according to one of the early systems of classification (Engler and Prantl system in North America and most European countries; Bentham and Hooker in the British Isles and Commonwealth). A few are organized by one or another of the recent systems (e.g., Cronquist). In some herbaria the families are organized alphabetically within the major groups (gymnosperms, monocots, dicots, etc.). Within the families, genera may be arranged alphabetically or in systematic order. Species are almost always organized alphabetically within the genera. Sometimes the specimens are segregated according to the geographical area where they have been collected. For instance in some large herbaria specimens from each continent are filed in separate (often color-coded) folders.

Before using an herbarium you should always check with the herbarium director or curator. Herbaria often have strict guidelines that regulate access to the specimens, how specimens are to be handled, etc. These are designed to prevent damage to the specimens. Two of the greatest dangers to an herbarium are fire and insect damage. Smoking is generally strictly prohibited. Many herbaria have sterile-entry regulations to keep plant specimens that might be contaminated with insects from entering the premises. If someone wants to bring specimens for study, they must be sent ahead to be fumigated. Finally, it is very important to put specimens back in the proper order or in a designated location where they can be refiled properly. A misfiled specimen is a lost specimen.

Herbarium specimens should be handled with care. The dry plant parts are often quite brittle and the adhesives used to attach the specimens to the paper sometimes break down in age. Avoid bending the paper on which the specimens are mounted. When examining a series of specimens that are filed in the same folder, *do not turn specimens upside down* as you go through the pile. This causes the specimens to break free from the paper and loose parts to fall off. Instead pick up the herbarium sheets with both hands and set them to the side, reversing the order of the sheets as you go through them. In many herbaria the order in which specimens are arranged within a species folder is of no importance. Check with the director or curator to determine if you should return the sheets to their original order. If you see loose pieces that have broken free from the paper place them into the fragment folder. If there is no fragment folder, one can be added to the specimen. If you find specimens with loose stems, evidence of insect infestation, or other serious damage, bring them to the attention of the curator.

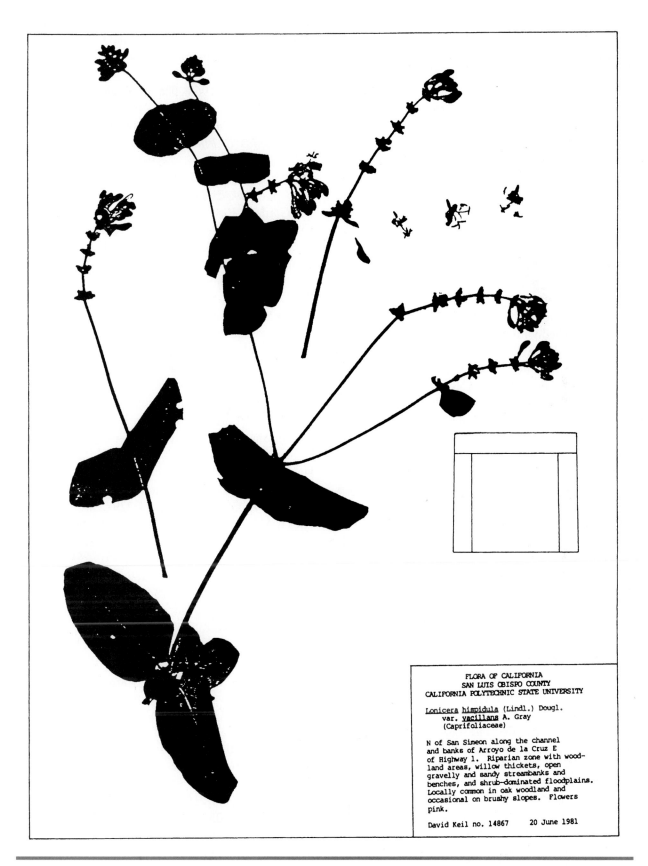

Inside the label:

FLORA OF CALIFORNIA
SAN LUIS OBISPO COUNTY
CALIFORNIA POLYTECHNIC STATE UNIVERSITY

Lonicera hispidula (Lindl.) Dougl.
    var. vacillans A. Gray
    (Caprifoliaceae)

N of San Simeon along the channel
and banks of Arroyo de la Cruz E
of Highway 1. Riparian zone with woodland areas, willow thickets, open
gravelly and sandy streambanks and
benches, and shrub-dominated floodplains.
Locally common in oak woodland and
occasional on brushy slopes. Flowers
pink.

David Keil no. 14867    20 June 1981

**Figure 7-4**  A fully processed mounted herbarium specimen.

As a general rule plant parts should not be permanently removed from the herbarium sheet to which they are attached. In some instances it may be necessary for you to dissect a flower or fruit. If this is the case you must obtain permission from the curator or director first. If parts must be dissected it is preferable for you to use structures already in a fragment folder (if any) before removing pieces of the mounted specimen. Herbarium regulations usually require you to place any dissected parts into the fragment folder upon the conclusion of your study. You may need to soften plant parts before you dissect them. Often a few drops of hot soapy water are sufficient to soften a plant part. Some structures may have to be boiled to soften them enough for a dissection. Softened parts should be dried and returned to the specimen's fragment folder when you are finished with them.

Occasionally when studying a series of plant specimens in an herbarium you may encounter a specimen that does not seem to belong where it is filed. It is not at all uncommon to find misidentified specimens in an herbarium. It is also possible that the specimen merely represents an extreme of variation of the taxon and is actually identified correctly. If the specimen has been misidentified and you are able to correct the misidentification (and you are certain that you are correct in your determination of the name), you should annotate the specimen. A specimen may be annotated by affixing a small label to the specimen with the corrected name and author, the name of the person who has reidentified the specimen, and the date (at least the year) of the annotation. The annotation label is generally glued to the herbarium sheet above or near the collection label. It should not permanently cover any part of the specimen itself. Many herbaria have blank annotation labels for use by staff and visitors. Do not write directly on the herbarium sheet. If you are not certain you should ask someone in charge of the collection what to do. The curator may elect to annotate the specimen himself (or herself) or to leave the specimen in place for a specialist to examine it later.

# References

Baker, M. A., M. W. Mohlenbrock and D. J. Pinkava. 1985. A comparison of two new methods of preparing cacti and other stem succulents for standard herbarium mounting. *Taxon* 34:118–121.

Benson, L. 1979. Preparation and preservation of plant specimens, pp. 423–444 *in Plant Classification,* 2nd Ed.

Brennan, J. P. M. 1968. The relevance of the national herbaria to modern taxonomic research, pp. 23–32 *in* V. H. Heywood (ed.), *Modern Methods in Plant Taxonomy.* Academic Press, London.

Bridson, D., and L. Forman. 1992. *The Herbarium Handbook,* 2nd edition (revised). Royal Botanic Gardens, Kew.

Croat, T. B. 1978. Survey of herbarium problems. *Taxon* 27:203–218.

Cronquist, A. 1968. The relevance of the national herbaria to modern taxonomic research in the United States of America, pp. 15–22 *in* V. H. Heywood (ed.), *Modern Methods in Plant Taxonomy.* Academic Press, London.

Dransfield, J. 1986. A guide to collecting palms. *Annals of the Missouri Botanical Garden* 73:166–176.

Fosberg, F. R., and M.-H. Sachet. 1965. Manual for tropical herbaria. *Regnum Vegetabile* No. 39. International Bureau for Plant Taxonomy and Nomenclature, Utrecht, The Netherlands.

Haynes, R. R. 1984. Techniques for collecting aquatic and marsh plants. *Annals of the Missouri Botanical Garden* 71:229–231.

Hicks, A. J., and R. M. Hicks. 1978. A selected bibliography of plant collection and herbarium curation. *Taxon* 27:63–99.

Irwin, H. S., W. W. Payne, D. M. Bates and P. S. Humphrey (eds.). 1973. *America's Systematics Collections: a National Plan.* Association of Systematics Collections.

Ketchledge, E. H. 1970. *Plant Collecting. A Guide to the Preparation of a Plant Collection.* State University College of Forestry, Syracuse University, Syracuse, N.Y. 21 pp.

Kobuski, C. E., C. V. Morton, M. Ownbey and R. M. Tryon. 1958. Report of the committee for recommendations on desirable procedures in herbarium practice and ethics. *Brittonia* 10:93–95.

Lawrence, G. H. M. 1951. Field and herbarium techniques. pp. 234–262 *in Taxonomy of Vascular Plants.*

McGaugh, M. H., and H. H. Genoways. 1976. State laws as they pertain to scientific collecting permits. *Museology, Texas Tech University* 2:1–81.

McNeill, J. 1968. Regional and local herbaria, pp. 33–44 *in* V. H. Heywood (ed.), *Modern*

*Methods in Plant Taxonomy.* Academic Press, London.

Metsger, D. A. and S. C. Byers. 1999. Managing the modern herbarium: an interdisciplinary approach. Society for the Preservation of Natural History Collections, Royal Ontario Museum, Ontario. 384 pp.

Porter, C. L. 1967. Field and herbarium methods, pp. 42–54 *in Taxonomy of Flowering Plants.*

Prance, G. T. 1984. Completing the inventory, pp. 365–396 *in* V. H. Heywood and D. M. Moore (eds.), *Current Concepts in Plant Taxonomy.* Systematics Association Special Publication no. 25. Academic Press, London.

Robertson, K. R. 1980. Observing, photographing, and collecting plants. *Illinois Natural History Survey Circular* 55. Urbana, Illinois. 62 pp.

Soderstrom, T. R., and S. M. Young. 1983. A guide to collecting bamboos. *Annals of the Missouri Botanical Garden* 70:128–136.

von Reis Altschul, S. 1977. Exploring the herbarium. *Scientific American* 236(5):96–104.

Womersley, J. S. 1981. *Plant Collecting and Herbarium Development.* FAO Plant Production and Protection Paper 33, Rome.

# Exercises

1. Your assignment is to prepare a plant collection from the area around the campus, your home or other nearby locations. Your instructor will determine how many plants to include in your collection. Certain areas are off-limits. **Do not collect in parks, nature reserves or private land where you do not have permission to collect.** If in doubt, check with the landowner or agency before collecting. If a plant is uncommon or rare, do not collect it.

2. Purchase a small notebook in which to record your field notes. Whenever you go out into the field to collect plants, record your field observations in this notebook. The field notebook is a record of your field activities. It is not to be rewritten though you may wish to add other information later. Even if you decide not to turn in some of the plants you have collected, the notes you wrote when you collected them should be remain in the notebook. **Number each plant as you record information in your notebook.** The number should be written both in your notebook and on the sheet of newsprint that contains the specimen.

3. Your class may share a set of plant presses or you may have one checked out to you for your own use. Place your specimens in the press that you are using. Do not take for granted that your press will be looked after properly; check it often! It is your responsibility to make sure that your specimens are pressed properly. You may have access to a plant press drier or you may have to dry the specimens on your own.

4. Determine the correct names for the plants (see Chapters 2 and 5). You may want to identify the plants before you press them or to keep some portions of the specimens fresh until you have had the opportunity to key them out.

5. After the specimens are dry, store them in the cabinet provided until laboratory time is available for identifying the plants (if you have not already determined the name). Labels, except for identification, can be prepared during this interval or labels can be prepared after the identification has been completed.

6. On the date designated by your instructor, turn in your field notebook and the properly labeled, unmounted specimens. (**Do not** glue down, tape down, or otherwise affix the labels or the specimens to the newspaper folder in which the plants have been placed.)

7. Allot yourself plenty of time for collecting, identifying, drying and labeling your plants. Do not wait until the last minute to start your collection. If you do, you will probably not get finished, your plants will not be dry and you will lose points on the assignment.

# Survey of the Vascular Plants

chapter 8

Land plants have been traditionally divided into "bryophytes" and vascular plants. Based on the fossil record, land plants first appear about 450 million years ago. These first land plants possessed a gametophytic and sporophytic stage of their life cycle. The sporophyte has a resting stage when it first forms and this has led scientists to name the land plants "embryophytes" (Kenrick and Crane 1997; Shaw and Renzaglia 2004).

The "bryophytes" are recognized as the mosses, liverworts, and hornworts. These all have a dominant gametophytic stage in the life cycle. The "bryophytes" are now considered to represent two or three ancient lineages of land plants. Although they have a waxy cuticle that reduces water loss and either pores (liverworts) or stomates (hornworts and mosses) for gas exchange, they do not have the specialized cells for water conduction called tracheids (Graham 1993).

True vascular plants appear in the fossil record about 420 million years ago. They possess a cuticle and stomates, but they also have elongated cells with specialized thickened cell walls. These cells are dead at maturity and conduct water from cell to cell by openings or pits. Much like the invention of the elevator allowed for the development of cities and skyscrapers, these water conducting cells allowed land plants to grow taller and taller to compete for light.

Vascular plants are the most important plants on earth today in terms of number of species and biomass. They dominate almost every terrestrial habitat on the earth's surface. They range from minuscule herbs to giant trees. They provide many of the raw materials that support our modern civilization. Agri-culture, man's oldest cultural breakthrough, is basically the management of vascular plants to produce food and fiber for man and his animals.

Taxonomists have been studying vascular plants for hundreds of years. For many taxa the relationships are now considered to be relatively well-established. Although there are several competing systems of classification available, a cursory survey of them will show a consistency of pattern for many of the taxa. However, the relationships and classification for other taxonomic groups remain controversial, especially at the highest and lowest levels in the taxonomic hierarchy. Various taxonomists have expressed their own ideas of relationships and how best to indicate these in a classification system. The level and number of subcategories differ from one classification system to another.

Figure 8-1 shows our latest understanding of the relationships of land plants or embryophytes. According to this phylogeny, we can recognized three main clades or groups of land plants. The Lycophytes consist of the three living families (Lycopodiaceae, Selaginellaceae, and Isoetaceae) that have leaves with a single vascular strand, called microphylls. Those plants with highly vascularized leaves, called megaphylls, represent the other two groups, the Monilophytes or the ferns and fern allies, and the Lignophytes, whose living members include the seed plants. (Judd et al. 2001; Palmer, Soltis and Chase 2004; Pryor et al. 2004). In the survey of the vascular plants that follows, we have chosen to use the following classification systems. For the Lycophytes and Monilophytes we follow the Flora of North America treatment (Vol 2, 1993). For the seed plants, we use the four traditional living

125

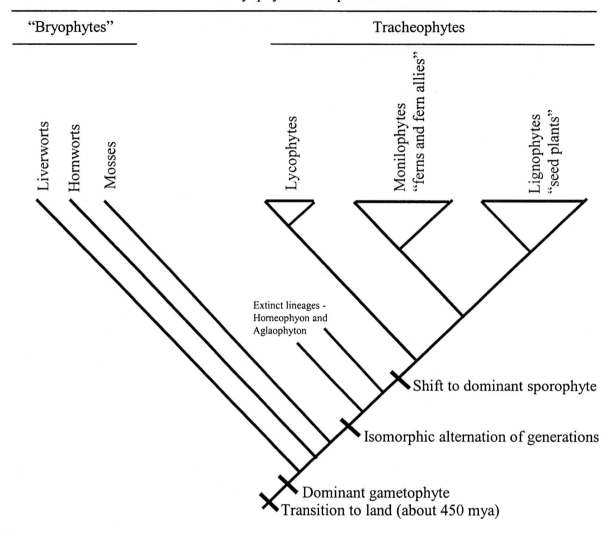

**Embryophytes/Land plants**

"Bryophytes"

Tracheophytes

Liverworts

Hornworts

Mosses

Lycophytes

Monilophytes "ferns and fern allies"

Lignophytes "seed plants"

Extinct lineages -
Horneophyon and
Aglaophyton

Shift to dominant sporophyte

Isomorphic alternation of generations

Dominant gametophyte
Transition to land (about 450 mya)

**Figure 8-1** Phylogeny of land plants, showing the polyphyletic "bryophytes" and the three major clades of Tracheophytes. It should be noted that the relationships of the three "bryophyte" lineages are poorly resolved. Character state changes indicate approximate shifts in the life history of early land plants. This figure is adapted from Judd et al. (2002) and Pryor et al. (2004).

orders (Cycadales, Coniferales, Ginkgoales, and Gnetales) that are currently recognized by Burleigh and Mathews (2004) as a monphyletic group and the dynamic treatment maintained by Peter Stevens (2001) for the fifth group of seed plants, the angiosperms or the phylum Anthophyta.

There are several older systems of classifications of flowering plants that offer tremendous insight into our understanding of angiosperm evolution (Cronquist, Dahlgren, Takhtajan, and Thorne). These systems are referenced at the end of the chapter. Previous editions of this textbook have used the Cronquist system of classification.

However, the exploration of the DNA sequences of plants over the past 15 years, coupled with new methods of data analysis, have radically changed our understanding of plant relationships. Given the availability of a more recent classification system on the Internet (Stevens 2001), it is appropriate to recognize these latest developments in our understanding of plant evolution. It should be noted, particularly by advanced students of systematics, that the systems of Cronquist and Tahktajan both recognized paraphyletic taxa. It should be of interest for such advanced students to compare these systems in order to understand different concepts of classification, as well as to understand where molecular data are in conflict with morphological and/or biochemical analyses.

# Organization of the Survey

The next 9 chapters present the survey of vascular plants with emphasis at the level of family. The family level is particularly appropriate for an introductory taxonomy class. There are 300,000+ species of vascular plants on the earth's surface today and many thousands of genera. It is beyond the capabilities of even well-trained professional taxonomists to gain familiarity with more than a small fraction of these taxa. On the other hand there are fewer than 400 families of vascular plants, many of which are represented by only a few genera and species. By concentrating on "major families" a student can gain a working knowledge of well over half of the vascular plants. With certain exceptions classification at the family level is considerably more stable than that of higher level taxa.

The families are presented in this text at three somewhat arbitrary levels of importance. For each family that we consider to be important from an ecological, economic, or evolutionary point of view, a technical description is presented along with illustrations and a general discussion of the family's importance. Consideration has been given to families well-represented in the native or cultivated floras of the continental United States. It is expected that students will learn to recognize the families of the first order importance by sight.

Families of somewhat lesser importance are represented by a statement of geographical range, an indication of the number of genera and species, and a one-paragraph technical description. A few illustrations are provided for some of these families. Some of these second-order families are well-represented in tropical regions or in the Old World but have comparatively few species in the U.S. Others are relatively small families.

Families that we have considered of least importance to students in the U.S. are merely listed with their range and number of genera and species. Some of these are very small families and others are restricted to the tropics or southern hemisphere areas, and are seldom encountered in cultivation.

# Synoptic Keys vs. Identification Keys

Keys can be used in two very different ways—to define the major characters of a taxon (**synoptic keys**) or to determine the taxonomic placement of individuals or groups. Because the synoptic keys often use anatomical or physiological characters that are not readily visible without special equipment or training, they often are not useful in field identification of a particular plant. A given plant species may differ widely from a typical representative of its family in one or more readily visible characteristics. Synoptic keys are used primarily to help a student conceptualize the major patterns of variation discussed in each chapter.

Keys can also be used for identification (Chapter 5). Such keys should be based on the most readily visible characters, whether or not they correlate with characters actually important in defining taxa. Since the same set of visible characters may be found in two completely unrelated taxa, identification keys must be constructed in an artificial manner. This means that two taxa that appear next to one another in such a key can not *a priori* be said to be related. All that can be stated is that they resemble each other in those characters used in the key.

An artificial identification key to many of the families of vascular plants found native to or cultivated in the contiguous United States can be found on p. 128. This key does not include those families considered to be of least taxonomic significance. This key does give a broader alternative to the keys in most local and regional floras so that cultivated plants can be determined at least the family level. It also gives an alternative route to identify difficult plants and thus can serve as check on the correctness of an identification.

# Synoptic Key[1] to the Major Taxa of Vascular Plants with Reference to Chapters in this Book

The following synoptic key defines the groups of vascular plants discussed in the following 9 chapters. Whenever you begin studying a new chapter, we recommend that you return to this key to establish the relationships of the plants discussed in the chapter to the rest of the vascular plants.

1. Plants microphyllous or megaphyllous, never producing seeds, flowers or fruits; primary terrestrial dissemination by one-celled drought resistant spores; plants heterosporous or homosporous; male gametes motile, produced in antheridia [Pteridophytes].

    2. Leaves present, megaphyllous; veins of leaves many and usually intricately branched, sporangia borne away from the stem on leaf blades or modified leaves.
    . . . . . . . . . . . . . . . . . . . . . . . . . . . . . . . . . . . . Division **POLYPODIOPHYTA** (Chapter 9)

    2′ Leaves absent or microphyllous; veins of leaves 1 and unbranched; leaves scale-like or linear.

        3. Stem jointed and hollow; sporangia borne in cones on the undersurface of peltate sporangiophores. . . . . . . . . . . . . . . . . . . . . . . . . . . . . Division **EQUISETOPHYTA** (Chapter 9)

        3′ Stem smooth and solid; sporangia associated with leaves or lateral cauline scales.

            4. Sporangia fused into groups of 2 or 3 (synangia) in axil of veinless scales or flattened, 1-veined leaflike structures; true leaves absent.
            . . . . . . . . . . . . . . . . . . . . . . . . . . . . . . . . . . . .Division **PSILOTOPHYTA** (Chapter 9)

            4′ Sporangia single, borne in the axil of a leaf with a single vein . . . . . . . . . . . . . . . . . . . . . . . . .
            . . . . . . . . . . . . . . . . . . . . . . . . . . . . . . . . . .Division **LYCOPODIOPHYTA** (Chapter 9)

1′ Plants megaphyllous, always producing true seeds, often producing flowers and fruits; primary terrestrial dissemination by seeds containing preformed multicellular embryos; heterosporous; microspores develop within microsporangia to form pollen; male gametes rarely motile, and if so, associated with pollen tube; megaspore develops inside an integumented megasporangium (ovule) forming an embryo sac.

    5. Ovules usually borne in simple or compound cones, rarely on stalks, exposed to air at the time of pollination; pollen carried to micropyle of ovule; microsporangia 1-many per microsporophyll; second sperm usually disintegrates after fertilization of egg cell by first sperm; leaves often scale, awl, or needle-like; trees or shrubs [Gymnosperms]. . . . . . . . . . . . .Division **PINOPHYTA** (Chapter 10)

    5′ Ovules enclosed within a carpel at time of pollination; pollen carried to stigma formed of carpellary tissue; microsporangia 2–4 per microsporophyll, forming anthers usually borne on a slender filament; second sperm fuses with polar nuclei of embryo sac forming polyploid endosperm; habit and leaves variable [Angiosperms] . . . . . . . . . . . . . . .Division **MAGNOLIOPHYTA** (Chapters 11–16)

        6. Cotyledons usually two; mature root system usually primary, adventitious or both; vascular bundles of stem in a ring surrounding a pith; inter- and intrafascicular cambium usually present; leaves mostly net-veined; floral parts when definite in number usually in sets of 5, less often 4, rarely 3; pollen typically triaperturate or of triaperturate derived types [Dicots].
        . . . . . . . . . . . . . . . . . . . . . . . . . . . . . . . . . .Class **MAGNOLIOPSIDA** (Chapters 12–16)

            7. Plants relatively primitive; perianth usually evident, polypetalous or apetalous; stamens usually numerous, centripetal; pollen always binucleate and often uniaperturate; gynoecium typically apocarpous or monocarpous; ovules betegmic and crassinucellate.
            . . . . . . . . . . . . . . . . . . . . . . . . . . . . . . . . . .Subclass **MAGNOLIIDAE** (Chapter 12)

---

[1]Key to subclasses of the Magnoliophyta used by permission; adapted from *Evolution and Classification of Flowering Plants,* 2nd ed., pp. 265–266; 457–458, by A. Cronquist, copyright 1988, The New York Botanical Garden.

7    Plants displaying characters more advanced than above in one or more respects; pollen never uniaperturate.

   8.  Plants mostly displaying adaptations to wind pollination; perianth reduced, usually apetalous, but calyx usually present in staminate flowers; flowers most imperfect and often in catkins; seeds usually 1 or 2 per fruit. . . . . . . . . . . . . . . . . . . . . . . . . . . . . . . . . .Subclass **HAMAMELIDAE** (Chapter 18)

   8′  Plants usually *not* displaying characteristics of wind pollination such as absent corolla, reduced calyx, catkin inflorescences, imperfect flowers, and 1 or 2 seeds per fruit.

      9.  Flowers perigynous to epigynous with a well developed hypanthium or fleshy nectar ring or both; stamens when many with a centripetal initiation sequence; corolla polypetalous or sympetalous, rarely apetalous; sympetalous members with stamens borne on the corolla.

         10.  Flowers polypetalous, rarely apetalous; ovules bitegmic and crassinucellate; commonly trees and shrubs. . . . . . . . .Subclass **ROSIDAE** (Chapters 12 and 13)

         10′  Flowers sympetalous; ovules unitegmic, tenuinucellate; carpels usually 2, sometimes 3, rarely more by the formation of false septa or seemingly 1; trees and shrubs relatively rare. . . . . . . . . . .Subclass **ASTERIDAE** (Chapters 14 and 15)

      9′  Flowers hypogynous or epigynous, hypanthia and nectar rings rare; stamens when many initiating centrifugally; corolla various, if sympetalous then stamens not epipetalous; placentation often basal, free central, or parietal.

         11.  Pollen usually binucleate (except Brassicaceae and Ericaceae); ovules various, but seldom with perisperm; no betalains; placentation often parietal, never free-central except in flowers with sympetalous corollas and opposite stamens. . . . . . . . . . . . . . . . . . . . .Subclass **DILLENIIDAE** (Chapter 16)

         11′  Pollen trinucleate; ovules bitegmic, crassinucellate, mostly campylotropous or amphitropous; seeds often with perisperm (nutritive tissue derived from nucellus); plants either with betalains instead of anthocyanins or with free-central to basal placentation or both; many succulent. . . . . . . . . . . . . . . . . . . . . . . .Subclass **CARYOPHYLLIDAE** (Chapter 17)

6′  Cotyledon one, rarely undifferentiated; leaves mostly with parallel venation; inter- and intrafascicular cambium lacking, as often are all cambia; vascular bundles generally scattered or in two or more rings; floral parts when definite in sets of three, seldom four and never five; pollen uniaperturate or uniaperturate derived types [Monocots]. . . . . . . . . . . . . . . . . . . . . . . . . . . . . . . . . . . . . . . . . . . . . .Class **LILIOPSIDA** (Chapters 19 and 20)

   12.  Plants not grass-like; flowers usually with conspicuous perianth, or if apetalous, then associated with a large, often colored infolding bract (spathe); plants herbaceous to woody. . . . . . . . . . . . . . . . . . . . . . . . . . . . . . . . . . . . . . . . . . . . . . . . . .**MONOCOTS I** (Chapter 19)

   12′  Plants mostly ± grass-like; flowers usually without perianth or with perianth small, brown or green or reduced to bristles or scales. . . . . . . . . . . . . . . . . .**MONOCOTS II** (Chapter 20)

# Tips for Learning Families

Each plant family has its own pattern of variation. No one species is truly representative of that pattern unless it is the only member of its family. Usually a family has several genera, each comprising one or more species. They are united as members of one family because they share a set of characteristics and because they are thought to share a common ancestry. The shared characteristics within a family are those features that were apparently present in the original members of the family and that have been passed on to most or all of the species and genera that make up that family. The genera and species differ among each other in features that have been modified through evolutionary change. Developing a "family concept" in your mind is a matter of learning to recognize the theme behind all the variation. The theme is the family concept and the features of genera and species are variations on that theme.

The set of features that characterizes a family is often a seemingly esoteric pattern. A feature that is uniform in one family may be highly variable in another. Each family has its own combination of "diagnostic features." Basically a diagnostic feature is one that can be used to distinguish one family from the others. Usually no one feature by itself is diagnostic. Each family is characterized by a combination of features. A family is made up of individuals, species and genera, each of which differs by varying degrees from the others. However, the members of the family generally share most if not all the defining set of characteristics.

The amount of time and effort that you put into studying family characteristics will probably determine how well you learn the families. Some families are easy to recognize. They have clearly discernible patterns of variation. You probably know a few of them even now. Others may be more difficult to learn because they have combinations of features that are not immediately evident to you. Sometimes two or more families share a set of conspicuous features and differ by characters that you have not yet learned to observe. It is your task as a student to develop a concept for each of the families you are assigned to learn. This concept should be broad enough that you will be able to recognize family members on sight, even if they are species or genera that you have not seen before.

Various procedures and study aids are available that can help you to determine the characteristics of a family. Not all necessarily will work for you. Some of these are listed below.

1. **Keying plants in laboratory (or on your own) to family.** This gives a set of characteristics that have been used to separate one family from all the others. Retrace your steps and make a list of the characters in the sequence that the author of the key used them to key the plant to family. If the route through the key is the same for all members of a family, this key summary provides an excellent set of characters for distinguishing the family from all the others. If different members of a family key along differing pathways, compare the routes to see what they have in common. The different routes represent variation on the theme of the family.

2. **Keying several members of a family to genus and species.** Each species has its own set of characteristics. By observing the features of several different members of a family you soon will begin to separate the family characteristics that they share from their own specific variations. Because variation in a given geographical area or at a particular time of year is likely to be restricted, you are unlikely to observe all the possible variations during one laboratory experience. Do not consider the few representatives of a family that you see in the lab to be representative of all of the variation in the family. Supplement your laboratory exposure with additional sources of information.

3. **Consult one or more of the many references or study aids available in the laboratory or library.** Each of these sources probably approaches the families from a slightly different point of view. Collectively they provide a broad perspective on family features and patterns of variation. Suggested sources of information include:

   a. this textbook. The descriptions, illustrations and commentary in this book are presented to help you to learn family characteristics.

   b. sets of study slides and the accompanying commentary that describes the features illustrated.

   c. *Study Guide to Selected Families of Plants* (Keil, 1986) with microfiche illustrations and accompanying commentary.

| Superior | Inferior | Both |
|---|---|---|
| | Grossulariaceae | Nymphaeaceae |
| ALL | Onagraceae | Rosaceae |
| | Myrtaceae | Saxifragaceae |
| THE | Apiaceae | Ericaceae |
| | Caprifoliaceae | Liliaceae |
| REST | Rubiaceae | |
| | Asteraceae | |
| OF THE | Bromeliaceae | |
| | Caetaceae | |
| FAMILIES | Aizoaceae | |
| | Betulaceae | |
| | Fagaceae | |
| | Iridaceae | |
| | Orchidaceae | |

**Table 8-1** A comparison of ovary position in some important families.

d. other textbooks, floras and manuals, and botanical reference books. Some suggested references are listed at the end of this chapter, and additional references are listed at the end of each chapter in the survey.

e. posters, charts, and pictures in books that illustrate representatives of the families.

4. **Prepare a set of one-character/family tables.** Characters take on their importance through their ability to discriminate among different families. A given character may be diagnostic for one family and highly variable in another. Character importance is best gauged by studying the families in comparison with one another. Table 8-1 is an example of a character state/family table for some families based on ovary position.

These tables have two advantages. First they show immediately which families have what traits. Secondly they cut down on the memory work. For the character Ovary Position, three columns are needed. Two of the columns are relatively short, while the third is very long. It seems logical to memorize the short lists, rather than the long ones. Note, if you know the short columns, by default you have also memorized the long list. If a family is not on one of the short lists, then it must be on the long one (if

you have been careful in 'constructing your table.')

5. **Make a family/character table.** Get a large sheet of posterboard and list characters across the top and families down the side. Use the same sequence of characters that is used in writing descriptions (see Chapter 5). This allows you to put the information on all the families you are learning into a single table. As you meet additional families, add them onto the table. Table 8-2 is an example of a small portion of such a table for some families of the Magnoliidae. This table has the advantage of making all the information easily available and very compact. It allows you to make comparisons quickly and easily. By scanning through such a table you can identify families that have features in common. This can help you to determine which groups of families are likely to give you trouble. Examples of potential trouble spots include families with

(a) 4-merous flowers,

(b) many stamens,

(c) inferior ovaries,

(d) apocarpous gynoecia,

(e) sympetalous corollas,

(f) hypanthia, etc.

Maybe the sympetalous dicots are a trouble area for you. From the table you can determine the features by which the families differ. The table also allows you to recognize families that are polymorphic for particular features.

6. **Make up a set of flash cards** for the families you are learning. Put the family name on one side and a description on the other. You may want to include drawings, floral diagrams, floral formulas, etc. Are there any terms that are restricted to members of a particular family? Include them on the card and make sure that you know what they mean. Put the cards face down on the table. Can you write a description for the family? Do you know how to separate the families?

7. **Review and study with friends.** Establish a study group with two or three of your classmates. Pool your knowledge. Each of you probably remembers something different from the lecture and lab. Compare your flash cards,

| FAMILY | CONDITION OF COROLLA | | | NUMBER OF STAMENS | | | |
|---|---|---|---|---|---|---|---|
| | Apetalous | Apopetalous | Sympetalous | Many | 2 × Petals | = Petals | < Petals |
| Magnoliaceae | — | X | — | X | — | — | — |
| Ranunculaceae | X | X | — | X | — | — | — |
| Papaveraceae | rare | X | — | X | — | — | — |
| Nymphaeaceae | — | X | — | X | — | — | — |
| Rosaceae | — | X | — | X | X | — | — |
| Solanaceae | — | — | X | — | — | X | — |
| Lamiaceae | — | — | X | — | — | — | X |
| Scrophulariaceae | — | — | X | — | — | — | X |
| Asteraceae | — | — | X | — | — | X | — |
| Other families | | | | | | | |

**Table 8-2** Portion of a family/character table for some families of dicots. The table can be expanded to include additional characters and more families.

family comparison tables, etc. with those constructed by your classmates. Resolve any discrepancies. Quiz each other using slides, microfiche, pictures from books or charts, or from your notes. Take a walk together on campus, in a park or in some other area where you can see a diversity of plants. Ask questions and discuss what you see. Ask your friends how to distinguish the families that are giving you trouble. If you are not sure, pool your efforts. If you can successfully tell your friends how to separate difficult families, you are learning the families. **CAUTION!!** If questions arise that you are unable to answer, do not ignore the problem. Look into reference books for answers or ask your instructor.

8. **Answer study questions.** At the end of each chapter of the plant survey in this book there is a set of study questions. Answer the questions on the families you have studied. Use the text, your notes, and other study aids to find out information that you do not recall. You may wish to compare your answers with those of your classmates or to check with your instructor if you are not sure of an answer.

9. **Write your own keys to families.** Pull out five flash cards at random and without looking at your character lists, write an indented dichotomous key to those families (see Chapter 5).

Check your key for accuracy by comparing it with descriptions, your family table, flash cards, etc. If your instructor is available, let him or her check the key. Shuffle the deck. Pull out five more cards . . .

10. **Use computer self-help programs.** Although not as good, computer programs are becoming available that will review and test your understanding. If such programs are available, they can aid you doing character association and recall. They have the advantage in that they force a correct response without allowing you to look ahead to the answer. They are also quite impersonal, but have infinite patience. They will sit and wait for the correct response as long as it takes you to come up with it.

# References

Bedell, H. G., and J. L. Reveal. 1982. An outline and index to Takhtajan's 1980 classification of the flowering plants. *Taxon* 31:211–232.

Benson, L. 1979. *Plant Classification,* 2nd ed. D. C. Heath & Co., Lexington, Mass.

Brummitt, R. K. 1992. *Vascular Plant Families and Genera.* Royal Botanic Gardens, Kew.

Burleigh, J. G. and S. Mathews. 2004. Phylogenetic signal in nucleotide data from seed plants: implications for resolving the seed plant tree of

| SYMMETRY | | HYPANTHIUM | | OVARY POSITION | | CONDITION OF GYNOECIUM | | |
|---|---|---|---|---|---|---|---|---|
| Radial | Bilateral | Present | Absent | Superior | Inferior | Apocarpous | Monocarpous | Syncarpous |
| X | — | — | X | X | — | X | — | — |
| X | X | — | X | X | — | X | rare | rare |
| X | X | — | X | X | — | — | — | X |
| X | — | — | X | X | X | — | — | X |
| X | — | X | X | X | X | X | X | X |
| X | rare | — | X | X | — | — | — | X |
| — | X | — | X | X | — | — | — | X |
| — | X | — | X | X | — | — | — | X |
| X | X | — | X | — | X | — | — | X |
| | | | | | | | | |
| | | | | | | | | |

Table   Continued
8-2

life. *American Journal of Botany* 91(10): 1599–1613.

Cronquist, A. 1988. *The Evolution and Classification of Flowering Plants,* 2nd ed. New York Botanical Garden, Bronx, N.Y.

———. 1981. *An Integrated System of Classification of Flowering Plants.* Columbia University Press, New York.

———, A. Takhtajan, and W. Zimmermann. 1965. On the higher taxa of Embryophyta. *Taxon* 15:129–134.

Dahlgren, R. 1975. A system of classification of the angiosperms to be used to demonstrate the distribution of characters. *Botaniska Notiser* 128:119–147.

———. 1980. A revised system of classification of the angiosperms. *Botanical Journal of the Linnaean Society* 80:91–124.

———. 1983. General aspects of angiosperm evolution and macrosystematics. *Nordic Journal of Botany* 3:119–149.

———, and H. T Clifford. 1982. *The Monocotyledons. A Comparative Study.* Academic Press, New York.

———, ———, and P. F. Yeo. 1985. *The Families of the Monocotyledons.* Springer Verlag, New York.

Flora of North America Editorial Committee. 1993. Flora of North America, North of Mexico. New York: Oxford University Press.

Graham, L. E. 1993. The origin of land plants. Wiley Press: New York.

Heywood, V. H. (ed.). 1993. *Flowering Plants of the World,* updated edition. Oxford University Press, New York.

Hutchinson, J. 1967. *The Genera of Flowering Plants.* Clarendon Press, Oxford. 2 vol.

———. 1973. *The Families of Flowering Plants,* 3rd ed. Clarendon Press, Oxford.

Jones, S. B., and A. E. Luchsinger. 1979. *Plant Systematics.* McGraw-Hill, New York.

Judd, W. S., C. S. Campbell, E. A. Kellogg, P. F. Stevens, M. J. Donoghue. 2002. Plant systematics: a phylogenetic approach. Sinauer Associates Inc.: Sunderland, MA.

Keil, D. J. 1986. *Study Guide to Selected Families of Plants.* Meckler Publ. Co., Westport, Connecticut.

Kenrick, P. and P. R. Crane. 1997. The origin and early evolution of plants on land. *Nature* 389: 33–39.

Lawrence, G. H. M. 1951. *Taxonomy of Vascular Plants.* Macmillan Co., New York.

Palmer J. D., Soltis D. E., and Chase, D. W. 2004. The plant tree of life: an overview and some points of view. *American Journal of Botany* 91(10): 1437–1445.

Pool, R. J. 1941. *Flowers and Flowering Plants,* 2nd ed. McGraw- Hill, New York.

Porter. C. L. 1967. *Taxonomy of the Flowering Plants,* 2nd ed. W. H. Freeman and Co., San Francisco.

Pryer, K. M., E. Schuettpelz, P. G. Wolf, H. Schneider, A. R. Smith, and R. Cranfill. 2004. Phylogeny and evolution of ferns (Moniolophytes) with a focus on the early leptosporangiate divergences. *American Journal of Botany* 91(10): 1582–1598.

Radford, A. E. 1986. *Fundamentals of Plant Systematics.* Harper and Row, New York.

Silba, J. 1986. Encyclopedia Coniferae. *Phytologia Memoirs* 8:1–217.

Shaw, J. and K. Renzaglia. 2004. Phylogeny and diversification of bryophytes. *American Journal of Botany* 91(10): 157–1581.

Smith, J. P. 1974. *Vascular Plant Families.* Mad River Press, Eureka, Ca.

Sporne, K. R. 1956. The phylogenetic classification of angiosperms. *Biological Review* 31:1–29.

———. 1965. *The Morphology of the Gymnosperms: the Structure and Evolution of Primitive Seed Plants.* Hutchinson & Co., London.

Stebbins, G. L. 1974. *Flowering Plants. Evolution Above the Species Level.* Belknap Press, Cambridge, Mass.

Stevens, P. F. (2001 onwards). Angiosperm Phylogeny Website, Version 5, May 2004 [and more or less continuously updated since]. http://www.mobot.org/MOBOT/research/APweb/.

Takhtajan, A. L. 1969. *Flowering Plants. Origin and Dispersal.* Oliver and Boyd, Edinburgh.

———. 1980. Outline of the Classification of Flowering Plants (Magnoliophyta). *Botanical Review* 46:225–359.

Thorne, R. F. 1976. A phylogenetic classification of the Angiospermae. *Evolutionary Biology* 9:35–106.

———. 1981. Phytochemistry and angiosperm phylogeny, a summary statement, pp. 233–295 in D. A. Young and D. S. Seigler (eds.). *Phytochemistry and Angiosperm Phylogeny.* Praeger, New York.

———. 1983. Proposed new realignments in the angiosperms. *Nordic Journal of Botany* 3:85–117.

Willis, J. C. 1973. *A Dictionary of Flowering Plants and Ferns,* 8th ed. Revised by H. K. Airy Shaw. University Press, Cambridge.

# Exercises

1. With the help of your instructor put together a family by character chart to fill in as the quarter progresses. Your instructor can provide you with a list of the families you are expected to learn. Discuss with your instructor and classmates which characters and character states you should include. You probably will not want to include all possible details. Discuss what you should do if a feature is variable within a family.

2. When you first encounter a family during the school term, fill in its features on the chart. Review the chart often to help you remember the families and make comparisons. Don't wait until the night before the exam to start filling out the chart.

3. You may wish to assemble a set of flash cards. Discuss with your instructor and classmates the kinds of information you should include or exclude.

# Lycophytes and Seed Free Plants

## chapter 9

Early land plants possessed several evolutionary innovations, such as a cuticle that reduces water loss, pores or stomates that are critical in gas exchange, and a symbiotic relationship with fungi. These characteristics, along with an alternation of multicellular **gametophytic** and **sporophytic** life forms in a **sporic life cycle,** are traits that are shared by all land plants. The living members of these early land lineages are liverworts, hornworts, and mosses, and these groups have a large gametophytic life phase and a more reduced sporophytic life phase. Based upon evidence from fossil groups such as *Aglaophyton* and *Rhynia,* we now understand that the early vascular plants had very similar gametophytes and sporophytes. Therefore, the lineage that gave rise to living vascular plants or **tracheophytes** about 420 mya (million years ago), made a transition from isomorphic alternation of generations to a reduction of the gametophytic stage and an elaboration of the sporophytic stage.

Based upon molecular and morphological data, we now recognize three main lineages of vascular plants, as shown in Figure 9-1. One, called Lycophytes, was represented by large trees during the Devonian Period (417–354 mya) and Carboniferous Period (354–290 mya), but today are represented by relatively small plants in three families that possess **microphylls,** or leaves with a single vascular strand (Lycopodiaceae, Isoetaceae, and Selaginellaceae). The two other main lineages, the "seed free plants" or Monilophytes and the "seed plants" or Lignophytes, are thought to have a common ancestry since they all possess highly vascularized leaves **(megaphylls),** differentiation of the main axis and side branches, multiflagellate sperm, and a large (30 kilobase) inversion in the chloroplast.

Chapter 9 will explore diversity in the Lycophyta and the Monilophyta. The Lycophytes are represented by 1,200 species placed within three families, the Lycopodiaceae divided into 15 genera that are all homosporous, and two heterosporous families, the Selaginellaceae and Isoetaceae. The Monilophytes have been traditionally divided into the Psilophyta, Equisetophyta, and the Pteridophyta with two eusporangiate groups (Marattiales and Ophioglossaceae) and one leptosporangiate group, the Polypodiales. The Pteridophyta (often considered the "true ferns") are similar in having large leaves that unfold in a spiral or "fiddlehead" fashion, while the Psilophyta (whisk ferns) have very reduced leaves called enations, and the Equisetophyta (scouring rushes) have hollow stems that are jointed, with reduced megaphylls whorled at the nodes.

## Life Cycles

The Lycophytes and the Monilophytes, together often called the "ferns and fern allies," undergo an alternation of free-living diploid and haploid phases in their life cycle. These two phases have markedly different morphologies.

### The Sporophyte Phase

The large green fern plant that one encounters is a **sporophyte** (spore-producing plant). Sporophytes are diploid (having 2 sets of homologous chromosomes per cell) and have well-developed leaves, roots, and often an underground stem or rhizome.

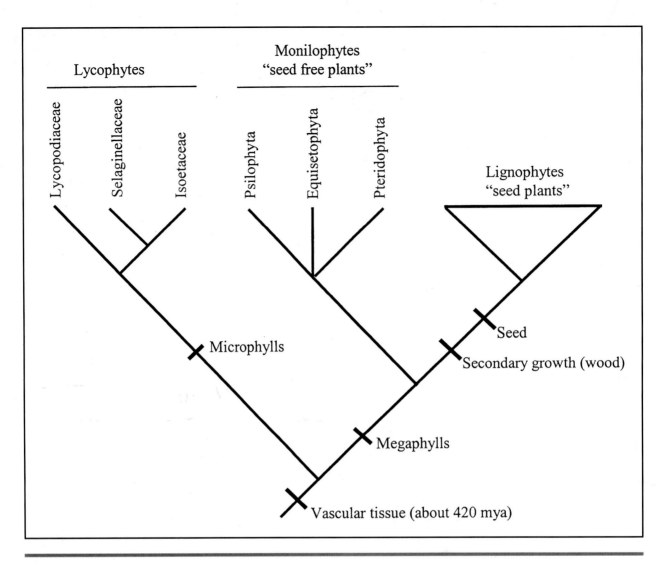

**Figure 9-1** Evolutionary history of "Ferns and Fern Allies." Character state changes emphasize the shifts in vascularization of photosynthetic structures. The figure is adapted from Judd et al. (2002) and Pryer et al. (2004).

Spores are produced as a result of meiosis in specialized multicelluar sacs called **sporangia.** A cluster of sporangia on the leaf surface is called a sorus (sori) **sorus** (pl., **sori;** Fig. 9-2 B, C). In ferns these structures are usually borne on the underside of leaves. In many ferns each sorus is covered by a flap of tissue called an **indusium.** In the fern allies the sporangia are borne in various other locations.

Each sporangium consists of an inner zone of sporogenous tissue and an outer layer of **jacket cells.** In many ferns each sporangium is borne on a stipe (stalk) consisting of one to several rows of elongated cells. The cells of the sporogenous tissue undergo meiosis, and the resulting haploid cells are the spores. Each sporangium generally produces numerous dust-like spores.

Most ferns and some of the fern allies are **homosporous;** all the spores are uniform in size and appearance. The life cycle of a *Nephrolepis* (Dryopteridaceae) illustrated in Figure 9-2, will serve as an example of the life cycle of a homosporous fern. A few ferns and some of the fern allies are **heterosporous,** producing two kinds of spores in two kinds of sporangia. The larger are **megaspores** and the smaller are **microspores.**

In many fern sporangia, in addition to the thin-walled jacket cells, there is also a zone or band of thick-walled cells called the **annulus.** In many ferns the annulus nearly encircles the sporangium, being interrupted only by the stalk. The cells of the annulus are hygroscopic, changing their dimensions in response to changes in the moisture content of their

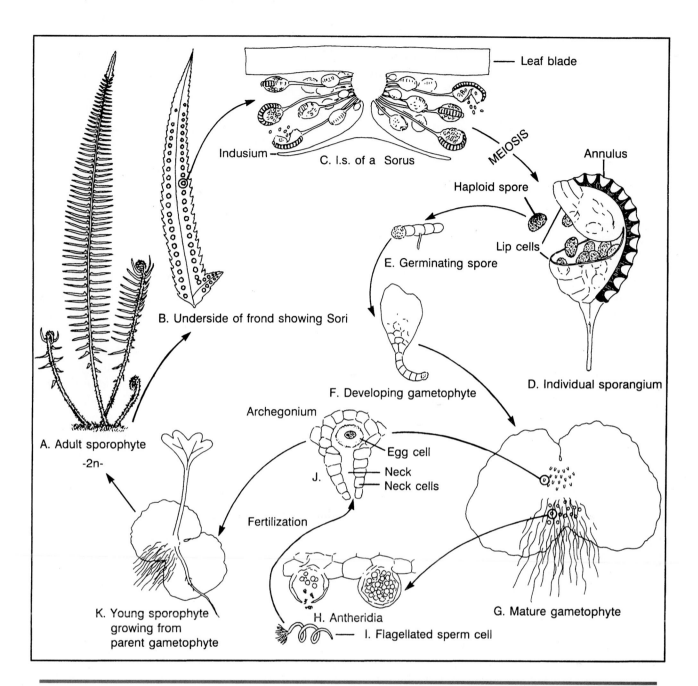

Leaf blade

Indusium

C. l.s. of a Sorus

MEIOSIS

Annulus

Haploid spore

Lip cells

E. Germinating spore

B. Underside of frond showing Sori

D. Individual sporangium

F. Developing gametophyte

Archegonium

Egg cell

J.

Neck

Neck cells

A. Adult sporophyte

-2n-

Fertilization

K. Young sporophyte
growing from
parent gametophyte

H. Antheridia

I. Flagellated sperm cell

G. Mature gametophyte

**Figure
9-2** Life cycle of a typical fern. A. Diploid sporophyte. B. Underside of a single leaflet with circular sori. C. Individual sorus (l.s.) with stalked sporangia and a peltate indusium (sterile covering). D. Single dehiscing sporangium showing the thick-walled annulus and thin-walled lip cells and the escaping haploid spores. E–F. Germination and early growth of the haploid gametophyte (prothallus). G. Undersurface of a mature gametophyte with rhizoids, archegonia (near notch) and antheridia (among rhizoids). H. Enlarged antheridia (l.s.) containing numerous flagellated sperm cells. I. Individual sperm cell. J. Enlarged archegonium with basal egg cell and elongated neck. A sperm cell swims through the neck of the archegonium and fertilizes the egg cell forming the zygote. K. Young sporophyte attached to gametophyte. Note developing root and leaf of the young sporophyte.

cell walls. When annulus cells dry, they contract and the sporangium splits open along a zone of weakness bordered by **lip cells,** releasing the spores. You may observe this happening as you observe the sporangia with the aid of a dissecting microscope. The heat of the lamp can dry out the sporangia, causing them to split open, flinging out spores.

## The Gametophyte Phase

Spores germinate if they happen to land in a suitably moist environment. They grow by mitotic divisions into small multicellular plantlets called **gametophytes** (gamete-producing plants). Each cell of a gametophyte is haploid, containing only one set of chromosomes. The gametophyte (also called the **prothallus**) of most ferns is a tiny, fragile, heart-shaped, free-living, photosynthetic plantlet, without differentiation into leaves, stem, or roots. It has single-celled, water absorbing structures **(rhizoids)** on the lower surface. [In some ferns the gametophyte is subterranean and non-photosynthetic and depends on a mycorrhizal association with a fungus for its nutrition.]

When mature, a prothallus can produce (via mitosis) one or two kinds of sex organs. The **antherdia** are spherical and produce sperm cells. The **archegonia** are vase-shaped and each produces a single egg within its swollen base. In homosporous ferns, the archegonia and antheridia are often produced on the same prothallus. In heterosporous ferns the prothalli formed from megaspores **(megagametophytes)** produce archegonia and the prothalli from microspores **(microgametophytes)** form antheridia.

In the presence of water (usually rain or dew) the antheridia release swimming sperm that are chemically attracted to the archegonia. The sperm cells of ferns have numerous cilia that propel the sperm through the water. Fertilization takes place within an archegonium. The resulting **zygote** is diploid and represents the first cell of the next sporophyte generation. By mitotic divisions the zygote forms an embryonic sporophyte that eventually grow to maturity as the familiar fern plant.

# Taxonomy

The identification of ferns and fern allies depends almost exclusively on the nature of the vegetative sporophyte and the form, location, and associated structures of the sporangia and spores. Although certain gametophytes are very distinctive, gametophytes are not generally used for identification because of their seasonality and the difficulty of locating and studying them. Their structure is important in fern classification, however.

Several features of the stems and leaves are commonly used in fern taxonomy. Most terrestrial ferns have a rhizome which may be short or elongated. The rhizome commonly bears hairs or scales. Fern leaves are commonly known as **fronds.** The petiole, often called a **stipe,** may be smooth or covered with scales or hairs. The blades of fern leaves vary from simple and entire to deeply lobed or 1–4 or more times pinnately compound. A **pinna** is a primary division of the frond. A **pinnule** is one of the ultimate (smallest) divisions of a 2 or more times compound frond. Pinnules may be lobed or incompletely divided into **segments.**

Most of the fern families have **circinate vernation.** The young leaves are coiled into a tight spiral, and as they develop, they unroll. Circinate vernation is the process by which the fronds uncoil. The immature leaf is called a **fiddlehead** (crosier).

The sporangia of most ferns are borne in sori on the undersurface of the leaf. A sorus is an aggregation of sporangia. The shape and location of the sori are important characters in fern taxonomy. Sori may appear as round dots or elongated streaks on the undersurface of the leaf or they may be borne at the ends of veins along the margin. Marginal sori may form a continuous band or may be small, discrete aggregations of sporangia. In a few ferns the sporangia are distributed across the undersurface of the leaf and do not form discrete sori.

In many ferns the sori are partially or wholly covered by a flap of leaf tissue called an indusium. The shape and position of attachment of the indusia are important features. The indusium may be attached to one side of the sorus (lateral indusium), beneath the sorus (inferior indusium), or borne above the sorus (superior indusium). In ferns with marginal sori the indusium may be formed by the reflexed or revolute margin of the leaf segments (false indusium). Some ferns (such as bracken fern) have a revolute margin as well as a true marginal indusium. Ferns without indusia have naked sori. In the majority of ferns there is little difference in structure between **fertile fronds** (those bearing sori) and **sterile fronds** (those without sori). In a few genera, however, the fronds are strongly dimorphic. Some ferns (e.g., the Ophioglossaceae) have leaves with dimorphic segments, some of which bear sporangia and some that are sterile.

The sporangia of most ferns (**leptosporangiate ferns**) are very small structures, generally with slender stalks. These sporangia typically contain 16–64 spores apiece. In a few fern families (**eusporangiate ferns**) the sporangia are considerably larger and contain hundreds to thousands of spores.

The ferns and fern allies have traditionally been viewed as a related group, with Tryon and Tryon (1982) grouping them all together in a single division, the Pteridophyta. Others, such as Cronquist, Takhtajan, and Zimmermann (1965) divided these plants into four divisions, the Lycophyta, Psilophyta, Equisetophyta and Pteridophyta. The recent work of Kenrick and Crane (1997), Doyle (1998), Pryer et al. (2001) have recognized the Lycophyta as a separate lineage and also recognized the Psilophyta, Equisetophyta and Pteridophyta as a monophyletic group. In fact, in spite of the differences among the Psilophytes, Equisetophytes, and Pteridophytes, it appears that the reduction in leaves and roots are derived conditions in the whisk ferns and scouring rushes that had previously obscured their close relationship to the "true ferns." In this book we will treat the Lycophytes as a separate group, and the Monilophytes as a group of three phyla (Psilophyta, Equisetophyta, and Pteridophyta).

There have been several attempts to establish a stable, universally accepted system of classification for the fern families. In many older treatments most ferns were placed into a large and polymorphic family, the Polypodiaceae. Most recent workers have divided the Polypodiaceae into smaller, more coherent families. A major problem has been the lack of consensus among experts on the limits of the segregate families. See Jermy et al. (1973), Nagar (1970) and Tryon and Tryon (1982) for discussions of some of the proposed systems. Unfortunately, there are now several competing classifications. As a student, you should be aware that family limits presented in the flora that you are using may differ from those used here. Here we accept the family limits used by Wagner and Smith (1993) and for *Flora North America*.

# Synoptic Key to the Major Taxa of Living Pteridophytes

1. Stem jointed and hollow; sporangia borne in cones on the undersurface of peltate sporangiophores. Division **EQUISETOPHYTA**

1′ Stem smooth and solid; sporangia associated with leaves or lateral cauline scales.

   2. Leaves absent or veins of leaves 1 and unbranched; leaves scale-like or linear.

      3. Sporangia fused into groups of 3 (synangia) in axil of veinless scales; true leaves absent. Division **PSILOTOPHYTA**

      3′ Sporangia single, born in the axil of a leaf with a single vein. Division **LYCOPHYTA**

         4. Plants homosporous, ligule absent. . . . . . . . . . . . . . . . . . . . . . . . . . . . .Class **LYCOPSIDA**

         4′ Plants heterosporous, ligule present. . . . . . . . . . . . . . . . . . . . . . . . .Class **ISOETOPSIDA**

  2′ Leaves present; veins of leaves many and usually intricately branched, sporangia borne away from the stem on leaf blades or modified leaves. . . . . . . . . . . . . . . . . . . .Division **POLYPODIOPHYTA**

     5. Annulus absent, sporangia often fused into synangia.

        6. Synangia many on the abaxial surface of a leaf; stem stout to massive. . . . . . . . . . . . . . . . . . . . . . . . . . . . . . . . . . . . . . . . . . . . . . .Subclass **MARATTIIDAE**

        6′ Synangia two at the apex of a spike or absent; stem small, subterranean. . . . . . . . . . . . . . . . . . . . . . . . . . . . . . . . . . . . . . . . . . . . . . .Subclass **OPHIOGLOSSIIDAE**

     5′ Annulus present, sporangia not fused into synangia.

7. Plants mostly terrestrial, homosporous; sporangia not enclosed in hollow sporocarps.
. . . . . . . . . . . . . . . . . . . . . . . . . . . . . . . . . . . . . . . . . . . . .Subclass **POLYPODIIDAE**

7′ Plants aquatic, heterosporous; sporangia enclosed in ± leathery sporocarps that extrude sporangia in a gelatinous matrix.

8. Plants floating or stranded on mud; leaves simple and entire or 2-lobed; sporocarps attached to lobes of underwater leaves. . . . . . . . . . . . . . . . . . . .Subclass **SALVINIIDAE**

8′ Plants rooted in mud; leaves linear and entire or with 2–4 leaflets; sporocarps borne on leaf-bases at or below surface of mud. . . . . . . . . . . . . . . . . .Subclass **MARSILEIDAE**

# FAMILIES

## Phylum Lycophyta

### Class Lycopodiopsida
### Order Lycopodiales

**Lycopodiaceae**—The Club-moss Family (Fig. 9-4). 10–15 genera, 350–400 species. Cosmopolitan

## Phylum Equisetophyta

**Equisetaceae**—The Horsetail or Scouring-rush Family (Fig. 9-12). 1 genus, 15 species. Cosmopolitan.

§See page 151 for description and illustrations.§

## Phylum Psilotophyta

**Psilotaceae**—The Whisk Fern Family (Fig. 9-3). 2 genera, 4–8 species. Tropical region.

Perennial herbs. Roots absent. Stems subterranean and aerial, the underground stems rhizomatous, dichotomously branched, the aerial stems erect or pendant, dichotomously branched, rounded or flattened, bearing alternate, scalelike, veinless leaves or flattened, 1-veined leaves (these perhaps not homologous with leaves in other divisions); Plants homosporous. Sporangia large, bilobed or trilobed, borne laterally along stem, subtended by a pair of veinless scales (enations); spores uniform in size, small. Gametophytes bisexual, subterranean or fleshy branched, rhizomelike, mycorrhizal, non-photosynthetic.

Herbs; stems densely leafy, usually dichotomously branched; stolons mostly dichotomous, arising directly from stems. Leaves alternate, opposite or whorled (rarely all basal), simple, 1-veined, linear or scale-like, without ligules. Plants homosporous, sporangia borne in axils or base of fertile leaves

**Figure 9-3** Examples of Psilotaceae. A. Branch of *Tmesipteris* sp. with flattened, leaflike structures and 2-lobed sporangia. B. Portion of dichotomously branched stem of *Psilotum nudum* with 3-lobed sporangia.

[sporophylls] that are sometimes aggregated into cone-like strobili; spores uniform in size, trilete, and small. Gametophytes bisexual, subterranean and mycotrophic or above ground and photosynthetic.

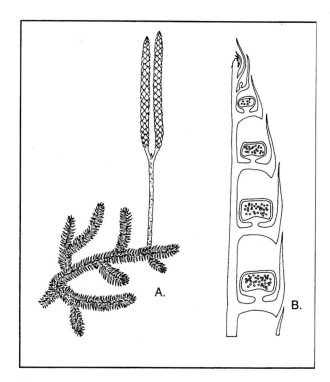

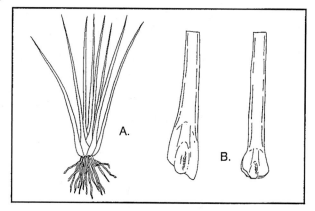

**Figure** Examples of Isoetaceae. A–B. *Isoetes* sp.
**9-5** A. Habit of plant with cluster of basal leaves attached to corm-like stem. B. Closeup of leaf-bases, each with one sporangium covered by membranous velum.

**Figure** Examples of Lycopodiaceae.
**9-4** A–B. *Lycopodium clavatum.* A. Leafy branch and pair of strobili. B. Longitudinal section of strobilus (only one side shown).

### Class Isoetopsida
#### Order Selaginellales
**Selaginellaceae**—The spike-moss Family (Fig. 9-11). 1 genus, 700 species. Cosmopolitan; most diverse in the tropics and subtropics.

#### Order Isoetales
**Isoetaceae**—The Quillwort Family (Fig. 9-5). 1 genus, 150 species. Widely distributed, especially in temperate regions.

Herbs, aquatic or semiaquatic with corm-like stem. Leaves tapering to apex, simple, linear, terete with internal air chambers, at base expanded with a membranous ligule and a large chambered sporangium. Plants heterosporous; microsporophylls usually alternate cycle to megasporophylls. Microspores very numerous. Megaspores 50–300, variously sculptured. Gametophytes unisexual, retained within spore walls, non-photosynthetic.

# Phyllum Polypodiophyta
## Class Polypodiopsida
### Subclass Ophioglossidae
#### Order Ophioglossales
**Ophioglossaceae**—The Adder's Tongue or Grape-fern Family (Fig. 9-6). 3–5 genera, 70–80 species. Widespread in tropics and temperate areas.

Herbs with short, erect, unbranched stems. Leaves solitary or alternate, with a simple to much dissected or compound sterile segment and a long-stalked spikelike or paniculiform fertile segment

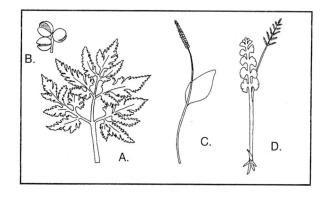

**Figure** Examples of Ophioglossaceae.
**9-6** A–B. *Botrychium multifidum.* A. Sterile frond. B. Closeup of sporangia. C. Fertile frond of *Ophioglossum vulgatum.* D. Plant of *Botrychium lunaria* with fascicled roots and fertile frond.

bearing numerous large sporangia; vernation not circinate. Plants homosporous. Sporangia without annulus; spores numerous, small. Gametophyles bisexual, fleshy, subterranean, mycotrophic. Lacks root hairs.

### Subclass Marattiidae
### Order Marattiales

**Marattiaceae.** 4–7 genera, 145 species. Pantropical.

### Subclass Polypodiidae
### Order Polypodiales

**Osmundaceae**—The Royal Fern or Cinnamon Fern Family. 3 genera, 16–36 species. Temperate North America and Eurasia.

Herbs with upright aerial or horizontal subterranean, dichotomously branched stems; scales absent, black fibrous roots. Leaves alternate, large, one or more times pinnate, either dimorphic or some leaves differentiated into fertile and sterile segments, glabrous or pubescent; scales absent; vernation circinate. Plants homosporous. Sporangia large, stalked, with annulus poorly differentiated, borne on the abaxial surface of leaf blade either in loose clusters or solitary, sori absent; spores uniform in size, small. Gametophytes bisexual, above ground, thin, green, photosynthetic.

**Gleicheniaceae**—The Forking Fern Family. 4 genera, 140 species. Tropics to subtropics.

Coarse herbs with long-creeping, scaly (hairy) rhizomes. Leaves large, many-times compound, repeatedly forking, indeterminate in growth, often scrambling over vegetation, pubescent with simple or stellate hairs. Sporangia sessile or subsessile, small, with complete annulus, borne in raised sori on undersurface of leaf segments; spores uniform in size. Gametophytes bisexual, above ground, thin, green, photosynthetic.

**Schizaeaceae**—The Curly-grass Family. 2–3 genera, 30 species. Widely distributed, especially in the tropics.

Herbs with short, erect or horizontal, stiffly pubescent stems. Leaves all basal, all fertile and once pinnately compound with long, slender petioles and much reduced tufts of revolute-margined pinnae, or sometimes dimorphic with pinnate fertile leaves and sterile bladeless leaves. Plants homosporous. Sporangia sessile, large, with annulus subapical, grouped on undersurface of pinnae; spores

uniform in size, small. Gametophytes bisexual, above ground, thin, green, photosynthetic or subterranean, thick, mycorrhizal, nonphotosynthetic.

**Lygodiaceae**—The Climbing Fern Family. 1 genus, 40 species. Widely distributed, principally in tropics.

Herbs with branched, subterranean, brownish pubescent stems. Leaves all basal, 2 or more times compound, indeterminate in growth with rachises twining, fertile leaves with ± dimorphic pinnae, the fertile pinnae generally toward apices of leaves, bearing sporangia toward tips of narrowed lobes. Plants homosporous. Sporangia sessile, large, with annulus apical, each covered by a hoodlike indusium; spores uniform in size, small. Gametophytes bisexual, above ground, thin, green, photosynthetic.

**Anemiaceae**—The Anemia family. 2 genera, 119 species. Widely distributed in tropical and subtropical areas.

Herbs with creeping, horizontal, pubescent rhizomes. Leaves all basal, 1–3 times pinnately compound, partially dimorphic with a pair of fertile pinnae with reduced blades arising below the sterile portion of leaf, or fully dimorphic, the fertile leaves with much reduced blades, the sterile leaves with well-developed blades. Plants homosporous. Sporangia sessile, large, with annulus apical, borne in 2 rows on ultimate leaf segments if fertile pinnae; spores uniform in size, small. Gametophytes bisexual, above ground, thin, green, photosynthetic.

**Parkeriaceae**—The Water Fern Family. 1 genus, 3–4 species, widespread in tropics and warm subtropics.

Herbs, aquatic and free-floating or subaquatic and rooted on mud, the stems short, erect, sparingly scaly. Leaves all basal, 1–4 times pinnately compound, dimorphic, the sterile leaves with broad, flat segments, commonly bearing plantlets along the margins that break free from parent plants, the fertile leaves with narrow, revolute-margined segments. Plants homosporous. Sporangia subsessile, densely crowded on undersurface of fertile segments, covered by revolute margins of segments; spores uniform in size, small. Gametophytes bisexual, terrestrial, aboveground, thin, green, photosynthetic.

**Stromatopteridaceae.** 1 genus, 1 species. New Caledonia.

**Matoniaceae**—2 genera, 4 species. Tropical Malaysia.

**Dipteridaceae**—1 genus, 8 species. Mountains of tropical Polynesia.

**Cheiropleuriaceae**—1 genus, 1 species. East Asia.

**Loxomataceae**—2 genera, 3 species. New Zealand and Central America.

**Hymnophyllopsidaceae**—6 genera, 650 species. Tropical South America.

**Plagiogyriaceae**—1 genus, 15 species. Mexico to northern South America.

**Dicksoniaceae**—The Dicksonia Family. 5–6 genera, 155 species. Tropics to warm temperate.

Shrubs or trees (rarely herbs) with persistent leaf bases. Leaves alternate, 2–4 times pinnate, very large with stout, densely woolly petioles; vernation circinate. Plants homosporous. Sporangia small, stalked, with an oblique continuous annulus, clustered in circular sori borne at tips of veinlets on leaf margins; indusium present, cup-shaped, sometimes with a lid-like cover; spores uniform in size, small. Gametophytes bisexual, aboveground, thin, green, photosynthetic.

**Lophosoriaceae**—1 genus, 1 species, Mexico to throughout South America.

**Metaxyaceae**—1 genus, 1 species. Southern Central America to northern South America.

**Cyatheaceae**—The Tree Fern Family. 1–6 genera, 480 species. Tropics and subtropics.

Shrubs or trees with erect trunks bearing persistent leaf circinate. Plants homosporous. Sporangia small, stalked, with an oblique continuous annulus, clustered in circular sori borne along veins on undersurface of leaves; indusium present, cuplike, enclosing the sorus, or sometimes absent; spores uniform in size, small. Gametophytes bisexual, aboveground, thin, green, photosynthetic.

**Pteridaceae**—The Maidenhair Fern Family (Figure 9-13). 40 genera, 1000 species. Cosmopolitan.

§See page 153 for description and illustrations.§

**Vittariaceae**—The Shoestring Fern Family. 10 genera, 100 species. Worldwide in tropics and subtropics.

Herbs with short-creeping, densely scaly rhizomes. Leaves simple, entire, linear, not differenti- ated into blade and petiole; vernation circinate. Plants homosporous; sporangia small, stalked, with an erect partial annulus, borne in linear sori in sub- marginal grooves parallel to midvein; indusium absent; spores uniform in size, small. Gametophytes bisexual, aboveground, ribbonlike, green, photosyn- thetic, sometimes forming long-lived, vegetatively reproducing clones.

**Hymenophyllaceae**—The Filmy Fern Family. 6 genera, 650 species. Tropics to warm temperate.

Herbs with erect to horizontal, sometimes much- branched, pubescent rhizomes. Leaves varying from minute scales to large and pinnate, the blades gener- ally a single cell thick between the veins; vernation circinate. Plants homosporous; sporangia small, stalked, with a complete, ringlike annulus, in mar- ginal sori, borne on a slender receptacle surrounded by or covered by a bivalved or tubular indusium; spores uniform in size, small. Gametophytes bisex- ual, aboveground, thin, filamentous or ribbonlike, much-branched, photosynthetic.

**Dennstaedtiaceae**—The Bracken Fern Family (Fig- ure 9-14). 17–20 genera, 400 species. Cosmopoli- tan, but especially abundant in the tropics.

§See page 155 for description and illustrations.§

**Thelypteridaceae**—The Marsh Fern Family. 1–30 genera, 900 species, depending on definition. Mostly tropical. In many treatments, genera here included in the Dryopteridaceae are placed in this family.

Herbs with erect or horizontal rhizomes, these scaly at tip. Leaves simple and 2–3 times pinnately divided or 1–2 times pinnately compound or simple and 2–3 times pinnately divided, monomorphic or more or less dimorphic, pubescent (scaly); verna- tion circinate. Plants homosporous. Sporangia stalked, small, with an erect partial annulus, borne on the veins away from margins on undersurface of leaf segments, variously organized into round or oblong (elongate) sori; indusium reniform and attached to one side of sorus or absent; spores uni- form in size, small. Gametophytes bisexual, above- ground, thin, green, photosynthetic.

**Blechnaceae**—The Chain Fern Family (Figure 9-7). 9–10 genera, 250 species. Mostly tropical and southern temperate; notable exception *Woodwardia*.

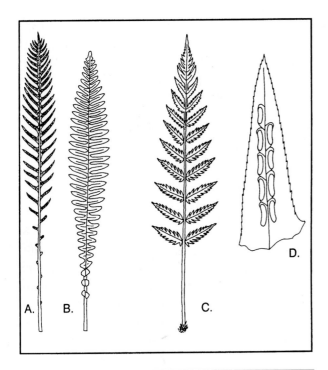

**Figure 9-7** Examples of Blechnaceae. A–B. *Blechnum spicant.* A. Fertile frond. B. Sterile frond. C–D. *Woodwardia fimbriata.* C. Frond. D. Leaf segment with sori in chainlike rows parallel to midvein.

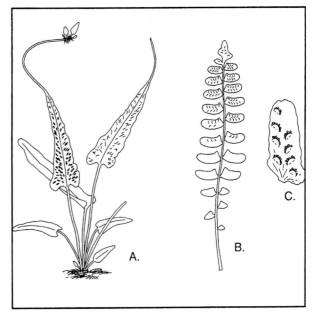

**Figure 9-8** Examples of Aspleniaceae. A. *Asplenium rhizophyllum* plant with plantlet at tip of entire leaf. B–C. *Asplenium platyneuron.* A. Pinnately compound frond. B. Closeup of pinnule with sori and laterally attached indusia.

Herbs with erect, or creeping scaly rhizomes. Leaves simple and pinnately lobed to once pinnately compound with deeply pinnately lobed pinnae, monomorphic or dimorphic, glabrous to pubescent or scaly; vernation circinate. Plants homosporous. Sporangia stalked, small, with an erect partial annulus, organized into linear or oblong sori parallel to midveins on undersurface of leaf segments; indusium linear or oblong, attached along 1 side; spores uniform in size, small. Gametophytes bisexual, aboveground, thin, green, photosynthetic.

**Aspleniaceae**—The Spleenwort Family. (Figure 9-8) 1–9 genera, 700 species. Cosmopolitan.

Herbs with erect, or more commonly, creeping, scaly rhizomes. Leaves simple and entire to variously 2–4 times pinnately compound, monomorphic or dimorphic, glabrous or glandular-puberulent, often sparingly scaly; vernation circinate. Plants homosporous. Sporangia stalked, small, with an erect partial annulus, organized into linear or crescent-shaped sori parallel to veins on undersurface of leaf blade; indusium present or absent; spores uniform in size, small. Gametophytes bisexual, aboveground, thin, green, photosynthetic.

**Davalliaceae**—9–10 genera, 200 species, Mostly tropical Asia.

**Dryopteridaceae**—The Wood Fern Family (Figure 9-15). 60 genera, 3000 species. Cosmopolitan.

§See page 157 for description and illustrations.§

**Grammitidaceae.** 10–14 genera, 500 species. Primarily American tropics and subtropics.

**Polypodiaceae**—The Polypody Family (Figure 9-16). 40–47 genera, 700 species.

**Subclass Marsileidae**
**Order Marsileales**
**Marsileaceae**—The Water-clover Family (Figure 9-9). 3 genera, 50 species. Widespread, tropics to temperate.

Herbs, usually aquatic or semiaquatic with creeping, hairy rhizomes. Leaves alternate, simple, bifoliolate or quadrifoliolate, sometimes bearing one or more swollen sporocarps (modified leaflets) near base; vernation circinate. Plants heterosporous. Microsporangia and megasporangia borne in same sori, embedded in gelatinous material within sporocarp. Microspores numerous; megaspore 1 per spo-

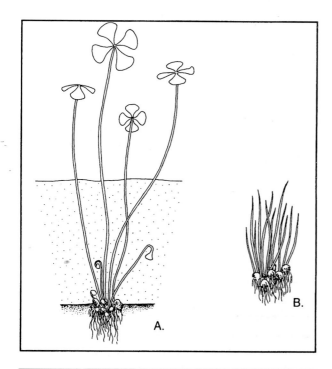

**Figure 9-9** Examples of Marsileaceae. A. Partially submersed plant of *Marsilea vestita* with emergent leaves and basal sorocarps. B. Plant of *Pilularia americana* with linear, entire leaves and spherical sorocarps.

**Figure 9-10** Example of Azollaceae. Floating plant of *Azolla filiculoides.*

lobe submersed, the other floating, the floating leaf lobes containing colonies of *Anabaena azollae,* a nitrogen-fixing, photosynthetic cyanobacterium. Plants heterosporous. Sporangia small, stalked, without an annulus, borne in sporocarps on submersed leaf segments. Microsporangia and megasporangia borne in different sporocarps, enclosed by spherical indusia; megasporangia 1 per megasporocarp; microsporangia many per microsporocarp. Megaspores 1 per sporangium; microspores 32 or 64 per sporangium. Gametophytes unisexual, protruding from spore walls; megagametophytes with *Anabaena* colonies.

rangium. Gametophytes unisexual, retained within spore walls, non-photosynthetic.

### Subclass Salviniidae
### Order Salviniales
**Salviniaceae**—The Floating Fern Family. 1 genus, 10 species. Mostly tropical.

Tiny aquatic floating herbs with short, branching, pubescent stems; roots absent. Leaves whorled with 2 leaves floating and entire, the third submersed and dissected, rootlike. Plants heterosporous. Sporangia small, stalked, without an annulus, borne in sporocarps on submersed leaves. Microsporangia and megasporangia borne in different sporocarps; megasporangia ca. 10 per megasporocarp; microsporangia many per microsporocarp. Spores of 2 different sizes; megaspores 1 per sporangium; microspores 64 per sporangium.

**Azollaceae**—The Mosquito Fern Family (Figure 9-10). 1 genus, 7 species. Tropics and temperate regions.

Tiny aquatic floating herbs with short, branching, pubescent stems that readily fragment; roots present, unbranched. Leaves alternate, bilobed with one

## References

Bierhorst, D. W. 1971. *Morphology of Vascular Plants.* MacMillan, New York, pp. 153–192, 243–326.

Bold, H. C., and J. W. La Claire. 1987. *The Plant Kingdom,* 5th ed. Prentice-Hall, Englewood Cliffs, N.J., pp. 98–173.

Bower, F. O. 1923–1928. *The Ferns (Filicales) Treated Comparatively with a View to their Natural Classification.* Cambridge, England. 3 vols.

Brummitt, R. K. 1992. *Vascular Plant Families and Genera.* Royal Botanic Gardens, Kew.

Cobb, B. 1956. *A Field Guide to the Ferns.* Houghton Mifflin Co., Boston.

Copeland, E. B. 1947. *Genera Filicum.* Chronica Botanica. Waltham, Mass.

Crabbe, J. A., A. C. Jermy, and J. M. Mickel. 1975. A New Arrangement for a Pteridophyte Herbarium. *Fern Gazette* 11:141–162.

Cronquist, A., A. Takhtajan, and W. Zimmermann. 1965. On the higher taxa of Embryophyta. *Taxon* 15:129–134.

Doyle, J. A. 1998. Phylogeny of the vascular plants. Annual Rev. Ecol. Syst. 29: 567–599.

Flora North America Editorial Committee. 1993. *Flora of North America, North of Mexico, Vol. 2: Pteridophytes and Gymnosperms.* Oxford University Press, Oxford, pp. 11–344.

Grillos, S. J. 1966. *Ferns and Fern Allies of California.* University of California Press, Berkeley.

Hauke, R. L. 1963. A taxonomic monograph of the genus *Equisetum,* subgenus *Hippochaete.* J. Cramer Publ. Co., Weinheim, Germany.

————. 1978. A taxonomic monograph of the genus *Equisetum,* subgenus *Equisetum. Nova Hedwigia* 30:385–455.

Holttum, R. E. 1949. The classification of ferns. *Botanical Review* 24:267–296.

Jermy, A. C., J. A. Crabbe, and B. A. Thomas (eds.). 1973. The phylogeny and classification of ferns. *Botanical Journal of the Linnaean Society* 67, Suppl. 1:1–283.

Kenrick, P. and P. R. Crane. 1997. The origin and early evolution of plants on land. Nature 389: 33–39.

Kubitzki, K. (ed.) 1990. *The Families and Genera of Vascular Plants.* Vol. 1. *Pteridophytes and Gymnosperms.* Edited by K. U. Kramer and P. S. Green, assisted by E. Götz. Springer Verlag, Berlin.

Lellinger, D. B. 1985. *A Field Manual of Ferns and Fern Allies of the United States and Canada.* Smithsonian Institution Press, Washington, D.C. pp. 466.

Nagar, B. K. 1970. A phylogenetic classification of the homosporous ferns. *Taxon* 19:229–236.

Pryer, K. M., H. Schneider, A. R. Smith, R. Cranfil, P. G. Wolf, J. S. Hunt, S. D. Sipes. 2001. Horsetails and ferns are a monophyletic group and the closest living relatives to seed plants. Nature 409: 618–622.

Scagel, R. F., R. J. Bandoni, G. E. Rouse, W. B. Schofield, J. R. Stein and T. C. Taylor. 1984. *An Evolutionary Survey of the Plant Kingdom.* Wadsworth Publ. Co., Belmont, California.

Sporne, K. R. 1979. *The Morphology of the Pteridophytes: the Structure of Ferns and Allied Plants.* Hutchinson & Co., London.

Thomas, B. A., and S. D. Brack-Hanes. 1984. A new approach to family groupings in the Lycophytes. *Taxon* 33:247–255.

Tryon, R. M., and A. F. Tryon. 1982. *Ferns and Allied Plants with Special Reference to Tropical America.* Springer-Verlag, New York.

# Selaginellaceae

## The Spike-Moss Family

Lycopodiophyta, Isoetopsida, Selaginellales ▪

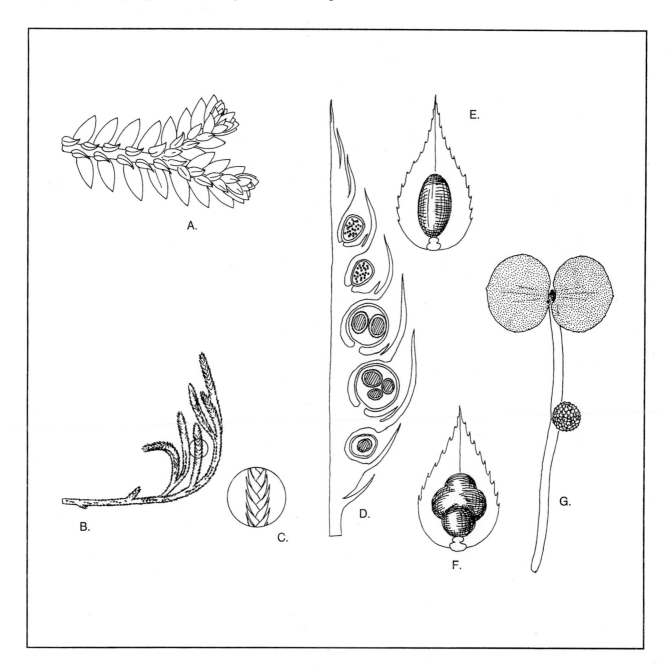

**Figure 9-11** Features of Selaginellaceae. A. Branch tip of *Selaginella kraussiana* illustrating leaves of different sizes. B-C. *Selaginella bigelovii*. B. Branch with all leaves the same size. C. Closeup of leaves. D. Diagrammatic l.s. of strobilus of *Selaginella*. E. Microsporophyll of *Selaginella*. F. Megasporophyll of *Selaginella*. G. Two-leaved sporeling (young sporophyte) of *Selaginella* still attached to female gametophyte (retained within megaspore wall).

# FAMILY DESCRIPTION

Herbs; stems densely leafy, sometimes dichotomously branched; roots borne on wiry rhizophores arising from forks in stem. Leaves alternate, opposite or whorled, simple, 1-veined; mostly scalelike, uniform in size and radially distributed (Fig. 9-11 B) or dimorphic with stem dorsiventral (Fig. 9-11 A); a scale-like ligule produced in leaf axil during early development but soon withering. Plants heterosporous. Sporangia borne in axils of fertile leaves (sporophylls) that are clustered, usually in 4 ranks, in a ± conelike strobilus (Fig. 9-11 C); microsporangia and megasporangia borne in the same or different strobili; microsporangia (Fig. 9-11 D) enclosing many tiny microspores; megasporangia enclosing 4 large megaspores (Fig. 9-11 E). Gametophytes unisexual, retained within spore walls, nonphotosynthetic.

# FAMILY NOTES

*Selaginella,* the only living genus in the Selaginellaceae, is nearly cosmopolitan in its distribution, although it is best represented in tropical areas. It comprises about 700 species. The present-day clubmosses (Selaginellaceae and Lycopodiaceae) are all herbaceous plants, but fossils indicate that the Lycopodiophyta was once much more diverse. The carbonized remains of long-extinct, tree-like relatives of the club-mosses are a major component of the coal fields of the eastern United States.

The gametophytes of *Selaginella* are very small, forming within the spore wall. Often they begin development before the spores are released from the sporangia. Later developmental stages and fertilization take place on the ground. Each microgametophyte forms a single antheridium that releases swimming sperm. A megagametophyte begins development by free-nuclear divisions; mitosis occurs without accompanying cell wall formation. Later, cell walls develop around the nuclei, and the spore wall is ruptured. Each megagametophyte produces a single egg within an archegonium. The embryo remains attached to the megagametophyte until it has developed its own roots and leaves (Figure 9-11 G).

*Selaginella lepidophylla,* a native of the Chihuahuan Desert of North America, is called resurrection plant. Under drought conditions it curls up and becomes brown and dry. When water is once more available it uncurls and becomes moist and green again. It is sometimes gathered and sold as a novelty at Easter. In some areas this harvesting has seriously depleted the plant's natural populations. Several other species of dryland habitats also are able to undergo extreme desiccation and revive when moisture is available. These often grow in rocky habitats where there is little soil to hold moisture.

Many of the tropical species of *Selaginella* are not at all drought-tolerant. Most species are epiphytes on the branches of trees and shrubs. Some are vines with long trailing stems and flattened, frondlike branches bearing dimorphic leaves.

Today a few species of *Selaginella* are grown as ornamentals or as novelties. Otherwise the family is of little economic importance.

# Equisetaceae

## The Scouring Rush or Horsetail Family
### Equisetophyta, Equisetopsida, Equisetales ■

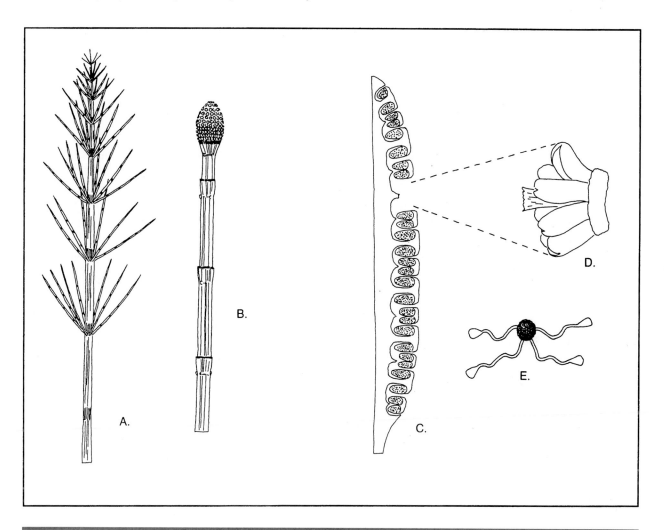

**Figure 9-12** Features of Equisetaceae. A. Branched vegetative stem of *Equisetum arvense*. B. Unbranched stem of *Equisetum laevigatum* with terminal strobilus (cone). C. Diagrammatic l.s. of strobilus of *Equisetum*. D. Individual peltate sporangiophore with sporangia. E. Spore with elaters.

## FAMILY DESCRIPTION

Herbs; stems hollow, jointed, longitudinally ridged, distinct nodes, unbranched or bearing whorled branches (irregularly branched when damaged), monomorphic and all green, or dimorphic with some stems green and branched and others pale and mostly non-photosynthetic and terminated by a strobilus (cone). Leaves whorled, connate, forming a tubular sheath, simple, scalelike, 1-veined, often non-green. Plants homosporous. Sporangia borne on undersurface of polygonal peltate sporangiophores that are aggregated on an axis forming a terminal strobilus; spores green, uniform in size, small, each bearing threadlike, hygroscopic elaters. Gameto-phytes unisexual, thin, dichotomously forking, aboveground, green, photosynthetic.

## FAMILY NOTES

The Equisetaceae is a monogeneric family with 35 living species. The single living genus, *Equisetum,* is found throughout the world in moist to wet soils. *Equisetum* is probably the most ancient living genus of vascular plants. It is the last survivor of a once widespread division. Huge tree-like relatives of the horse-tails dominated some of the coal producing floras of the Carboniferous period and their carbonized remains are burned today as coal.

*Equisetum* is a very distinctive genus that is easily recognized by the combination of jointed, ribbed, hollow stems and whorled, basally connate scale leaves. Plants in this genus often occur in shallow water. The hollow stems may be an adaptation to aquatic habitats. The hollowness is produced by a system of channels through the cortex plus a hollow central region. In some species the stems are unbranched (Figure 9-12 B), whereas in others there are whorled branches at each node (Figure 9-12 A).

The sporangia are borne in ovoid cone-like strobili that are terminal on the stems (Figure 9-12 B). A strobilus consists of a central axis to which are attached numerous short-stalked, peltate, polygonal sporangiophores (Figure 9-12 C.) Each sporangio-phore bears on its undersurface a ring of oblong sporangia (Figure 9-12 D), each containing numerous tiny green spores. The spores have 4 hygroscopic, threadlike elaters (Figure 9-12 E) that coil and uncoil in response to changes in moisture.

In some species stems that bear strobili are similar in appearance to those that do not. Other species have dimorphic stems. The stems that bear strobili in these species are non-pigmented, unbranched and ephemeral, withering soon after the spores have been released. Other stems of the same plants are green and photosynthetic, much-branched and lack strobili. They carry on photosynthesis throughout the growing season. The two types of stems are joined together by a system of underground rhizomes.

*Equisetum* has little economic importance. Some species are toxic to cattle and are weedy in poorly drained pastures. The epidermal cells of *Equisetum* contain silicon dioxide crystals, and some are rough to the touch. The giant scouring rush *E. hyemale,* has sometimes been used to scour pots and pans. Stems of this species are also sold as a tea in health food stores.

# Pteridaceae

## The Maidenhair Family
### Polypodiophyta, Polypodiopsida, Polypodiales ■

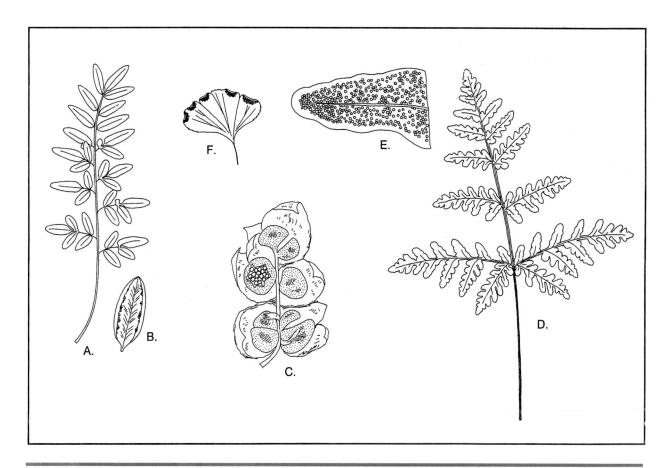

**Figure 9-13** Features of the Pteridaceae. A–B. *Pellaea glabella.* A. Frond. B. Closeup of pinnule with continuous recurved margin forming false indusium. C. Closeup of beadlike pinnules of *Cheilanthes* sp. with one pinnule turned over to show incurled margin and sporangia; the underside of the leaf is covered by membranous scales. D–E. *Pentagramma triangularis.* D. Frond. E. Closeup of undersurface of frond showing scattered sporangia not clustered into discrete sori. F. Pinule of *Adiantum* sp. with marginal lobes reflexed over sori as discontinuous false indusium.

## FAMILY DESCRIPTION

Herbs with erect, or more commonly, creeping, pubescent and/or scaly rhizomes. Leaves simple or 1–6 times compound, monomorphic or dimorphic, glabrous to pubescent and/or scaly; vernation circinate. Plants homosporous. Sporangia stalked, small, with an erect partial annulus, variously organized into sori or scattered on undersurface of leaf blade; indusium, present or absent, when present formed by reflexed margin of pinnules; spores uniform in size, small. Gametophytes bisexual, aboveground, thin, green, photosynthetic.

# FAMILY NOTES

The Pteridaceae is a cosmopolitan family of about 40 genera and 1000 species. Many members of the Pteridaceae occur in dry habitats and show a wide range of adaptations to drought. Some occur in deserts and other extremely arid environments. Many grow in rock crevices or beneath overhanging ledges where the moisture supply is moderately enhanced. Common genera in the U.S. include *Adiantum* (maidenhair fern), *Aspidotis* (lace fern), *Cheilanthes* (lip fern), *Cryptogramma* (parsley fern), *Notholaena* (cloak fern), *Pellaea* (cliffbrake), and *Pentagramma* (goldback fern).

Most Pteridaceae are small ferns only 1–3 dm tall although the giant swamp fern *(Achrostichum danaeifolium)* has fronds up to 5 meters long. The stipes and rachises of the fronds are often slender and wiry and commonly are dark brown to nearly black. In many Pteridaceae the leaves are 2 or more times compound. The leaf blades are frequently much reduced and in most genera have revolute margins. These conditions reach an extreme condition in some species of *Cheilanthes* in which the many pinnules are reduced to tiny bead-like structures (Figure 9-13 C). Fronds of arid-land species can withstand extreme desiccation and still revive when moist conditions return. In others the plant dies back to a rhizome during the dry season and forms a new set of leaves with the onset of moist conditions.

Sori are marginal in most genera, and are concealed beneath the revolute or reflexed margins which form a false indusium (Figure 9-13 A–B). In a minority of genera (e.g., *Achrostichum, Penta-gramma*) the sporangia are distributed over the undersurface of the frond and are only loosely if at all aggregated into sori (Figure 9-13 D–E).

Although it is a large family, the members of the Pteridaceae are economically unimportant. Various members of the family are cultivated as ornamentals including species of *Achrostichum, Adiantum, Cheilanthes, Cryptogramma, Doryopteris, Hemionitis, Onychium, Pellaea, Pentagramma, Pityrogramma,* and *Pteris*. Some of these are easy to cultivate and others require exacting conditions and much patience. The various species of *Adiantum* (maidenhair fern) are particularly prized because of their delicate foliage.

Taxonomic treatments of the ferns included here as Pteridaceae vary widely. In older classifications and in many floras the ferns we treat as Pteridaceae were placed into a highly polymorphic Polypodiaceae (sensu lato). We have accepted the family limits as defined for the *Flora North America* treatment (Flora North America Editorial Committee. 1993). Tryon and Tryon (1982) treated the Pteridaceae slightly more broadly than this, including the aquatic genus *Ceratopteris* which we recognize as a member of the monogeneric family Parkeriaceae. Most of the ferns here included in the Pteridaceae were treated as Sinopteridaceae by Lellinger (1985) who recognized the Pteridaceae as a separate smaller family. Brummitt (1992) treated them in a similar fashion but used the name Adiantaceae instead of Sinopteridaceae.

# Dennstaedtiaceae

## The Bracken Fern Family
### Polypodiophyta, Polypodiopsida, Polypodiales ■

**Figure 9-14** Features of the Dennstaedtiaceae. A–B. *Pteridium aquilinum;* A. Frond attached to creeper rhizome. B. Leaf segment illustrating revolute leaf margin, continuous marginal sori, and linear marginal indusium. C. Leaf segment of *Dennstaedtia punctilobula* with discrete marginal sori and cup-like indusia formed from reflexed marginal teeth fused to true indusium.

# FAMILY DESCRIPTION

Herbs with erect, or more commonly, long-creeping pubescent (scaly) rhizomes. Leaves simple or 1-several times pinnately compound, monomorphic, pubescent (scaly); vernation circinate. Plants homosporous. Sporangia stalked, small, with an erect partial annulus, variously organized into continuous or discontinuous marginal or submarginal sori on the undersurface of the leaf blade; indusium marginal (absent), linear or cup-shaped, the sori often also covered by reflexed leaf margin; spores uniform in size, small. Gametophytes bisexual, aboveground, thin, green, photosynthetic.

# FAMILY NOTES

The Dennstaedtiaceae comprises 17–20 genera and 400 species. The family has a cosmopolitan distribution, but is especially abundant in the tropics. They are best known in the U.S. from the widespread *Pteridium aquilinum* (bracken fern) and *Dennstaedtia punctilobula* (hay-scented fern) of the eastern forests.

Bracken fern is the most widespread and weedy of all the ferns. It occurs on all continents and on many islands as well. It occurs from the boreal forests to the tropics in diverse habitats ranging from shade to full sun. It can become an invasive weed in pastured areas, sometimes forming dense pure stands. It spreads by its long, creeping rhizomes and the large leaves shade out smaller plants. Under favorable growing conditions the fronds of bracken fern sometimes grow to a height of 3 meters or more.

Although the young fiddleheads of bracken fern are sometimes harvested for human consumption, the plant is known to be somewhat toxic. In addition to the presence of a neurotoxin in its tissues, it is carcinogenic to animals that consume it in large quantities and has been implicated in cases of human cancer as well. Mature leaves are unpalatable to livestock. Bracken fern contains compounds mimicking insect molting hormones and is largely free of insect predation.

A few Dennstaedtiaceae are grown as ornamentals, principally outdoors where their coarse, rhizomatous habit allows them to serve as ground cover plants. These include *Dennstaedtia punctilobula* (hay-scented fern), *Hypolepis repens* (creeping bramble fern), *Microlepia* spp., and *Pteridium aquilinum* (bracken fern).

# Dryopteridaceae

## The Wood Fern Family
Polypodiophyta, Polypodiopsida, Polypodiales ■

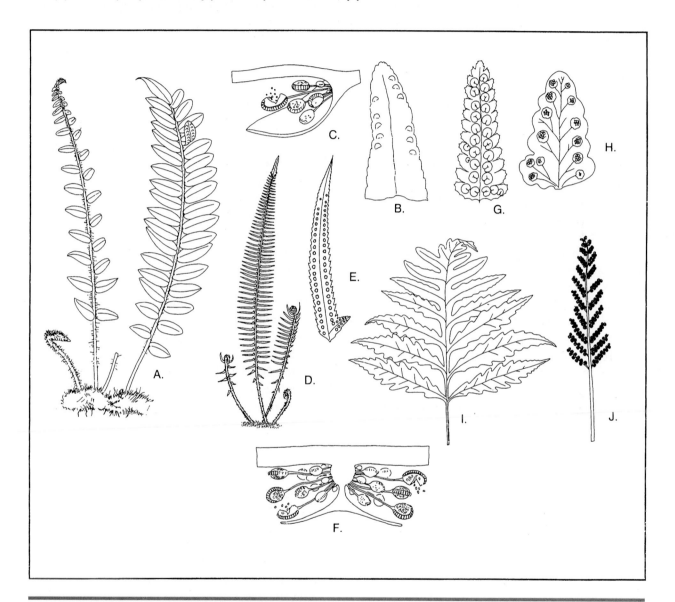

**Figure 9-15** Features of the Dryopteridaceae. A–C. *Nephrolepis exaltata*. A. Mature sporophyte. B. Pinna with laterally attached indusia covering the sori. C. Diagrammatic section of sorus (cluster of sporangia) on lower surface of leaf partially covered by laterally attached indusium (flap of tissue). D–F. *Polystichum munitum*. D. Mature sporophyte. E. Pinnule with circular sori. F. Diagrammatic section of sorus with centrally attached indusium. G. Pinnule of *Dryopteris arguta* illustrating round sori covered by deeply cordate indusia. H. Pinnule of *Woodsia* illustrating sori with inferior, cup-shaped indusia. I–J. Dimorphic sterile and fertile fronds of *Onoclea sensibilis*.

# FAMILY DESCRIPTION

Herbs with erect or creeping scaly rhizomes. Leaves simple or 1–5 times pinnately compound, monomorphic or dimorphic, more or less scaly, at least at petiole base; vernation circinate. Plants homosporous. Sporangia stalked, small, with an erect partial annulus, variously organized into round to oblong or linear sori on undersurface of leaf blade away from margin or continuously distributed across undersurfaces of leaf blades; indusium linear to round, attached centrally or on 1 side of sorus (absent); spores uniform in size, small. Gametophytes bisexual, aboveground, thin, green, photosynthetic.

# FAMILY NOTES

The Dryopteridaceae is a large, cosmopolitan family of 60–78 genera, and about 3000–3500 species. These ferns are well represented in both temperate and tropical regions. The family includes some of the most common woodland ferns of temperate North America. Common genera in the U.S. include *Athyrium* (lady fern), *Cystopteris* (fragile fern), *Dryopteris* (wood fern), *Gymnocarpium* (oak fern), *Matteuccia* (ostrich fern), *Onoclea* (sensitive fern), *Polystichum* (sword fern, holly fern, Christmas fern), and *Woodsia* (cliff fern).

In most Dryopteridaceae the sori are well-defined and are located on the undersurface of the frond away from the margin. In most there is a well-developed indusium. In some genera the indusium is round and shield-like, either peltate and centrally attached as in *Polystichum* or attached by a notch on one side as in *Dryopteris*. It may be laterally attached as in *Nephrolepis* or *Athyrium* or inferior and attached below and more or less enclosing the sorus as in *Woodsia*. In *Elaphoglossum* the sporangia are spread across the undersurface of the frond and do not form distinct sori.

Many members of the Dryopteridaceae are cultivated as ornamentals. Many have handsome foliage and may be grown either indoors or outdoors. Cultivated members of the Dryopteridaceae include *Dryopteris* (wood fern), *Polystichum,* (sword fern, holly fern, Christmas fern), *Athyrium* (lady fern), *Cyrtomium* (holly fern), *Cystopteris* (fragile fern), *Gymnocarpium* (oak fern), *Matteuccia* (ostrich fern), *Onoclea* (sensitive fern), *Woodsia* (cliff fern), *Nephrolepis* (Boston fern), *Diplazium* (twin-sorus fern), and *Tectaria* (button fern).

*Matteuccia struthiopteris, Diplazium esculentum* (vegetable fern), and some other members of the Dryopteridaceae are edible and are sometimes commercially grown and harvested. The fiddleheads of *M. struthiopteris* are the state vegetable of Vermont.

Taxonomic treatments of the ferns included here as Dryopteridaceae vary widely. In older classifications and in many floras the ferns we treat as Dryopteridaceae were placed into a highly polymorphic Polypodiaceae (sensu lato). Some authors in the past have used the illegitimate name Aspidiaceae instead of Dryopteridaceae. We have accepted the family limits of Dryopteridaceae as defined for the *Flora North America* treatment (Flora North America Editorial Committee, 1993). The treatment of the Dryopteridaceae by Tryon and Tryon is similar to the *Flora North America*. Some authors (Lellinger, 1985; Brummitt, 1992) place some of the genera into additional families including the Lomariopsidaceae, Oleandraceae, and Woodsiaceae. The commonly cultivated *Nephrolepis* (Boston fern) is variously placed. Tryon and Tryon (1982) and Lellinger (1985) placed *Nephrolepis* into the Davalliaceae. Brummitt treated it as a member of the Oleandraceae.

# Polypodiaceae

## The Polypody Family

Polypodiophyta, Polypodiopsida, Polypodiales ■

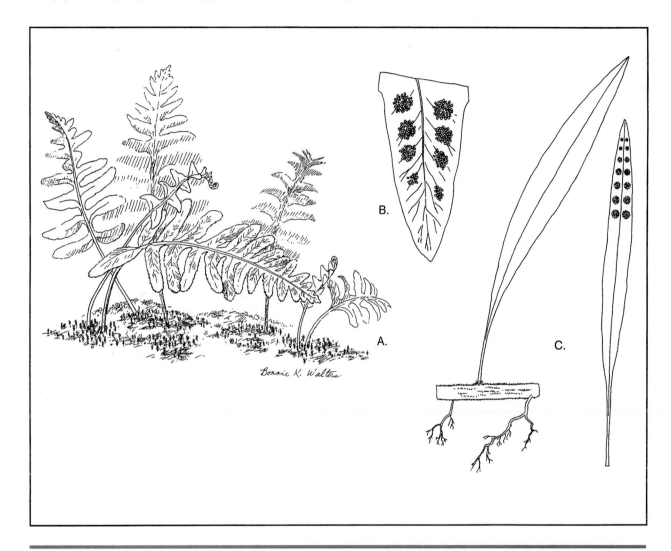

**Figure** Features of the Polypodiaceae. A. Mature sporophyte of *Polypodium californicum*. B. Lobe of frond of *Poly-*
**9-16** *podium californicum* illustrating naked sori (without indusia). C. Frond of *Pleopeltis macrocarpa.*

# FAMILY DESCRIPTION

Herbs with erect, or more commonly, creeping, scaly rhizomes. Leaves simple and entire to pinnately lobed or once compound (rarely 2 or more times compound), monomorphic or dimorphic, glabrous to pubescent or scaly; vernation circinate. Plants homosporous. Sporangia stalked, small, with an erect partial annulus, organized into round or oblong sori or distributed over the undersurface of leaf blade; indusium absent; spores uniform in size, small. Gametophytes bisexual, aboveground, thin, green, photosynthetic.

# FAMILY NOTES

The Polypodiaceae as treated here comprise 40–47 genera and about 700 species, widely distributed throughout the tropics and subtropics with a limited presence in temperate regions. In most of temperate North America the family is represented only by *Polypodium* (polypody). In Florida there are several additional genera and one of these, *Pleopeltis* (shielded sorus ferns) occurs throughout much of the southeastern U.S. Many of the tropical members of the family are epiphytic.

Various members of the Polypodiaceae are cultivated as ornamentals. In addition to various species of *Polypodium* (polypody), cultivated members of the family include *Drynaria* (oak-leaf fern), *Lemmaphyllum,* and *Platycereum* (staghorn fern). The stag-horn ferns are peculiar fleshy epiphytes with sessile, heart-shaped sterile fronds and erect, often dichotomously forking fertile fronds that are not differentiated into petiole and blade.

In older treatments, the Polypodiaceae was more broadly defined as an enormous cosmopolitan family of about 180 genera and 7000 species. The large number of species, interesting morphology and wide diversity prompted much work in the family. This work left little doubt that the Polypodiaceae as traditionally recognized was an artificial assemblage of smaller families. Some of the genera formerly included in the Polypodiaceae are now placed in the following families: Parkeriaceae, Pteridaceae, Vittariaceae, Dennstaedtiaceae, Thelypteridaceae, Blechnaceae, Aspleniaceae, Dryopteridaceae, Grammitidaceae.

The generic limits of Polypodiaceae are likewise controversial. *Polypodium* has often been very broadly defined to include most or all of the species in the family. In recent treatments this genus has been divided into a number of smaller genera, some of which are very distinctive and some separated on seemingly minor characters.

Polypodiaceae are notable by their uniform lack of an indusium. In many the sori are comparatively large and round. The leaves are usually simple and vary from entire to merely lobed. The petiole has a joint near the base and breaks free leaving a clean scar.

# Exercises

## ■ Study Questions

### Ferns and Fern Allies—General

1. Equisetaceae and Selaginellaceae are called "fern allies."

   In what way(s) are they similar to ferns?

   In what ways are they different?

2. For each of the following, indicate where it occurs, by what processes it is formed, and its function:

   Sporangium

   Spore

   Antheridium

   Archegonium

### Equisetaceae

3. *Equisetum* is often described as a "living fossil."

   Why?

4. Describe the sporophyte of an *Equisetum*.

   What do the stems look like?

   What do the leaves look like?

   How are the leaves arranged?

5. Where are the spores of *Equisetum* produced?

   What color are they?

   What do they grow into?

6. The leaves of *Equisetum* are called microphylls.

   Does this mean more than just small leaves?

   What are the features of a microphyll?

   What are the differences between microphylls and megaphylls?

7. *Equisetum* and most ferns are homosporous.

   What does this mean?

## Selaginellaceae

8. Describe the sporophyte of a *Selaginella*.

   What do the stems look like?

   Where are the roots produced?

   How are the leaves arranged?

   Are the leaves microphylls or megaphylls?

9. Where are the spores of *Selaginella* produced?

   What color are they?

   Is *Selaginella* homosporous or heterosporous?

   What do the spores grow into?

## Ferns

10. The fern plants that are familiar to everyone are sporophytes.

    Why are they called sporophytes?

    Are they diploid or haploid?

11. Are fern leaves megaphylls or microphylls.

12. What is a sporangium?

    What process takes place in a sporangium?

    What is an annulus?

13. What is a sorus?

    Where would you look for sori?

14. What is an indusium?

    What is its relationship to a sorus?

    Do all ferns have indusia?

15. What happens to the spores that are produced by a fern plant?

16. How does a fern gametophyte differ from a fern sporophyte?

    Where would you look in nature to find a fern gametophyte?

17. Where are the gametes of a fern plant produced?

    By what process are they produced?

18. Where does fertilization take place in ferns?

    How do sperm and eggs come together?

    What is the product of fertilization?

    How does this change into a new sporophyte?

19. List the families and species of ferns and fern allies that you identified in the lab.

_____     _____

_____     _____

_____     _____

_____     _____

_____     _____

# Gymnosperms

chapter 10

The seed plants are represented by five living groups;

1. Cycadales,

2. Coniferales,

3. Ginkgoales,

4. Gnetales

5. angiosperms or flowering plants.

The "gymnosperms" is a term that has been applied to the four seed bearing lineages that lack the specialized features that characterize the angiosperms (flowering plants). The name, gymnosperm, means naked seed. In some gymnosperms the ovules are borne on the margins of modified leaves; in others they occur at the tips of specialized short branches. In many they are located in various types of **strobili** (cones). Gymnosperms lack the folded, marginally sealed carpels that characterize the flowering plants. Gymnosperm ovules are exposed to the air at the time of pollination. The pollen-receptive structures are the ovules rather than stigmatic portions of carpels.

The fossil record indicates that the gymnosperms are less abundant and diverse now than in the past. The decline of the gymnosperms has been accompanied by the diversification of the angiosperms. Fossil evidence indicates that the cycads, ginkgos and conifers first appeared in the Paleozoic, the Gnetales and modern conifers evolved in the Triassic to Jurassic, and angiosperms in the Mesozoic (Stewart and Rothwell 1993, Crane 1996). Several lineages of ancient gymnosperms are now wholly extinct. Others persist as a few relict genera and species. A

few families are still well represented, but the only major exception to the general decline of the gymnosperms is the Pinaceae, a family that is well represented in modern plant communities in much of the northern hemisphere.

The gymnosperms have been classified in various ways, yet all the systems tend to recognize the same general categories. All of the classification systems used today recognize at least four major extant lineages and three fossil ones. None of them agree totally at what level (order, class, division) the taxa should be recognized.

It is not at all certain that all of the gymnosperms represent a single evolutionary lineage. The modern gymnosperms may all be derived from the seed ferns, a wholly extinct group, or they may have two or more separate origins. Some morphologists have suggested that true seeds evolved independently several times during the history of the earth. There is no consensus as to the correct interpretation of the fossil record of the gymnosperms. At least four very distinct evolutionary lines are represented among modern gymnosperms. These are the cycads, ginkgos, conifers, and the gnetophytes. The ginkgos, cycads and the conifers have long fossil records. The gnetophytes, on the other hand, have a very limited fossil record and do not closely resemble any other living taxa. All four lineages are so distinctive that any shared ancestors are very remote in time.

The gymnosperms are not a monophyletic group in the cladistic sense (Doyle et al, 1994; Nixon et al, 1994; see Chapter 23). The angiosperms apparently originated from a lineage that, if it were alive today, would be classified as a gymnosperm. If the gymnosperms represent a single lineage they are

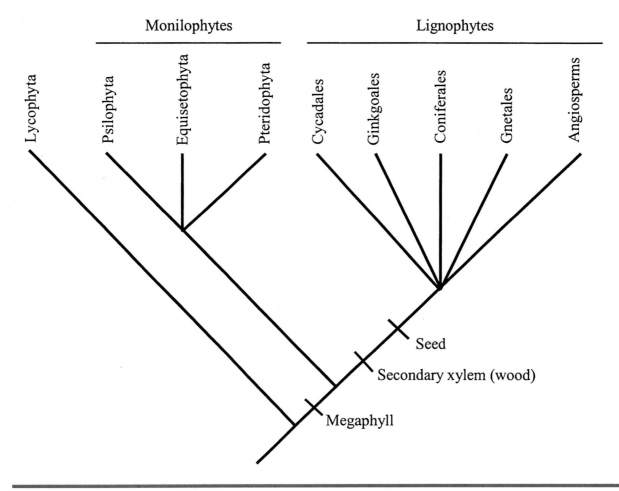

**Figure 10-1** Phylogeny of the tracheophytes (vascular plants). Time is on the vertical axis. Note that the vascular tissue of xylem and phloem is a shared derived character (synapomorphy) that unites the Tracheophytes. The megaphyll is a synapomorphy that unites the Euphyllophytes. The presence of a seed and secondary growth are synapomorphies that unite the Lignaphytes or seed plants.

paraphyletic. If gymnosperms have evolved independently more than once, they are polyphyletic. They are at best a "grade taxon" defined by having greater complexity than the ferns and fern allies and by their lack of angiosperm characteristics. Recent work by Burleigh and Mathews (2004) indicates that there is no clear resolution of relationships within the seed plants. However, there is some evidence suggesting that the Gnetales may share a common ancestry with the Pinaceae. If this is the case, then the Coniferales would be a paraphyletic group. In light of the lack of consensus about the relationships of seed plants, they may best be considered an unresolved polytomy (see Figure 10-1). In the discussion that follows we are treating each of the four lineages as separate orders.

# Gymnosperm Life Cycles (Fig. 10-2)

Gymnosperms and angiosperms are grouped together as seed plants. Both have **ovules.** An ovule is a specialized **megasporangium** that is enclosed by one or more outer layers (integuments) and is retained on the parent plant until it matures as a seed. Within the ovule meiosis takes place, a **megagametophyte** develops, fertilization occurs and an embryo develops. Both gymnosperms and angiosperms have **pollen.** Pollen grains are tiny **microgamethophytes,** produced in pollen sacs **(microsporangia).** Gymnosperms and angiosperms differ in the location of the ovules and pollen sacs on the plant, in the method of pollination, and in the events that take place within the ovules as seed-formation occurs.

An ovule of a gymnosperm has several layers. The outermost layer **(integument)** covers, and to some extent, protects the tissues within. At maturity the integument forms the seed coat. At one end of an ovule there is a pore that penetrates the integument; this is the **micropyle.** In *Pinus* and some other gymnosperms the micropyle is flanked by a pair of hornlike projections. Within the integument is the megasporangium, consisting of the **nucellus** (a nutritive tissue) and a **megasporocyte** (a cell that undergoes meiosis, forming megaspores). After meiosis three of the megaspores degenerate and the fourth, the **functional megaspore,** undergoes repeated mitotic divisions forming a megagametophyte (female gametophyte). In most gymnosperms the megagametophyte is a multicellular structure that produces several **archegonia,** each of which contains an egg. Development of the female gametophyte takes place over a period of several months.

Pollen grains are the dispersal stage of the male gametophytes (microgametophytes) of angiosperms and gymnosperms. In gymnosperms they are formed in microsporangia that are located in various types of strobili (cones) or stamen-like structures. The structures that bear the pollen sacs range in size from massive cones 50 cm or more long in some cycads to tiny cones only a few mm long in some conifers. Each microsporangium contains numerous microsporocytes (microspore mother cells). Meiotic division of each microsporocyte produces a group of four microspores. Subsequent mitotic divisions and formation of a characteristic cell-wall transform the microspores into pollen grains. The mature pollen grain usually consists of two or several cells enclosed together within an outer wall (exine). The exines of pollen grains are variously sculptured, and the pollen grains of different kinds of plants are often easy to recognize. The pollen grains of *Pinus* and several other genera have bladder-like wings. Wind pollination is characteristic of most gymnosperms.

In gymnosperms the micropyle of the ovule is the pollen-receptive structure. In some gymnosperms the micropyle exudes a droplet of liquid to which pollen grains adhere. As moisture evaporates from the pollination drop (or is resorbed by the ovule tissue), surface tension draws pollen grains through the micropyle into a pollen chamber within the ovule. A pollen grain that arrives within a receptive micropyle germinates, forming a pollen tube. Development of the pollen tube varies considerably among the gymnosperms. Sperm in some gymnosperms (e.g., cycads and ginkgos) are motile and swim from the pollen tube into the archegonia where fertilization takes place. In other gymnosperms a pollen tube grows through the nucellar tissue into the neck of an archegonium where sperm are released as free nuclei.

Several eggs may be fertilized within a single ovule, and cells that form from the first mitotic divisions of the zygotes following fertilization may all begin development as independent embryos. Although numerous embryos may develop initially within an ovule, only one usually survives when the seed is mature. The remainder die and are absorbed either by the surviving embryo or by the maturing female gametophyte.

A mature gymnosperm seed consists of a seed coat (ripened integument), a thin nucellar layer, the ripened female gametophyte and the embryo. The seed coat of some gymnosperms is fleshy and berrylike and in others it is dry and hard or thin and membranous. The principal food storage tissue in most gymnosperm seeds is the female gametophyte. Note that gymnosperms [except *Ephedra*] do not have double fertilization and the seeds lack endosperm; these are features restricted to the angiosperms. The embryo of a gymnosperm may have as few as two cotyledon or it may have several. Upon maturation the seed is presented for dispersal. Presentation can be done a number of ways. Most commonly, the seeds are borne on the top of scale-like elements of a woody cone, but they may also be borne on stalks, surrounded by fleshy tissue, or on the edge of modified leaves. Some gymnosperm seeds are winged and wind-dispersed. Others are fleshy or hard and nutlike and are commonly dispersed by animals, especially birds. Still others have no evident dispersal mechanism.

Gymnosperms possess many complicated specializations in leaf features, ovulate cone structure, and branching pattern. The leaves of *Gnetum* (Gnetaceae) are very similar to those of dicotyledonous angiosperms. Plants of the monotypic genus *Welwitschia* (Welwitschiaceae), an extreme xerophyte of South African deserts produce only two leaves during the life of the plant. The strap-shaped, leathery leaves grow from the base, split into strips, and weather away at the tips.

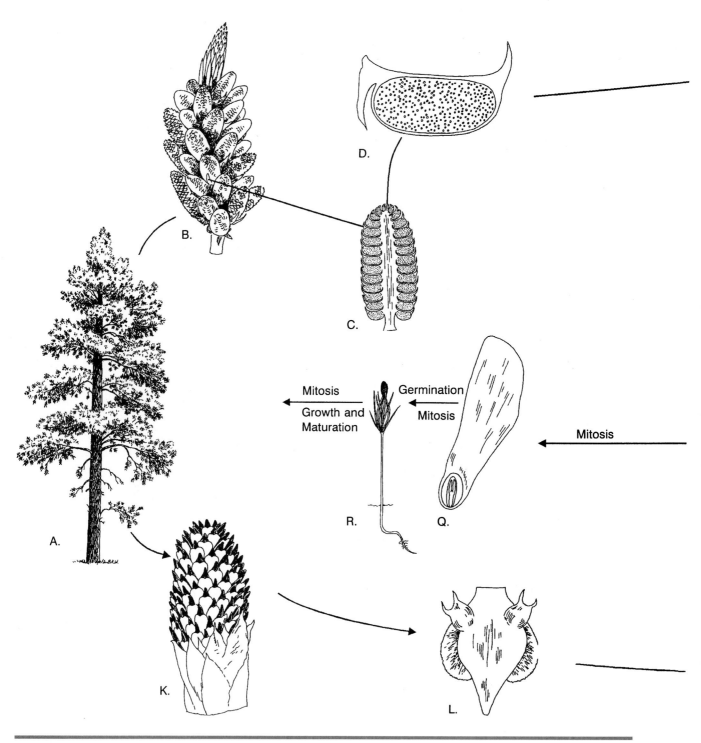

Mitosis
Growth and Maturation
Germination
Mitosis
Mitosis

D.

C.

B.

A.

R.

Q.

K.

L.

**Figure 10-2** Pine life cycle. A. Adult sporophyte. B. Clusters of microsporangiate (pollen) cones. C. Single microsporangiate cone (l.s.). D. Enlarged l.s. of microsporophyll with microsporangium containing many microsporocytes (microspore mother cells) that undergo meiosis. E-J. Stages in development of microgametophyte. E. Single microsporocyte. F. Tetrad of microspores produced by meiosis of one microsporocyte. G. Microgametophyte (pollen grain) at time of dispersal for pollination with degenerating prothalial cells and the functional tube and generative cells. H-J. Germinating pollen grain and growing pollen tube showing the eventual division of the generative cell into the stalk cell and body cell which then divides into two sperm cells.

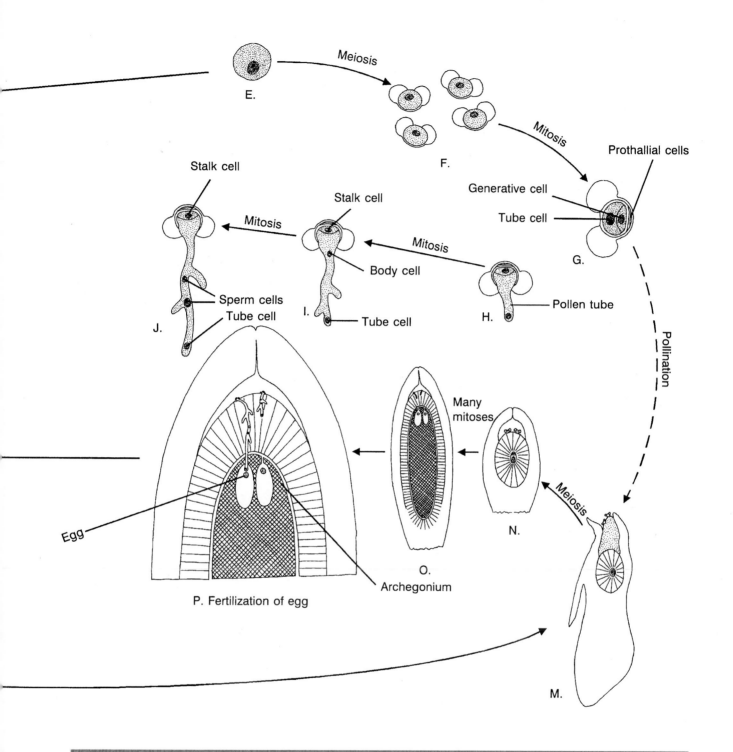

**E.**

Meiosis

**F.**

Mitosis

Prothallial cells

Generative cell

Tube cell

**G.**

Stalk cell

Mitosis

Stalk cell

Mitosis

Body cell

Tube cell

Pollen tube

**H.**

**I.**

Sperm cells

Tube cell

**J.**

Pollination

Many mitoses

Meiosis

**N.**

Archegonium

**O.**

Egg

**P. Fertilization of egg**

**M.**

K. Immature ovulate cone. L. Upper (adaxial) surface of ovulate cone-scale with two ovules (bearing horn-like appendages) and shaded subtending bract. M-P. Stages in development of megagametophyte. M. Single developing ovule at time of pollination with micropyle filled with sticky pollination drop and megasporangium (nucellus) with single megasporocyte (megaspore mother cell). N. Developing ovule with pollen grains within the pollen chamber and a linear tetrad of 4 megaspores, 3 of which are disintegrating. O. Maturing ovule with enlarged integument sealing the micropyle over the pollen grains and pollen tubes growing through the nucellus toward the megagametophyte with its 2 egg-containing archegonia. P. Mature ovule just before fertilization by the first sperm cell to enter the archegonia. Q. Developing seed (l.s.) with mature embryo. R. Young seedling produced after seed dispersal and germination.

Most gymnosperms lack vessel elements in their xylem and have tracheids as their water-conducting and structural support cells. However, vessel elements are present in the members of the Gnetales.

Microsporangia in most gymnosperms are borne in strobili. These are massive in the cycads (over 50 cm long in some cases) and generally not more than a few centimeters long in other groups. In most families the microsporangia are sessile, but in the Ginkgoaceae, the Gnetales, and some others the microsporangia are stamen-like with filaments and anthers.

# Synoptic Key to the Orders of Living Gymnosperms

1. Sperm cells motile, swimming from pollen tube to the archegonia.

    2. Leaves pinnately (rarely bipinnately) compound; plants commonly with a short, stout, few-branched trunk or a subterranean corn-like trunk; ovules borne on margins of megasporophylls that are often aggregated into simple strobili; microsporophylls aggregated in dense strobili; pollen sacs borne in sorus-like clusters on abaxial surfaces of microsporophylls. . . . . . . . . . . . . . . . . . . . . **CYCADALES**

    2′ Leaves simple, fan-like with dichotomous venation; stems openly branched; ovules borne in terminal pairs on slender, axillary, bractless stalks; microsporangia borne in slender, catkinlike strobili; microsporophylls stamenlike with 2 pollen sacs borne at end of slender filament-like stalk. . . . . . . . . . . . . . . . . . . . . . . . . . . . . . . . . . . . . . . . . . . . . . . . . . . . . . . . . . . . . . . . . **GINKGOALES**

1′ Sperm cells non-motile, delivered by pollen tube as free nuclei.

    3. Microsporangiate strobili simple. . . . . . . . . . . . . . . . . . . . . . . . . . . . . . . . . . **CONIFERALES**

    3′ Microsporangiate strobili compound; resin canals absent from stems and leaves. . . . . . . . . . . . . . . . . . . . . . . . . . . . . . . . . . . . . . . . . . . . . . . . . . . . **GNETALES**

# FAMILIES OF GYMNOSPERMS

## Order Cycadales

**Cycadaceae**—The Cycad Family (Figure 10-6). 1 genus, 47 species. Tropics and subtropics from Australia to S.E. Asia.

§See page 174 for description and illustrations§

**Zamiaceae**—The Zamia Family (Figure 10-6). 8 genera, 164 species. Tropical & subtropical North and South America, sub-Saharan Africa, and Australia.

§See page 174 for description and illustrations§

**Stangeriaceae**—2 genera, 4 species. Subtropical Coasts of Eastern Australia and Coastal South Africa.

## Order Ginkgoales

**Ginkgoaceae**—The Ginkgo Family (Figure 10-7). 1 genus, 1 species. China.

§See page 177 for description and illustrations§

## Order Coniferales

**Taxaceae**—The Yew Family (Figure 10-3). 5 genera, 17–20 species. Widely distributed, mostly in northern hemisphere.

Shrubs or trees. Leaves alternate, simple, linear, needle-like, decurrent on twigs. Plants dioecious (rarely monoecious). Pollen borne in cones; ovules solitary or in pairs, borne in reduced cones. Pollen cones small, each with several flattened sterile basal scales and 4–32 spirally arranged or whorled, peltate or ± flattened sporangiophores; pollen sacs 2–9 per sporangiophore, linear. Ovulate cones terminal on short shoots; ovules 1–2, subtended by opposite scale-like bracts; seeds solitary, surrounded by a green or red fleshy aril.

**Cephalotaxaceae**—The Plum-yew Family. 1 genus, 3 species. Temperate eastern Asia.

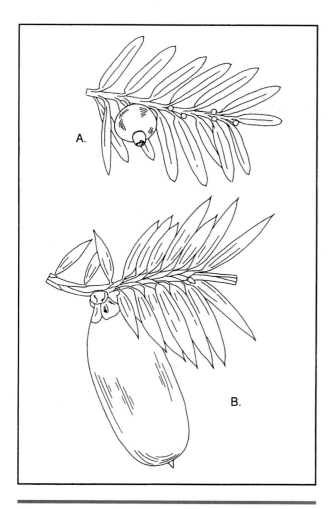

**Figure 10-3** Examples of Taxaceae A. Twig of *Taxus baccata* (yew) showing the alternate, flat, needle-shaped leaves and seed with fleshy aril (red when mature) that does not fully cover the ovule. B. Twig and seed of *Torreya californica* (California-nutmeg).

Shrubs or trees. Leaves opposite or whorled, simple and linear. Plants dioecious. Pollen and ovules both borne in cones. Pollen cones with many spirally arranged scales; sporophylls 1–8, stamenlike, pollen sacs 3–5 on undersurface of scale. Ovulate cones very reduced; bracts few, opposite, without cone scales, each subtending 2 short-stalked ovules; seeds fleshy, drupelike, normally only one maturing per cone.

**Pinaceae**—The Pine Family (Figures 10-2 and 10-8). 9–12 genera, 210 species. Northern hemisphere, particularly at high latitudes and in high mountains.

§See page 179 for description and illustrations§

**Cupressaceae** (including Taxodiaceae)—The Cypress Family (Figure 10-9). 28 genera, 120 species. Worldwide in distribution with several areas of high endemism.

§See page 181 for description and illustrations§

**Araucariaceae**—The Araucaria Family. 2 genera, 39 species. Tropical East Indies and southern temperate zone of South America and Australian region.

Trees. Leaves alternate or opposite, simple, entire, needle-like or awl-like to ovate, sessile on leafy branchlets that are sometimes deciduous in age. Plants dioecious (rarely monoecious). Pollen and ovules both borne in cones. Pollen cones elongate with many spirally arranged scales; pollen sacs many, elongate, borne on undersurface of scale. Ovulate cone large, woody; cone scales many, flattened, spirally arranged, wholly or partly adnate to subtending bracts; ovules 1(2), on upper surface of cone scale or covered by a flap of tissue and appearing to be embedded in scale; cones disintegrating at maturity; seeds either winged and breaking free from cone scales or wingless and falling together with the cone scale-bract complex as a unit.

**Podocarpaceae**—The Podocarp Family (Figure 10-4). 13–77 genera, 172 species. Widespread in south temperate and tropical regions.

Shrubs or trees (rarely root parasites). Leaves alternate (rarely opposite), simple, scalelike, needle-like or with flattened blades (rarely branchlets forming lobed or toothed cladodes). Plants monoecious or dioecious. Pollen borne in cones; ovules borne in

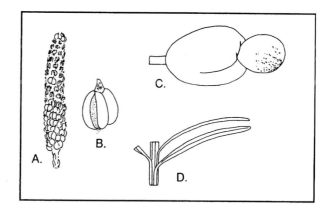

**Figure 10-4** Examples of Podocarpaceae. A–D. *Podocarpus* sp. A. Pollen cone. B. One microsporophyll enlarged. C. Terminal ovule subtended by large fleshy cone axis. D. Leafy twig.

cones or 1–2 on ends of branchlets. Pollen cones with many spirally arranged scales; pollen sacs 2 per scale, borne on undersurface of scale. Ovulate cone spikelike, or reduced and comprising 1–2 terminal ovules, each ± adnate to a cone scale and bract, and to a swollen fleshy cone axis with 1 or more reduced cone scales; seeds dry or fleshy and berry-like, sometimes enveloped by a fleshy aril.

## Order Gnetales

**Gnetaceae**—The Gnetum Family. 1 genus, 30 species. Pantropical.

Shrubs, woody vines or trees. Leaves opposite, simple and entire, pinnately veined, petiolate. Plants dioecious (monoecious). Pollen and ovules borne in flower-like structures. "Flowers" whorled in spikelike clusters. "Staminate flowers" very numerous; "perianth" tubular, "stamen" 1 with a filament and 1 or 2 pollen sacs. "Ovulate flowers" with a tubular "perianth" surrounding a solitary ovule; seeds drupelike with "perianth" forming a fleshy outer layer.

**Welwitschiaceae**—The Welwitschia Family. 1 genus, 1 species. Extreme deserts of South Africa.

Herbs with fleshy underground stem and taproot. Leaves 2, opposite, straplike, leathery, growing from base and functioning throughout life of plant. Plants dioecious. Pollen and ovules borne in flower-like structures grouped in strobili on dichasially branching stalks. "Staminate flowers" individually subtended by bracts (cone scales) and a pair of bractlets; "perianth scales" 2, connate; "stamens" 6, surrounding a non-functional ovule; filaments basally connate; anther sacs 3 per "stamen." "Ovulate flowers" subtended by a bract and a pair of reduced bractlets; "perianth scales" 2, connate around ovule, dry and wing-like at maturity; seeds dry.

**Ephedraceae**—The Mormon-tea Family (Figure 10-5). 1 genus, 40 species. Desert and semidesert regions of Northern Hemisphere and South America.

Shrubs or woody vines (rarely trees) with jointed green stems. Leaves opposite or whorled, simple, scale-like, often deciduous. Plants dioecious. Pollen and ovules borne in flower-like structures borne in strobili. "Staminate flowers" subtended by scale-like bracts, grouped in spikelike strobili; "perianth scales" 2; "stamens" 1–8 with 1–several anther sacs per filament. "Ovulate flowers" 1–3 per strobilus,

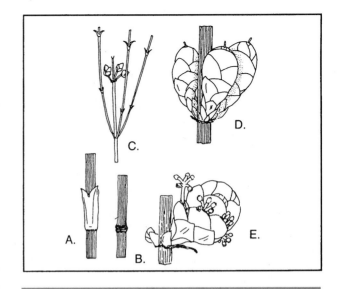

**Figure 10-5** Examples of Ephedraceae. A–B. *Ephedra californica.* A. Closeup of vertically ridged young branchlet with opposite, connate, scale leaves. B. Older twig after leaves have withered and fallen. C–E. *Ephedra viridis.* C. Branchlets with ovulate strobili and opposite scale leaves. D. Whorl of ovulate strobili with perianth-like scales subtending terminal ovules. E. Pollen cone showing many spiraling scales and branched microsporophylls with terminal microsporangia (pollen sacs).

subtended by 2–4 pairs or whorls of scale-like bracts; "perianth" in 2 series, the outer of 4 basally connate scales, the inner of 2 slender connate scales forming a beaklike tube, basally adnate to ovule; ovule 1 per "flower"; seeds dry or seeds and subtending structures fleshy and ± coalescent.

## References

Beck, C. B. (ed.). 1988. *Origin and Evolution of Gymnosperms.* Columbia Univ. Press, New York.

Bierhorst, D. W. 1971. *Morphology of Vascular Plants.* MacMillan, New York.

Brummitt, R. K. 1992. *Vascular Plant Families and Genera.* Royal Botanic Gardens, Kew.

Burleigh, J. G. and S. Mathews. 2004. Phylogenetic signal in nucleotide data from seed plants; implications for resolving the seed plant tree of life. *American Journal of Botany* 91(10): 1599–1613.

Crane, P. R. 1996. Fossil history of Gnetales. *International Journal of Plant Sciences* 157:S50–S57.

Dittmer, H. J. 1964. *Phylogeny and Form in the Plant Kingdom.* D. Van Nostrand Co., Princeton, N.J.

Doyle, J. A., M. J. Donoghue, and E. A. Zimmer. 1994. Integration of morphological and ribosomal RNA data on the origin of angiosperms. *Annals of the Missouri Botanical Garden* 81:419–450.

Eckenwalder, J. E. 1976. Re-evaluation of Cupressaceae and Taxodiaceae: a proposed merger. *Madroño* 23:237–256.

Flora North America Editorial Committee. 1993. *Flora of North America, North of Mexico, Vol. 2: Pteridophytes and Gymnosperms.* Oxford University Press, Oxford. Pp. 345–434.

Florin, R. 1955. The systematics of the gymnosperms, pp. 323–403 in *A Century of Progress in the Natural Sciences,* California Academy of Sciences, San Francisco.

Foster, A. S., and E. M. Gifford. 1974. *Comparative Morphology of Vascular Plants.* Freeman, San Francisco.

Hart, J. 1987. A cladistic analysis of conifers: preliminary results. Journal of the Arnold Arboretum 68:269–307.

Jones, D. L. 1993. *Cycads of the World.* Smithsonian Institution Press, Washington D.C.

Kubitzki, K. (ed.) 1990. *The Families and Genera of Vascular Plants.* Vol. 1. *Pteridophytes and Gymnosperms.* Edited by K. U. Kramer and P. S. Green, assisted by E. Götz. Springer Verlag, Berlin.

Meyen, S. V. 1984. Basic features of gymnosperm systematics and phylogeny as evidenced by the fossil record. *Botanical Review* 50:1–111.

Nixon, K. C., W. L. Crepet, D. Stevenson, and E. M. Friis. 1994. A reevaluation of seed plant phylogeny. *Annals of the Missouri Botanical Garden* 81:484–533.

Scagel, R. F., R. J. Bandoni, G. E. Rouse, W. B. Schofield, J. R. Stein and T. M. C. Taylor. 1984. *An Evolutionary Survey of the Plant Kingdom.* Wadsworth Publ. Co., Belmont, California.

Silba, J. 1986. Encyclopaedia coniferae. *Phytologia Memoirs* 8:1–217.

Sporne, K. R. 1965. *The Morphology of the Gymnosperms: the Structure and Evolution of Primitive Seed Plants.* Hutchinson & Co., London.

Stewart, W. N. and G. W. Rothwell. 1993. Paleobotany and the evolution of plants, 2nd edition. Cambridge University Press: Cambridge, UK.

# Cycadaceae and Zamiaceae

## The Cycad Families

Pinophyta, Cycadicae, Cycadopsida, Cycadales ■

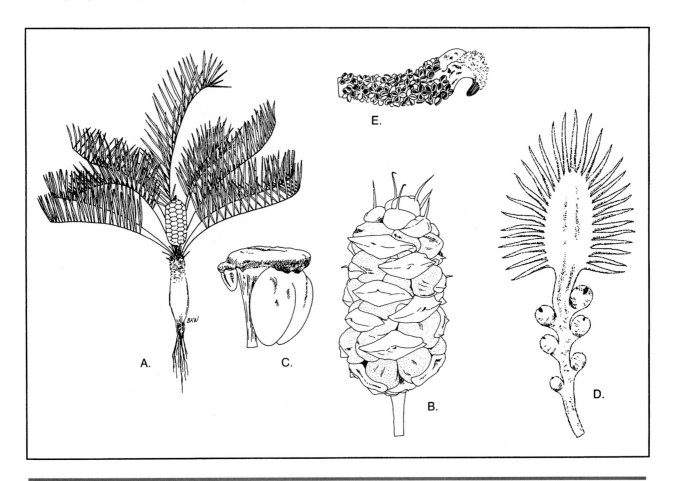

**Figure 10-6** Features of Cycadales. A. Habit of *Zamia pumila* (Zamiaceae) with ovulate strobilus (seed cone). B–C. *Macrozamia* sp. (Zamiaceae). B. Ovulate strobilus. C. Peltate megasporophyll. D–E, *Cycas revoluta* (Cycadaceae). D. Leaflike megasporophyll with marginal ovules. E. Microsporophyll with densely aggregated microsporangia (pollen sacs).

# CYCADACEAE–FAMILY DESCRIPTION _____

Shrubs or palmlike trees. Leaves tightly alternate, forming a dense terminal rosette; pinnate, with circinate vernation; leaf bases persistent and woody. Plants dioecious. Pollen cones large with many spirally arranged scales. Pollen borne on lower surface of flattened microsporophylls in terminal or subterminal cones; pollen sacs 6–many on undersurface of scales. Ovulate cones loosely aggregated with ovules 2–8 on margins of leaflike sporophylls with pinnately toothed or lobed terminal appendages; seeds large, drupelike.

# ZAMIACEAE—FAMILY DESCRIPTION

Shrubs or palmlike trees (sometimes with underground trunks). Leaves tightly alternate, forming a dense terminal rosette; pinnate, sometimes with circinate vernation; leaf bases usually persistent and woody. Plants dioecious. Pollen cones large with many spirally arranged scales. Pollen borne on lower surface of flattened microsporophylls in terminal or subterminal cones; pollen sacs 6–many on undersurface of scales. Ovulate cones small or very large with many spirally arranged peltate scales, each bearing on its undersurface a pair of large ovules; seeds large, drupelike.

# FAMILY NOTES

The cycads of today are a small group of tropical and subtropical plants totaling about 11 genera and 185 species. The greatest diversity is in the southern hemisphere. They represent the last survivors of an ancient lineage that was much more widespread and diverse during the Mesozoic Era. They are not closely related to any other modern gymnosperms.

In older treatments all the members of the order Cycadales were placed into one family, the Cycadaceae. Recent work has indicated that three or four lineages can be recognized. In most recent treatments these lineages have been recognized as separate families. These may be distinguished in the following key.

1. Ovules 2–8, borne laterally along the petioles of megasporophylls with pinnately lobed terminal blades; megasporophylls borne in a whorl or tight spiral on main stem axis, not forming well-organized strobili.
. . . . . . . . . . . . . . . . .Cycadaceae *(sensu stricto)*

1′ Ovules usually 2, borne on peltate or imbricate bladeless megasporophylls; megasporophylls grouped along short, determinate axes, forming strobili.

   2. Leaves lacking the pair of stipules, or if present, then stipules lacking veins; ovules attached above the sporophyll stalk.
. . . . . . . . . . . . . . . . . . . . . . . .Zamiaceae

   2′ Leaves subtended by pair of stipules that bear at least a single vascular bundle; ovules attached below the sporophyll stalk.
. . . . . . . . . . . . .Stangeriaceae *(sensu lato)*

      3. Leaves once pinnate, thin, fernlike; stems subterranean.
. . . . . . . .Stangeriaceae *(sensu stricto)*

      3′ Leaves bipinnate, stiff; plants evidently woody above the ground.
. . . . . . . . . . . . . . . . . . . .Boweniaceae

The Cycadaceae *(sensu stricto)* is composed of only one genus, *Cycas,* and about 47 species, distributed in the tropics and subtropics from Australia to S.E. Asia. The Stangeriaceae is composed of two genera, *Stangeria,* a monotypic genus from South Africa, and *Bowenia;* with three species from northern Australia. *Stangeria* and *Bowenia* are sometimes placed into separate monogeneric families. The remaining eight genera and 164 species comprise the Zamiaceae, which is distributed in tropical and subtropical North and South America, sub-Saharan Africa, and Australia. The only cycad native to the U.S. is *Zamia integrifolia.*

Most cycads are relatively small with the trunk rarely over 3 meters tall, and are very slow growing. In some, specialized retractile tissues pull the developing trunks underground. The leaves are pinnately compound, a very unusual feature among gymnosperms, and in some species the blades expand in a manner resembling the circinate vernation characteristic of most ferns. Since most do not branch and there is very little secondary growth in the trunk, cycads strongly resemble small palm trees (Figure 10-6 A). *Stangeria* is unusual in having thin, herbaceous leaves that resemble those of ferns.

Cycads are dioecious plants. Pollen is produced in massive terminal strobili that in *Cycas* may reach a length of over 50 cm. Each pollen cone consists of a central axis with many spirally arranged cone scales. The undersurface of the scale bears masses of sessile pollen sacs (Figure 10-6 E) in sorus-like clusters. The cones produce copious amounts of pollen. Although most older references indicate that cycads are wind-pollinated, recent studies have demonstrated that most, if not all, are insect-pollinated.

Cycadaceae and Zamiaceae differ in the position of the ovules. Ovules in *Cycas* are borne along the petioles of more or less leaflike megasporophylls.

The megasporophylls have a reduced, often densely pubescent, pinnately lobed blade (Figure 10-6 D). They are loosely clustered at the apex of the trunk and do not form a cone-like structure. The terminal bud alternates between production of vegetative leaves and megasporophylls. In the Zamiaceae and Stangeriaceae the megasporophylls are aggregated in conelike strobili. Each megasporophyll is a peltate, or flattened, scale-like structure with little or no blade development and bears a pair of ovules on its undersurface or margin (Figure 10-6 C). The strobili develop from axillary buds.

Sexual reproduction of cycads is rather primitive. The large ovules mature slowly. Pollen grains germinate into branched pollen tubes that grow for prolonged periods of time. The sperm cells are flagellated and swim from the pollen tubes to the archegonia within the ovules. The only other modern gymnosperms with flagellated sperm are the ginkgos.

Several members of the Cycadales are grown as ornamentals in warm climates and for the production of "palm-fronds" for the Christian celebration of Palm Sunday. Cycads are very popular with collectors of unusual horticultural items and this, unfortunately, is bringing many species to the edge of extinction. All species of cycads must be considered poisonous as all contain the toxins **cycasin** and **macrozamin** throughout the plant. Therefore the reports that *Zamia* and *Cycas* seeds are edible or that the pith of species of *Cycas* can be used as source of sago (or palm) starch must be tempered with a warning. It is true that native peoples in Africa and Australia did use cycad products locally for food, but each people had elaborate leaching and cooking procedures that eliminated the toxins.

# Ginkgoaceae

## Maidenhair Tree Family

### Pinophyta, Pinicae, Pinopsida, Ginkgoales ■

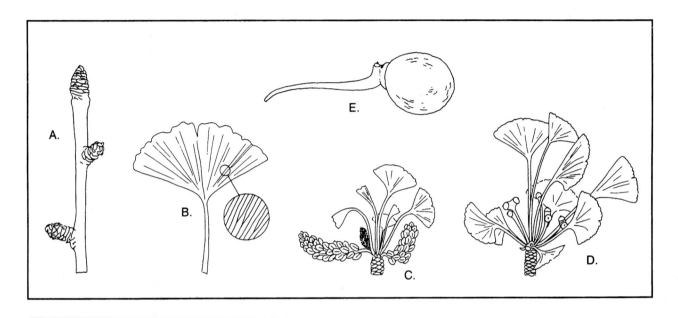

**Figure 10-7** Features of Ginkgoaceae. A–E. *Ginkgo biloba.* A. Twig in deciduous dormant condition illustrating long and short shoots. B. Fan-shaped leaf with close-up of dichotomous venation. C. Short shoot from male tree bearing catkin-like microsporangiate strobikli (pollen cones). D. Short shoot from female tree bearing paired stalked ovules. E. Stalk with one seed developed and one undeveloped ovule.

## FAMILY DESCRIPTION

Trees. Leaves alternate, or fascicled on short shoots, simple, fan-shaped with many dichotomous veinlets, petiolate. Plants dioecious. Pollen borne in axillary spike-like clusters of stamen-like sporangiophores; pollen sacs 2 per sporangiophore, reflexed. Ovules 2 (3 or more) naked, borne at ends of forking peduncle; seeds large, drupelike with foul-scented fleshy outer layer.

## FAMILY NOTES

*Ginkgo biloba,* the only extant member of the Ginkgoaceae, can be described as a living fossil. The fossil record indicates that ginkgos and their relatives were widespread in the Mesozoic Era. All species but *G. biloba* apparently became extinct long ago. Even this lone survivor is not known to occur in the wild. This species was in cultivation in monastery gardens in China when European botanists first encountered it. It is now widely grown in temperate regions as an ornamental.

Ginkgos are openly branched trees with distinctly different long shoots and short shoots (Figure 10-7 A).

Most leaves are borne on the stubby short-shoots (spur branches). The slender-petiolate, fan-shaped leaves are unlike those of any other living gymnosperms, with many dichotomously branched veins (Figure 10-7 B).

The pollen-bearing structures strongly resemble the stamens of angiosperms. The paired pollen sacs are attached to a filament-like stalk. These are aggregated in slender, catkinlike clusters (Figure 10-7 C). The naked ovules are borne in pairs on slender stalks (Figure 10-7 D). The reproductive structures are borne on the short shoots among the leaves. The pollen tubes of ginkgos, like those of the cycads, release swimming sperm within the micropyles of the ovules.

Male trees are preferred as street trees to female trees. The outer seed coat of ginkgos is fleshy and develops a very unpleasant odor as it ripens. The odor of a tree full of ripe ginkgo seeds can last for several weeks. Despite their odor the seeds are edible and in parts of the Orient the seeds are considered to be a delicacy.

# Pinaceae

## The Pine Family
### Coniferales ■

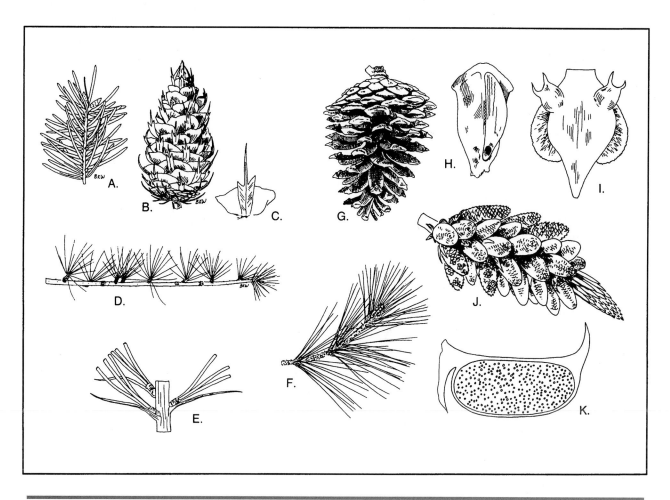

**Figure 10-8** Features of Pinaceae. A–C. *Pseudotsuga menziesii.* A. Twig bearing solitary needle-like leaves. B. Ovulate cone with exserted three-lobed bracts. C. Bract and undersurface of ovulate cone-scale. D. Long shoot of *Larix* bearing short shoots with fascicled needle-like leaves. E. Young long shoot of *Pinus* bearing scale-leaves and axillary short shoots with fascicled needle-like leaves. F. Branch of *Pinus* bearing mature fascicles of needle-like leaves. G. Ovulate cone of *Pinus* with cone scales longer than and concealing the bracts. H. Upper surface of mature ovulate cone scale of *Pinus* with winged seed. I. Immature bract and ovulate cone scale of *Pinus* bearing ovules on upper surface. J. Branchlet of *Pinus* with cluster of pollen cones. K. Cone-scale of pollen cone of *Pinus* (l.s) with pollen sac on lower surface.

# FAMILY DESCRIPTION

Shrubs or trees. Leaves alternate, or closely fascicled on determinate or indeterminate short shoots, simple, linear to needle-like. Plants monoecious. Pollen and ovules both borne in cones. Pollen cones with many spirally arranged scales; pollen sacs 2 on undersurface of scale. Ovulate cones woody at maturity; cone scales many, flattened, spirally arranged, free or nearly so from subtending bracts; bracts mostly shorter than scales (less commonly the bract tips exserted from between scales); ovules 2 on upper surface of each cone scale; cone scales separating or cones disintegrating at maturity; seeds usually winged (rarely wingless and nutlike).

# FAMILY NOTES

Members of the Pinaceae are common forest trees in the northern hemisphere, particularly at high latitudes and in high mountains. The family consists of 9–12 genera and about 210 species. Pines, firs, spruces, larches and other members of the family are often dominant members of the communities in which they grow. The Pinaceae is the most successful of the gymnosperm families.

Vegetative features of the Pinaceae are important in their ecological success. Most species have a strongly dominant central axis and grow with a characteristic spire-like habit (Figure 10-2 A). Although their stems have no vessels in the xylem their tracheids are efficient in water conduction. The stems, leaves and cones have resin ducts that contain substances that seal damaged surfaces and that are somewhat insect-resistant. The needle-like leaves have a thick epidermis and sunken stomata that retard water loss. Many species are tolerant of freezing. Members of all but one genus *(Larix)* are evergreen.

The degree of differentiation of short shoots and long shoots varies from genus to genus. In some (e.g., *Abies, Picea, Pseudotsuga* and *Tsuga*) there is little difference between the two (Figure 10-8 A). In other genera (e.g., *Cedrus* and *Larix*) the short shoots are stubby lateral branches with indeterminate growth that produce fascicles of leaves year after year (Figure 10-8 D). In *Pinus* the short-shoots are determinate, each producing a fascicle with a set number of leaves (Figure 10-8 E, F).

Members of the Pinaceae have small, ephemeral pollen cones and small to large, more or less woody ovulate cones. A pollen cone consists of a central axis and numerous small flat microsporophylls, each with two microsporangia on its lower surface (Figure 10-8 J, K). The ovulate cones are more complicated. The cone has a woody axis to which are attached spirally arranged cone scales (ovuliferous scales), each of which is subtended by a scale-like bract. Two ovules are attached to the upper surface of each scale. These mature in most members of the family as winged seeds that break free from the cone scale (Figure 10-8 H) at the time that the woody cone scales dry and separate (Figure 10-8 B, G).

Economically the Pinaceae is the most important gymnosperm family. Many species are important lumber trees. In much of the world the Pinaceae is the principal source of wood pulp for making paper. Some are economically valuable as sources of resins, turpentine and other extracts. In many areas young trees are harvested as Christmas trees. Seeds of some species of pine (e.g., pinyon pines) are edible and were used as a major food source by some tribes of American Indians.

# Cupressaceae
# [Including Taxodiaceae]
## The Cypress Family
Coniferales ■

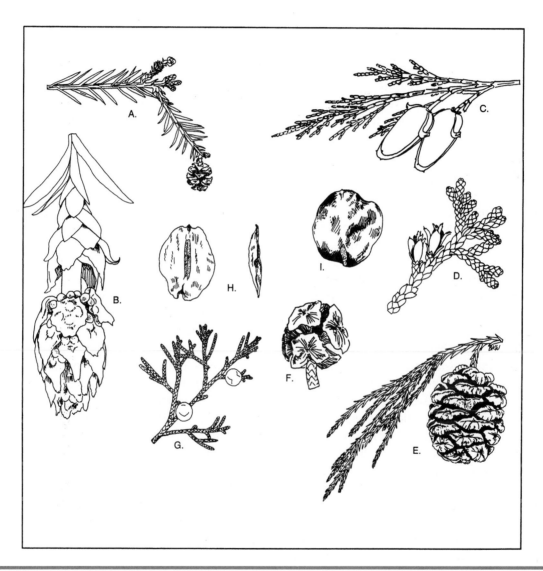

**Figure 10-9** Features of Cupressaceae (*sensu lato* including Taxodiaceae). A–B. *Sequoia sempervirens.* A. Branchlets bearing alternate needle-like leaves and woody ovulate cone. B. Pollen cone bearing several microsporangia (pollen sacs) per microsporophyll. C. Flattened branchlets of *Calocedrus decurrens* with opposite scale-like leaves and woody cones with flattened cone scales. D. Flattened branchlets of *Platycladus orientalis* with opposite scale-like leaves and woody cones with flattened cone scales. E. Branchlets of *Sequoiadendron giganteum* with alternate awl-like leaves and woody ovulate cone with peltate cone scales. F. Ovulate cone of *Cupressus* sp. with peltate cone scales. G. Branchlets of *Juniperus* sp. with whorled scale-like leaves and fleshy ovulate cones. H. Winged seeds of *Sequoia.* I. Non-winged seeds of *Cupressus.*

# FAMILY DESCRIPTION

Shrubs or trees. Leaves alternate, opposite, or whorled, simple, scale-like, awl-shaped or linear, in one genus connate side-by-side in pairs; leafy branchlets sometimes deciduous in age. Plants monoecious or dioecious. Pollen and ovules borne in cones. Pollen cones usually small with scales spirally arranged, opposite or whorled; pollen sacs 2–9 per scale, borne on undersurface of scale. Ovulate cones mostly woody at maturity (or fleshy and berry-like in one common genus); cone scales several to many (1–3), flattened or peltate, alternate, opposite or whorled, adnate almost to apex to subtending bract; ovules 2–many (1) per cone scale; cone scales separating at maturity or cones disintegrating (cones indehiscent and fleshy); seeds small, wingless or narrowly winged, falling free from cone scales.

# FAMILY NOTES

We are combining here the plants treated by many authors as two families, the Cupressaceae and Taxodiaceae. Although representatives of the Taxodiaceae and Cupressaceae are quite distinctive in North America, the families are not readily separable when considered on a worldwide scope. See Eckenwalder (1976) for a full discussion of the rationale for combining these two families. Recent cladistic analyses (e.g., Hart, 1987) have supported the combination of the two families. Some recent authors (e.g., Brummitt, 1992), however, continue to recognize Taxodiaceae as distinct from Cupressaceae.

The Cupressaceae has a comparatively large number of genera but relatively few species. Sixteen of the 28 genera are monotypic and most of these are narrowly endemic. Over 60 percent of the species are in only three genera, *Juniperus* (50), *Callitris* (14), *Cupressus* (13). The family is worldwide in distribution with several areas of high endemism. The fossil record indicates that some genera were formerly much more widely distributed than at present. Several genera have major disjunctions in their ranges.

Leaves in the Cupressaceae vary from short appressed scales (Figure 10-9 C, D, G) to ascending or stiffly spreading awl-shaped structures (Figure 10-9 E) or short needles (Figure 10-9 A, B). They are opposite or whorled (Figure 10-9 A–C) in the genera traditionally assigned to the Cupressaceae *(sensu stricto)* and alternate (Figure 10-9 A, B, E) in most genera assigned to the Taxodiaceae. However, the Taxodiaceous genus *Metasequoia* has opposite leaves.

Pollen cones in the Cupressaceae bear two to several pollen sacs per cone scale (Figure 10-9 B). The pollen cones are generally smaller and have fewer scales than in the Pinaceae.

Ovulate cones in the Cupressaceae have cone bracts and cone scales wholly adnate. The cones in genera such as *Sequoia, Sequoiadendron* and *Cupressus* have peltate cone scales (Figure 10-9 A, E, F) with numerous ovules scattered on the undersurface. In genera such as *Calocedrus* and *Platycladus* the cones (Figure 10-9 C, D) have flattened scales and only two ovules per scale. In *Juniperus* the cone scales are fleshy and completely enclose the few seeds in a berrylike aggregate at maturity (Figure 10-9 G).

Members of the Cupressaceae are widely grown as ornamentals and some are important lumber trees, especially where decay-resistant wood is needed. The wood tends to be full of tannins and/or aromatic compounds that repel insects. The wood of some species is used to make cedar chests. The California coast redwood is renowned for its attractive and decay-resistant wood. Berries of juniper are used to flavor gin. Important genera include *Platycladus* and *Thuja* (both called arbor vitae), *Juniperus* (juniper), *Calocedrus* (incense-cedar), *Taxodium* (bald-cypress), *Cupressus* (cypress), *Metasequoia* (dawn-redwood), *Sequoia* (coast redwood) and *Sequoiadendron* (Sierra redwood).

# Exercises

## ■ Study Questions

### Gymnosperms—General

1. What is meant by the name, "gymnosperm"?

   What do angiosperms have that gymnosperms lack?

   What do both angiosperms and gymnosperms have that ferns and fern allies lack?

2. Where are the sporangia of gymnosperms located?

   Where are the sporangia of angiosperms located?

   Are gymnosperms and angiosperms homosporous or heterosporous?

3. What do we call the microgametophyte of a gymnosperm or an angiosperm?

   By what process does it change from a microspore?

4. Where would we find the megagametophyte of a gymnosperm or an angiosperm?

   By what process does it change from a megaspore?

   What structures are present in gymnosperm megagametophytes that are absent in angiosperms?

5. What is the pollen-receptive surface in a gymnosperm?

   What receives pollen in an angiosperm?

6. Where does fertilization take place in most gymnosperms? (Be specific.)

   What process takes place when fertilization occurs in angiosperms that generally does not take place in gymnosperms?

   What special tissue do angiosperm seeds have that gymnosperm seeds never produce?

### Cycadaceae and Zamiaceae

7. Describe the sporophyte of a member of the Cycadaceae.

   What do the stems look like?

   How are the leaves arranged?

   Are the leaves microphylls or megaphylls?

   In what ways do the leaves differ from those of other gymnosperms?

8. Where is pollen produced in cycads?

9. Describe the location of the ovules in the Cycadaceae.

   Describe the location of the ovules in the Zamiaceae.

10. How do sperm get to the eggs in cycads?

    How does this differ from the situation in most other gymnosperms?

11. Cycads are often described as "living fossils." Why?

# Ginkgoaceae

12. Describe the sporophyte of a member of the Ginkgoaceae.

    What do the stems look like?

    How are the leaves arranged?

    What is the leaf venation?

    Describe what the leaves look like.

13. Where is pollen produced in *Ginkgo?*

14. Where does *Ginkgo* produce ovules?

    Describe the appearance of the ovule.

    What are the features of *Ginkgo* seeds?

15. How do sperm get to the eggs in *Ginkgo?*

    What do the sperm cells of *Ginkgo* have in common with those of cycads?

16. How do the diversity and geographic distribution of the Ginkgoaceae today compare with earlier periods of earth history?

# Pinaceae

17. Describe the sporophyte of a typical member of the Pinaceae.

    How are the leaves arranged?

    Describe a typical Pinaceae leaf.

18. Some members of the Pinaceae have well-differentiated "short shoots" and "long shoots."

    What is the difference between a long shoot and a short shoot?

19. Members of the genus *Pinus* have different kinds of leaves on the two types of shoots. What kind of leaves are produced on the long shoots?

    What kind are produced on the short shoots?

    What is a fascicle of needles of a pine?

    What is the sheath at the base of the fascicle composed of?

20. Where is pollen produced in members of the Pinaceae?

    How is pollen disseminated?

    What is the pollen-receptive surface?

21. Describe the structure of an ovulate cone in the pine family.

    What is the relationship between cone scales and cone bracts?

    Where are ovules located?

22. Ovulate cones in the Pinaceae are described as complex cones whereas those in the Cycadaceae are simple cones. What are the differences?

23. Describe a seed of a member of the Pinaceae.

    What is the difference between one of these seeds and a samara?

    What is the storage tissue in one of these seeds?

    Why is the "pinyon nut" a misnomer?

# Cupressaceae

24. What kinds of habit do we find in the Cupressaceae?

    Leaf arrangement?

    Describe the different kinds of leaves that occur in this family.

25. Where is pollen produced in members of the Cupressaceae?

    How is pollen disseminated?

26. In what ways does a cone of the Cupressaceae differ from a cone in the Pinaceae?

27. How do ovulate cones of *Juniperus* differ from those of other genera in the Cupressaceae?

28. List the families and species of gymnosperms that you identified in the lab.

_____       _____

_____       _____

_____       _____

_____       _____

_____       _____

_____       _____

# Introduction to the Flowering Plants

## chapter 11

The most important plants on the surface of the earth today are the flowering plants. Angiosperms dominate most terrestrial and semiaquatic habitats. They are the base for most terrestrial food chains, and many kinds of animals are wholly dependent on them for food, cover, and other aspects of their lives. Their stems and leaves provide much of the shelter, paper, and vegetable foods used by humans and domestic animals. The beauty of flowers provides us with aesthetic pleasure.

Many substances that we use in our daily lives are plant products. Numerous medicines were first isolated from species of flowering plants, and essentially all of the spices and culinary herbs are products of flowering plants. Many of the drugs and flavorings that are now synthesized in laboratories were originally discovered in extracts of flowering plants. Many industrial products such as rubber, paint bases, nonpetroleum oils, resins and sizing starches are derived from flowering plants.

Most important of all are the edible plant products that are the food base of human culture. The earliest civilizations are correlated with the domestication of one or another of the grasses such as wheat, rice, or corn. The food bases for various tropical and alpine cultures are roots, tubers, corms and bulbs produced by various unrelated kinds of flowering plants. So important and all pervasive are flowering plants and their products that it would be a difficult exercise to imagine a world without them. Certainly, the earth could not support more than a tiny fraction of the existing human population without flowering plants. It is unlikely that humans would have evolved in the absence of the angiosperms.

## Origin and Relationships of Angiosperms

The earliest flowering plants apparently evolved near the beginning of the Cretaceous period (about 100 million years ago). At the time that the angiosperms first made their appearance, the gymnosperms and ferns were the dominant plants on the land. There is very little tangible evidence for what the immediate ancestors of the flowering plants looked like. No angiosperm fossils have been found in sediments deposited prior to the Cretaceous period, and the earliest known angiosperms apparently already possessed all the characteristic angiosperm features. The ancestors of the angiosperms were plants that, if alive today, would probably be classified as gymnosperms. However no surviving gymnosperm family is considered to be a likely ancestor or even a close relative for the angiosperms. By the end of the Cretaceous the angiosperms had become very widely distributed and soon became the dominant plants that they are today.

The angiosperms are generally grouped together as monocotyledons and dicotyledons. The earliest known angiosperms were apparently dicots, but well before the end of the Cretaceous Period, monocots had evolved and began to diversify. Most of the very early angiosperm fossils represent lineages that are apparently extinct or have descendants so changed that their relationships are uncertain. However, some monocot and dicot families (e.g., Arecaceae and Fabaceae) are very ancient and were present during the Cretaceous. Many families are apparently much more recent.

Many evolutionary novelties have arisen since the angiosperms first appeared on the earth's surface. The earliest known angiosperms had features that are retained by some members of the subclass Magnoliidae.

# The Angiosperm Life Cycle

A characteristic feature of the angiosperms is the grouping of reproductive structures with sterile auxiliary ones into a single unit known as the **flower.** The original form of the flower remains a subject of much discussion among taxonomists and paleobotanists. Most consider the flower to represent a shortened branch bearing several kinds of modified leaves. The outermost of these, the sepals and petals are sterile (nonreproductive) and the inner two whorls, stamens (microsporophylls) and carpels (megasporophylls) are reproductive (producing spores through the process of meiosis). There is a large gap between the structure of any known fossil or living gymnosperm and that of the flower. However, some gymnosperms have single features that approach but do not exactly equal angiosperm features.

The reproductive life cycle of flowering plants (Figure 11-1) involves several unique features. The structure for which the name angiosperm (covered seed) was coined is the **carpel.** Only angiosperms have carpels. All extant gymnosperms bear ovules that are exposed to the air at the time of pollination (see Chapter 10), and the micropyle of the ovule itself is the pollen receptive area. This was probably the case in the gymnosperm lineage that gave rise to the angiosperms. At the time of pollination in the angiosperms, however, the ovules are completely enclosed within a hollow chamber formed by the folded carpel and a specialized portion of the carpel (the stigma) serves as the pollen receptive site.

Carpels are usually differentiated into three regions: the pollen receptive **stigma,** a slender connecting **style,** and the hollow **ovary** that contains one or more ovules. The ovule is a megasporangium **(nucelleus)** enwrapped by one or more additional tissue layers **(integuments).** Within the nucellus, one diploid cell, the **megasporocyte** (megaspore mother cell), undergoes meiosis producing four haploid megaspores. The next stages are variable among the various groups of angiosperms, and only the most basic sequence is presented here. Three of the four megaspores generally degenerate soon after meiosis. Three mitotic divisions [usually without cell wall formation] then take place within the remaining **functional megaspore** resulting in a large 8-nucleate cell, the embryo sac, a structure unique to the angiosperms. The embryo sac is the female gametophyte phase of an angiosperm's life cycle. Two of the eight nuclei, the **polar nuclei,** cluster together near the center of the embryo sac while another nucleus, the **egg nucleus,** is located

---

**Figure 11-1** Life cycle of a typical angiosperm. A. Branch of an adult sporophyte bearing a perfect flower. B–E. Stages in development of pollen grains (microgametophytes). B. Individual stamen bearing anther composed of 4 microsporangia. C. Cross section of anther with 4 microsporangia, each containing numerous microsporocytes (microspore mother cells). D. A single microsporangium after meiosis but before the individual microspores have split from their tetrads. E. Release of pollen grains (microgametophytes) from anther. F. Cross section of individual 2-nucleate pollen grain at the time it is released from the anther. G–M. Stages in the development of the embryo sac (megagametophyte). G. Longitudinal section of pistil with one basal ovule. H. Pistil with l.s. of ovule showing the single megasporocyte (megaspore mother cell) within the nucellus or megasporangium (shaded area) surrounded by 2 integumentary layers. I–M. Ovule (l.s.). I. Linear tetrad of megaspores formed by meiosis. J. Degeneration of three megaspores leaving one functional megaspore. K–M. Formation of 8-nucleate embryo sac (megagametophyte) by three mitotic divisions of the functional megaspore. N. Pistil (l.s.) after pollination with pollen on the stigma and pollen tubes growing through the style. Note that the generative cell has divided by mitosis forming two sperm nuclei. O–Q. Fertilization and development of seed. O. Ovule and embryo sac (l.s.) at time of double fertilization with one sperm uniting with the egg cell forming the zygote and the second sperm uniting with the two polar nuclei forming the triploid fusion nucleus. P. Developing seed containing triploid endosperm formed by mitosis from the fusion nucleus and the young diploid embryo formed by mitosis from the zygote. Q. Mature seed (l.s.) at time of dispersal with mature embryo embedded in nutritive endosperm surrounded by the hardened integuments (seed coats) and thin layer of perisperm (mature nucellus).

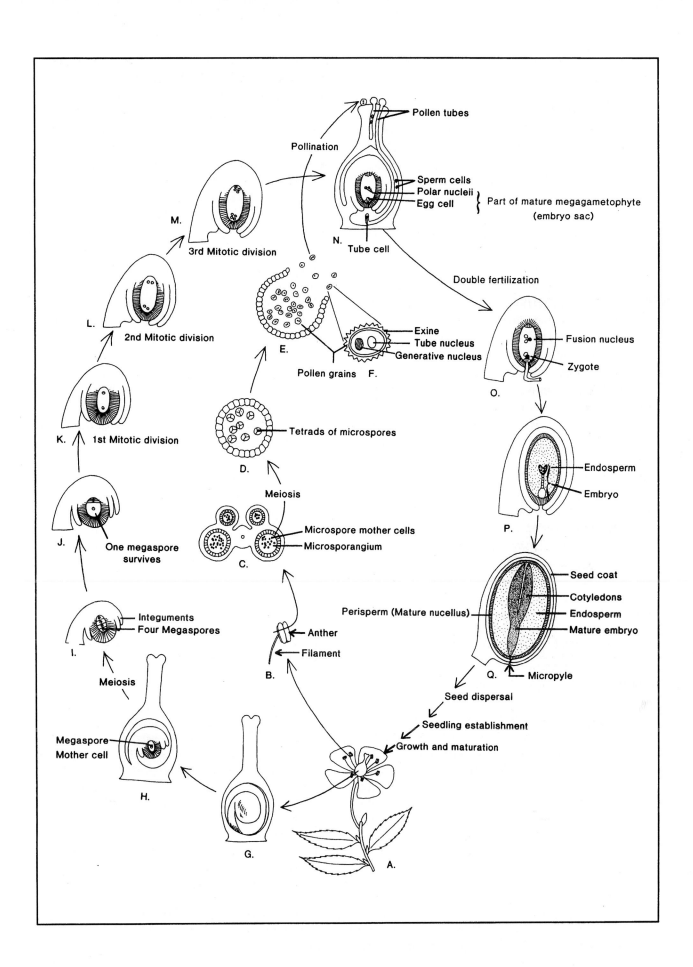

Pollen tubes

Pollination

Sperm cells
Polar nucleii
Egg cell
} Part of mature megagametophyte (embryo sac)

N.
Tube cell

M.
3rd Mitotic division

Double fertilization

L.
2nd Mitotic division

Exine
Tube nucleus
Generative nucleus

E.

Pollen grains   F.

Fusion nucleus

Zygote

O.

K.    1st Mitotic division

Tetrads of microspores

Endosperm

Embryo

D.

P.

Meiosis

J.

One megaspore
survives

Microspore mother cells
Microsporangium

Seed coat

Cotyledons

Perisperm (Mature nucellus)

Endosperm

Mature embryo

C.

Integuments
Four Megaspores

Q.    Micropyle

I.

Anther

Seed dispersal

Filament

Seedling establishment

Meiosis

B.

Growth and maturation

Megaspore
Mother cell

H.

G.

A.

near the micropyle of the ovule. The ovule is usually at this stage when the pollen tube arrives at the micropyle.

Pollen production occurs within **anthers** (hollow sac-like parts of the stamens). Each anther comprises one to four microsporangia. A microsporangium contains several to many diploid **microsporocytes (microspore mother cells),** each of which undergoes meiosis, resulting in the production of four haploid microspores. The nucleus of each microspore then undergoes a mitotic division forming a **tube nucleus** and a **generative nucleus.** This two celled structure is a **pollen grain;** it is the male gametophyte of an angiosperm. A thickened outer cell wall or **exine** is deposited as the pollen grain matures. Usually the anther opens at this stage and releases the pollen to be transported to the stigma **(pollination).** The pollen grain remains in this **2-nucleate** condition until it attaches to the stigma whereupon the generative cell divides to form two sperm cells. Sometimes the generative nucleus divides immediately while the pollen is still in the anther and the pollen is shed in a **3-nucleate** condition.

The transport of pollen from anther to stigma of the same species is essential to reproductive success. This process **(pollination)** often involves some external agent that transports the pollen. The various configurations of flower parts are considered to be adaptations that have evolved in response to one or another pollination mechanism. Pollination may be brought about by such abiotic agents as wind and water currents or by various groups of animals including many different kinds of insects, birds, bats and even rodents. Some plants are self-pollinating with the pollen transferred directly from stamen to stigma without the intervention of any outside agent.

Upon arriving at the appropriate stigma, a **pollen tube** emerges through the pollen grain wall and grows through the stigma and style into the locule of the ovary where the ovules are located. Nutrients are provided to the developing pollen tube by the tissues of the stigma and style. The growth of the pollen tube through the style may be very rapid (as much as 3 cm in an hour). When a pollen tube passes through the micropyle of an ovule, the wall of the pollen tube breaks down and the sperm nuclei are released.

The next stage in the life cycle, **double fertilization,** is unique to the angiosperms. One sperm nucleus from the pollen tube unites with the egg cell forming a diploid zygote, and the second sperm nucleus unites with the two polar nuclei producing a triploid fusion nucleus. The fusion nucleus immediately undergoes a rapid sequence of mitotic divisions forming a nutritive tissue unique to the angiosperms, **endosperm.** As the endosperm develops, the zygote also divides mitotically. The cells derived from the zygote differentiate into an **embryo** or tiny sporophyte. The developing endosperm and embryo are supplied with nutrients by the surrounding nucellar tissues. The embryo, endosperm, and hardened integuments mature as a seed. In some seeds the nucellus also matures as a layer of storage tissue, the **perisperm.**

Concurrently with seed formation, the carpellary tissue develops into the last of the uniquely angiospermous features, the **fruit.** Fruits are extremely variable. Some are hard and dry whereas others are soft and fleshy. Some break open, releasing seeds into the surrounding environment. Others develop air foils or buoyancy chambers that allow the fruits to float on air or water currents. Some hitch a ride on passing animals by sharp hooks and prickles. Seed dispersal mechanisms are as diverse and as effective as pollination mechanisms and help account for the fact that angiosperms are the most common and wide-spread group of land plants on the earth today.

# Survey of Flower and Fruit Terminology

The flowering plants are the most numerous, most diverse and most important plants over much of the earth's land surface. An understanding of their reproductive structures is basic to any introductory plant taxonomy class. It is very important for taxonomy students to become familiar with the terms used to describe flowers and fruits. This chapter presents a working vocabulary of the most important terms. The list of terms introduced here is not meant to be comprehensive, but it includes most of the terms commonly encountered in identification keys.

## The Flower

A flower is generally defined as a reduced branch system with the leaves modified into **sepals, petals, stamens,** and **carpels.** The reduced stem axis to

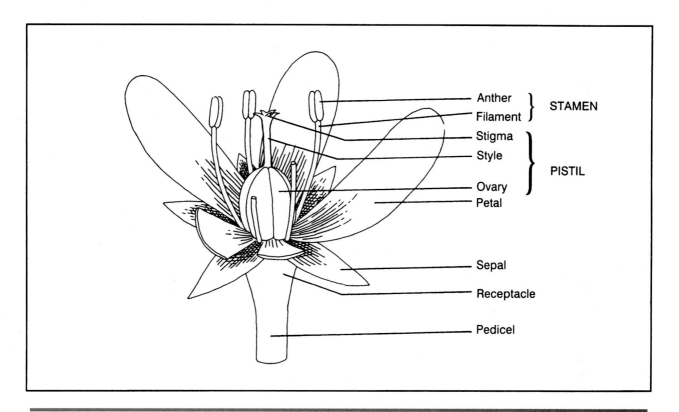

**Figure 11-2** Diagrammatic representation of a complete, 5-merous flower.

which these parts are attached is the **receptacle.** Often a flower contains **nectaries** (nectar-secreting glands) as well. The four principal types of flower parts are defined primarily by their position and only secondarily by their function within a flower. Figure 11-2 shows a diagrammatic representation of a single **complete** flower (one that has all four whorls of parts present). If one or more of these whorls are absent, the flower is **incomplete.**

Each of the four principal types of flower parts, the sepals, petals, stamens and carpels, is highly variable. Among the 300,000+ species of flowering plants there is much variation in number, size, shape, form, position, and fusion of the parts of each whorl as well as relationships among the parts of adjacent whorls. Because fusion is common and because we often need to refer to all the members of a whorl of flower parts together as a group or unit, we sometimes use collective terms. Table 11-1 lists the individual types of flower parts and the terms that describe whorls of the individual parts. For example, the calyx is the collective term for all the sepals of a flower.

Another way to look at a flower is from a functional point of view. A flower that produces both

| Individual parts | Whorl(s) of parts | |
|---|---|---|
| Sepals _____ | Calyx | Perianth |
| Petals _____ | Corolla | |
| Stamens _____ | Androecium | |
| Carpels _____ | Gynoecium | |

**Table 11-1** Principal flower parts and the collective terms that describe the whorls.

pollen (male) and ovules (female) is said to be **perfect.** If one of these is absent, then the flower is **imperfect.** Imperfect flowers are **staminate** if only stamens are functional (pollen-producing), and **pistillate** if only carpels are functional (ovule-producing). [Imperfect flowers may have non-functional carpels or non-functional stamens, or these parts may be absent entirely]. A plant with imperfect flowers is **dioecious** if staminate and pistillate flowers are borne on different individuals and **monoecious** if staminate and pistillate flowers are borne on the same individual. If all the flowers of a plant are perfect, the plant may be described as

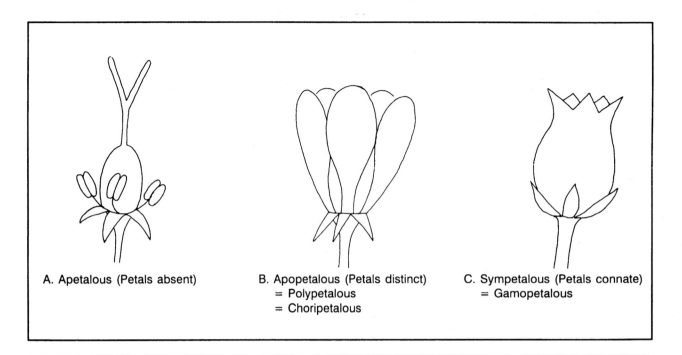

A. Apetalous (Petals absent)

B. Apopetalous (Petals distinct)
= Polypetalous
= Choripetalous

C. Sympetalous (Petals connate)
= Gamopetalous

**Figure 11-3** Terms describing the presence or absence and degree of connation of the petals.

**synoecious.** It is **polygamous** if there is a mixture of imperfect and perfect flowers.

One of the common trends in flowers is fusion of parts. Two or more parts of the same embryonic origin that are united are **connate.** Parts of the same embryonic origin that are not joined or connote are **distinct.** Connate petals, for instance, are fused side-by-side forming a ring or tube, and distinct petals are separate from each other and individually attached to the receptacle.

Parts of different origin that are fused together are **adnate.** Unlike parts that are not joined together are **free** from each other. For example, stamens may be adnate to the petals or free from them.

## The Perianth

The perianth is the set of sterile (nonreproductive) flower parts that surround the stamens and carpels. In many cases the perianth is **biseriate,** consisting of an outer ring of sepals and an inner ring of petals. The calyx is the outermost whorl of the flower parts. In some cases the perianth is **uniseriate,** consisting of only one whorl. A general working assumption that taxonomists often apply in this situation is that if only one ring of perianth parts is present, the parts present are sepals and the missing parts are petals. This assumption is occasionally found to be inaccurate in specific cases, but it is a good starting point.

A flower that lacks petals is **apetalous.** Sometimes perianth parts are spirally arranged rather than in whorls or are represented by several whorls. In such a flower there may be intergradation in color and texture among the different series or whorls of perianth. In these cases the decision as to which parts are sepals and which are petals may be rather arbitrary.

Sepals are most commonly green and leaf-like **(sepaloid)** but it is not unusual for them to have the color and texture usually associated with petals **(petaloid).** Sepals may be distinct or variously connate. A calyx of connate sepals is **synsepalous.** The fused portion of a synsepalous calyx is a **calyx tube,** and the separate tips are **calyx lobes.** Sepals may be entirely missing in some cases. They may fall off as the flower opens **(caducous sepals)** or persist through various lengths of time of flowering and fruiting.

The corolla is the inner of the two sterile whorls when both whorls are present. Petals often are brightly colored but they are sepaloid in some flowers. Three different synonymous terms are used to describe a corolla of distinct petals: **apopetalous, choripetalous** and **polypetalous.** If the petals are connate into a ring or tube and fall together as a unit, the corolla may be described as either **sympetalous** or **gamopetalous.** Generally if the number of sepals and petals is the same, the petals alternate

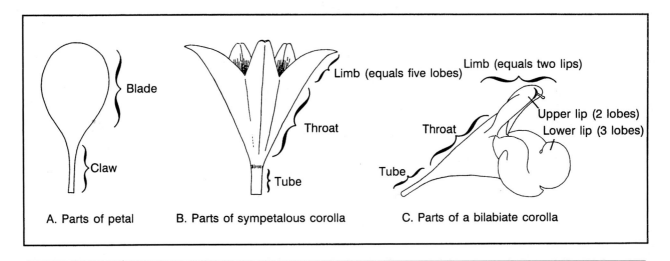

A. Parts of petal     B. Parts of sympetalous corolla     C. Parts of a bilabiate corolla

**Figure 11-4**  Terms describing corolla parts.

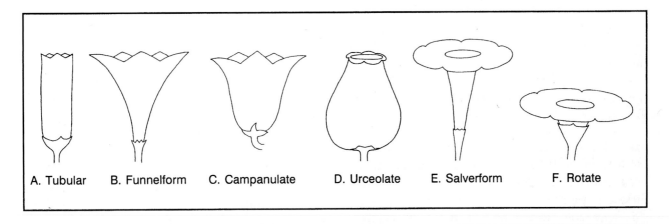

A. Tubular   B. Funnelform   C. Campanulate   D. Urceolate   E. Salverform   F. Rotate

**Figure 11-5**  Terms describing shapes of sympetalous corollas. These terms can also be used to describe the shape of a calyx or a hypanthium.

with the sepals. This means that a line drawn from the center of a flower through the middle of a petal will pass between two sepals.

Because the corolla is often the largest and most conspicuous whorl in the flower, it usually has the most to do with determining the overall appearance of the flower. The shape of a corolla depends on the number, relative size, placement, and shape of the petals. The number of petals can be combined with the suffix, **"-merous"** to describe the corolla. For example, monocots most commonly have a 3-merous corolla; dicot corollas most commonly are 4-merous or 5-merous. The shape of individual petals is extremely variable. Some petals have a thin stalk-like basal portion called a **claw** and an expanded **blade** (Figure 11-4 A); others are sessile. Sympetalous corollas are often differentiated into three sections (Figure 11-4 B): a cylindrical basal **tube,** a gradually expanding **throat** and a flaring **limb.** The limb is usually lobed. Any of these portions can be present, absent or scarcely developed. The relative proportions of these sections determine the overall shape of the corolla (Figure 11-5).

The appearance of a flower is strongly influenced by the **symmetry** of the corolla (Figure 11-6). Taxonomists recognize two common symmetry types. A flower is **radial (actinomorphic, regular)** if all the petals are of equal size and shape and equally spaced around the axis of the flower. A radially symmetrical flower can be divided into two mirror images by drawing a line from the tip of any petal through the middle of the flower. A **bilaterally symmetrical** flower **(zygomorphic, irregular)** has only one line of symmetry. Only a line drawn vertically

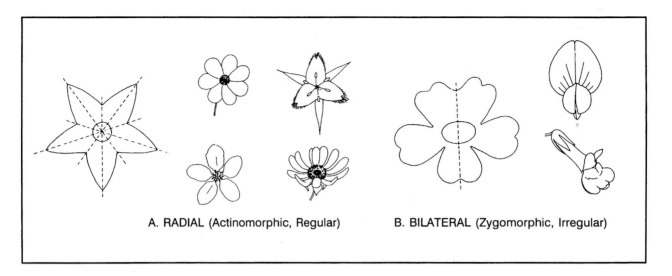

A. RADIAL (Actinomorphic, Regular)  B. BILATERAL (Zygomorphic, Irregular)

**Figure 11-6** Diagrams and examples of floral symmetry. The flowers can be divided along the dashed lines to form mirror images.

through the flower can divide the flower into mirror images. This line of symmetry divides the flower into a left and a right side. A sympetalous corolla with bilateral symmetry is often **bilabiate** (two-lipped). From the side it looks like an open mouth with an **upper lip** and a **lower lip** (Figure 11-4 C).

## The Androecium

The androecium consists of the stamens of a flower, usually arranged in a spiraled or whorled pattern. Stamens are located around the gynoecium if carpels are present and in the center of the flower if the gynoecium is absent. Most stamens consist of a narrow thread-like **filament** (stalk) bearing a terminal, 2–4-lobed sac-like **anther** filled with **pollen** (Figure 11-7 A). Occasionally the filaments are absent or nearly so and the anthers are attached directly to the receptacle. Rarely stamens are broad and leaf-like **(laminar).** In most plants, an anther is composed of 2–4 pollen sacs joined by a sterile region called the **connective.**

Stamens may be fertile or sterile. Fertile stamens produce pollen. **Staminodes** (sterile stamens) have lost the ability to make pollen and are extremely variable in form and size. These structures sometimes can be recognized only by their position within the androecial whorl. Staminodes occasionally are petaloid in color and texture. In fact, the structures commonly considered to be petals in some families probably originated as staminodes. Staminodes may secrete nectar or may be reduced to minute remnants with no apparent function.

Stamens may be distinct and free or variously fused to each other or to other parts of the flower. If the filaments of all the stamens are connate into a ring, the stamens are **monadelphous** (united in one group; Figure 11-7 B). In many members of the pea family the stamens are **diadelphous** (in two groups); in such plants there are ten stamens, nine of which are connate by their filaments into a U-shaped tube and one which is distinct. In members of several other families stamens are connate in bunches **(polyadelphous),** often with several to many stamens each; these bunches may be alternate with or opposite the petals. In the very large sunflower family (Chapter 15) and several smaller families, the filaments are distinct but the anthers are connate, forming a tube that surrounds the style.

Stamens may be attached directly to the receptacle or may be adnate to other parts of the flower. Stamens with filaments adnate to the petals are **epipetalous** (Figure 11-7 C). The adnation of the bases of all three outer whorls of flower parts (sepals, petals, and stamens) results in a structure called a **hypanthium** (Figure 11-10 B). In flowers with an inferior ovary the bases of the stamens and the perianth parts are adnate to the ovary.

Whether they are joined or distinct, the stamens of a flower do not necessarily all look alike. Filaments may be of different lengths (Figure 11-7 C), or anthers may be of different sizes or shapes within the same flower.

The number of stamens in a flower is extremely variable. Identification keys often give the number

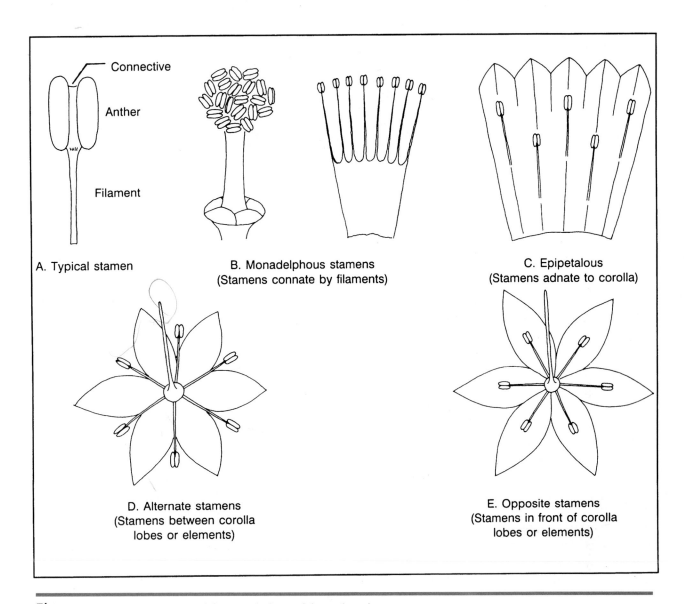

A. Typical stamen

B. Monadelphous stamens
(Stamens connate by filaments)

C. Epipetalous
(Stamens adnate to corolla)

D. Alternate stamens
(Stamens between corolla
lobes or elements)

E. Opposite stamens
(Stamens in front of corolla
lobes or elements)

**Figure**  Common structure and form variations of the androecium.
11-7

of stamens relative to the number of corolla elements. For example stamens may be fewer than, equal to, twice as many as, or more than twice as many as the petals. In the last case the stamens are often considered to be **numerous** or **many.** Stamens are also considered to be numerous when the number of both petals and stamens is too high to be conveniently counted. When the stamens are as many as or fewer than the petals they are **alternate** with the petals (Figure 11-7 D) if they attach between the petals or **opposite** if they are directly in front of the petals (Figure 11-7 E). In a similar fashion stamens may be alternate with or opposite the sepals.

## The Gynoecium

In all flowers that have a gynoecium, the carpels form the terminal or centermost whorl of parts on the floral axis. The carpels are the basic units of the gynoecium. Carpels are thought to have originated many millions of years ago as flat, leaflike structures that bore ovules on their margins. In the ancestor to the angiosperms these structures became folded or inrolled on themselves so as to form a hollow structure, the ovary, that enclosed one or more ovules. The margins of the carpels became fused together and the enclosed ovules remained attached to them (Figure 11-8 A–C). Almost all modern angiosperms have fully closed carpels. A pea pod is an example of a carpel.

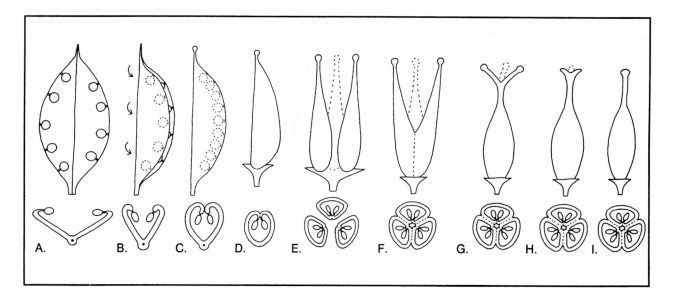

**Figure 11-8** Variation in fusion and closure of carpels. A–C. Origin of the carpel from a flat, leaflike megasporophyll. D. Flower with a monocarpous gynoecium. E. Flower with an apocarpous gynoecium. F–I. Flowers with syncarpous gynoecia illustrating varying degrees of carpellary fusion. Note: the number of carpels and locules in E through I is shown as three for convenience only. The number of carpels in an apocarpous or syncarpous gynoecium ranges from two to many.

The number of carpels in a flower and the degree of carpellary fusion is quite variable. Some angiosperms have only a single carpel per flower. These have a **monocarpous gynoecium** (Figure 11-8 D). Others have two or more carpels. If the carpels are distinct from each other the gynoecium is **apocarpous** (Figure 11-8 E). If the carpels are connate into a single unit, the gynoecium is **syncarpous** (Figure 11-8 F–I). Another term often used to describe the gynoecium is **pistil** (Figure 11-2). A pistil is the visual unit of the gynoecium and consists of an **ovary,** one or more **stigmas** (pollen receptive surfaces), and usually one or more **styles** (slender connections between stigma and ovary). A pistil may consist of only one carpel or of two or more connate carpels. If the gynoecium is monocarpous, it comprises a single one-carpellate pistil. A one-carpellate pistil is a **simple pistil.** An apocarpous gynoecium consists of two or more simple pistils. When two or more carpels are connate, forming a syncarpous gynoecium, the resulting unit structure is a **compound pistil.** A compound pistil has a **compound ovary.** Often only the lower parts of the carpels are fused. A compound pistil may have two or more styles, style branches or stigmas (Figure 11-8 F–I).

It is often necessary to determine the number of carpels that comprise a pistil. There are five criteria that can be used. Not all of them are necessarily present or equally useful in a given flower.

1. Number of stigmas.

2. Number of styles or style-branches.

3. Number of lobes of the ovary.

4. Number of locules (chambers in ovary).

5. Number of placentae. (Note: if actual seed rows can be counted, then the number of carpels is generally half as many as the number of seed rows.)

## Placentation

**Placentation** is the attachment of **ovules** (immature seeds) within the interior of an ovary. The placement of the ovules is quite variable and is dependent on the number of carpels, their degree of fusion and evolutionary modifications of the interior of the ovary and its contents. Figure 11-9 illustrates common placentation types. Note that all of these patterns are discernible only by sectioning the ovary. A cross section is best for observing marginal, parietal, axile and free-central placentation. Basal and apical placentation are best observed in a longitudinal section. Both sections may be needed to confirm the placentation type.

Placentation type is based on the nature of the ovule attachment within the ovary and on whether the ovary is simple or compound. If the ovary is

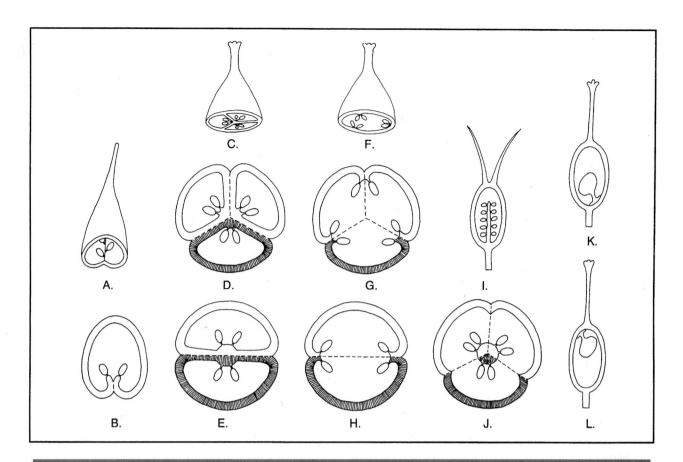

Figure 11-9 Common types of placentation and terms describing the structures associated with the attachment of ovules to the ovary wall. In examples of compound ovaries the dotted lines indicate carpel boundaries and shading indicates carpel tissue from one of the component carpels. A–B. Marginal placentation. C–E. Axile placentation. F–H. Parietal placentation. I–J. Free-central placentation. K. Basal Placentation. L. Apical placentation.

simple, the attachment is usually **marginal** with ovules attached in two vertical rows that correspond to the fused carpel margins. Compound ovaries have several different types of placentation. An ovary with **axile** placentation is divided by **septa** (partitions) into two to many **locules** (chambers) and the ovules are borne along the central axis. The axis comprises the fused margins of the component carpels of the compound ovary. An ovary with **parietal placentation** typically has no septa and the ovules are borne on the inner wall of the ovary. The carpels of an ovary with parietal placentation are fused together in an "open" position around a single central locule. Ovaries with free-central placentation are also one-loculed; the several to many ovules are attached to an elongate, sometimes free-standing placenta located in the middle of the locule. **Apical** and **basal** placentation occur in both simple and compound ovaries; the one or few ovules are restricted to the top or bottom of the locule respectively.

## Ovary Position and Insertion Points

The ovary is generally attached directly to the receptacle. The other flower whorls are also attached to the receptacle, but their *apparent place of attachment* may be to the rim of a hypanthium or to the sides or top of the ovary. The apparent point of attachment of the outer floral parts relative to the ovary is their place of **insertion** (Figure 11-10).

If the bases of the perianth parts and stamens are attached directly to the receptacle, the insertion is **hypogynous** (below the ovary; Figure 11-10 A). The ovary of such a flower is **superior** (free from the bases of the outer parts of the flower).

If the flower has a **hypanthium** (floral tube), formed by adnation of the bases of sepals, petals and stamens, the apparent point of attachment of the outer flower parts is the rim of the hypanthium rather than the receptacle. The shape of a hypanthium can resemble a flat disc, a bowl, a tube or even a flask. If the hypanthium is free from the ovary, the insertion

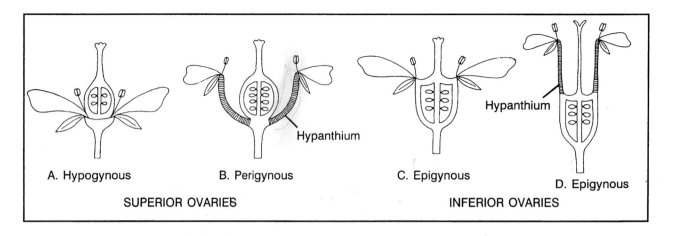

**Figure 11-10**   Ovary position and insertion of flower parts.

is **perigynous** (around the ovary; Figure 11-10 B) and the ovary is superior. Even if an ovary is completely surrounded by a tubular or flask-shaped hypanthium, the ovary is superior if its walls are free from the surrounding hypanthium.

If the bases of the perianth parts and stamens are adnate to the ovary wall (either directly or as a hypanthium), the insertion is **epigynous** (on top of the ovary; Figure 11-10 C, D). The apparent point of attachment of the outer flower parts is the top or sides of the ovary rather than the receptacle. In such a flower the ovary is **inferior** to the point of insertion of the outer flower parts. A flower with an inferior ovary *always* has epigynous insertion. A hypanthium may extend beyond the top of an inferior ovary (Figure 11-10 D).

## Nectaries

Flowers that are animal-pollinated often bear **nectaries.** These are secretory glands that exude a solution of sugars or other substances that are gathered by the animal as a food source. In the process of gathering nectar the animal is likely to come into contact with stamens and stigmas. As it moves from flower to flower gathering nectar, it effects pollination. Nectaries may be located in several different positions in a flower. Their structure and size are extremely variable. Some flowers have a nectary disc that sits below or surrounds the base of the ovary. Nectar glands may be located on the surface of a petal or sepal. In certain plants sepals or petals produce nectar inside **spurs** (hollow projections that jut from the surface of the perianth part). Staminodes may be modified into nectar-secreting glands.

In many monocots the nectaries are located on the septa of the ovaries. Non-floral structures such as bracts may occasionally secrete nectar. The chemical composition of nectar varies depending upon the nature of the pollinator that characteristically visits the flowers. Some nectars are rich in sugars and others contain significant quantities of amino acids.

## Inflorescences

Flowers may be borne individually on a plant or its branches or in groups called **inflorescences** (Figure 11-13). The flowers of an inflorescence may be very scattered or aggregated together into very tight clusters. Inflorescences may consist of only a few flowers or in some cases may contain thousands of flowers. An inflorescence (Figure 11-11 A) may be borne on a **peduncle** (inflorescence stalk) or it may be sessile. The flowers may be borne on **pedicels** (flower stalks) or they may be sessile. An elongated inflorescence has a central **rachis** (inflorescence axis). In addition to flowers inflorescences often contain **bracts** (modified leaves) in association with the flowers. A bract may differ from foliage leaves of the plant in size, shape, color, texture or in various other ways. In some plants an **involucre** (a ring or cup composed of several to many bracts) surrounds the flowers (Figure 11-11 B). A bract (or involucre) **subtends** a flower (or cluster of flowers) when it is attached directly below the flower.

The form of an inflorescence is dependent upon the pattern of its development. Inflorescences form as a result of the formation and differentiation of buds. An inflorescence begins its development from the activities of a terminal bud which produces bract

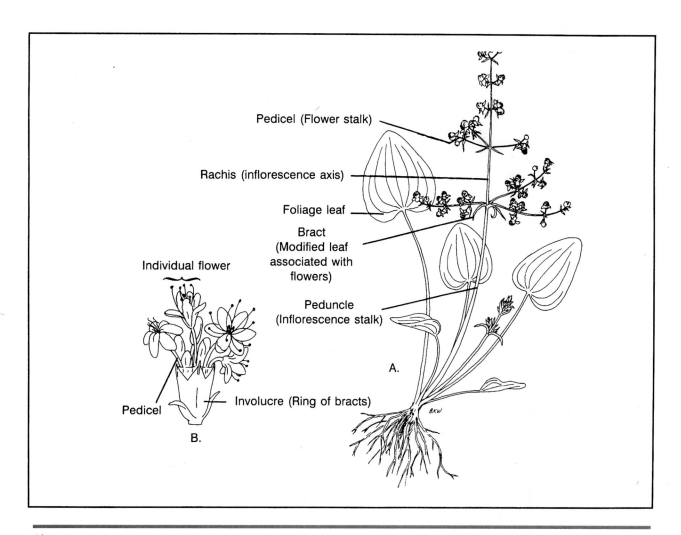

Pedicel (Flower stalk)

Rachis (inflorescence axis)

Foliage leaf

Bract
(Modified leaf
associated with
flowers)

Peduncle
(Inflorescence stalk)

Individual flower

A.

Pedicel

Involucre (Ring of bracts)

B.

**Figure** Terms used to describe the parts of an inflorescence. A. Plant of *Echinodorus rostratus* (Alismataceae).
**11-11** B. Involucrate cluster of flowers of *Eriogonum fasciculatum* (Polygonaceae).

primordia and bud primordia. There are two different patterns of inflorescence development.

If the apical meristem of the terminal bud from which the inflorescence is initiated remains active as the inflorescence develops and all of the flowers form from lateral buds, **indeterminate** (racemose) inflorescences of unrestricted growth result (Figure 11-12 A). These often are elongated inflorescences with a **rachis** (central axis) formed by elongation of cells produced by the terminal bud. Generally the first lateral buds to differentiate in such an inflorescence are the first to mature as flowers. As a result, the oldest flowers or buds are those located farthest from the terminal bud, at the base or the outside of the inflorescence.

**Determinate** (cymose) inflorescences have a different pattern of development (Figure 11-12 B). Soon after the inflorescence begins its development,

the terminal bud differentiates into a flower. Once a bud has differentiated as an immature flower, the stem it terminates is blocked from future growth in length and additional flowers can develop only from buds that have already formed. Since such an inflorescence is of limited growth it is determinate. In a determinate inflorescence the most mature flower or bud is located at the end of the stem or in the center of the inflorescence.

Determinate inflorescences sometimes are able to form additional flowers when the lateral buds that form before the terminal bud differentiates remain active long enough to form bracts and associated lateral buds of their own. When this happens the inflorescence becomes progressively more complex. In some cases a determinate inflorescence can *appear* to be indeterminate in the pattern of its growth. However, the *apparent rachis* of such an

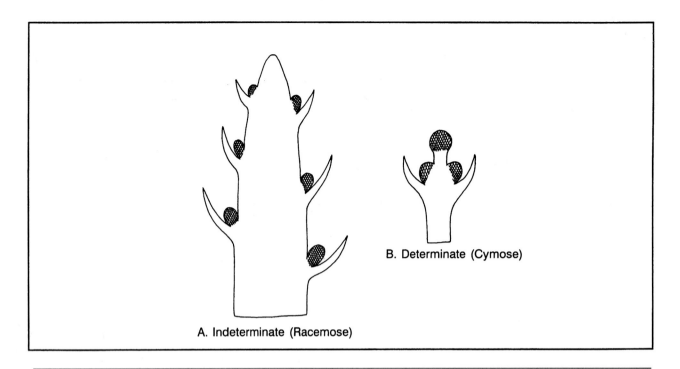

**Figure** Origin of indeterminate and determinate inflorescence types. Shaded areas are flower buds. The relative sizes
**11-12** of the buds indicate their ages and order of origin.

B. Determinate (Cymose)

A. Indeterminate (Racemose)

inflorescence (e.g., a scorpioid cyme) is **sympodial,** resulting from the combined activities of several to many different buds, each producing a section of the axis. A true rachis is the product of a single bud. Some types of **compound** (branching) inflorescences are truly intermediate in their development with some parts of the cluster developing in a determinate fashion and others in an indeterminate pattern [e.g., a paniculiform cyme].

## Key to Some Common Inflorescence Types

1. Terminal bud(s) active throughout development of inflorescence; the most mature flowers located at base of or to the outside of inflorescence or its branches. **[INDETERMINATE INFLORESCENCES]**
    2. Rachis present, inflorescence elongate.
        3. At least some flowers or their pedicels attached to secondary or tertiary rachises . . . . . . .**Panicle**
    3′ Flowers or their pedicels attached to primary rachis.
        4. Flowers sessile . . . . . . . . . . . . . . . . . . . . . . . . . . . . . . . . . . . . . . . . . . . . . . . . . . .**Spike**
        4′ Flowers borne on pedicels.
            5. Upper and lower pedicels equal in length . . . . . . . . . . . . . . . . . . . . . . . . . . . .**Raceme**
            5′ Upper pedicels much shorter than lower . . . . . . . . . . . . . . . . . . . . . . . . .**Corymb**
    2′ Rachis absent, inflorescence short.
        6. Flowers sessile . . . . . . . . . . . . . . . . . . . . . . . . . . . . . . . . . . . . . . . . . . . . . . . . . . . . . .**Head**
        6′ Flowers borne on pedicels
            7. Flowers attached directly to primary branch of inflorescence . . . . . . . . . . . . . .**Simple umbel**
            7′ Flowers in secondary clusters at ends of primary branches of
                inflorescence (rays) . . . . . . . . . . . . . . . . . . . . . . . . . . . . . . . . . .**Compound umbel**

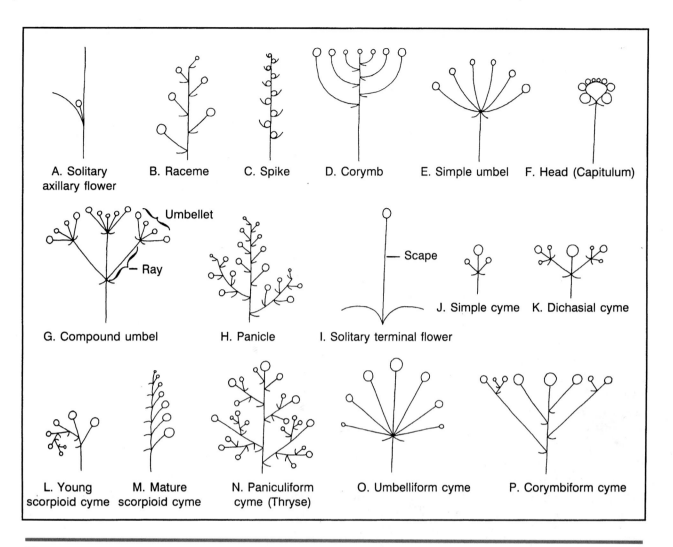

A. Solitary axillary flower  B. Raceme  C. Spike  D. Corymb  E. Simple umbel  F. Head (Capitulum)

G. Compound umbel — Umbellet, Ray  H. Panicle  I. Solitary terminal flower — Scape  J. Simple cyme  K. Dichasial cyme

L. Young scorpioid cyme  M. Mature scorpioid cyme  N. Paniculiform cyme (Thryse)  O. Umbelliform cyme  P. Corymbiform cyme

**Figure 11-13**  Common types of inflorescences. Relative age and maturity of flower buds is indicated by size.

1′ Terminal bud(s) used up in forming first flower(s) of inflorescence or branches of inflorescence; oldest flowers at top or center of inflorescence or its branches.   **[DETERMINATE INFLORESCENCES]**

   8.  Inflorescence one-sided, uncoiling as it develops, becoming elongate in age; the apparent rachis of the inflorescence sympodial . . . . . . . . . . . . . . . . . . . . . . . .**Scorpioid cyme**

   8′  Inflorescence not one sided; not uncoiling.

      9.  Inflorescence few flowered; consisting of a cluster of 3–4 flowers . . . . . . . . . . . . .**Simple cyme**

      9′  Inflorescences several- to many-flowered.

         10.  Rachis present, well developed; inflorescence elongate, pyramid shaped **Paniculiform cyme**

         10′  Rachis absent or very short; inflorescence short, rounded or flat-topped.

            11.  Inflorescence repeatedly forked . . . . . . . . . . . . . . . . . . . . . . . . . . . . . .**Dichasial cyme**

            11′  Inflorescence ± flat-topped or broadly rounded, with several elongated branches.

               12.  Branches of inflorescence all arising from the same point . . . . .**Umbelliform cyme** (cymose umbel)

               12′  Branches of inflorescence arising at different levels from a short rachis . . . . . . . . . . . . . . . . . . . . . . . . . . . . . . . . . . . . . . . . . . . . . . . . .**Corymbiform cyme**

# Fruits

The word "fruit" has different meanings to different people. To a layman it may carry the connotation of a soft, sweet plant product. To a botanist, however, the word has a more precise definition. Botanically, a fruit is a mature ovary together with any floral or vegetative structures that are attached to it and that become enlarged and ripen with it. Not all structures that a layman would consider to be a fruit fit the botanical definition, and many structures that a botanist considers to be fruits are neither sweet nor good to eat. In the discussion that follows, the botanical definition will be used.

Seed dispersal is the primary function of fruits. The pattern of ripening of the **pericarp** (ovary wall) determines the fruit's mature features. In some fruits the pericarp becomes very juicy as it matures. Often such a fruit has a high sugar content and is readily consumed by animals. In many cases, though, the seeds are hard and indigestible and are regurgitated or deposited in the animal's feces. This is an effective seed dispersal mechanism for many plants. The pericarp in many other plants dries as it matures. Often the pericarp is **dehiscent,** splitting open and releasing the seeds that it contains. In such plants the seeds are the units of dispersal. In other plants the fruit is **indehiscent;** the seeds remain enclosed within pericarp tissue and the fruit is the unit of dispersal. In certain plant families the pericarp breaks into pieces that retain the seeds on the inside. The pericarp may develop flattened, wing-like airfoils or tufts of long hair or bristles that enable the fruit to flutter or float through the air away from the parent plant. Fruits may be armed with bristles, hooks or barbs that cling to the fur or feathers of an animal. Seeds may have similar features.

The pericarp is often differentiated into layers (Figure 11-14 B), particularly in fleshy fruits. The **exocarp** is the outer skin-like layer. The **mesocarp** is the middle layer that often becomes soft and fleshy. Only certain kinds of fruits have a third layer, the **endocarp.** This innermost layer closely surrounds the seed or seeds. It may be hard and stonelike as in an olive pit or stiff and papery as in an apple core. Many fleshy fruits have no endocarp (e.g., a tomato).

Taxonomists and morphologists have developed a usable classification of fruits based primarily on features such as texture, nature of dehiscence, number of seeds, presence or absence of attached parts, and the morphology of the ovary from which the fruits were derived. The outline presented on the next page is not a dichotomous key, but it can be used in a similar fashion. For example, if a particular specimen has a simple fruit that is fleshy at maturity, select group IA and then decide which of the specific types it is. Note that if all of the features of the more specific type are not present, then the

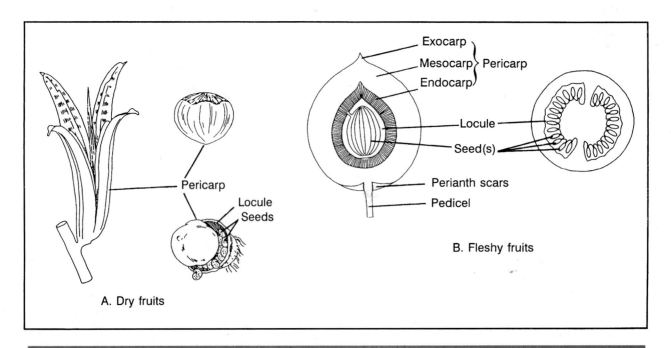

**Figure** Terms used to describe the parts of fruits.
**11-14**

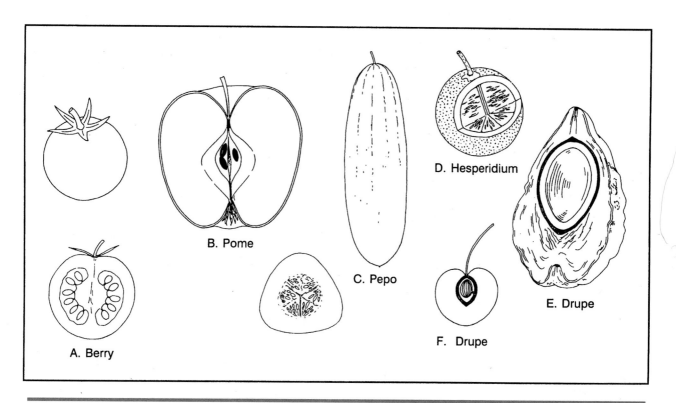

**Figure 11-15** Examples of fleshy fruits. A. Berry of *Lycopersicon* (tomato). B. Pome of *Malus* (apple). C. Pepo of *Cucumis* (cucumber). D. Hesperidium of *Citrus* (orange). E. Drupe of *Cocos* (coconut). F. Drupe of *Prunus* (cherry)

fruit is considered to be the generalized type. For example, if a fleshy fruit does not have the particular features of a drupe, pome, pepo or hesperidium then it is considered to be an unspecialized berry.

## Outline of Common Types of Fruits[1]

I. Fruit derived from a monocarpous or syncarpous gynoecium. (SIMPLE FRUITS)

  A. Fruit fleshy at maturity.

    1. **BERRY**—Entire pericarp soft; generally with two or more seeds. Derived from either superior or inferior ovaries (Figure 11-15 A). Examples: tomato, banana, grape.

      a. **Pome**—Derived from an inferior ovary and surrounded by a fleshy hypanthium. Restricted to the Rosaceae subfamily Maloideae (Figure 11-15 B). Examples: pear, apple, quince.

      b. **Pepo**—Derived from an inferior ovary with a leathery rind. Generally restricted to Cucurbitaceae (Figure 11-15 C). Examples: watermelon, pumpkin, squash, cucumber.

      c. **Hesperidium**—Derived from a superior ovary with a leathery pericarp having oil glands. The carpels are easily separated as segments that are lined with fleshy hairs (Figure 11-15 D). Restricted to *Citrus* and its close relatives (Rutaceae). Examples: orange, lemon, grapefruit.

    2. **DRUPE**—Pericarp distinguishable as three layers: exocarp, mesocarp, and endocarp. The endocarp is a "stone" that encloses the seed (Figure 11-15 E, F). Examples: peach, cherry, coconut, olive.

  B. Fruit dry, one-seeded, not breaking apart at maturity.

    3. **ACHENE**—Extremely variable in form. Seed and pericarp attached only by the funiculus (ovule stalk; Figure 11-16 A, D). Examples: buttercup, buckwheat.

[1]Prepared by Rhonda Riggins. Used by permission.

a. **Samara**—A winged achene (Figure 11-16 H–J). Examples: elm, ash, birch.

b. **Cypsella**—An achene with adnate calyx (pappus; Figure 11-16 B, C). Examples: members of the Asteraceae such as the dandelion, aster.

c. **Utricle**—A bladder achene with the wall loosely surrounding the seed (Figure 11-16 E). Example: pigweed.

4. **CARYOPSIS**—Seed coat and pericarp completely adnate (Figure 11-16 F, G). Restricted to Poaceae. Examples: maize, wheat, rye, barley, sorghum.

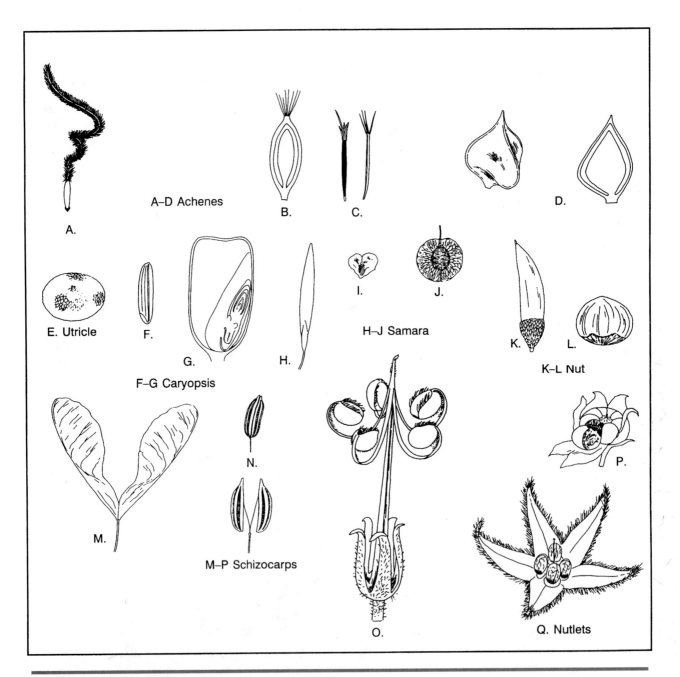

**Figure 11-16** Examples of single-seeded indehiscent dry fruits. A–D. Achenes. A. *Cercocarpus* (mountain-mahogany). B. *Sonchus* (sow-thistle). C. *Tagetes* (marigold), D. *Rumex* (dock). E. Utricle of *Chenopodium* (goosefoot). F–G. Caryopses. F. *Triticum* (wheat) G. *Zea* (maize, l.s.). H–J. Samaras. H. *Fraxinus* (ash). I. *Betula* (birch). J. *Ptelea* (hop-tree). K–L. Nuts. K. *Quercus* (oak). L *Corylus* (hazelnut). M–P. Schizocarps. M. *Acer* (maple). N. *Foeniculum* (fennel), O. *Erodium* (filaree). P. *Lavatera* (tree mallow). Q. Nutlets of *Borago* (borage).

5. **NUT**—Exocarp hard; the fruit often subtended by an involucre. Derived from a syncarpous gynoecium, but one-seeded by abortion of ovules (Figure 11-16 K, L). Examples: acorn, filbert.

C. Fruit fleshy or dry, at maturity breaking into one- or few-seeded segments, the seeds remain completely surrounded by the pericarp.

6. **SCHIZOCARP**—Fruit breaking into one-carpellate segments (mericarps; Figure 11-16 M–P). Examples: members of the Apiaceae such as fennel, dill, and celery and scattered examples in other families such as hollyhock, cheeseweed, and madder. Schizocarps may be winged [samaroid schizocarps, Figure 11-16 M.] as in maple or fleshy as in madder.

7. **NUTLET**—Fruit deeply four-lobed, breaking into one-seeded half-carpellate segments; used especially for fruits of borage, vervain and mint families (Figure 11-16 Q). Examples: sage, borage.

D. Fruit dry, at maturity dehiscing either lengthwise, or by pores, teeth or a lid and exposing the seeds.

8. **CAPSULE**—Syncarpous and typically several- to many-seeded. The most common and variable of the dehiscent fruits. Capsules can be distinguished by the way they dehisce.

   a. **Septicidal Capsule**—opens lengthwise along the septa (zigadene; Figure 11-17 A).

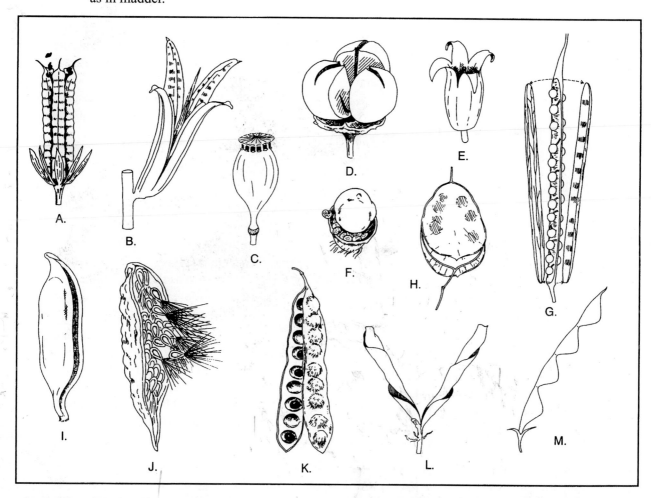

**Figure 11-17** Examples of dry dehiscent fruits. A. Septicidal capsule of *Zigadenus*. B. Loculicidal capsule of *Oenothera*. C. Poricidal capsule of *Papaver*. D. Schizocarpic capsule of *Croton*. E. Denticidal capsule of *Lychnis*. F. Circumscissile capsule of *Portulaca*. G. Silique of *Brassica*. H. Silicle of *Lunaria*. I. Follicle of *Brachychiton*. J. Follicle of *Asclepias*. K. Legume of *Albizia*. L. Legume of *Lupinus*. M. Loment of *Desmodium*.

b. **Loculicidal Capsule**—opens lengthwise between the septa and directly into the locule (evening primrose, iris; Figure 11-17 B).

c. **Poricidal Capsule**—opens by pores near the top (poppy; Figure 11-17 C).

d. **Schizocarpic Capsule**—breaks completely apart at the septa into one-seeded segments that then dehisce through the locules (castorbean, croton; Figure 11-17 D).

e. **Denticidal Capsule**—opens apically by a ring of teeth (chick-weed, *Eucalyptus;* Figure 11-17 E).

f. **Circumscissile Capsule**—[pyxis] opens horizontally in a ring so that a lid is formed (purslane, plantain; Figure 11-17 F).

g. **Silique**—A long, narrow, 2-carpellate capsule (more than twice as long as wide) whose outer walls break away from a central papery partition (replum; Figure 11-17 G). Restricted to Brassicaceae. Example: mustard.

h. **Silicle**—A short silique (twice as long as wide or less) (Figure 11-17 H). Restricted to Brassicaceae. Examples: shepherd's purse, sweet alyssum.

9. **FOLLICLE**—Unicarpellate, dehiscing along one suture (Figure 11-17 I, J). Examples: milkweed, larkspur.

10. **LEGUME**—Unicarpellate, dehiscing along two sutures (Figure 11-17 K, L). Restricted to the Fabaceae. Examples: bean, pea.

a. **Loment**—a legume that has constrictions between the seeds and breaks apart at these constrictions (Figure 11-17 M). The segments may be indehiscent or dehiscent. Example: beggar's-ticks.

II. Fruit derived from more than one pistil.

A. Fruit derived from a flower with an apocarpous gynoecium.

11. **AGGREGATE**—The ovary of each carpel matures into one of the simple fruits described in section I. Thus *Magnolia* produces an aggregate of follicles (Figure 11-18 A); a raspberry is an aggregate of small drupes (Figure 11-18 B).

a. **Accessory**—Tissue other than or in addition to the ovary enlarges and becomes fleshy. Examples: strawberry and blackberry in which the receptacle becomes fleshy (Figure 11-18 C).

b. **Hip**—The hypanthium of a perigynous flower enlarges and becomes fleshy at maturity but does not become adherent to the carpels (Figure 11-18 D, E). Known only from certain Rosaceae, especially *Rosa.* The ovaries usually mature into achenes.

B. Fruit derived from the gynoecia of several to many separate flowers of an inflorescence.

12. **MULTIPLE FRUITS**—The individual pistils of each flower develop into one of the simple fruits described in section I (Figure 11-18 F). Accessory tissue such as the calyx of the mulberry or the inflorescence axis of the pineapple may become fleshy.

a. **SYCONIUM**—a special kind of multiple fruit restricted to *Ficus* (fig) where a hollow fleshy inflorescence axis surrounds the flowers (Figure 11-18 G). The individual ovaries mature as achenes.

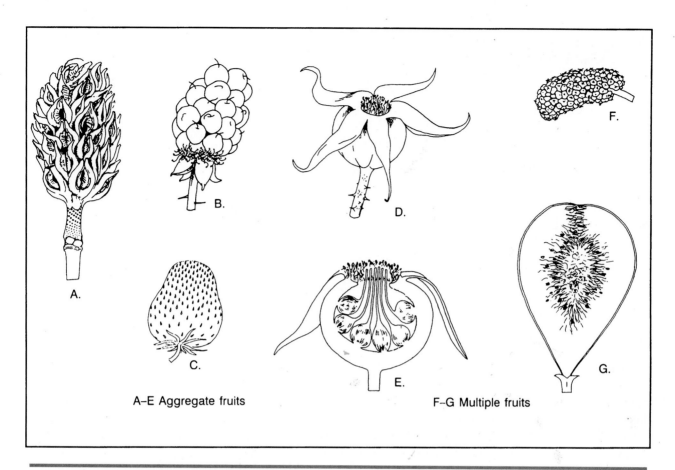

A–E Aggregate fruits          F–G Multiple fruits

**Figure 11-18** Examples of aggregate and multiple fruits. A. Aggregate of follicles of *Magnolia*. B. Aggregate of drupelets of *Rubus* (raspberry). C. Aggregate of achenes of *Fragaria* (strawberry) with accessory tissue derived from fleshy receptacle. D. Intact hip (aggregate of achenes surrounded by fleshy hypanthium) of *Rosa* (rose). E. Hip of *Rosa* (l.s.). F. Multiple accessory fruit of *Morus* (mulberry). The individual fruits are achenes surrounded by fleshy sepals. G. Syconium of *Ficus* (fig). The individual fruits are achenes surrounded by a hollow inflorescence axis.

# References

Borror, D. J. 1960. *Dictionary of Word Roots and Combining Forms.* Mayfield Publ. Co., Palo Alto, California.

Brown, R. W. 1956. *Composition of Scientific Words.* Smithsonian Institution Press, Washington, D.C. (1991 reprint).

Harrington, H. D., and L. W. Durrell. 1957. *How to Identify Plants.* Swallow Press, Chicago.

Harris, J. G., and M. W. Harris. 1994. *Plant Identification Terminology. An Illustrated Glossary.* Spring Lake Publishing, Payson, Utah.

Henderson, I. F., and W. D. Henderson. 1957. *A Dictionary of Scientific Terms,* 6th ed., revised by J. H. Kenneth. Oliver and Boyd, Edinburgh.

Jaeger, E. C. 1950. *A Source-book of Biological Names and Terms,* 2nd ed. Charles C. Thomas, Publisher, Springfield, Illinois.

Lawrence, G. H. M. 1955. *An Introduction to Plant Taxonomy.* MacMillan, New York.

Little, R. J., and C. E. Jones. 1980. *A Dictionary of Botany.* Van Nostrand Reinhold Co., New York.

Radford, A. E., W. C. Dickison, J. R. Massey and C. R. Bell. 1974. *Vascular Plant Systematics.* Harper & Row, New York.

Woods, R. S. 1944. *The Naturalist's Lexicon.* Abbey Garden Press, Pasadena, California.

# Exercises

You will be provided with a variety of flowers, inflorescences, and fruits for examination or dissection. Fill out a copy of Worksheet 1 for each flower that you dissect. This will help you to gain familiarity with floral features and the terms that describe them, and practice in preparing floral formulas, floral diagrams, and diagrammatic longitudinal sections.

**Equipment needed:**

- 2 dissecting needles,

- a sharp razor blade or scalpel,

- a pair of forceps,

- a dissecting microscope or hand lens.

When you dissect a flower you should count the parts in the various whorls. Determining the number of sepals, petals, and stamens is usually easy (though sometimes the number of stamens is inconveniently large and variable from flower to flower). Determining the number of carpels may require the use of external or internal clues (see p. 000). To obtain internal clues you will have to make a cross section through the ovary. A good, clean cross section is essential. This will enable you to determine locule number and placentation. It is often best to make several thin slices through the ovary rather than cutting it apart in just one location. One of the slices may be better at showing the features you are looking for than any of the others.

You also should make a longitudinal section of the flower to determine the points of attachment of each floral whorl. Cut the flower lengthwise as close to the center as possible through the receptacle and the various floral whorls. This will enable you to see whether the ovary is superior or inferior, whether the flower has a hypanthium, etc. You should compare your section with the illustrations in Figure 11-10 to determine the type of flower insertion. After you have made the dissection, prepare a **diagrammatic longitudinal section** (Figure 11-19) of each flower. All parts should be drawn in their proper size and insertion relationship to each other. It is important that all parts be labeled as there are no conventions that will identify the structures drawn.

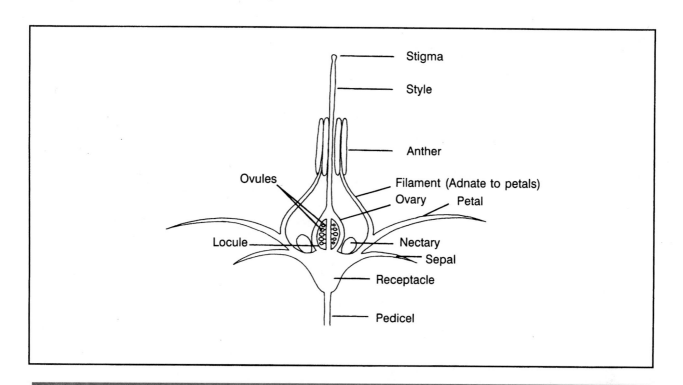

**Figure 11-19** An example of a diagrammatic longitudinal section of a potato flower. Note: since there are no standard symbols, and the usefulness of these drawings depends somewhat on the artistic skill of the student, explanatory labels are very important.

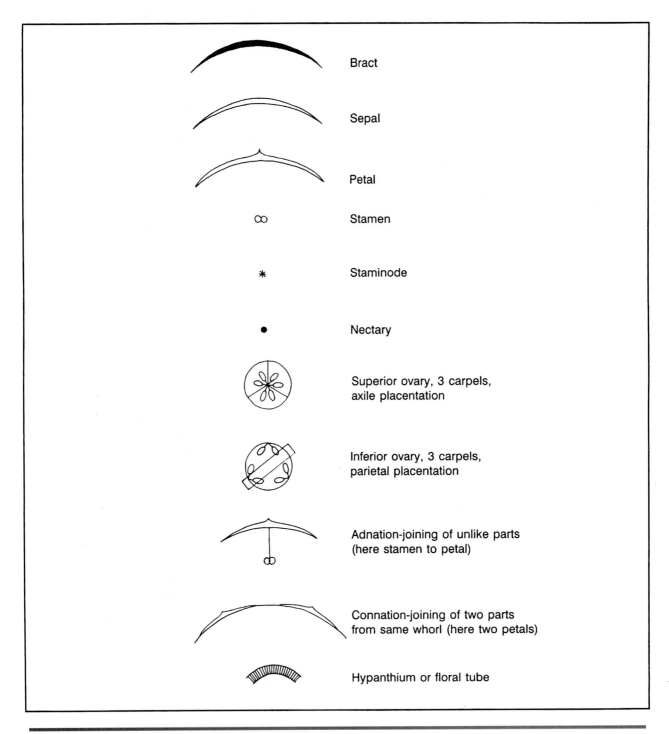

Bract

Sepal

Petal

Stamen

Staminode

Nectary

Superior ovary, 3 carpels,
axile placentation

Inferior ovary, 3 carpels,
parietal placentation

Adnation-joining of unlike parts
(here stamen to petal)

Connation-joining of two parts
from same whorl (here two petals)

Hypanthium or floral tube

**Figure** Commonly used symbols for constructing floral diagrams.
**11-20**

A **floral diagram** is an idealized cross-section of a flower with standardized symbols for the parts of each whorl. Figure 11-20 shows the symbols most commonly used, and Figure 11-21 shows examples of completed diagrams for several different flowers. A floral diagram illustrates not only the parts present in the flower but also their spatial relationship to each other. Stamens have a spatial relationship to the petals, for instance. They may be opposite (directly in front of) the petals or alternate with (in the notches between the petals). You should prepare a floral diagram for each of the flowers you dissect. Check with your instructor to see if you are preparing the floral diagrams correctly.

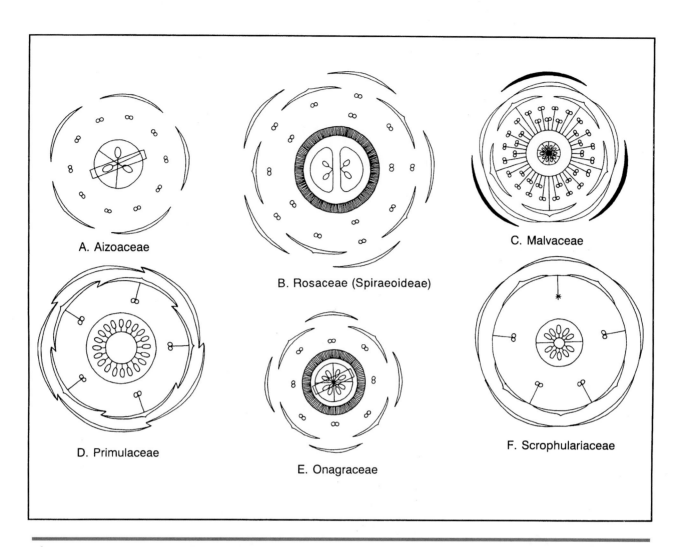

**Figure 11-21** Sample floral diagrams. A. 5-merous apetalous flower with 3-carpellate inferior ovary with axile placentation (Aizoaceae). B. 5-merous flower with perigynous flower and 2-carpellate apocarpous gynoecium (Rosaceae). C. 5-merous flower closely subtended by 3 bracts; sepals connate, petals distinct, adnate at base to connate filaments of the many stamens; ovary superior, 5-carpellate with axile placentation (Malvaceae). D. 5-merous flower with connate sepals and petals, 5-epipetalous stamens opposite the corolla lobes and a superior ovary with free-central placentation (Primulaceae). E. 4-merous flower with hypanthium and inferior 4-carpellate ovary and axile placentation (Onagraceae). F. 5-merous zygomorphic flower with connate sepals, connate petals, 4 epipetalous stamens and a staminode, alternate with the corolla lobes, and a 2-carpellate superior ovary with axile placentation (Scrophulariaceae).

**Floral formulas** are used to summarize the structure and special features of a flower or plant by abbreviations similar to those used in chemical formulas. The symbols and their explanation are presented in Table 11-2. Each flora formula should include the symbols for calyx (**Ca**), corolla (**Co**), androecium (**A**), and gynoecium (**G**). Each of these should be accompanied by a superscript number that indicates the number of individual parts. For instance, **Ca**$^5$ indicates that the calyx comprises five sepals. A circle drawn around a number (e.g., **Co**$^④$) indicates that the parts are connate; the lack of a circle indicates parts that are distinct. A floral formula that looks like a fraction with one symbol above another is indicative of adnate fusion. The part or parts represented by the symbols above the line are adnate to the part below the line. Any flower can be symbolized by one of the four basic floral formulas presented in Figure 11-22.

**Ca = Calyx, sepals**

| | |
|---|---|
| Ca$^0$ | Sepals none; calyx absent |
| Ca$^x$ | Sepals indefinite in number |
| Ca$^{3-\infty}$ | Sepals 3 to many, distinct |
| Ca$^{4-5}$ | Sepals 4–5, distinct |
| Ca$^{\underline{5}}$ | Sepals 5, cennate at base |
| Ca$^{⑤}$ | Sepals 5, connate; calyx actinomorphic (radial) |
| Caz$^{⑤}$ | Sepals 5, connate; calyx zygomorphic (bilateral) |

**Co Corolla, petals**

| | |
|---|---|
| Co$^0$ | Petals none; corolla absent |
| Co$^x$ | Petals present, indefinite in number |
| Co$^{\underline{5}}$ | Petals 5, slightly connate at base |
| Co$^{⑤}$ | Petals 5, connate; corolla actinomorphic (radial) |
| Coz$^{⑤}$ | Petals connate; corolla zygomorphic (bilateral) |

**A Androecium, stamens**

| | |
|---|---|
| A$^\infty$ | Stamens many, distinct |
| A$^{\overline{\infty}}$ | Stamens many, connate (monadelphous) |
| A$^{⑤}$ | Stamens 5, connate |
| $\frac{A^5}{Co^5}$ | Stamens epipetalous, opposite petals |
| $\frac{A^5}{Co^{⑤}}$ | Stamens epipetalous, alternate with the petals |
| A$^{2+4}$ | Stamens in 2 series, 2 in one series, 4 in the other |

**G Gynoecium, Carpels**

| | |
|---|---|
| G$^1$ | Gynoecium monocarpous; carpel 1 |
| G$^\infty$ | Gynoecium apocarpous; carpels many, distinct |
| G$^3$ | Gynoecium apocarpous; carpels 3, distinct |
| G$^{③}$ | Gynoecium syncarpous; carpels 3, connate |

**Additional features**

| | |
|---|---|
| HYP | Hypanthium |
| BER | Berry |
| DRP | Drupe |
| ACH | Achene |
| SAM | Samara |
| SCHIZ | Schizocarp |
| CAPS | Capsule |
| FOL | Follicle |
| AGG | Aggregate |
| MONOEC | Plants monoecious |
| DIOEC | Plants dioecious |
| AXILE | Axile placentation |
| PARIETAL | Parietal placentation |
| MARGINAL | Marginal placentation |
| FREE-CNT | Free central placentation |
| APICAL | Apical placentation |
| BASAL | Basal placentation |
| TEND | Tendrils |
| SUC | Foliage or stems succulent |
| INVOL | Involucre |
| SPATHE | Spathe |

**Table 11-2**    Symbols and abbreviation commonly used for construction of floral formulas.

| | | | | | | | | | **A** | | | | | | **A** | |
|---|---|---|---|---|---|---|---|---|---|---|---|---|---|---|---|---|
| **Ca** | **Co** | **A** | **G** | | **Ca** | **Co** | **G** | **Ca** | **Co** | **A** | | **Ca** | **Co** | | |
| | | | | | | | | | | **G** | | | | **G** | |
| | **A.** | | | | | **B.** | | | **C.** | | | | **D.** | | |

**Figure 11-22** Basic floral formulas. One of these four basic formulas can be used for any flower. A. Any flower with a superior ovary and stamens free from the petals. B. Any flower with a superior ovary and stamens adnate to the petals (A above Co). C. Any flower with an inferior ovary (below the line) and stamens free from the petals. D. Any flower with an inferior ovary and stamens adnate to the corolla.

Additional features can be indicated under a line drawn beneath the floral formula. Symbols for additional features are usually the first two or three letters of the term describing the feature to be represented. You are not limited to the features in Table 11-2. You might, for instance, want to include the inflorescence type, leaf venation, flower color, or other noteworthy features of the plant. Some sample completed formulas are presented in Figure 11-23.

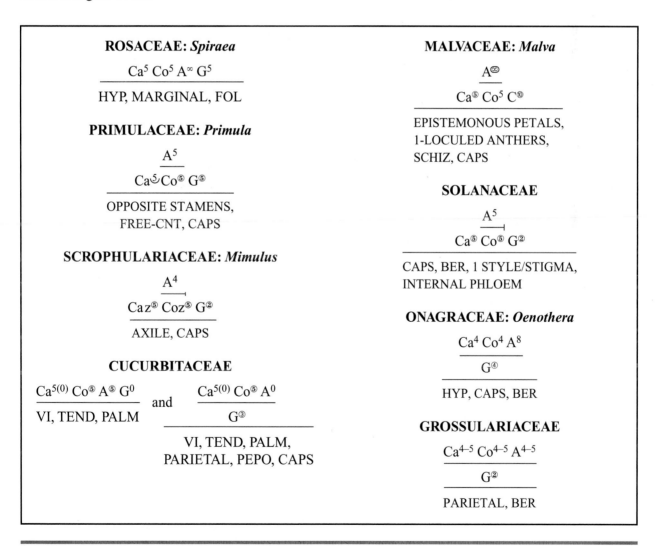

**Figure 11-23** Floral formulas for members of some common plant families.

# WORKSHEET 1

_____     _____
              name of plant                              your name

1. How many sepals? _____

2. Are the sepals distinct or connate? _____

3. What is the symmetry of the calyx? _____

4. How many petals? _____

5. Are the petals distinct or connate? _____

6. What is the symmetry of the corolla? _____

7. How many stamens? _____

8. Are the stamens distinct or connate? _____

9. Are the stamens free from the petals or are they adnate to the corolla (epipetalous)? _____

10. If the stamens are fewer than or equal in number to the petals, are they alternate with or opposite the petals. _____

11. Does the flower have a hypanthium? _____

12. How many carpels? _____
    What is your source of evidence for carpel number? _____

13. If the flower has more than one carpel, are they distinct or connate? _____

14. Placentation type? _____

15. In the space to the right make a drawing *(not a diagram)* of a cross section through the ovary. Label the pericarp, septa, placentae, ovules, and locules.

16. Ovary position? _____

17. Is the insertion of flower parts hypogynous, perigynous or epigynous? _____

18. Are the flowers perfect or imperfect? _____

19. Are the flowers complete or incomplete? _____

20. Is the plant monoecious, dioecious or synoecious? _____

21. Based on your answers to the questions above, fill out the correct floral formula.

$$Ca\ Co\ A\ G \qquad Ca\ Co\ \underline{A} \ G \qquad Ca\ Co\ A \atop \overline{G} \qquad Ca\ Co \atop \overline{G} $$

22. In the space below prepare a diagrammatic longitudinal section of the flower. Label the pedicel, receptacle, hypanthium (if any), sepals, petals, filaments, anthers, ovary or ovaries, style, stigma, locules, placentae, and ovules.

23. In the space below, prepare a floral diagram. Do not make your diagram too small. It is recommended that you start with the gynoecium and work your way outward.

# WORKSHEET 1

______________________________ name of plant

______________________________ your name

1. How many sepals? _____
2. Are the sepals distinct or connate? __________
3. What is the symmetry of the calyx? __________
4. How many petals? _____
5. Are the petals distinct or connate? __________
6. What is the symmetry of the corolla? __________
7. How many stamens? _____
8. Are the stamens distinct or connate? __________
9. Are the stamens free from the petals or are they adnate to the corolla (epipetalous)? __________
10. If the stamens are fewer than or equal in number to the petals, are they alternate with or opposite the petals. __________
11. Does the flower have a hypanthium? __________
12. How many carpels? _____

    What is your source of evidence for carpel number? __________
13. If the flower has more than one carpel, are they distinct or connate? __________
14. Placentation type? __________
15. In the space to the right make a drawing *(not a diagram)* of a cross section through the ovary. Label the pericarp, septa, placentae, ovules, and locules.
16. Ovary position? __________
17. Is the insertion of flower parts hypogynous, perigynous or epigynous? __________
18. Are the flowers perfect or imperfect? __________
19. Are the flowers complete or incomplete? __________
20. Is the plant monoecious, dioecious or synoecious? __________
21. Based on your answers to the questions above, fill out the correct floral formula.

$$\text{Ca Co A G} \qquad \frac{\text{A}}{\text{Ca Co G}} \qquad \frac{\text{Ca Co A}}{\text{G}} \qquad \frac{\frac{\text{A}}{\text{Ca Co}}}{\text{G}}$$

22. In the space below prepare a diagrammatic longitudinal section of the flower. Label the pedicel, receptacle, hypanthium (if any), sepals, petals, filaments, anthers, ovary or ovaries, style, stigma, locules, placentae, and ovules.

23. In the space below, prepare a floral diagram. Do not make your diagram too small. It is recommended that you start with the gynoecium and work your way outward.

# WORKSHEET 1

_____   _____
               name of plant                                              your name

1. How many sepals? _____

2. Are the sepals distinct or connate? _____

3. What is the symmetry of the calyx? _____

4. How many petals? _____

5. Are the petals distinct or connate? _____

6. What is the symmetry of the corolla? _____

7. How many stamens? _____

8. Are the stamens distinct or connate? _____

9. Are the stamens free from the petals or are they adnate to the corolla (epipetalous)? _____

10. If the stamens are fewer than or equal in number to the petals, are they alternate with or opposite the petals. _____

11. Does the flower have a hypanthium? _____

12. How many carpels? _____

    What is your source of evidence for carpel number? _____

13. If the flower has more than one carpel, are they distinct or connate? _____

14. Placentation type? _____

15. In the space to the right make a drawing _(not a diagram)_ of a cross section through the ovary. Label the pericarp, septa, placentae, ovules, and locules.

16. Ovary position? _____

17. Is the insertion of flower parts hypogynous, perigynous or epigynous? _____

18. Are the flowers perfect or imperfect? _____

19. Are the flowers complete or incomplete? _____

20. Is the plant monoecious, dioecious or synoecious? _____

21. Based on your answers to the questions above, fill out the correct floral formula.

$$Ca\ Co\ A\ G \qquad Ca\ Co\ \underline{A}\ G \qquad Ca\ Co\ \underline{\underline{A}}\ \underline{G} \qquad Ca\ Co\ \underline{\underline{A}}\ \underline{G}$$

22. In the space below prepare a diagrammatic longitudinal section of the flower. Label the pedicel, receptacle, hypanthium (if any), sepals, petals, filaments, anthers, ovary or ovaries, style, stigma, locules, placentae, and ovules.

23. In the space below, prepare a floral diagram. Do not make your diagram too small. It is recommended that you start with the gynoecium and work your way outward.

# Early Evolution of Flowering Plants: Basal Angiosperms, Magnoliids, and Basal Eudicots

chapter 12

Enormous controversy surrounds our understanding of relationships within the flowering plants. Cronquist (1988), Thorne (1992), and Takhtajan (1997) considered the "Magnoliidae" to represent the ancestral lineages of flowering plants. This group includes magnolias, water lilies, and black peppers, and varies from trees and shrubs to aquatic herbs, and from very large terminal flowers to very small flowers in distinct inflorescences. Recently, this concept has been challenged and clarified by several molecular studies (Mathews and Donoghue 1999, Qui et al. 1999, Soltis et al. 1999, Barkman et al. 2000, Zanis et al. 2002, Soltis and Soltis 2004). Recent discovery of the fossil genus *Archaefructus* (Archaefructaceae) pushes the origin of angiosperms to at least 125 million years ago (Sun et al. 2002).

Three groups of flowering plants consistently appear as basal lineages in recent studies (Amborellales, Nymphaeales, and Austrobaileyales). *Amborella* exists as a single species in New Caladonia, and is shrubby, with small flowers and tracheids for water conduction. This would suggest that early flowering plants did not have vessels (Feild et al. 2000, Feild et al. 2003). The Nymphaeales or water lilies form a second basal lineage, and these plants are aquatic herbs with large showy flowers. The

Austrobaileyales contains four families of shrubs and vines with showy flowers. Based upon these three groups, we can speculate that early flowering plants had alternate simple leaves, and terminal or axillary flowers with multiple whorls of tepals, stamen undifferentiated into stamen and filament, and several to many separate carpels.

These early angiosperms had carpels that did not close completely in development, but were sealed by secretions from within the carpel. This supports that concept that carpels were derived from leaves. The "core angiosperms" have carpels sealed by epidermal layers, suggesting a shift in carpel development in the ancestor that gave rise to this group (Endress and Igersheim 2000). Within this "core angiosperm" group, we can identify four lineages, the Magnoliids with four orders, the Chloranthales, the monocots and the eudicots.

Figure 12-1 depicts our latest understanding of the relationships within the angiosperms or flowering plants. The Magnoliids includes the Magnoliales, Laurales, and Canellales that are generally trees or shrubs with alternate simple leaves, but also includes the Piperales (with Piperaceae, Sauraceae, and Aristolochiaceae) that are mostly vines or herbs. The Piperales have often been placed basal to the monocots. The Chloranthales had often been placed

221

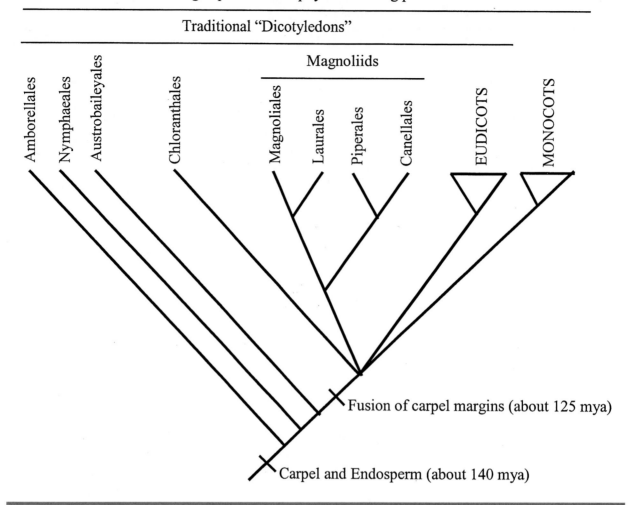

Angiosperms/Anthophyta/flowering plants

Traditional "Dicotyledons"

Magnoliids

Amborellales
Nymphaeales
Austrobaileyales
Chloranthales
Magnoliales
Laurales
Piperales
Canellales
EUDICOTS
MONOCOTS

Fusion of carpel margins (about 125 mya)

Carpel and Endosperm (about 140 mya)

**Figure 12-1** Summary of relationships within the flowering plants, focusing on the basal lineages. The Eudicots and Monocots are major clades that will be addressed in Chapters 12–16. This figure is adapted from Zanis et al. (2002) and Judd et al. (2002).

in the Magnoliids, but recent evidence does not support this placement.

The eudicots represent more than half of all flowering plants (160,000 species) that have three colpi or furrows in the pollen and the flower parts tend to be in multiples of fours or fives. The monocots represent about 65,000 species, with orchids and grasses being the dominant families in this clade. The monocots can be identified by having a single cotyledon or seed leaf, and they tend to have flower parts in threes and parallel leaf venation.

The eudicots are currently recognized as two major lineages, the rosids and the asterids. However, the relationships of these two lineages to those eudicots that are considered basal is not well understood. These basal eudicots include the buttercups,

poppies, sycamore, barberry, and mayapple. Some of these groups of basal eudicots have traditionally been placed in the Hammamelidae (such as Platanaceae) and others were evidently misplaced, such as the Proteaceae that had been considered a member of the rosid clade. Figure 12-2 presents our understanding of relationships within the eudicots.

Chapter 12 will provide an overview of the three basal lineages of angiosperms, the Magnoliids, the Chloranthales and the basal eudicots. Chapters 13 (Caryophyllanae), 14 (Rosids), and 15 (Asterids) will cover the eudicots and Chapter 16 will describe the diversity within the monocots. The families and orders generally follow the treatments in Stevens (2001 onwards) and Judd et al. 2002.

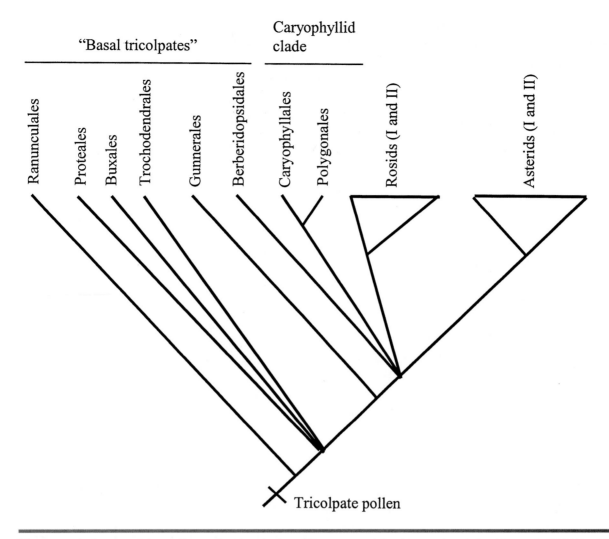

Figure
12-2
Phylogeny of the eudicots, showing the basal lineages, and the three major clades (Caryophyllids, Rosids and Asterids). This figure is adapted from Stevens (2001 onward).

# BASAL ANGIOSPERMS

## 1. Order Amborellales

**1.A    Amborellaceae**—1 genus, 1 species. New Caledonia.

## 2. Order Nymphaeales

**2.A    Nymphaeaceae**—The Waterlily Family (Figure 12-9) 5 genera, 50 species. Widely distributed in tropical and temperate regions.

§ See page 230 for descriptions and illustrations §

**2.B    Barclayaceae**—1 genus, 4 species. Southeastern Asia to New Guinea in aquatic habitats.

**2.C    Cabombaceae**—The Water-Shield Family. 2 genera, 7 species. North-temperate and tropical latitudes of New and Old World.

Aquatic herbs with submersed rhizome and leafy submersed or floating stems. Leaves alternate, opposite or whorled, short- to long-petiolate; blades simple and floating or submersed and deeply dissected, estipulate. Plants synoecious. Inflorescence of solitary, axillary, long-peduncled flowers. Flowers perfect, regular. Sepals 2–4, distinct. Petals 0. Stamens 3–many, distinct, arranged in whorls; filaments slightly flattened. Carpels (1) 2–18, distinct; ovaries superior with 1 locule and 1–3 laminar ovules; styles short. Fruit an aggregate of achenes or follicles.

## 3. Order Austrobaileyales

**3.A    Austrobaileyaceae**—1 genus, 1 species. Australia.

**3.B    Illiciaceae**—The Star-anise Family. 1 genus, 40 species. Subtropical southeastern Asia and southeastern North America.

Shrubs or trees. Leaves alternate, simple, estipulate, punctuate. Plants synoecious Inflorescence of solitary flowers. Flowers perfect, ± regular. Tepals 7–many, ± imbricate, the outermost ± sepaloid, distinct. Stamens many, distinct, spirally attached, developing centripetally. Carpels 5–20, distinct; whorled; ovaries superior, unilocular with 1 basal-marginal ovule per carpel; styles 1 per carpel. Fruit a cluster of spreading follicles.

**3.C    Trimeniaceae**—1–2 genera, 5 species. Southwestern Pacific region.

**3.D    Schisandraceae**—2 genera, 50 species. Temperate Asia and eastern North America.

# THE MAGNOLIIDS

## 4. Order Magnoliales

**4.A    Myristacaceae**—15–18 genera, 300 species. Pantropical.

**4.B    Magnoliaceae**—The Magnolia Family (Figure 12-10) 7–12 genera, 220 species. SE United States to southern Brazil, and from the Himalayas to Japan, SE Asia and the East Indies.

**4.C    Degeneriaceae**—1 genus, 1–2 species. Fiji.

**4.D    Himantandraceae**—1 genus, 3 species. Australia, Molucca Islands and New Guinea.

**4.E    Eupomataceae**—1 genus, 2 species. New Guinea and Australia.

**4.F    Annonaceae**—The Custard-apple Family (Figure 12-3). 125–130 genera, 2300 species. Pantropical with a few temperate species.

Shrubs or trees. Leaves alternate, 2-ranked, simple and entire, estipulate. Plants synoecious. Inflorescence of solitary flowers. Flowers perfect, regular. Tepals 6–9 in two or 3 whorls, distinct or basally connate. Stamens many, distinct, spiraled, developing centripetally. Carpels 3 to many, distinct (rarely connate and adnate to receptacle), spiraled; ovaries superior with 1 locule and 1–many marginal ovules per carpel; styles 3–many or stigmas sessile. Fruit a cluster of stipitate or sessile berries, these sometimes aggregated into fleshy syncarps.

## 5.    Order Laurales

**5.A    Calycanthaceae**—The Spicebush Family. 4 genera, 6 species. Widely disjunct in temperate China, northern Australia, and in eastern and western North America.

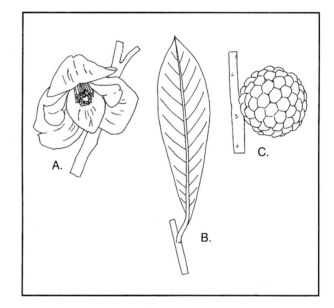

**Figure 12-3**    Examples of Annonaceae: A–B. *Asimina triloba.* A. Flower. B. Leaf. C. Fruit of *Annona cherimola.*

Shrubs. Leaves opposite, simple, estipulate. Plants synoecious. Inflorescence of solitary flowers. Flowers perfect, regular. Tepals many, grading from sepaloid to petaloid, distinct. Stamens 5–many, distinct, when many spirally attached, developing centripetally. Carpels many, distinct on a concave receptacle; ovaries superior with one locule and one or two marginal ovules, styles distinct. Fruit an aggregate of achenes surrounded by the enlarge fleshy receptacle.

**5.B    Gomortegaceae**—1 genus, 1 species. Chile.

**5.C    Monimiaceae**—30–39 genera, 450 species. Pantropical, especially well represented in southern hemisphere.

**5.D   Hernandiaceae**—4 genera, 60 species. Pantropical.

**5.E   Lauraceae**—The Laurel Family (Figure 12-11). 30–50 genera, 2000+ species. Pantropical with a few temperate species.

§ See page 234 for description and illustrations §

# 6.  Order Cannellales

**6.A   Canellaceae**—6 genera, 20 species. West Indies to tropical South America; disjunct in tropical eastern Africa and Madagascar.

**6.B   Winteraceae**—9 genera, 100 species. Circum-Pacific.

# 7.  Order Piperales

**7.A   Piperaceae**—The Pepper Family (Figure 12-4). 10 genera, 1400–2000 species. Pantropical.

Herbs, shrubs or trees. Leaves alternate (less commonly opposite or whorled), simple, stipulate or estipulate. Plants monoecious, dioecious, or synoecious. Inflorescence a spike. Flowers very reduced, perfect or imperfect, subtended by bractlets. Sepals absent. Petals absent. Stamens 1–10, distinct. Carpels 2–5, connate; the ovary superior, 1-loculed with 1 basal ovule; stigma 2–5, sessile. Fruit a drupe or drupelet.

**7.B   Saururaceae**—The Lizard-tail Family. 5 genera, 7 species. Eastern Asia, western and eastern North America.

Herbs. Leaves alternate, simple and entire, stipulate. Plants synoecious. Inflorescence a raceme or spike, sometimes with petaloid bracts. Flowers perfect, naked. Sepals absent. Petals absent, Stamens 3–8, distinct and free or ± adnate to ovary. Carpels 3–4, distinct, the ovaries superior with 2–4 marginal ovules, *or* carpels 3–4, connate, the ovary superior or inferior with 1 locule and 6–8 parietal ovules per placenta or with 3–4 locules and 1 axile ovule per placenta; styles 1 per carpel. Fruit a capsule or cluster of achenes.

**7.C   Lactoridaceae**—1 genus, 1 species. Juan Fernandez Islands.

**7.D   Aristolochiaceae**—The Birthwort Family (Figure 12-5), 8–10 genera, 600 species. Mostly tropical, few temperate.

Herbs, shrubs or woody vines. Leaves alternate, simple, estipulate. Plants synoeciuos. Inflorescence of solitary or clustered axillary flowers, racemes, or cymes. Flowers perfect, regular to very irregular. Sepals 3(4), petaloid, connate and basally adnate to

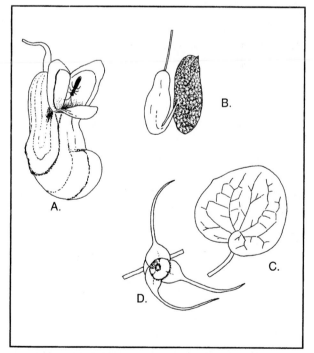

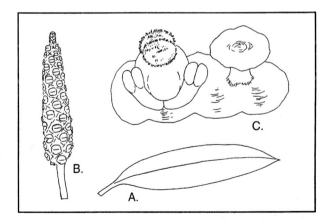

**Figure 12-4**   Examples of Piperaceae. A–B. *Piper* sp. A. Leaf. B. Spike of minute flowers. C. *Peperomia* sp., naked flower with peltate, scale-like bract.

**Figure 12-5**   Examples of Aristolochiaceae. A. Flower of *Aristolochia californica*. B. Lateral view of flower of *Aristolochia* sp. C–D. *Asarum caudatum*. C. Leaf. D. Flower.

ovary, sometimes contorted into bizarre shapes. Petals absent (rarely 3, vestigial). Stamens 8–many, distinct or basally connate and basally adnate to calyx and ovary *or* wholly adnate to stigmas forming a gynostegium. Carpels 3–6, connate; ovary inferior with 3–6 locules and axile ovules; style 1 with 3–6 stigmas. Fruit a capsule.

**7.E** Hydnoraceae—2 genera, 10 species. Africa.

## 8. Order Chloranthales

**8.A** **Chloranthaceae**—4–5 genera, 75 species. Tropics and subtropics.

# BASAL TRICOLPATES

## 9. Order Ranunculales

**9.A** **Lardizabalaceae**—The Lardizabala Family. 9 genera, 31 species. Southeastern Asia and Chile.

Shrubs, sometimes climbing (rarely trees). Leaves alternate, palmate (rarely pinnate), estipulate. Plants monoecious or dioecious (synoecious). Flowers imperfect (perfect), regular. Sepals 3 or 6, distinct, petaloid. Petals 6 (0), distinct. Stamens 6, distinct or connate. Carpels 3–15, distinct; ovaries superior, unilocular with many (1–few) marginal or laminar ovules; stigmas sessile or subsessile. Fruit an aggregate of follicles, berries, or drupes.

**9.B** **Circaeasteraceae**—1–2 genera, 1–2 species. Southeastern Asia.

**9.C** **Menispermaceae**—The Moonseed Family. 70–73 genera, 400 species. Tropics to subtropics, few temperate.

Shrubs or woody vines (less commonly trees). Leaves alternate, simple, estipulate. Plants dioecious. Inflorescences various, cymose or paniculiform. Flowers imperfect, regular. Sepals 6 in two series (4–8), distinct. Petals 6–9 in 2–3 series, distinct. Stamens 6–many, distinct. Carpels 2–6 (many), distinct; ovaries superior with 1–2 marginal ovules; style 1 per carpel or stigmas sessile. Fruit an aggregate of drupes.

**9.D** **Berberidaceae**—The Barberry Family (Figure 12-12). 16–18 genera, 600 species. Mostly north temperate with a few in tropical mountains and south-temperate regions.

§ See page 236 for description and illustrations §

**9.E** **Ranunculaceae**—The Buttercup Family (Figure 12-13). 50–52 genera, 2000 species. Cosmopolitan.

§ See page 238 for description and illustrations §

**9.F** **Papaveraceae** (including Fumariaceae)—The Poppy Family (Figure 12-14). 34–41 genera, 600 species. Mostly north-temperate.

§ See page 240 for description and illustrations §

**9.G** **Sabiaceae**—1–3 genera, 60 species. Southeastern Asia and tropical America.

**9.H** **Eupteleaceae**—1 genus, 2 species. Southeast Asia.

## 10. Order Proteales

**10.A** **Nelumbonaceae**—The Water-Lotus Family (Figure 12-6). 1 genus, 2 species. Eastern North America, Australia and Eurasia.

Aquatic herbs. Leaves simple, long-petiolate, peltate, estipulate, floating or emersed, arising from submersed rhizome. Plants synoecious. Inflorescence of solitary, axillary, long-peduncled flowers. Flowers perfect, regular. Sepals 2, distinct. Petals 20–30, spirally attached. Stamens many, distinct, spirally attached, developing centripetally; filaments slender. Carpels 12–40, distinct, sunken into pits in flat-topped corky receptacle; ovaries superior with 1 locule and 1 apical ovule; style absent; stigma sessile. Fruit an aggregate of nuts loose in the cavities of the raft-like fruiting receptacle.

**10.B** **Platanaceae**—The Sycamore Family (Figure 12-7). 1 genus, 6–7 species. North temperate zone to tropical southeast Asia.

Trees. Leaves alternate, simple and palmately lobed, stipulate. Plants monoecious. inflorescence of spherical staminate and pistillate heads. Flowers imperfect, regular, hypogynous. Sepals 3–8, very small, distinct or in age connate. Petals 3–8, very small, distinct or in age connate. Stamens 3–8, connate by stamen connectives. Carpels 6–9 (3), distinct; ovaries superior with 1 locule and 12 apical

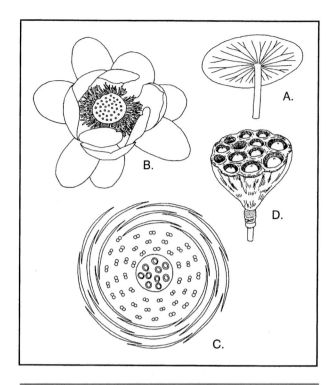

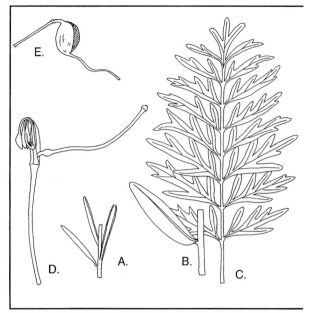

**Figure 12-6** Examples of Nelumbonaceae. A–D. *Nelumbo nucifera.* A. Peltate leaf. B. Flower. C. Floral diagram. D. Fruiting receptacle with nuts.

**Figure 12-8** Examples of Proteaceae. A–C. Leaves of *Grevillea noellii, Protea* sp., and *Grevillea robusta.* D. Flower of *Grevillea* sp. E. Follicle of *Grevillea* sp.

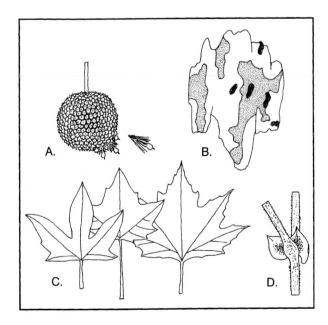

**Figure 12-7** Examples of Plantanaceae. A–E. *Platanus* spp. A. Fruiting head releasing achenes with bristly perianth. B. Bark with flat plates. C. Leaves of *P. racemosa, P. x acerifolia,* and *P. occidentalis.* D. Twig and base of leaf with stipules.

ovules; style 1 per carpel. Fruit of dense heads of achenes with long bristles.

**10.C Proteaceae**—The Protea Family (Figure 12-8). 69–75 genera, 1000 species. Temperate southern hemisphere.

Shrubs or trees (very rarely herbs). Leaves alternate (rarely opposite or whorled), simple to deeply divided, or compound, estipulate. Plants synoecious or dioecious (rarely polygamous). Inflorescence of solitary flowers, racemes, spikes or heads, sometimes involucrate. Flowers perfect (imperfect), regular or irregular. Sepals 4, petaloid, basally connate. Petals absent or represented by 2–4 tiny scales. Stamens 4, opposite the sepals with filaments adnate to sepals (rarely free). Carpel 1, sometimes stipitate; ovary superior with 1 locule and 1–many apical or marginal ovules; style 1. Fruit a follicle, achene, drupe or nut.

# 11. Order Buxales

**11.A Buxaceae**—The Boxwood Family. 5 genera, 60 species. Cosmopolitan.

Shrubs or trees (rarely herbs). Leaves alternate or opposite, simple, estipulate. Plants monoecious or dioecious (polygamous). Inflorescence of spikes, racemes, cymes or 1–few axillary flowers. Flowers imperfect (rarely perfect), regular. Sepals 4–5 (6 or more) (0), distinct or basally connate. Petals absent. Stamens 4–6 (8–many), distinct. Carpels 3 (2–6), connate; ovary superior with 3 (2–6) locules and 2 apical-axile ovules per locule; styles 3 (2–6), distinct. Fruit a capsule or drupe.

**11.B Didymelaceae**—1 genus, 2 species. Madagascar.

# 12. Order Trochodendrales

**12.A Trochodendraceae**—1 genus, 1 species. Southeast Asia.

# 13. Order Gunnerales

**13.A Myrothamnaceae**—1 genus, 2 species. Africa and Madagascar.

**13.B Gunneraceae**—The Gunnera Family. 1 genus, 50 species. Southern hemisphere to southern Mexico.

Herbs. Leaves alternate or all basal, simple, stipulate, often very large. Plants synoecious, monoecious, dioecious or polygamous. Inflorescence a scapose spike, head or spicate panicle. Flowers perfect or imperfect, regular or irregular. Sepals 0, 2 or 3, distinct or connate. Petals 0 or 2, distinct. Stamens 1–2, distinct; filaments sometimes adnate to petals. Carpels 2, connate; ovary inferior with 1 locule and 1 apical ovule; styles 2. Fruit a drupe.

# 14. Order Berberidopsidales

**14.A Aextoxicaceae**—1 genus, 1 species. Chile.

# References

Angiosperm Phylogeny Group II. 2003. An update of the Angiosperm Phylogeny Group classification for the orders and families of flowering plants: APG II. *Botanical Journal of the Linnean Society* 141:399–436.

Barkman T. J., G. Chenery, J. R. McNeal, J. Lyons-Weiler, W.J. Ellisens, G. Moore, A. D. Wolfe, and C.W. dePamphilis. 2000. Independent and combined analyses of sequences from all three genomic compartments converge on the root of flowering plant phylogeny. Proceedings of the National Academy of Sciences 97(24):13166–13171.

Cronquist, A. 1981. *An Integrated System of Classification of Flowering Plants.* Columbia University Press, New York. Pp. 23–149.

————. 1988. *The Evolution and Classification of Flowering Plants,* 2nd ed. New York Botanical Garden, Bronx, N.Y. pp. 261–292.

Endress, P. K., and A. Igersheim. 2000. Gynoecium structure and evolution in basal angiosperms. *Int. J. Pl. Sci.* 161 (Suppl.): S211–S223.

Ernst, W. R. 1962. The genera of Papaveraceae and Fumariaceae in the southeastern United States. *Journal of the Arnold Arboretum* 43:315–343.

————. 1964. The genera of Berberidaceae, Lardizabalaceae, and Menispermaceae in the southeastern United States. *Journal of the Arnold Arboretum* 45:1–35.

Feild, T. S., M. A. Zwieniecki, and N. M. Holbrook. 2000. Winteraceae evolution: an ecophysiological perspective. *Ann. Missouri Bot. Gard.* 87:323–334.

Feild T.S., N. C. Arens, and T. E. Dawson. 2003. The ancestral ecology of angiosperms: emerging perspectives from extant basal lineages. *International Journal of Plant Sciences* 164(3 Suppl.):S129–S142.

Goremykin V. V., K. I. Hirsch-Ernst, S. Wolfl, and F. H. Hellwig. 2003. Analysis of the *Amborella trichopoda* chloroplast genome sequence suggests *Amborella* is not a basal angiosperm. *Molecular Biology and Evolution* 20:1499–1505.

Heywood, V. H. (ed.). 1993. *Flowering Plants of the World,* updated edition. Oxford University Press, New York. Pp. 27–55.

Hilu K.W., T. Borsch, K. Müller, D. E. Soltis, P. S. Soltis, V. Savolainen, M. W. Chase, M. P. Powell, L. A. Alice, R. Evans, H. Sauquet, C. Neinhuis, T. A. B. Slotta, J. G. Rohwer, C. S. Campbell, and L. W. Chatrou. 2003. Angiosperm phylogeny based on *mat*K sequence information. *American Journal of Botany* 90(12):1758–1776.

Judd W. S., C. S. Campbell, E. A. Kellogg, P. F. Stevens, and M. J. Donoghue. 2002. *Plant Systematics: A Phylogenetic Approach.* 2nd ed. Sunderland, MA: Sinauer Associates.

Kubitzki, K. (ed.) 1990. *The Families and Genera of Vascular Plants.* Vol. 2. *Flowering Plants - Dicotyledons: Magnoliid, Hamamelid and Caryophyllid families* Edited by K. Kubitzki, J. G. Rohwer, and V. Bittrich. Springer Verlag, Berlin.

Li, H. L. 1955. Classification and phylogeny of Nymphaeaceae and allied families. *American Midland Naturalist* 54:33–41.

Mathews S., and M. J. Donoghue. 1999. The root of angiosperm phylogeny inferred from duplicate phytochrome genes. *Science* 286:947–950.

Parkinson C. L., K. L. Adams, and J. D. Palmer. 1999. Multigene analyses identify the three earliest lineages of extant flowering plants. Current Biology 9:1485–1488.

Qiu, Y.-L., J. Les, F. Bernascond-Quadroni, D. E. Soltis, P. S. Soltis, M. Zanis, E. A. Zimmer, Z. Chen, V. Savolainen, and M. W. Chase. 1999. The earliest angiosperms: evidence from the mitochondria, plastic and nuclear genomes. *Nature* 402:404–407.

Qiu Y.-L., F. Bernasconi-Quadroni, D. E. Soltis, P. S. Soltis, M. J. Zanis, E. A. Zimmer, Z. Chen, V. Savolainen, and M. W. Chase. 2000. Phylogeny of basal angiosperms: analyses of five genes from three genomes. *International Journal of Plant Sciences* 161:3–27.

Qiu Y.-L., J. Lee, B. A. Whitlock, F. Bernasconi-Quadroni, and O. Dombrovska. 2001. Was the ANITA rooting of the angiosperm phylogeny affected by long-branch attraction? *Molecular Biology and Evolution* 18(9):1745–1753.

Soltis, D. E., and P. S. Soltis. 2004. *Amborella* not a "basal angiosperm"? Not so fast. *American J. of Botany* 91(6): 997–1001.

Soltis P. S., D. E. Soltis, and M. W. Chase. 1999. Angiosperm phylogeny inferred from multiple genes as a research tool for comparative biology. Nature 402:402–404.

Soltis P. S., D. E. Soltis, M. J. Zanis, and S. Kim. 2000. Basal lineages of angiosperms: relationships and implications for floral evolution. *American Journal of Botany* 161:97–107.

Stefanovic S., D. W. Rice, and J. D. Palmer. 2004. Long branch attraction, taxon sampling, and the earliest angiosperms: Amborella or monocots? BioMed Central Evolutionary Biology Online. URL http://www.biomedcentral.com/content/pdf/1471-2148-4-35.pdf

Stevens, P. F. (2001 onwards). Angiosperm Phylogeny Website, Version 5, May 2004 [and more or less continuously updated since]. http://www.mobot.org/MOBOT/research/APweb/.

Sun G., Q. Ji, D. L. Dilcher, S. Zheng, K. C. Nixon and X. Wang. 2002. Archaefructaceae, a new basal angiosperm family. *Science* 296:899–904.

Takhtajan, A. 1997. *Diversity and classification of flowering plants.* Columbia University Press, New York.

Thorne, R. F. Classification and geography of flowering plants. *Bot. Rev.* 58:225–348.

Wood, C. E. 1958. The genera of the woody Ranales in the southeastern United States. *Journal of the Arnold Arboretum* 39:296–346.

———. 1959. The genera of Nymphaeaceae of the southeastern United States. *Journal of the Arnold Arboretum* 40:94–112.

———. 1971. The Saururaceae in the southeastern United States. *Journal of the Arnold Arboretum* 52:479–485.

Zanis M. J., D. E. Soltis, P. S. Soltis, S. Mathews, and M. J. Donoghue. 2002. The root of the angiosperms revisited. Proceedings of the National Academy of Sciences 99:129–150.

# Nymphaeaceae

## The Waterlily Family

### Nymphaeales ■

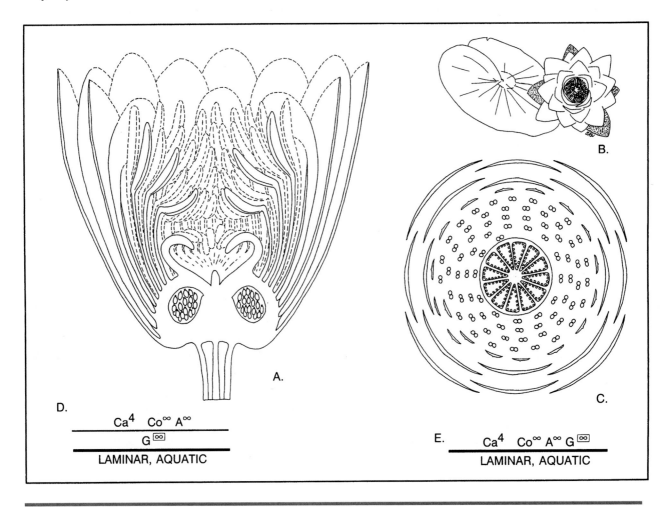

$$\frac{Ca^4 \quad Co^\infty \quad A^\infty}{G^{\boxed{\infty}}}$$

LAMINAR, AQUATIC

E. $\underline{Ca^4 \quad Co^\infty \quad A^\infty \quad G^{\boxed{\infty}}}$

LAMINAR, AQUATIC

**Figure 12-9** Features of Nymphaeaceae. A. Flower of *Nymphaea* (l.s.) with numerous perianth parts, stamens and carpels. Note the inferior ovary. B. Leaf and flower of *Nymphaea*. C. Floral diagram of *Nuphar*. Note the superior ovary. D. Floral formula for *Nymphaea*. E. Floral formula for *Nuphar*.

## FAMILY DESCRIPTION

Aquatic herbs with submersed rhizomes. Leaves large, alternate, simple, long-petiolate and floating or emersed, estipulate. Plants synoecious. Inflorescence of long-peduncled solitary flowers. Flowers perfect, regular. Sepals 4–many, distinct or basally connate. Petals 8–many (0), distinct or basally connate and adnate to sepals and stamens, sometimes grading into petaloid stamens. Stamens numerous, spirally arranged, developing centripetally, distinct or basally connate and adnate to sepals and petals; filaments flattened and often petaloid. Carpels 5–many, connate; style absent or very short; stigmas distinct; ovary superior to inferior with 5–many locules and numerous laminar ovules. Fruit a spongy berry.

# FAMILY NOTES

The waterlilies are a distinctive group of aquatic herbs. As treated here the Nymphaeaceae consists of 5 genera and about 50 species, widely distributed in tropical and temperate latitudes. Members of the family are all rhizomatous perennials that lack both secondary growth and vessel elements in their xylem. The leaves are all long-petiolate with leaf blades ranging from deeply cordate to peltate. The leaves may float on the water surface (Figure 12-9 B) or extend a short distance above the surface. *Victoria amazonica,* the waterlily of the Amazon basin, has very large raft-like leaf-blades that are sometimes 2 meters or more in diameter. The leaves of *Victoria* have upturned edges and numerous stout riblike veins. The petioles of waterlilies generally have extensive systems of air chambers that apparently aid in gas exchange between the leaves and the submersed rhizomes.

The flowers of waterlilies are large and usually have several to many petals (Figure 12-9 A, B). The petals are very conspicuous in *Euryale, Nymphaea,* and *Victoria,* inconspicuous and hidden by the stamens in *Nuphar* and absent in *Ondinea.* The stamens are often petaloid and it is probable that the extra petals are staminodal in origin. The bases of the perianth parts and stamens vary from free and distinct in *Nuphar* to connate and basally adnate in *Euryale* and *Victoria.* The ovary is compound and multilocular with numerous ovules spread over the inner surfaces of the locules. Ovaries in the Nymphaeaceae range from superior in *Nuphar* (Figure 12-9 C) to partially inferior in *Nymphaea* (Figure 12-9 A) and wholly inferior in *Euryale* and *Victoria.*

The Nymphaeaceae are considered here in a restricted sense. In some references several additional genera are included in the family. Most botanists now place *Nelumbo* (water-lotus) in its own family, and evidence is accumulating that it belongs its own order as well. Members of the Barclayaceae and Cabombaceae are also sometimes classified in the Nymphaeaceae.

Evolutionarily the Nymphaeaceae and its close relatives are considered to be very isolated. The species have several features in common with the monocots. The stems have scattered vascular bundles and lack secondary growth. Vessel-less xylem is common in some monocot groups. There is also a tendency toward reduction of the cotyledons in the Nymphaeaceae. Some botanists have suggested that the monocots originated from an ancestor similar to the modern Nymphaeaceae, and a few have even suggested that the family should be classified as monocots.

Economically the Nymphaeaceae are of limited significance. Species of *Euryale, Nuphar, Nymphaea* and *Victoria* are grown in pools and water gardens as ornamentals. Their large leaves and showy flowers are very attractive. Many species and hybrids of *Nymphaea* are in cultivation. The rhizomes of some species of *Nymphaea* are edible.

# Magnoliaceae

## The Magnolia Family
### Magnoliales ■

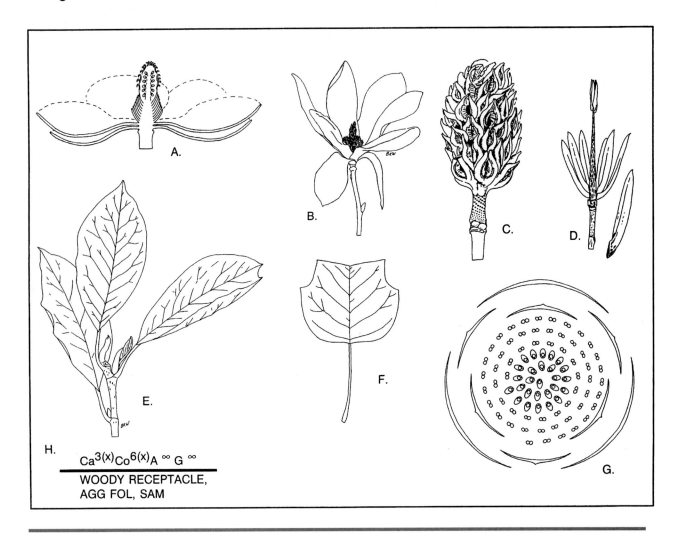

**Figure 12-10** Features of Magnoliaceae. A. Flower of *Magnolia grandiflora* (l.s.). B. Flower of *Magnolia soulangiana*. C. Aggregate of mature follicles of *Magnolia grandiflora*. D. Mature samaras of *Liriodendron tulipifera* breaking free from woody receptacle. E. Leafy twig of *Magnolia*. F. Individual leaf of *Liriodendron tulipifera*. G. Floral diagram of *Magnolia*. H. Generalized floral formula.

## FAMILY DESCRIPTION

Trees or shrubs. Leaves alternate, simple, stipulate. Plants synoecious. Inflorescence of solitary flowers. Flowers large, perfect, regular. Tepals in 3 to several whorls or sometimes spiraled, distinct, the outermost sometimes sepaloid. Stamens many, distinct, spiraled, developing centripetally. Carpels many, distinct, or weekly connate, spiraled on elongated receptacle; ovaries superior, unilocular with 1–5 marginal ovules; styles 1 per carpel. Fruit an elongated aggregate of woody follicles, samaras, or berries.

# FAMILY NOTES

The Magnoliaceae is a relict family of 7–12 genera and about 220 species. The family has a disjunct distribution with New World species occurring from the southeastern United States to southern Brazil, and Old World species ranging from the Himalayas to Japan throughout southeast Asia to the East Indies. Fossils indicate that the family once occurred in western North America. All three of the genera that occur in the New World are also represented by Old World species.

Most species are small to large trees. The leaves in all the genera but one are entire; in *Liriodendron* the leaves are pinnately few-lobed. Stems and leaves of the Magnoliaceae contain one-celled oil glands.

The flowers are solitary and generally large and showy (Figure 12-10 A, B). The receptacle is elongated and becomes more or less woody as the fruits mature. The perianth parts are spirally attached to the receptacle in some members of the family and in three or more whorls of three in others. These may be ± differentiated into sepals and petals or all may be of similar petaloid color and texture. The stamens are spirally attached to the receptacle. They are somewhat flattened and often there is not a clearly differentiated filament. The carpels are spirally attached to the receptacle. They range from completely distinct as in *Michelia* to somewhat coherent as in *Magnolia* or basally connate in *Talauma*.

Fruits in the Magnoliaceae are quite varied. The most common fruit type is an aggregate of follicles. (Figure 12-10 C). The follicles are unusual in that they do not split along the margins, but along the midvein of the follicle. Thus the follicles open away from the floral axis rather than toward it as in the follicles of other families. In genera with carpels connate the fruit is capsular. In genera with dehiscent fruits the seeds have a fleshy aril (outer seed coat) and dangle from the opened fruit by a silk-like thread. Some Magnoliaceae have indehiscent fruits. The pericarp in some members of the family is fleshy, and the ovaries ripen as aggregates of berries. In *Liriodendron* the fruit is an aggregate of one-seeded samaras (Figure 12-10 D).

Members of the Magnoliaceae are extensively planted for the beauty of their flowers and for shade. Many species and hybrids are cultivated as ornamentals. The tulip tree *(Liriodendron tulipifera)* is a valuable lumber tree. Some Indian tribes used its wood for dug-out canoes since it often has very long straight trunks.

# Lauraceae

## The Laurel Family
### Laurales ■

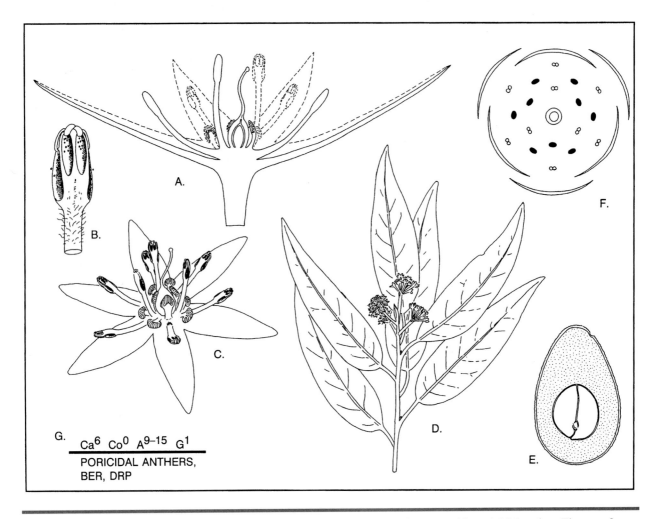

**Figure 12-11** Features of Lauraceae. A. Flower of *Persea americana* (l.s.). B. Stamen with poricidal anther. Flower of *Persea*. D. Leafy twig of *Umbellularia californica* showing the umbelliform inflorescences. E. Single-seeded berry of *Persea americana* (avocado; l.s.). F. Floral diagram of *Laurus nobilis*. G. Generalized floral formula.

## FAMILY DESCRIPTION

Shrubs or trees (in 1 genus twinning parasitic herbs). Leaves alternate or opposite, simple, estipulate (rarely reduced and scale-like or absent), usually punctate and strongly scented. Plants synoecious (rarely dioecious). Inflorescence of paniculiform or umbelliform cymes, racemes, spikes or heads, sometimes involucrate. Tepals 6, sepaloid, in 2 whorls of 3, distinct or basally connate. Stamens 9–15 in whorls of 3, some often reduced to staminodal nectaries, distinct, free or filaments basally adnate to tepals; anthers with oblong pores opening by uplifted lids. Carpel 1, ovary superior (inferior); with 1 locule and 1 apical ovule; style 1. Fruit a berry or drupe.

# FAMILY NOTES

The Lauraceae, with 30–50 genera and over 2000 species, are widespread and common in tropical and subtropical areas throughout the world. The family is especially common in Brazil and southeast Asia. A few genera range into temperate regions. Species in temperate North America include *Sassafras albidum* (sassafras), *Lindera benzoin* (spicebush), and *Persea palustris* (swamp bay) in the eastern U.S. and *Umbellularia californica* (California bay-laurel) in California and Oregon.

Except for *Cassytha,* a genus of leafless parasitic herbs, members of the Lauraceae are all trees and shrubs.The leaves are usually entire and scented with aromatic oils. *Sassafras,* a temperate genus of eastern North America, is unusual in having palmately lobed leaves.

The flowers are small and usually arranged in umbelliform (Figure 12-11 D) or paniculiform cymes. The six perianth parts (Figure 12-11 C) are not clearly differentiated into sepals and petals and are usually rather small. They are usually all treated as sepals. Generally there is a short hypanthium (Figure 12-11 A). The anthers have a characteristic manner of opening (Figure 12-11 B); flaps of tissue fold upward, forming 2–4 pores through which the pollen is released. Paired staminodal nectaries are often borne at the base of each filament.

The laurel family produces several economically important products. Cinnamon is produced from the bark of *Cinnamomum zeylandicum* and several other species of *Cinnamomum.* Other spices are produced by *Laurus nobilis* (bay leaf) and *Umbellularia californicum* (California bay-laurel). Edible fruits are produced by *Persea americana* (avocado; Figure 12-11 E). The roots of *Sassafras albidum* have been used as a flavoring and for herb teas, but recent studies indicate that some of the chemicals are potentially carcinogenic. Camphor is extracted from the foliage of *Cinnamomum camphora.* The family includes several valuable tropical timber trees, including species that produce aromatic cabinet woods. Various members of the family are cultivated as ornamentals. In ancient Greece a wreath made from the foliage of *Laurus nobilis* was used as a symbol of honor.

# Berberidaceae

## The Barberry Family
### Ranunculales ■

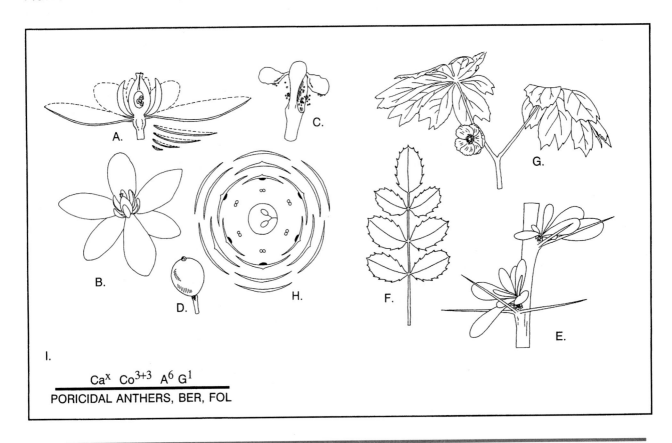

$$Ca^x \ Co^{3+3} \ A^6 \ G^1$$

**PORICIDAL ANTHERS, BER, FOL**

**Figure 12-12** Features of Berberidaceae. A. Flower of *Nandina domestica* (l.s.). B. Flower of *Nandina*. C. Stamen of *Berberis* with anther dehiscing by uplifted lids. D. Berry of *Nandina*. E. Branch of *Berberis* bearing a spiny primary leaf on main stem and foliage leaves on short secondary branches. F. Leaf of *Mahonia* sp. G. Stem, leaves, and flower of *Podophyllum peltatum*. H. Floral diagram of *Berberis*. I. Generalized floral formula.

## FAMILY DESCRIPTION

Herbs or shrubs. Leaves alternate (rarely opposite) or all basal, simple to quadripinnate, estipulate (rarely stipulate). Plants synoecious. Inflorescence of solitary flowers or variously clustered in cymes, racemes or panicles. Flowers perfect, regular. Sepals 4–many (0), distinct, often spiraled or in several whorls of 3 (2, 4), sometimes poorly differentiated from petals. Petals 4–many (0), distinct, sometimes in several whorls of 3 (2, 4). Stamens 3–6 (–18), distinct, anthers opening by pores with lids (rarely by lateral slits), Carpel 1; ovary superior with 1 locule and 2–many marginal ovules or 1–few basal (apical) ovules (rarely 2 locules by placental intrusion); style 1 or stigma sessile. Fruit a berry or follicle (rarely a legume-like pseudocapsule).

<hr />

# FAMILY NOTES

The Berberidaceae is a relatively small family of 16–18 genera and 600 species. Most occur in mesic habitats in the north temperate zone with a few in the mountains in the tropics and in south-temperate areas. The herbaceous species are all north temperate in distribution whereas woody members of the family occur in both the northern and southern hemispheres.

There is much variation in vegetative features of the barberry family. Some are herbaceous perennials and others are stiffly woody shrubs. Leaves range from the simple, peltate leaves of *Podophyllum* (Figure 12-12 G) to the spiny primary leaves of some *Berberis* species (Figure 12-12 E), the once-pinnate spiny-toothed leaves of *Mahonia* (Figure 12-12 F), and the several-times pinnate leaves of *Nandina*. Some botanists have divided the family into several smaller, more coherent groups.

Reproductive features are more uniform, however. Most have several to many relatively small flowers with several whorls of perianth parts, usually with three parts per whorl (Figure 12-12 B). The outermost two or three whorls are more or less sepaloid in color and texture. The middle two whorls are often petaloid but lack nectaries, and the innermost are petaloid and usually bear nectar-glands. Some authors treat the nectary-bearing structures as petals and others consider them to be staminodes. In most genera there are six stamens with anthers that dehisce by uplifted lids (Figure 12-12 C), somewhat like those of the Lauraceae. Certain individual genera deviate in one respect or another from these generalizations. The gynoecium is uniformly monocarpous.

Stamens of *Berberis and Mahonia* are sensitive to touch. In an open flower they are spread against the adjacent petals. When the filament base is touched by a visiting insect, the stamens spring upright and dust the body of the insect with pollen.

Members of the Berberidaceae have both positive and negative ecological and economic aspects. The fleshy fruits produced by species in several genera are eaten by birds and other wildlife. May-apples *(Podophyllum peltatum)* are sometimes eaten by humans (the pericarp is edible but the seeds are poisonous). The main economically important species are ornamental shrubs such as *Nandina domestica* (heavenly-bamboo) and species of *Berberis* (barberries) and *Mahonia* (Oregon-grapes). *Berberis vulgaris,* once widely planted as an ornamental, has been deliberately exterminated in many areas because it is the alternate host for wheat rust, economically a very serious plant disease.

# Ranunculaceae

## The Buttercup Family
### Ranunculales ◼

Ca$^x$ Co$^0$ A$^\infty$ G $^{3-\infty}$
PETALOID SEPALS, ACH, FOL

Ca$^5$ Co$^5$ A$^\infty$ G $^{3-\infty}$
ACH, FOL

**Figure 12-13** Features of Ranunculaceae. A–C. *Ranunculus californicus.* A. Flower (l.s.). B. Intact flower with many petals. C. Aggregate of achenes. D. Five-merous flower of *Ranunculus* sp. E. Twig and flower of *Anemone* with petaloid sepals and palmately compound leaves. F. Flower with spurred petals and aggregate of follicles of *Aquilegia eximia*. G. Zygomorphic flower of *Delphinium* sp. with spurred sepal. H–I. *Thalictrum fendleri*. H. Aggregate of achenes. I. Staminate flower. J. Floral diagram of *Ranunculus*. K. Generalized floral formula of apetalous Ranunculaceae. L. Generalized floral formula of apopetalous Ranunculaceae.

## FAMILY DESCRIPTION

Herbs, shrubs or woody vines. Leaves alternate, opposite or whorled, simple to several times compound, estipulate (rarely stipulate). Plants synoecious (rarely dioecious). Inflorescences various, determinate or indeterminate or flowers solitary.

Flowers perfect (rarely imperfect), regular (sometimes irregular). Sepals 4–many, distinct, often petaloid. Petals 0 or 4–5 (sometimes many), distinct. Stamens many (sometimes 5–10), distinct, spiraled, developing centripetally. Carpels 5–many (rarely

1–3), distinct (rarely 5, connate); ovaries superior with 1 locule and 1–many basal or marginal ovules (rarely ovary superior, 5–locular with axile ovules); styles 1 per carpel, distinct, undivided. Fruit a cluster of achenes or follicles (rarely a berry or capsule).

# FAMILY NOTES

The Ranunculaceae is a cosmopolitan family that is particularly well represented in temperate zones and montane environments, especially of the northern hemisphere. It is a medium-sized family of about 50–52 genera and 2000 species.

The Ranunculaceae are a polymorphic group of plants with great vegetative and floral diversity. Most are perennial herbs although *Xanthorrhiza* is a softly woody shrub and some species of *Clematis* are woody vines. Foliar morphology is extremely variable with leaves ranging from simple and entire to palmately or pinnately compound, bipinnate, tripinnate or even more complex. Inflorescences vary from solitary flowers to cymes or elongated racemes.

The ancestral floral form in the family was probably a perfect, radially symmetric flower with a flat or bowl-shaped uniseriate perianth of petaloid sepals; it had numerous spirally arranged stamens and an apocarpous gynoecium that matured as an aggregate of follicles. Flowers with these features occur in various modern members including *Caltha, Helleborus, Trautvetteria* and *Trollius.* The ancestral flowers most likely were insect-pollinated. Such flowers today are visited by beetles, flies, or bees.

Today there are assorted deviations from this floral form within the family. Many of the changes in floral morphology from the basic Ranunculaceous pattern are correlated with apparent adaptations related to pollination. Petals have evolved in several lineages (e.g., *Ranunculus,* Figure 12-13 A, B). The petals in Ranunculaceae apparently originated as staminodes. The conspicuous petals are important in attracting pollinators. The petal surfaces in various genera secrete nectar. Flowers of some genera are strongly three dimensional and/or zygomorphic and are visited by bees, moths, or birds. Petals in *Aquilegia* (Figure 12-13 F) and *Delphinium;* (Figure 12-13 G) have long, tubular, nectar-bearing spurs. Both *Aconitum* and *Delphinium* have bilaterally symmetric flowers.

Imperfect flowers have evolved independently in several genera. Flowers of some *Thalictrum* species (Figure 12-13 H–I) are small, imperfect and wind-pollinated. In staminate individuals of these species the flowers are inverted, the perianth is reduced and greenish and the stamens dangle in the air at the ends of slender filaments; in pistillate individuals the pollen receptive surfaces of the stigmas are enlarged. Dioecious species of *Clematis,* on the other hand, are insect-pollinated and have conspicuous flowers.

Some modifications are related to seed dispersal. Achenes have evolved one or more times from the ancestral follicle. In *Clematis* and some *Anemone* species the styles become elongated and are covered with long spreading hairs that aid in wind-dispersal of the achenes. The achenes of some species of *Ranunculus* bear hooks or prickles that cling to the fur of passing animals. The gynoecium of *Actaea* has been reduced to a single carpel that ripens as a berry. The berries are eaten by animals that disperse the seeds (however, they are poisonous to humans). The fruit of *Hydrastis* is an aggregate of berries. In *Nigella* the carpels are connate and the fruit is a capsule.

Various members of the family are grown as ornamentals for their showy flowers. These include species of *Anemone* (Figure 12-13 E) *Clematis* (virgin's bower), *Delphinium* (larkspur), *Aquilegia* (columbine, Figure 12-13 F), *Ranunculus* (buttercup), and *Nigella* (love in a mist).

Although they have beautiful flowers, species of *Delphinium* and *Aconitum* (monkshood) are poisonous and sometimes cause death to livestock. Ranchers regard these plants as weeds and have taken measures to eradicate them in some western states.

# Papaveraceae

## The Poppy Family

### Ranunculales ■

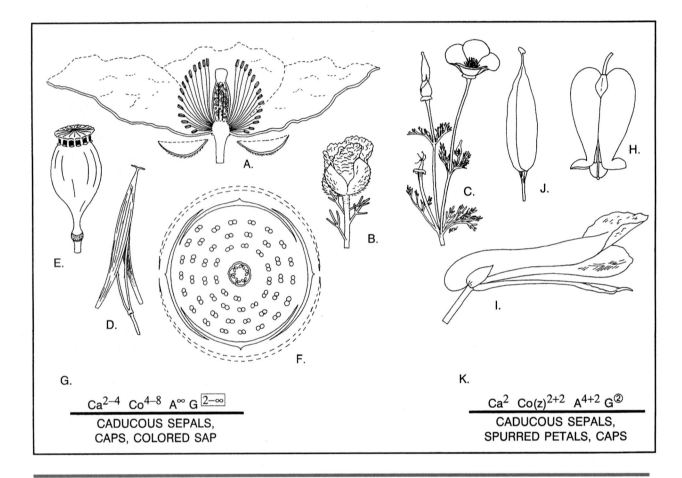

$$Ca^{2-4} \quad Co^{4-8} \quad A^{\infty} \quad G\boxed{2-\infty}$$

CADUCOUS SEPALS,
CAPS, COLORED SAP

$$Ca^2 \quad Co(z)^{2+2} \quad A^{4+2} \quad G^{②}$$

CADUCOUS SEPALS,
SPURRED PETALS, CAPS

**Figure 12-14** Features of Papaveraceae. A–G. Subfamily Papaveroideae. A. Flower of *Romneya californica* (l.s.); note deciduous sepals. B. Bud of *Romneya* with crumpled petals (sepals still present). C. Flowering branch of Eschscholzia *californica* with connate sepals deciduous as a cap. D. Longitudinally dehiscent capsule of Dendromecon. E. Poricidal capsule of Papaver *somniferum*. F. Floral diagram of *Papaver*. G. Generalized floral formula. H–K. Subfamily Fumarioideae. H. Isobilateral flower of *Dicentra*. I. Bilateral flower of *Corydalis*. J. Capsule of *Dicentra*. K. Generalized floral formula.

## FAMILY DESCRIPTION

Herbs or shrubs (rarely trees) often with milky or colored sap (sometimes only in the roots). Leaves alternate or opposite (whorled), simple to pinnately much dissected, estipulate, sometimes with twining petioles. Plants synoecious. Inflorescences various, determinate or indeterminate or flowers solitary.

Flowers perfect, irregular, regular or isobilateral [4 petals in 2 dissimilar sets]. Hypanthium absent (present in one genus). Sepals 2 or 3, distinct (in one genus connate to the tip), in many genera falling as the flower opens. Petals 4, 6 or 8 (12–16) (0), distinct rarely basally connate, often crumpled.

Stamens 4–12 or many, distinct, spirally attached, developing centripetally; filaments sometimes weakly adnate at base to petals. Carpels 2–many, connate (rarely nearly distinct); ovary with 1 locule and many parietal ovules (rarely locules 6–12 and ovules many, axile *or* locules 3–many and ovules apparently marginal); styles 1–many or stigma 1, sessile and ± lobed. Fruit a capsule (in 1 genus a cluster of indehiscent follicles; rarely a nut or indehiscent fruit breaking transversely into one-seeded pieces).

# FAMILY NOTES

The Papaveraceae is a mostly temperate family of about 34–41 genera and 600 species. The greatest diversity is in areas with Mediterranean climates. The two principal centers of distribution are western North America and eastern Asia.

The Papaveraceae consists of two well-marked subfamilies, the Papaveroideae and the Fumarioideae (Table 12-1). These are often treated as separate families. There are a couple of genera, however, that combine the features of the two subfamilies to such an extent that their subfamilial assignment is in doubt. Cronquist chose to recognize the Fumariaceae as a family distinct from the Papaveraceae and more or less arbitrarily assigned the intermediate genera to the former. We have chosen to combine the two families.

Most members of subf. Papaveroideae are herbs or shrubs. *Bocconia* sometimes forms small, weak-stemmed trees. The leaves vary from simple and entire to variously lobed or dissected. Most have milky or colored sap (which in some genera is restricted to the roots). The color of the sap ranges from creamy white or pale yellow to blood-red. The bitter-tasting sap generally contains significant quantities of mildly toxic to very poisonous alkaloids. Because of the presence of these distasteful or poisonous substances, the foliage of poppies is seldom eaten by herbivores.

The sepals characteristically fall from the flower as the petals and stamens expand (Figure 12-14 A, B). As a result open poppy flowers lack sepals. In *Eschscholzia* the sepals are connate to the tip and fall as a conical cap (Figure 12-14 C). The flowers generally have twice as many petals (4 or 6) as sepals (2 or 3) and in several genera the petals have a distinctive crumpled appearance. *Sanguinaria* is unusual in having 8–16 petals; it has only 2 sepals. Most genera have numerous stamens. The petals commonly spread apart and the open flower has the shape of a disk or bowl.

Members of subf. Fumarioideae are all herbaceous. None of them produce latex. In most cases the leaves are deeply dissected. The flowers in subf Fumarioideae have a distinctly different appearance from those of subf. Papaveroideae. The sepals are very small, commonly appearing as a pair of tiny triangular scales. These are usually caducous but sometimes persist in open flowers. The petals are

| Character | subf. Papaveroideae | subf. Fumarioideae |
|---|---|---|
| Habit | Annual or perennial herbs, shrubs, rarely small trees | Annual or perennial herbs |
| Latex | Present | Absent |
| Sepals | 2–3, large, enclosing flower bud | 2, small and bractlike, much shorter than flower bud |
| Symmetry | Radial or nearly so, fully distinct | Bilateral or isobilateral (4 petals in 2 unequal pairs), the inner cohering at tip |
| Spurs | Absent | Present on 1 or 2 petals |
| Stamens | (4–) many, equal | 6, tetradynamous |

**Table 12-1** Comparison of Papaveraceae subfamilies Papaveroideae and Fumarioideae.

paired. One or both of the outer pair have a pouch or spur. Flowers with one spurred petal are bilaterally symmetric. Those with two spurred petals and two unspurred petals have isobilateral (biradial) symmetry. These flowers have two lines of symmetry. The inner pair cling together at the tip over the stigma. There are only six stamens. Two of these have 4 pollen sacs and four have one pollen sac apiece.

The superior ovary in the poppy family matures as a capsule that may dehisce in various ways. In *Papaver* the capsule opens by pores that develop below the flattened, persistent stigma (Figure 12-14 D). In most other genera the capsule splits lengthwise. In *Chelidonium* (subf. Papaveroideae) and many Fumarioideae it matures as a silique much like those in the Brassicaceae. In *Platystemon* the carpels are only weakly connate, and as the fruit matures it separates into a group of follicle-like segments that break transversely into seed-bearing joints.

The poppy family includes species with considerable economic importance. *Papaver* includes several species of ornamental garden poppies as well as the opium poppy *(P. somniferum)*. The alkaloids of *P. somniferum* are the source of raw opium which is refined to produce narcotic drugs ranging from the medically important codeine and morphine to the illegal street drug, heroin. Other alkaloids in the family are also medicinally important. *Papaver somniferum* is also one of the sources of the poppy seeds used in cooking and of poppyseed oil, but the seeds lack alkaloids.

Several of North America's most conspicuous wildflowers are members of the Papaveraceae. Members of the Papaveroideae include *Eschscholzia californica* (California poppy), *Dendromecon rigida* (bush poppy), *Platystemon californicus* (creamcups), *Argemone* spp. (prickly poppy), and *Sanguinaria canadensis* (bloodroot). Members of the Fumarioideae include species of *Dicentra* (Dutchman's breeches, squirrel-corn, bleeding hearts) and *Corydalis.*

# Exercises

## ■ Study Questions

### Nymphaeaceae

1. What vegetative features of Nymphaeaceae are unusual?

2. What type(s) of perianth do we find in the Nymphaeaceae?

3. Describe the androecium in Nymphaeaceae flowers?

4. What kind of gynoecium characterizes the Nymphaeaceae?
   How many carpels?
   Placentation?

5. What is the ovary position in *Nymphaea?*
   in *Nuphar?*
   Fruit type?

### Magnoliaceae

6. Are there any herbaceous members of the Magnoliaceae?

7. Describe the perianth in the Magnoliaceae?
   What are the petals attached to?
   Floral symmetry?

8. How many stamens are there in Magnoliaceae flowers?
   What are they attached to?
   What do they look like?

9. What kind of gynoecium characterizes the families of the Magnoliaceae?
   Number of carpels?
   Ovary position?
   Placentation?

10. What are the fruit types in the Magnoliaceae?

### Lauraceae

11. What kinds of habit occur in the Lauraceae

12. Describe the perianth in the Lauraceae.
    Insertion of flower parts?

13. What is unusual about the stamens of the Lauraceae?

14. Describe the gynoecium of the Lauraceae.
    Fruit types?

# Berberidaceae

15. Describe the perianth of the Berberidaceae.

    How many sepals?

    How many petals?

    What is unusual about the petals?

16 What features of the stamens are unusual?

    Is the insertion of stamens hypogynous, perigynous, or epigynous?

17. Describe the gynoecium of the Berberidaceae.

    Fruit type?

# Ranunculaceae

18. Are vegetative features very useful in characterizing the Ranunculaceae? Explain.

19. Is it useful to try to characterize the Ranunculaceae by their inflorescences?

20. Describe the perianth in the Ranunculaceae.

    What are the petals and sepals attached to?

21. How many stamens are typical in Ranunculaceae flowers?

    What are they attached to?

    Do members of the Ranunculaceae have a hypanthium?

    What term describes the insertion (attachment) of the outer parts of the flower relative to the gynoecium?

22. What kind of gynoecium characterizes most Ranunculaceae?

    Ovary position?

    Fruit types?

    What is unusual about the fruits of *Clematis?*

23. What is the floral symmetry in most Ranunculaceae?

    In *Aconitum* and *Delphinium?*

    What specialized floral features occur in *Aconitum, Delphinium,* and *Aquilegia?*

# Papaveraceae

24. How many sepals do Papaveraceae flowers have?

    What happens to the sepals when a flower opens?

    What feature of the sepals of *Eschscholzia* (California poppy) are unusual?

25. How many petals do Papaveraceae flowers have?

    Would you expect poppies to have 5-merous flowers?

26. What is the flower symmetry in subf. Papaveroideae?

    In subf. Fumarioideae?

27. How many stamens are there in most Papaveroideae flowers?

In subf. Fumarioideae?

What is unusual about these stamens?

What is the insertion of the stamens?

28. What kind of gynoecium characterizes the Papaveraceae?

How many carpels do we find per flower?

Placentation?

Ovary position?

Fruit type?

29. List the families and species of subclass Magnoliidae that you identified in the lab.

_____        _____

_____        _____

_____        _____

_____        _____

_____        _____

_____        _____

# The Caryophyllid Clade

## chapter 13

The Caryophyllidae compromises four orders, 29 families, and about 15,000 species. Seventeen of the families and about 60 percent of the species are members of a single order, the Caryophyllales.

The Caryophyllid clade appears to be monophyletic on the basis of several characters such as seed coat anatomy, simple perforations in the vessel elements, pollen shed in 3-nucleate condition, and several types of sequence data (*rbcL, atpB, matK,* and 18S rDNA). The group includes two large orders (Caryophyllales and Polygonales) and two smaller orders (Santalales and Dilleniales). Many of these families had previously been placed in the subclass Dillenidae or the Rosidae.

There is clearly great support for the monophyly of the group, but how this group is related to the Rosids and Asterids is uncertain at this time. The Angiosperm Phylogeny Group treats this group as a single order (Caryophyllales *sensu lato*)

Many members of the Caryophyllidae are herbaceous, and the woody members of the subclass apparently all have anomalous types of secondary growth. Cronquist (1981) supported the hypothesis that the Caryophyllidae originated from herbaceous ancestors. Succulents occur in several families (e.g., Aizoaceae, Cactaceae, Portulacaceae).

In most members of the subclass the flowers are apetalous or have distinct petals. The lack of petals appears to be primitive within the subclass. Petals have apparently originated from staminodes in some families (e.g., Aizoaceae) and from bracts or sepals in others (e.g., Cactaceae). In some families the structures traditionally treated as sepals and petals probably originated as bracts and sepals respectively (e.g., Didiereaceae, Portulacaceae). Those members of the Caryophyllidae with numerous stamens have centrifugal stamen development.

Several of the families of the Caryophyllidae have ovaries with free-central or basal placentation. This gave rise to an old name for this group, the Centrospermae. Some families of the Caryophyllidae, however, have axile or parietal placentation. The ovules have two integuments and a thick nucellus. Food is stored in the seeds in the form of starch.

A chemical character present in some (but not all) families of the Caryophyllidae (Table 13-1) is the production of **betalain** pigment; instead of anthocyanins (Figure 13-1). Betalains are blue to red water-soluble pigments located in the cell sap. Although not all members of the Caryophyllidae produce these pigments, the only angiosperms to

| Betalain producers | Anthocyanin producers |
|---|---|
| Caryophyllales | Caryophyllales |
| Phytolaccaceae | Molluginaceae |
| Achatocarpaceae | Caryophyllaceae |
| Nyctaginaceae | |
| Cactaceae | Polygonales |
| Aizoaceae | Polygonaceae |
| Didiereaceae | Plumbaginaceae |
| Chenopodiaceae | |
| Amaranthaceae | |
| Portulacaceae | |
| Basellaceae | |

Table 13-1   Betalain and anthocyanin producing families of the Caryophyllidae.

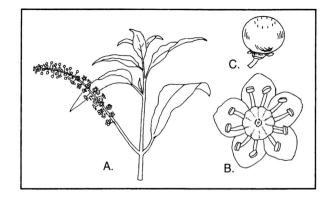

**Figure** Comparison of chemical structures of an anthocyanin pigment and a betalain pigment.
**13-1**

produce them are members of the Caryophyllidae. The families in which betalain pigments are produced are all members of the Caryophyllales. Even within this order there are two families that produce anthocyanins. The evolutionary significance of these pigments is unknown.

# FAMILIES OF THE CARYOPHYLLID CLADE

## 1. Order Caryophyllales

**1.A   Caryophyllaceae**—The Pink Family (Figure 13-10). 66–75 genera, 2000 species. Cosmopolitan, but most in temperate and warm-temperate regions in the northern hemisphere; greatest diversity in the Mediterranean region.

§See page 254 for description and illustrations§

**1.B   Phytolaccaceae**—The Pokeberry Family (Figure 13-2). 15–19 genera, 125 species. Mostly tropical and subtropical, especially in New World.

Herbs, shrubs or trees. Leaves alternate, simple and entire, estipulate (rarely stipulate). Plants synoecious (rarely monoecious). Inflorescence of racemes. Flowers perfect (rarely imperfect), regular. Sepals 4–5, distinct or connate. Petals absent. Stamens 4–5 (many), distinct. Carpels 1-many, distinct with ovaries superior (rarely ovary 1 and inferior), one-locular with 1 basal ovule per carpel *or* carpels

**Figure** Examples of Phytolaccaceae. A–C. *Phytolacca*
**13-2**    *americana.* A. Leafy branch with raceme.
           B. Flower. C. Berry.

2-many, connate, with ovary superior, 2-many locular with 1 axile ovule per locule; styles 1 per carpel. Fruit an aggregate of drupelets or achenes, a berry, nut, schizocarp or samara.

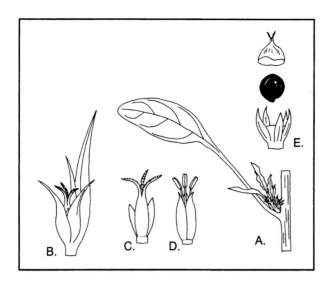

**Figure 13-3** Examples of Amaranthaceae. A–D. *Amaranthus californicus.* A. Node with leaf and axillary inflorescence. B. Pistillate flower with subtending bract. C. Pistillate flower with front sepals removed. D. Staminate flower. E. One-seeded circumscissile capsule of *Amaranthus blitoides.*

**1.C  Nyctaginaceae**—The Four O'clock Family (Figure 13-11). 30–38 genera, 300 species. Warm temperate to tropical regions of Old and New World with greatest diversity in New World

§See page 257 for description and illustrations§

**1.D  Amaranthaceae**—The Amaranth Family (Figure 13-3). 65–70 genera, 900 species. Mostly tropical or subtropical.

Herbs or shrubs (rarely trees or woody vines). Leaves alternate or opposite, simple, estipulate. Plants synoecious, monoecious, dioecious or polygamous. Inflorescence of solitary flowers or cymose clusters, often aggregated into spikelike, headlike, racemiform or paniculiform clusters. Flowers perfect or imperfect, regular (irregular), often closely subtended by bractlets. Sepals 4–5 (1–3), distinct or basally connate, opposite the sepals. Carpels 2–3, connate; ovary superior or inferior with 1 locule and 1-many basal ovules; styles 2–3, distinct, or 1 and entire or 2–3 branched. Fruit an achene, a utricle, or a circumscissile capsule (rarely a drupe or a berry).

**1.E  Chenopodiaceae**—The Goosefoot Family (Figure 13-12). 100–113 genera, 1500 species. Mostly temperate and subtropical in both Old and New Worlds.

§See page 259 for description and illustrations§

**1.F  Aizoaceae**—The Ice Plant Family (Figure 13-13). 100–150 genera, 2500 species. Mostly southern hemisphere, especially in South Africa and Australia, some tropical, few north temperate.

§See page 261 for description and illustrations§

**1.G  Portulacaceae**—The Purslane Family (Figure 13-14). 20–27 genera, 500 species. Nearly cosmopolitan distribution with greatest diversity in North America and Africa.

§See page 263 for description and illustrations§

**1.H  Cactaceae**—The Cactus Family. 50–150 genera, 1500–2000 species (Figure 13-15). New World, mostly warm-temperate to tropical.

§See page 265 for description and illustrations§

**1.I  Simmondsiaceae**—The Jojoba Family. 1 genus, 1 species. Southwestern deserts of North America.

Shrubs. Leaves opposite, simple and entire, estipulate. Plants dioecious. Inflorescence of peduncled axillary heads of staminate flowers or solitary axillary pistillate flowers. Flowers imperfect, regular. Sepals 4–6, distinct or basally connate. Petals absent. Stamens 10 (8–12), distinct. Carpels 3, connate; ovary superior with 3 locules and 1 apical-axile ovule per locule; style absent, stigmas 3, elongate, sessile. Fruit a one-seeded, capsule with 2 empty locules.

**1.J  Achatocarpaceae**—2 genera, 8 species. Texas to Argentina.

**1.K  Basellaceae**—The Madeira Vine Family. 4–5 genera, 15–20 species. Mostly New World tropics, few in Old World tropics.

Herbaceous vines. Leaves alternate, simple, estipulate, ± fleshy. Plants synoecious or monoecious. Flowers perfect or imperfect, regular. Hypanthium present, short. Sepals 2, distinct. Petals 5, distinct or basally connate. Stamens 5, opposite the petals, filaments adnate to corolla. Carpels 3, connate; ovary superior with 1 locule and 1 basal ovule; style 1. Fruit a drupe.

**1.L  Molluginaceae**—The Carpet-weed Family. 13–15 genera, 100 species. Tropical and subtropical, especially Africa.

Herbs (shrubs). Leaves opposite, alternate or whorled, simple, entire, estipulate. Plants synoe-

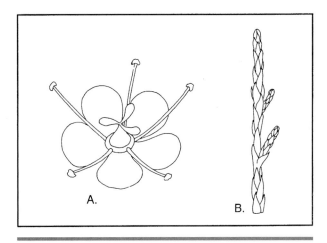

Figure 13-4 Examples of Tamaricaceae. A–B. *Tamarix* sp. A. Flower. B. Branchlet with scale leaves.

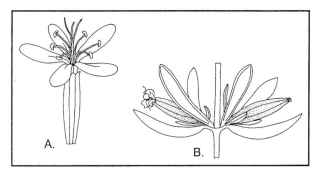

Figure 13-5 Examples of Frankeniaceae. A–B. *Frankenia salina*. A. Flower. B. Node with opposite leaves, axillary fascicles, and calyces with developing fruits.

cious (monoecious). Inflorescence variously cymose, or flowers solitary, axillary. Flowers perfect (imperfect), regular. Sepals 5 (4), distinct (rarely basally connate). Petals absent (4–5, distinct or connate). Stamens (2–) 5–10 (many), distinct or filaments connate. Carpels (1) 2–5, connate; ovary superior with (1) 2–5 locules and 1-many axile ovules, styles (1) 2–5. Fruit a capsule.

**1.M Tamaricaceae**—The Tamarisk Family (Figure 13-4). 4–5 genera, 100 species. Eurasia and Africa.

Shrubs or trees. Leaves alternate, simple, scale-like, estipulate. Plants synoecious. Inflorescence of solitary or racemose flowers. Flowers perfect, regular. Sepals 4–5, distinct. Petals 4–5 distinct. Stamens 4–10 (many), distinct or filaments basally connate. Carpels 2–5, connate; ovary superior with 1 locule and few-many basal or parietal ovules; styles 2–5, distinct or connate to near tip. Fruit a capsule with comose seeds.

**1.N Frankeniaceae**—The Frankenia Family (Figure 13-5). 4–5 genera, 80 species. Worldwide in warm-temperate to tropical salt marshes and salt deserts. Herbs or shrubs. Leaves opposite, simple and entire, estipulate. Plants synoecious (rarely monoecious or dioecious). Inflorescence of solitary flowers or variously cymose. Flowers perfect (imperfect), regular. Sepals 4–7, connate.

Petals 4–7, distinct. Stamens 4–7 (many) distinct or filaments basally connate. Carpels 2–4, connate; ovary superior with 1 locule and 2-many parietal ovules; style 1 with 2–4 lobes. Fruit a capsule.

**1.O Dioncophyllaceae**—3 genera, 3 species. Tropical Africa.

**1.P Ancistrocladaceae**—1 genus, 15–20 species. Southeast Asia.

**1.Q Didiereaceae**—4 genera, 11 species. Madagascar.

# 2. Order Polygonales

**2.A Polygonaceae**—The Buckwheat Family (Figure 13-16). 30–49 genera, 1000 species. Mostly north temperate with a few tropical genera and a single southern hemisphere genus; particularly common in western North America.

§See page 268 for description and illustrations§

**2.B Droseraceae**—The Sundew Family (Figure 13-6 A, B). 4 genera, 103 species. Very widespread in both the northern and southern hemispheres, particularly in boggy habitats.

Herbs (rarely subshrubs). Leaves alternate or all basal, simple, stipulate or estipulate, covered with long-stipitate insect-trapping glandular hairs (in one genus glabrous and jaw-like with a hinged midrib). Plants synoecious. Inflorescence of spiciform to paniculiform cymes. Flowers perfect, regular. Sepals 4–5 (6–8), distinct or connate. Petals 4–5 (6–8), distinct. Stamens 4–20, distinct (occasionally basally connate), free (rarely adnate to petals). Carpels 3–5, connate; ovary superior to half-inferior with 1 locule and numerous parietal ovules (rarely with few basal ovules or with 3–5 locules and 1-many axile ovules); styles 1–5, often

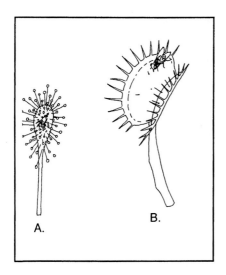

**Figure 13-6** Examples of Droseraceae. A. *Drosera rotundifolia* B. *Dionaea muscipula.*

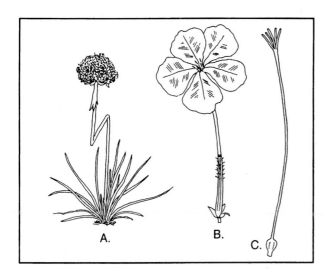

**Figure 13-7** Examples of Plumbaginaceae. A. *Armeria maritima,* plant with basal rosette and scapose inflorescence. B–C. *Plumbago auriculata.* B. Flower with subtending bractlets. C. Pistil.

**2.C  Plumbaginaceae**—The Leadwort Family (Figure 13-7). 12–25 genera, 400 species. Widely distributed, especially Eurasia.

Herbs or shrubs. Leaves alternate or all basal, simple, estipulate (rarely appearing stipulate). Plants synoecious. Inflorescence variously cymose or racemose, sometimes heads or spikes, sometimes involucrate. Flowers perfect, regular. Sepals 5, connate. Petals 5, connate or nearly distinct. Stamens 5, distinct, opposite the petals; filaments adnate to corolla. Carpels 5, connate; ovary superior with 1

locule and 1 basal ovule; styles 5 or 1 and 5-lobed. Fruit an achene or a utricle.

**2.D  Nepenthaceae**—1 genus, 75 species. East Indies to Madagascar.

# 3. Order Santalales

**3.A  Santalaceae**—The Sandalwood Family. 35–37 genera, 400 species. Cosmopolitan, especially tropics and subtropics.

Herbs, shrubs or trees, usually hemiparasitic on roots (or rarely on branches) of host plants. Leaves alternate or opposite, simple, estipulate. Plants synoecious, monoecious or dioecious. Inflorescences variously cymose or racemose or flowers solitary. Flowers perfect or imperfect, regular. Sepals 3–6, distinct or connate, sometimes petaloid. Petals absent. Stamens 3–6, distinct, opposite and filaments usually adnate to the sepals. Carpels 3–5 (2), connate; ovary superior to inferior with 1 locule and 1–5 basal or free-central ovules; style 1 and undivided or 3–5 lobed. Fruit a nut or drupe.

**3.B  Loranthaceae**—The Showy Mistletoe Family (Figure 13-8). 60–77 genera, 700 species. Mostly tropics, especially well-represented in South America.

Shrubby parasites anchored to the branches of woody hosts *or* root-parasitic shrubs, woody vines and trees, mostly green and photosynthetic. Leaves

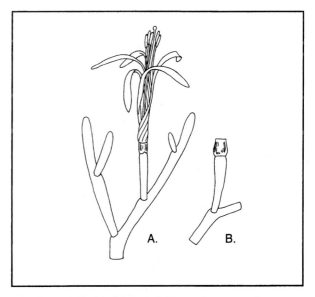

**Figure 13-8** Examples of Loranthaceae. A–B. *Psittacanthus* sp. A. Branchlet with flower. B. Branchlet with developing berry.

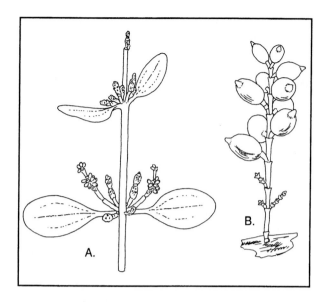

**Figure 13-9** Examples of Viscaceae. A. Branchlet of *Phoradendron flavescens* with developing berries. B. Branchlet of *Arceuthobium* sp. attached to host; note mature berries.

opposite (whorled), simple and entire, sometimes reduced to bladeless scales, estipulate. Plants synoecious (rarely monoecious or dioecious). Inflorescence of cymes, umbels, spikes, racemes or heads. Flowers perfect (imperfect), regular or irregular. Sepals 4–6, very short, connate, sometimes reduced to an unlobed rim. Petals 3–9, distinct or connate. Stamens 3–9, opposite the petals, free from or adnate to corolla. Carpels 3–4, connate; ovary inferior with 1 locale and 4–12 free-central or basal ovules (rarely locules 4 and placentation axile); style 1, undivided or stigma sessile. Fruit a berry or drupe (samara).

**3.C Viscaceae**—The Common Mistletoe Family (Figure 13-9). 7–8 genera, 350 species. Cosmopolitan but best represented in the tropics.

Herbaceous or shrubby parasites anchored to branches of woody hosts, mostly green and photosynthetic. Leaves opposite (rarely alternate), sometimes very reduced (or even absent), simple and entire, estipulate. Plants monoecious or dioecious. Inflorescence of small cymes or spikes. Flowers imperfect, regular: Sepals 2–4, connate, adnate to filaments in staminate flowers, adnate to ovary in pistillate flowers. Stamens 2–4 distinct, opposite the sepals, anthers sometimes multilocular. Carpels 3–4; ovary inferior, unilocular with 2 ovules not differentiated from basal placenta; style 1, short or stigma sessile. Fruit a berry with viscid seeds.

**3.D Olacaceae**—25–30 genera, 250 species. Tropics and subtropics.

**3.E Misodendraceae**—1 genus, 10 species. Temperate South America.

**3.F Opiliaceae**—9–10 genera, 50 species. Tropics and subtropics.

**3.G Eremolepidaceae**—3 genera, 12 species. Tropical America.

## 4. Order Dilleniales

**4.A Dilleniaceae**—The Dillenia Family. 10–11 genera, 350 species. Tropics and subtropics, especially Australia.

Trees, shrubs, woody vines or herbs. Leaves alternate (opposite), simple, entire to deeply lobed, stipulate or estipulate. Plants synoecious (rarely monoecious). Inflorescence of cymes, racemes or panicles or flowers solitary. Flowers perfect (rarely imperfect), regular. Sepals 5 (3–20), spirally overlapping, distinct or basally connate. Petals (2-) 5, distinct. Stamens many, centrifugal, often connate in groups. Carpels 4-many, distinct; ovaries superior, unilocular with 1-many marginal ovules (rarely carpels connate and locules 1 per carpel); styles 1 per carpel. Fruit a cluster of follicles or fruits indehiscent and enclosed by fleshy calyx.

## 5. Order Unknown

The following three families are poorly known and their placement in the Eudicots is uncertain. They have traditionally been placed in the Santalales.

**5.A Balanophoraceae**—17–19 genera, 45 species; Tropics and subtopics.

**5.B Medusandraceae**—1 genus, 1 species. Tropical west Africa.

**5.C Dipentodontaceae**—1 genus, 1 species. China and Burma.

## References

Angiosperm Phylogeny Group II. 2003. An update of the Angiosperm Phylogeny Group classification for the orders and families of flowering plants: APG II. *Botanical Journal of the Linnean Society* 141:399–436.

Behnke, H.-D. and T. J. Mabry (eds.). 1994. *Caryophyllales: Evolution and systematics.* Springer-Verlag, Berlin.

Blackwell, W. H. 1977. The subfamilies of the Chenopodiaceae. *Taxon* 26:395–397.

Bogle, A. L. 1969. The genera of Portulacaceae and Basellaceae of the southeastern United States. *Journal of the Arnold Arboretum* 50:566–598.

———. 1970. The Genera of Molluginaceae and Aizoaceae of the Southeastern United States. *Journal of the Arnold Arboretum* 51:431–462.

———. 1974. The Genera of Nyctaginaceae of the Southeastern United States. *Journal of the Arnold Arboretum* 55:1–37.

Channell, R. B., and C. E. Wood. 1959. The Genera of Plumbaginaceae in the Southeastern United States. *Journal of the Arnold Arboretum* 40:391–397.

Cronquist, A. 1981. *An Integrated System of Classification of Flowering Plants.* Columbia University Press, New York, pp. 231–285.

———. 1988. *The Evolution and Classification of Flowering Plants,* 2nd ed. New York Botanical Garden, Bronx, N.Y., pp. 172–186.

Eckardt, T. 1976. Classical morphological features of Centrospermous families. *Plant Systematics and Evolution* 126:5–25.

Graham, S. A, and C. E. Wood. 1965. The Genera of Polygonaceae of the Southeastern United States. *Journal of the Arnold Arboretum* 46:91–121.

Heywood, V. H. (ed.). 1993. *Flowering Plants of the World,* updated edition. Oxford University Press, New York, pp. 63–79.

Judd W. S., C. S. Campbell, E. A. Kellogg, P. F. Stevens, and M. J. Donoghue. 2002. *Plant Systematics: A Phylogenetic Approach.* 2nd ed. Sunderland, MA: Sinauer Associates.

Kubitzki, K. (ed.) 1990. *The Families and Genera of Vascular Plants.* Vol. 2. *Flowering Plants-Dicotyledons: Magnoliid, Hamamelid and Caryophyllid families* Edited by K. Kubitzki, J. G. Rohwer, and V. Bittrich. Springer Verlag, Berlin.

Mabry, T. J. 1977. The Order Centrospermae. *Annals of the Missouri Botanical Garden* 64:210–220.

Reveal, J. L. 1978. Distribution and Phylogeny of Eriogonoideae (Polygonaceae). *Great Basin Naturalist Memoirs* 2:169–190.

Rodman, J., M. Oliver, R. Nakamura, J. McClammer and A. Bledsoe. 1984. A taxonomic analysis and revised classification of Centrospermae. *Systematic Botany* 9:297–323.

Rogers, G. K. 1985. The genera of Phytolaccaceae in the southeastern United States. *Journal of the Arnold Arboretum* 66:1–37.

Soltis, D. E., P. S. Soltis, M. W. Chase, M. E. Mort, D. C. Albach, M. Zanis, V. Savolainen, W. H. Hahn. S. B. Hoot, M. E. Fay, M. Axtell, S. M. Swenson, L. M. Prince, W. J. Kress, K. C. Nixon, and J. S. Farris. 2000. Angiosperm phylogeny inferred from 18S rDNA, *rbcL,* and *atpB* sequences. *Bot. J. Linn. Soc.* 133: 381–461.

Stevens, P. F. (2001 onwards). Angiosperm Phylogeny Website, Version 5, May 2004 [and more or less continuously updated since]. http://www.mobot.org/MOBOT/research/APweb/.

Wohlpart, A., and T. J. Mabry. 1968. The distribution and phylogenetic significance of the betalains with respect to the Centrospermae. *Taxon* 17:148–152.

# Caryophyllaceae

## The Pink or Carnation Family
### Caryophyllales ■

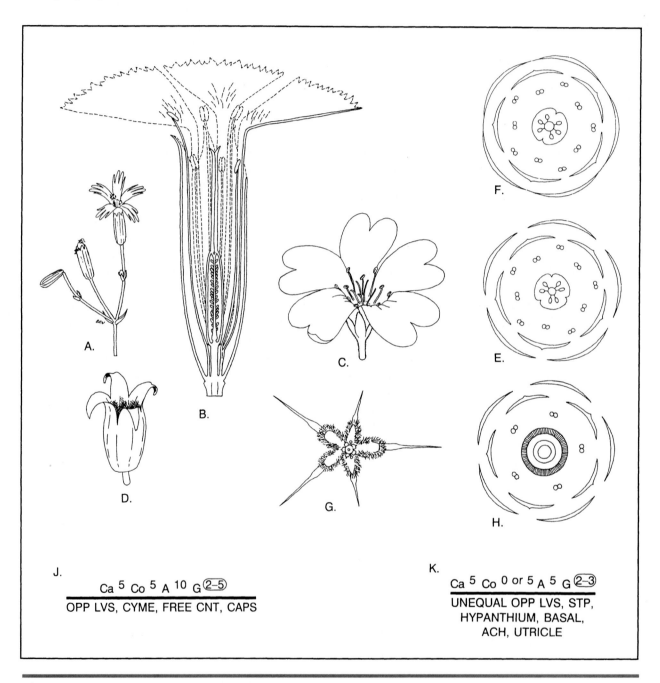

F.

C.

E.

A.

D.

B.

G.

H.

J.

$$Ca\ 5\ Co\ 5\ A\ 10\ G\ \widehat{(2\text{-}5)}$$

OPP LVS, CYME, FREE CNT, CAPS

K.

$$Ca\ 5\ Co\ 0\ or\ 5\ A\ 5\ G\ \widehat{(2\text{-}3)}$$

UNEQUAL OPP LVS, STP,
HYPANTHIUM, BASAL,
ACH, UTRICLE

**Figure 13-10** Features of Caryophyllaceae. A. Cyme of *Silene*. B. Flower of *Dianthus* (l.s.). C. Flower of *Stellaria* showing the notched petals and distinct sepals. D. Denticidal capsule of *Dianthus* with calyx removed. E. Floral diagram of *Stellaria*. F. Floral diagram of *Dianthus*. G. Flower of *Cardionema*. H. Floral diagram of *Cardionema*. J. Generalized floral formula for subf. Alsinoideae. K. Generalized floral formula for tribe Paronychieae.

# FAMILY DESCRIPTION

Herbs (very rarely shrubs). Stems often with swollen nodes. Leaves opposite (very rarely alternate), simple, entire, stipulate or estipulate. Plants synoecious (dioecious). Inflorescence variously cymose, often dichasial or flowers solitary. Flowers perfect (rarely imperfect), regular. Sepals 5 (4) distinct or connate. Petals 5 (4) (0), distinct, often notched or lacerate. Stamens (3-) 10, distinct, free or in some apetalous genera adnate to sepals. Hypanthium present in a few genera. Carpels 2–5, connate; ovary superior with 1 locule and few–many free-central ovules or 1 basal ovule (rarely with 2–5 locules with numerous axile ovules); styles 2–5 (rarely 1). Fruit a capsule or an achene.

# FAMILY NOTES

The Caryophyllaceae has a cosmopolitan distribution but most of the 66–75 genera and 2000 species occur in temperate and warm-temperate regions in the northern hemisphere. The greatest diversity occurs in the Mediterranean region.

The flowers of most members of the Caryophyllaceae have several distinctive features. They are generally borne in cymes of one sort or another (Figure 13-10 A). The calyx may be composed of distinct (Figure 13-10 E) or strongly connate sepals (Figure 13-10 B, H). The 5 petals are frequently differentiated into a basal claw and an expanded blade, and in some genera each petal bears a pair of scale-like appendages at the junction of the claw and blade (Figure 13-10 B). Although the petals of some genera are entire, it is common to find members of the Caryophyllaceae with petals that are notched at the apex (Figure 13-10 C), toothed ("pinked"; Figure 13-10 B), or variously lobed (Figure 13-10 A). There are usually 10 stamens attached directly to the receptacle. The ovary has free-central placentation in most members of the family although septa may be present when the ovary is immature. The free placenta extends the length of the ovary in some genera and is reduced to a short free-standing post in others. The number of carpels ranges from 2 to 5 and can easily be determined by counting the number of styles. The fruit matures as a denticidal capsule (Figure 13-10 D). Some genera differ in one or another feature from the conditions described above.

One group of genera, the tribe Paronychieae, differs from these features in several respects. These plants have small, often apetalous flowers with 3–10 stamens that are adnate to the sepals, forming a hypanthium (Figure 13-10 G, H). The number of ovules is reduced to only one which is basally attached. In addition, their leaves, though usually opposite as in other Caryophyllaceae are often very unequal in size. These genera are sometimes is segregated as a separate family, the Illecebraceae. The relationship of these plants to the remaining members of the Caryophyllaceae has been questioned by some taxonomists. A comparison among the infrafamilial taxa of the Caryophyllaceae is presented in Table 13-2.

The Caryophyllaceae includes colorful ornamentals and wildflowers and numerous weedy species. Economically, the most important member of the family is *Dianthus caryophyllus* (carnation). Various other members of this genus are also cultivated. Other ornamentals include species of *Gypsophila* (baby's breath), *Lychnis* (campion, Maltese cross), *Saponaria* (bouncing bet) and *Silene* (catchfly). Many others are suitable for rock-gardens. Weedy species include *Stellaria media* (common chickweed), *Cerastium* spp. (mouse-ear chickweed), *Spergula arvensis* (spurry), *Spergularia* spp. (sand spurry), and various others.

**subf. Alsinoideae (Fig. 13-10 C, E, I)**

Leaves of a pair equal, estipulate; flowers hypogynous; sepals distinct; petals entire or notched, unappendaged, sessile; stamens 10, free; ovary many-seeded; fruit a capsule.

**subf. Silenoideae (Fig. 13-10 A, B, E)**

Leaves of a pair equal, estipulate; flowers hypogynous sepals connate; petals entire, notched, fringed or deeply lobed, often appendaged and clawed; stamens 10, free or basally adnate to petals; ovary many-seeded; fruit a capsule.

**subf. Paronychioideae tribe Sperguleae**

Leaves of a pair equal, stipulate; flowers hypogynous; sepals distinct; petals sessile, unappendaged; stamens 10, free; ovary many-seeded; fruit a capsule.

**subf. Paronychioideae tribe Paronychieae [Illecebraceae] (Fig. 13-10 G, H, J)**

Leaves of a pair often unequal or leaves alternate, stipulate; flowers perigynous; sepals connate; petals very small or absent; stamens (3-) 5–10, adnate to sepals; ovary 1-seeded, maturing as an achene or a utricle

**Table 13-2**  Comparison of subfamilies and tribes of the Caryophyllaceae.

# Nyctaginaceae

## The Four O'clock Family
### Caryophyllales ■

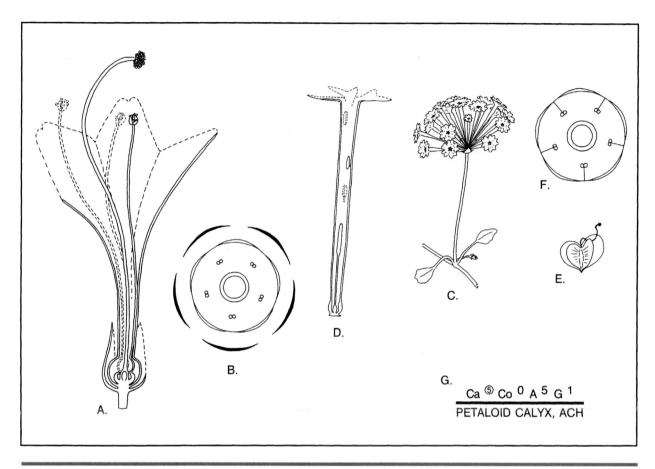

**Figure 13-11** Features of Nyctaginaceae. A. Flower of *Mirabilis jalapa* (l.s.) with the sepal-like bracts and showy, synsepalous, petaloid calyx. B. Floral diagram of *Mirabilis jalapa*. C. Branch of *Abronia umbellata* with involucrate head of flowers. D. Flower of *Abronia* with episepalous stamens (l.s.). E. Anthocarp (fused calyx base and enclosed achene) of *Abronia*. F. Floral diagram of *Abronia*. G. Generalized floral formula.

## FAMILY DESCRIPTION

Herbs, shrubs (rarely trees). Leaves opposite or alternate, simple and entire, estipulate. Plants synoecious (rarely monoecious or dioecious). Inflorescence of cymes, umbels, heads, racemes or panicles, sometimes involucrate (involucre sometimes 1-flowered and calyx-like). Flowers perfect (rarely imperfect), regular (irregular). Sepals 5, petaloid, connate, often constricted above the ovary. Petals absent. Stamens 1–30, distinct or filaments basally connate, free or adnate to calyx tube. Carpel 1; ovary superior with 1 locule and 1 basal ovule; style 1. Fruit an achene, often surrounded by the hardened base of the calyx tube.

# FAMILY NOTES

The Nyctaginaceae comprises 30–38 genera and about 300 species of herbs, shrubs and trees. The greatest diversity of the family is in tropical and subtropical regions. A few taxa occur in temperate latitudes.

Members of the Nyctaginaceae have a tabular petaloid calyx that strongly resembles a sympetalous corolla (Figure 13-11 A, D). The showy upper portion of the calyx withers after anthesis. The lower portion of the calyx often tightly enwraps the ovary and is persistent around the fruit as an **anthocarp** (Figures 13-11 F). The calyx base plus the enclosed fruit is the unit of dispersal. In some members of the Nyctaginaceae the persistent calyx base bears wing-like projections or sticky glands that aid in wind or animal dispersal of the fruits.

Among the most familiar members of the Nyctaginaceae is *Mirabilis jalapa,* the garden four o'clock, so named because its showy flowers open in late afternoon. Most members of the family, however, have flowers open at other times of the day.

*Mirabilis jalapa* has an involucre of connate bracts that encloses a single flower (Figure 13-11 A, B). The involucre almost exactly simulates a synsepalous calyx and the petaloid calyx looks exactly like a sympetalous corolla. If there were not other species of *Mirabilis* in which the involucre encloses several flowers we almost certainly would consider the garden four o'clock to have complete flowers.

The sand-verbenas, members of the genus *Abronia,* are among the showiest wildflowers of the family (Figures 13-11 C, D). These plants have involucrate heads of brightly colored flowers. Some members of the family, however, such as some species of *Boerhaavia,* have small, inconspicuous flowers.

The family has rather limited economic importance. Species of *Bougainvillea,* cultivated in warm areas, have colorful bracts that subtend their 3-flowered inflorescences. Various species of *Mirabilis* (four o'clock) are also cultivated. The leaves of some *Pisonia* species are used as vegetables in tropical areas.

# Chenopodiaceae

## The Goosefoot Family
### Caryophyllales ■

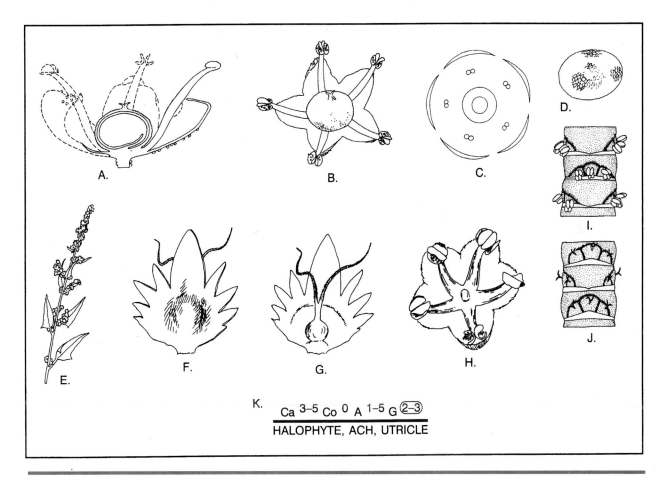

Ca 3-5 Co 0 A 1-5 G (2-3)

HALOPHYTE, ACH, UTRICLE

**Figure 13-12** Features of Chenopodiaceae. A–D. *Chenopodium* sp. A. Flower (l.s.). B. Intact flower. C. Floral diagram. D. Utricle. E–H. *Atriplex* sp. E. Stem of *Atriplex* with leaves and fruits. F–G. Pistillate flower with pair of bracts enclosing gynoecium (F) and with one bract removed (G). H. Staminate flower. I–J. Stems and inflorescences of *Salicornia virginica*. I. Inflorescence of staminate flowers. J. Inflorescence of pistillate flowers. K. Generalized floral formula.

## FAMILY DESCRIPTION

Herbs and shrubs, sometimes succulent. Leaves alternate (less commonly opposite), simple, estipulate. Plants synoecious, monoecious, dioecious or polygamous. Inflorescence of solitary or cymosely clustered flowers or often aggregated into racemiform, paniculiform or spike-like clusters. Flowers perfect or imperfect, regular or ± irregular; pistillate flowers sometimes enclosed by paired bracts. Sepals 3–5 (1–2) (0), distinct or connate, sometimes much reduced. Petals absent. Stamens 1–5, distinct. Carpels 2 (3) connate; ovary superior (inferior) with 1 locule and 1 basal ovule; styles 2 (3). Fruit an achene or utricle, sometimes surrounded by adnate or adherent calyx or by paired bracts; seed with spirally coiled embryo.

# FAMILY NOTES

The Chenopodiaceae consists of 100–113 genera and about 1500 species, distributed mostly in temperate and subtropical areas. Most members of the family occur in saline habitats. Chenopods are common and often dominant in coastal areas such as ocean beaches, coastal bluffs and salt marshes and in saline/alkaline desert areas where salts have been concentrated by evaporation. Species of *Atriplex* (saltbush) and other chenopods cover extensive areas in some desert regions. Some of the salt-tolerant members of the family are very succulent, accumulating a concentrated salt solution in their cell sap. This enables them to absorb water from salt solutions concentrated enough to kill most plants.

Flowers in the Chenopodiaceae are generally small and individually inconspicuous (Figure 13-12 A. B). The sepals are small and usually scale-like. *Atriplex* and several other genera are monoecious or dioecious. The staminate flowers consist of a calyx and a whorl of stamens (Figure 13-12 H). The pistillate flowers lack a corolla and are enclosed by a pair of bractlets (Figure 13-12 F, G). Often only the style tips are exposed at the time of pollination. The two bracts and the enclosed fruit serve as the unit of dispersal. The inflorescence of some members of the family is reduced and spikelike. The flowers of *Salicornia* (Figure 13-12 I, J) are actually arranged in tiny sessile 3-flowered cymes sunken into the fleshy stem axis. The leaves, bracts and sepals are all bladeless and fleshy. Despite this, ovaries are superior.

The herbage of many (but by no means all) members of the Chenopodiaceae is covered by farinose (mealy) pubescence (Figure 13-12 B). The surface of a leaf or other structure covered by these peculiar sessile water-filled vescicles may appear to be coated with tiny beads. If the trichomes dry out the surface appears to be covered with a dusting of flour.

The only economically important members of the Chenopodiaceae are *Spinacia oleracea* (spinach) and *Beta vulgaris* (beet). Some races of the latter species are cultivated as vegetables (garden beets, chard) but it is most important as a source of sugar (sugar beets). Minor crops include *Chenopodium quinoa* (quinoa) which is used as a grain and leaf-vegetable in parts of South America and *C. ambrosioides* (Mexican-tea), which is used as a condiment in some Mexican foods. The young leaves of various members of the family are edible as potherbs. Some *Atriplex* species are cultivated as ornamentals in arid regions. Various members of the Chenopodiaceae are agricultural weeds, especially in areas with salty soils. Russian thistle or tumbleweed *(Salsola iberica)* is a particularly troublesome weed in the western parts of the United States.

# Aizoaceae

## The Ice Plant Family
### Caryophyllales ■

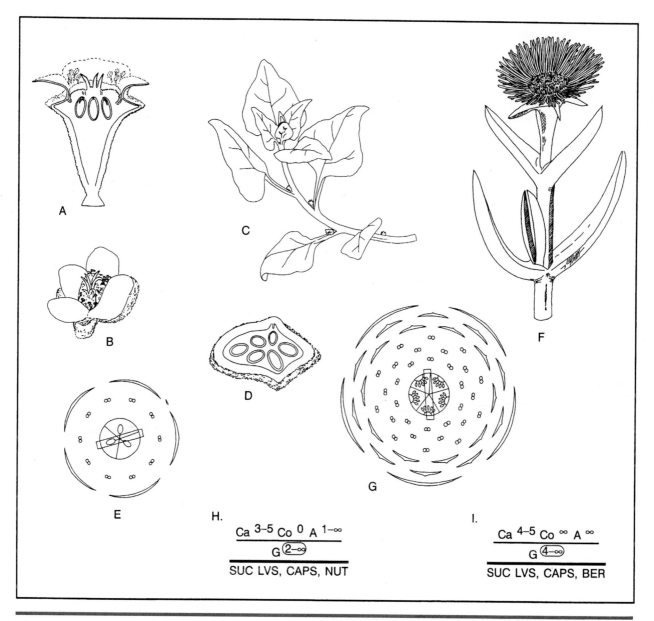

$$\text{Ca } ^{3-5} \text{ Co } ^0 \text{ A } ^{1-\infty}$$
$$\overline{\text{G } \boxed{2-\infty}}$$
SUC LVS, CAPS, NUT

$$\text{Ca } ^{4-5} \text{ Co } ^\infty \text{ A } ^\infty$$
$$\overline{\text{G } \boxed{4-\infty}}$$
SUC LVS, CAPS, BER

**Figure 13-13** Features of Aizoaceae. A–E. *Tetragonia tetragonioides*. A. Flower (l.s.). B. Intact flower. C. Leafy stem showing the weakly succulent leaves and small axillary flowers. D. Nutlike fruit (x.s). E. Floral diagram. F. Leafy stem of *Carpobrotus* showing thick, succulent leaves and large showy flowers with biseriate perianth. G. Floral diagram of *Carpobrotus*. H. Generalized floral formula of apetalous Aizoaceae. I. Generalized floral formula of petal-bearing Aizoaceae.

# FAMILY DESCRIPTION

Herbs or shrubs, often succulent. Leaves alternate or opposite, simple, usually estipulate. Plants synoecious (rarely monoecious). Inflorescence of solitary flowers or variously cymose. Flowers perfect (imperfect) regular. Hypanthium present (sometimes absent), often adnate to ovary. Sepals (3-) 4–5 (6–8) distinct or basally connate. Petals absent or numerous, distinct. Stamens (1-) 4–5 (8–10) or numerous, distinct or basally connate. Carpels 2-many, connate; ovary inferior (less commonly superior) with 2-many (1) locules with numerous (1) axile or parietal ovules per locule; styles 2-many or 1 and 2-many lobed or stigmas sessile. Fruit a capsule or berry (rarely a nut).

# FAMILY NOTES

The Aizoaceae is a largely southern hemisphere family of 100–150 genera and about 2500 species. Members of the Aizoaceae are particularly well represented in Africa and Australia. Most are strongly succulent with very fleshy leaves. Their succulence is accompanied in most genera by CAM (Crassulacean Acid Metabolism) photosynthesis. Members of the family produce betalain pigments.

Taxonomists differ in the interpretation of generic limits in the Aizoaceae. One large, mostly African group is sometimes treated as a single polymorphic genus, *Mesembryanthemum,* with about 2000 species. In place of this genus, some botanists instead recognize as many as 125 genera! Nomenclature in the group is complicated by the fact that many species are grown as ornamentals and numerous minor segregates and mutants have been given formal recognition.

Some of the South African species are commonly known as "living stones." Species of *Lithops* are small, ground-hugging succulents with leaves that strongly resemble pebbles. They occur in desert environments where water is in very short supply. They are very inconspicuous in stony desert soils. Their appearance changes greatly when they produce colorful, many-petaled flowers.

Many members of the Aizoaceae have large, very showy flowers with numerous petals (Figure 13-13 F). The flowers of other members of the family lack petals (Figure 13-13 A, B). The Aizoaceae are apparently primitively apetalous, and the petals of the showy species have most likely been derived from staminodes.

Fruits in the Aizoaceae are rather variable. Many produce many-seeded capsules. These are usually hygroscopic, opening and closing in response to moisture conditions. The adnate hypanthium tissues in *Carpobrotus* (sea-fig, Hottentot-fig) become very fleshy and the fruit is a berry. In other genera the fruit is indehiscent and nutlike (Figure 13-13 D).

Many of the South African succulent species are grown outdoors as ornamentals in California and other warm areas, and as greenhouse curiosities in other regions. *Tetragonia tetragonioides* (New Zealand-spinach; Figure 13-13 D), an apetalous species with edible leaves, has been introduced into coastal regions in various parts of the world. The fleshy fruits of *Carpobrotus edulis* (Hottentot-fig) and some other species are edible.

# Portulacaceae

## The Purslane Family
### Caryophyllales ◼

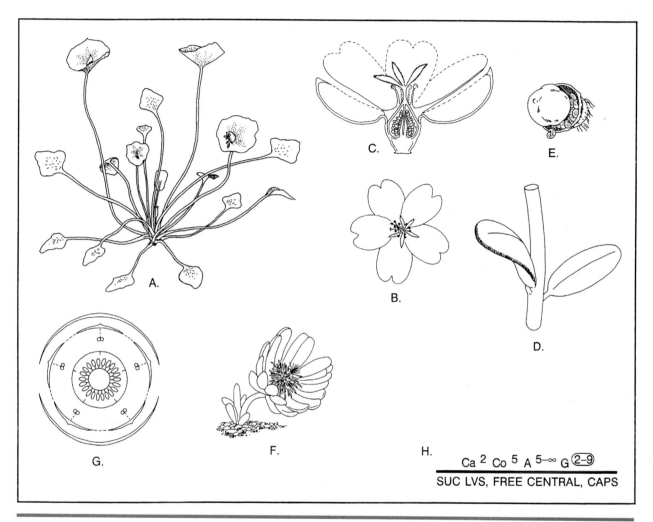

**Figure 13-14** Features of Portulacaceae. A. Habit of *Claytonia perfoliata.* B–E. *Portulaca oleracea.* B. Intact flower. C. Flower (l.s.). D. Stem with succulent leaves. E. Circumscissile capsule. F. Habit of *Lewisia reduviva.* G. Floral diagram of *Claytonia.* H. Generalized floral formula.

## FAMILY DESCRIPTION

Herbs or shrubs (rarely small trees). Leaves alternate or opposite or all basal, simple and entire, stipulate or estipulate, often succulent. Plants synoecious. Inflorescence of solitary flowers or variously cymose, racemose or headlike. Flowers perfect, regular (rarely slightly irregular). Sepals 2 (3–9), distinct (basally connate). Petals 5 (2–18) (0), distinct or basally connate. Stamens as many as and opposite the petals to twice as many or numerous (1–3), distinct or when numerous sometimes fascicled; filaments sometimes adnate at base to petals. Carpels 2–9, connate; ovary superior (rarely partially inferior) with 1 locule and 2-many basal or free-central ovules; styles 2–9 or 1 and lobed. Fruit a capsule, sometimes circumscissile (rarely a nut).

# FAMILY NOTES

The Portulacaceae comprises 20–27 genera and about 500 species. It has a nearly cosmopolitan distribution in a wide variety of habitats with the greatest diversity in North America and Africa.

Most members of the Portulacaceae can easily be recognized by their succulent leaves, two-merous calyx and 5-merous corolla (Figure 13-14 B, F). The paired sepals apparently originated as bracts comparable to the involucral bracts found in many Nyctaginaceae. Petals in the Portulacaceae are probably actually petaloid sepals. In the Nyctaginaceae where involucral bracts usually subtend more than one flower, the petaloid structures have traditionally been treated as sepals. In the Portulacaceae, how-ever, where the "involucre" is always one-flowered, the petaloid structures have been considered to be petals and the bracts to be sepals.

Several members of the purslane family are cultivated for the beauty of their flowers. These include *Portulaca grandiflora* (rose-moss), *Lewisia* spp. (Figure 13-14 F), *Talinum* spp., and various others. Various other members of the family are cultivated by collectors of succulents. A few, such as *Claytonia perfoliata* (miner's-lettuce; Figure 13-14 A), may be eaten in early spring as salad greens. The common purslane, *Portulaca oleracea* is sometimes grown as a pot herb, but more often is considered a weed of cultivated ground.

# Cactaceae

## The Cactus Family

Caryophyllales ■

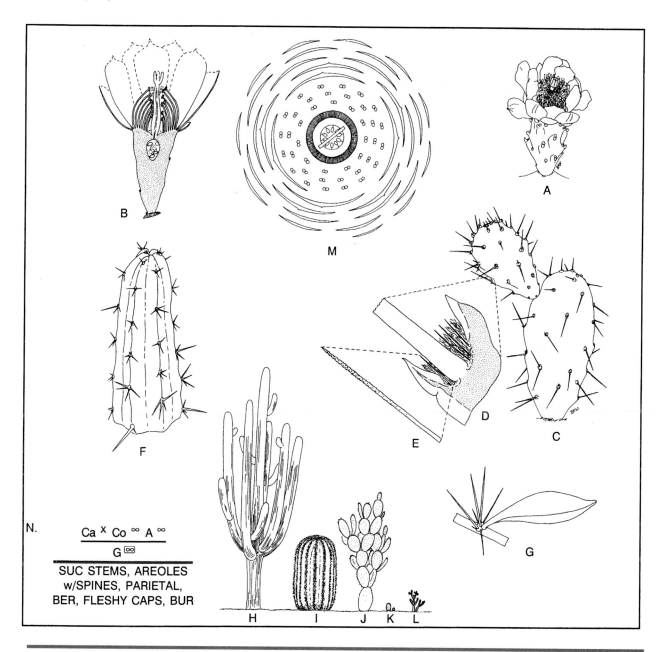

N.

$$\frac{Ca \ ^X \ Co \ ^\infty \ A \ ^\infty}{G \ ^{\overline{\infty}}}$$

SUC STEMS, AREOLES
w/SPINES, PARIETAL,
BER, FLESHY CAPS, BUR

**Figure 13-15** Features of Cactaceae. A. Flower of *Opuntia*. B. Flower of *Opuntia* (l.s.). C. Jointed, flattened stems of *Opuntia phaeacantha*. D. Details of areole of *Opuntia* with primary spines and glochids. E. Glochid of *Opuntia* (x 15). F. Cylindrical stem of *Lemairocereus*. G. Node of *Pereskia* sp. showing large foliage leaf and areole. H–L. Various forms of succulent stems found in the Cactaceae. H. *Cereus giganteus*. I. *Ferocactus* sp. J. *Opuntia* sp. (prickly-pear). K. *Mammilaria* sp. L. *Opuntia* sp. (cholla). M. Generalized floral diagram. N. Generalized floral formula.

# FAMILY DESCRIPTION

Shrubs, trees, or small to very large stem-succulents, sometimes epiphytic. Leaves alternate, simple, often highly reduced and early deciduous or absent, estipulate; axillary buds flattened and pad-like [areoles], often bearing 1-many spines. Plants synoecious (rarely dioecious). Inflorescence of solitary flowers emerging from areoles or bursting through epidermis (rarely cymose in one primitive genus). Flowers perfect (rarely imperfect), regular or less commonly irregular. Hypanthium sometimes well-developed. Tepals numerous, spirally arranged, the outer usually sepaloid, the inner petaloid. Stamens numerous, distinct, basally adnate to innermost tepals. Carpels 2-many, connate; ovary very inferior, sunken into areole-bearing stem tissue (rarely superior); locule 1 with many parietal ovules; style 1 with 2-many stigma lobes. Fruit a berry, a fleshy capsule, or an indehiscent spiny bur.

# FAMILY NOTES

The Cactaceae is an almost exclusively New World family of 50–150 genera and 1500–2000 species. Although cacti range well into the temperate zones of both North and South America, the greatest diversity occurs in the subtropical and tropical latitudes. Cactaceae are remarkable in the diversity of their bizarre growth forms. They range in size from tiny unbranched cylindrical succulents a few centimeters tall to giant candelabra-like "trees" as much as 20 meters tall. Some cacti are shrubby and intricately branched. Some have their stems sunken into stony dry ground and others grow as epiphytes in the branches of rain forest trees. They span an elevational range from below sea level to over 3000 meters.

Members of the cactus family are easy to distinguish from other succulents (such as members of the Aizoaceae, Crassulaceae, Euphorbiaceae and Asclepiadaceae). They are stem-succulents, unlike the Aizoaceae and Crassulaceae in which leaf-succulence is the norm. Of the stem succulents, only the Cactaceae have **areoles** (Figure 13-15 D). These are unusual pad-like axillary buds that in most cacti bear spines. The cactus-like Euphorbias and Asclepiadaceae lack areoles and are restricted to the Old World, particularly Africa.

The flowers of the cacti resemble those of some Aizoaceae in having inferior ovaries and many perianth parts but differ in gynoecium morphology and in the nature of the perianth parts. In both families the ovary is inferior but in the cacti it is sunken into a stem. The outer surface of a cactus ovary often bears areoles and spines. Petals in the cacti are spirally inserted and grade into the sepals which in turn often grade into bracts (Figure 13-15 A, B). Petals in the Aizoaceae are apparently derived from staminodes. The ovaries in both families commonly have parietal placentation, but in the Aizoaceae there often are "false" septa that divide the ovary into several to many locules, and the number of styles or style branches equals the number of carpels. Cacti generally have a one-locular ovary with diffuse parietal placentae that often bear ovules over the entire inner surface of the ovary. The style is generally unbranched but is tipped by as many stigmas as carpels.

Members of the Cactaceae comprise three subfamilies. The most primitive of these is the **Pereskioideae.** These plants are leafy trees and shrubs with scarcely succulent stems. In the axils of the flattened leaves are typical cactus areoles. The flowers of some species of *Pereskia* are arranged in cymes and have superior ovaries. Subfamily **Opuntioideae** consists of plants with reduced, usually early-deciduous leaves and jointed stems (Figure 13-15 C). Each stem-joint represents a season's growth. Members of this subfamily produce **glochids** (Figure 13-15 D, E). These structures are tiny deciduous barbed spines that are borne together with larger spines from the areoles. Seeds in the Opuntioideae have a bony aril, a feature lacking in the other subfamilies. In the **Cactoideae** leaves are extremely reduced, in many cases not developing beyond the primordium stage. The stems are generally not jointed; growth of one season is continuous with that of the preceding seasons. Glochids are absent. This subfamily comprises two tribes. Stems in the **Cereeae** are vertically ribbed with the areoles located on the ribs. Flowers emerge from old areoles. This tribe includes the giant columnar cacti. Most members of the **Cacteae** are small plants with ribbed or tuberculate stems. Flowers emerge from new areoles.

Cactus nomenclature is rather chaotic. Estimates of the number of species and genera vary widely because of variation in the interpretation of taxonomic boundaries in the family. Some of the most widely followed references on the family were written by splitters who recognized numerous microgenera, particularly in the Cactoideae. Amateur botanists have described many poorly defined taxa. Cactus specimens are difficult to prepare and are very bulky. Consequently cacti are poorly represented in museum collections. Documentation of published names is sometimes difficult because so many of the taxa have been described from cultivated specimens or fragmentary specimens.

Cacti are important economically in several ways. Their bizarre growth forms, their beautiful flowers and their hardiness have fascinated laymen and professionals for many years. This has led to a lucrative and in some cases illegal trade in exotic cacti. Collecting and raising cacti is a popular pastime for many amateurs. Some cactus species are in danger of extinction because of unscrupulous collectors who have removed them from their native habitats. Removal of native cacti from the wild is against the law in several states and countries. Some cacti, such as *Opuntia ficus-indica,* have edible fruits. Some are weedy in areas outside the range of their natural parasites. In Australia an introduced *Opuntia* nearly destroyed that continent's range lands.

# Polygonaceae

## The Buckwheat Family
### Polygonales ■

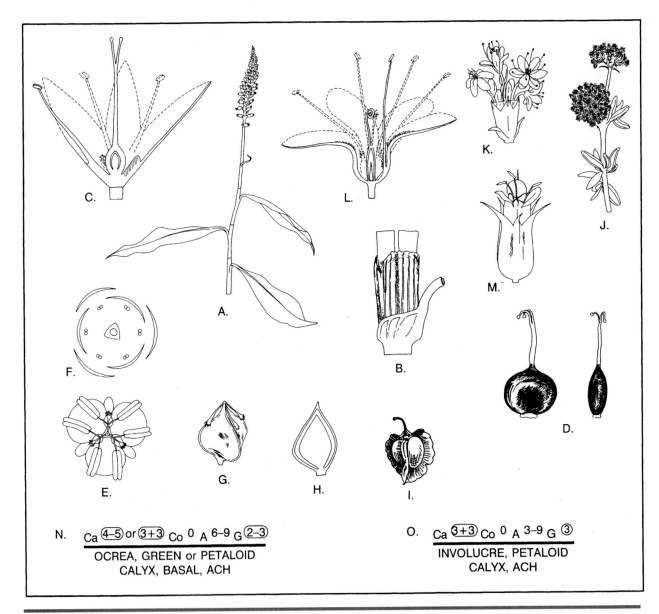

**Figure 13-16** Features of Polygonaceae. A–D. *Polygonum coccineum*. A. Leafy stem and inflorescence. B. Ocrea (stipular sheath). C. Flower (l.s.). D. Biconvex (lenticular) achenes. E–I. *Rumex crispus*. E. Flower. F. Floral diagram. G. Triangular achene with deflexed styles. H. Achene (l.s.). I. Ripened calyx with swollen callosities on the "valves" (inner sepals). J–L. *Eriogonum fasciculatum*. J. Leafy stem and inflorescences. K. Involucrate umbel. L. Flower (l.s.). M. One-flowered involucre of *Chorizanthe*. N. Generalized floral formula of subf. Polygonoideae. O. Generalized floral formula of subf. Eriogonoideae.

# FAMILY DESCRIPTION

Herbs, shrubs (rarely woody vines or trees). Leaves alternate, opposite, whorled or all basal, stipulate or estipulate; stipules (when present) connate and sheathing forming an ocrea. Plants synoecious (rarely monoecious or dioecious). Inflorescence of solitary or clustered axillary flowers, racemiform, paniculiform or spikelike clusters, involucrate umbels or solitary involucrate flowers. Flowers perfect (rarely imperfect), regular, sometimes subtended by a calyx-like involucre. Hypanthium often present though sometimes small. Sepals 2–5 in one series or 6 (4) in 2 similar or dissimilar whorls of three (two), distinct or basally connate, often petaloid, rarely the inner whorl adnate to ovary. Petals absent. Stamens 3–9, distinct, free or filaments adnate to sepals. Carpels 2–3 (4), connate; ovary superior with 1 locule and 1 basal ovule; styles 2–3 (4). Fruit a lenticular or triangular achene or small nut (enveloped by fleshy, berrylike calyx in 1 genus).

# FAMILY NOTES

The Polygonaceae comprises 30–49 genera and about 1000 species, most of them north temperate in distribution. There are a few tropical genera and a single southern hemisphere genus. Many of the genera are of restricted distribution. One of the centers of distribution is western North America where the genus *Eriogonum* and several of its close relatives are common.

Leaves in most genera of the Polygonaceae have a peculiar type of stipule called an **ocrea** (Figure 13-16 B). The stipules of a pair are united into a membranous sheath that enwraps the stem above the node. These structures are most easily observed on young stems. As lateral buds break dormancy the ocrea of the subtending leaf is generally ripped apart and may fall away entirely.

The perianth of most members of the Polygonaceae consists of two whorls of three tepals each. Traditionally both whorls have been treated as sepals. In some members of the family the outer perianth parts are clearly differentiated from the inner (Figure 13-16 E), and in others all the parts are structurally similar (Figure 13-16 K). In many members of the Polygonaceae the perianth elements are fused at the base in what appears to be a single ring. The perianth of *Polygonum* and *Coccoloba* comprises a single whorl of four or five sepals (Figure 13-16 C). This condition apparently is derived from ancestors with six perianth parts in two whorls. One sepal of the outer whorl became wholly fused to one from the inner whorl, reducing the apparent number from six to five.

In *Rumex* both whorls of perianth are sepaloid in color and texture. The outer sepals are much smaller than the inner sepals (Figure 13-16 E). The inner sepals (valves) closely enwrap the triangular achene and fall together with it at maturity. The valves often bear swollen "callosities" that are often mistaken for seeds. These structures function as floats in water dispersal of the achenes.

Several genera are characterized by an involucre that subtends one or more flowers. In *Eriogonum,* a large genus of herbs and shrubs, the involucre subtends an umbel (Figure 13-16 K). The umbels are distributed in various types of secondary clusters. The flowers have a petaloid perianth of 6 sepals (Figure 13-16 J). The outer three are sometimes differentiated from the inner three. In *Chorizanthe* and several smaller genera, the involucre is one-flowered and calyx-like (Figure 13-16 M). This, combined with the petaloid perianth, simulates a complete flower. *Eriogonum, Chorizanthe* and their relatives have estipulate leaves.

Comparatively few members of the Polygonaceae have economic importance. The petioles of *Rheum rhaponticum* (rhubarb) are edible (although the leaf blades are poisonous). The achenes of *Fagopyrum esculentum* (buckwheat) are ground to make buckwheat flour. *Coccoloba uvifera* (seaside grape) has fleshy sepals closely surrounding the achene forming an edible berrylike accessory fruit. Some species of *Antigonon, Eriogonum, Polygonum,* and *Rheum* are cultivated as ornamentals. Various members of the family are weedy.

# Exercises

## ■ Study Questions

### Caryophyllaceae

1. What is the characteristic leaf arrangement of the Caryophyllaceae?

   Leaf margin?

2. The inflorescence of a member of the Caryophyllaceae is generally some kind of _____.

3. What feature is characteristic of the sepals of subf. Silenoideae?

4. Caryophyllaceae flowers are usually ____-merous.

   What type of corolla do they have?

5. How many stamens are present in most Caryophyllaceae flowers?

6. What kind of gynoecium characterizes most Caryophyllaceae?

   Placentation?

   Ovary position?

   How can you determine the number of carpels?

7. In what features is tribe Paronychieae different from the other subfamilies and tribes?

8. What is the usual fruit type in the Caryophyllaceae?

   What is the fruit type in tribe Paronychieae?

9. Do the members of the Caryophyllaceae produce anthocyanins or betalains?

### Nyctaginaceae

10. Why are the flowers of the Nyctaginaceae often mistakenly interpreted as sympetalous?

11. What kind of gynoecium does the Nyctaginaceae have?

    How many locules?

    Placentation?

    How many ovules?

    Ovary position?

12. What is the dispersal unit of the Nyctaginaceae?

13. Does this family produce anthocyanins or betalains?

### Chenopodiaceae

14. What flower parts are characteristically lacking in the Chenopodiaceae?

    These flowers are usually ____-merous.

15. How many stamens typically occur in Chenopodiaceae flowers?

    What are they attached to?

16. What kind of gynoecium characterizes the Chenopodiaceae?

    How many carpels do we find per flower?

    What kind of placentation?

    Ovary position?

    How can you determine the number of carpels?

17. What unusual fruit do most members of the Chenopodiaceae produce?

    What is the difference between one of these fruits and an achene?

18. In what habitats would you expect to encounter members of the Chenopodiaceae?

19. Do the members of the Chenopodiaceae produce anthocyanins or betalains?

## Aizoaceae

20. How would you decide of a succulent plant is a member of the Aizoaceae or Cactaceae?

21. What kind of gynoecium occurs in the Aizoaceae?

    Number of carpels?

    Placentation types?

    Ovary position?

    Fruit types?

22. From what structures have the petals of Aizoaceae apparently evolved?

23. Does this family produce anthocyanins or betalains?

## Portulacaceae

24. How may sepals do flowers of most Portulacaceae have?

    Some botanists consider these to be modified _____.

25. Flowers in most Portulacaceae are ___-merous.

26. What kind of placentation is characteristic of Portulacaceae?

    How can the number of carpels be determined?

27. Do the members of the Portulacaceae produce anthocyanins or betalains?

## Cactaceae

28. What are the modified axillary buds characteristic of all Cactaceae called?

29. What are the sharp-pointed structures that radiate from cactus stems?

    From what part of the plant are they modified?

    What is a glochid?

30. Some succulent species of *Euphorbia* look remarkably like cacti. Can you explain the similarity between these very distantly related kinds of plants?

    Where are cacti native?

    Where are most succulent Euphorbias native?

31. How can you decide if a succulent is a cactus or a *Euphorbia?*

    What are the sharp pointed structures on the stems of succulent Euphorbias?

    How do these differ from the sharp-pointed structures of cacti?

32. Describe the perianth of a cactus flower.

    How would you decide where sepals stop and petals start?

    How many stamens do cactus flowers produce?

    How do cactus flowers differ from *Euphorbia* flowers?

33. What kind of gynoecium characterizes the Cactaceae?

    What is the ovary position?

    Placentation?

    How can you determine the number of carpels?

    What evidence do we have that cactus ovaries are sunken into stem tissue?

34. What kind of fruits are produced in the Cactaceae?

35. Do the cacti produce anthocyanins or betalains?

# Polygonaceae

36. What is the membranous stipular sheath found in many Polygonaceae called?

    Members of which subfamily of the Polygonaceae characteristically produce these structures?

    Members of which subfamily characteristically lack these structures?

37. Flowers in the Eriogonoideae are usually enclosed or closely subtended by _____.

38. Describe the perianth of most Polygonaceae.

39. How many stamens do Polygonaceae flowers have?

40. What kind of gynoecium characterizes the Polygonaceae?

    How many carpels?

    Placentation?

    Ovary position?

    How can you determine the number of carpels?

41. What kind of fruit do we find in the Polygonaceae?

    How does fruit appearance change with carpel number?

42. Do the members of the Polygonaceae produce anthocyanins or betalains?

43. List the families and species of subclass Caryophyllidae that you identified in the lab.

_____     _____

_____     _____

_____     _____

_____     _____

_____     _____

_____     _____

# Rosids

chapter 14

The Rosids are a large group of flowering plants. Its 18 orders, 114 families, and over 58,000 species comprise over one third of the dicotyledonous families. The number of species is approached only by the Asteriidae. Almost 75 percent of the species are classified in five orders—Fabales, Euphorbiales, Myrtales, Rosales, and Sapindales.

A group this large cannot be succinctly defined. Most character states found in the dicots can be found in one member or another of the Rosidae. The basal members are comparable to the primitive Magnoliidae whereas the most advanced share features with members of the Asteriidae. Members of the Rosidae tend to have perigynous or epigynous flowers. Those with hypogynous flowers usually bear a characteristic fleshy nectar ring associated with the ovary. Corollas most commonly are apopetalous; less commonly the petals are connate or absent. The number of stamens varies from numerous to twice as many as the petals, or less frequently equal in number to or fewer than the petals. When numerous, the stamens develop in a centripetal sequence. Gynoecia range from monocarpous or fully apocarpous to syncarpous, often with 3 or more carpels; ovules are usually borne on marginal or axile placentae and often are only one or two per locule. Exceptions occur in some families to almost all of these general trends. Members of the subclass occur in nearly all terrestrial ecological situations.

The Rosids are a diverse group. Based upon the work of the Angiosperm Phylogeny Group and Soltis et al. (2000) there appear to be two major clades in this group (Figure 14-1). This chapter divides the Rosids into these clades, based upon molecular characteristics.

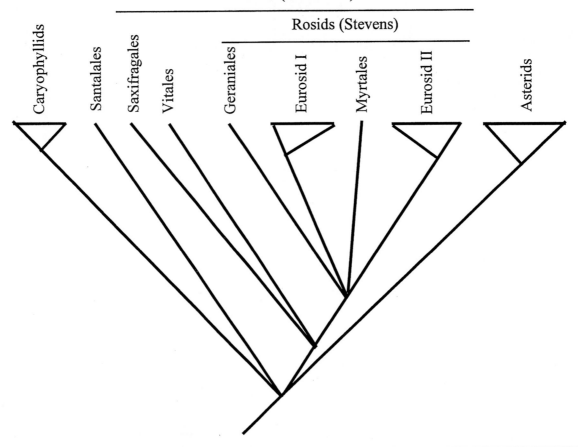

Rosids (Judd et al.)

Rosids (Stevens)

Caryophyllids    Santalales    Saxifragales    Vitales    Geraniales    Eurosid I    Myrtales    Eurosid II    Asterids

**Figure 14-1** Phylogeny of the rosid clade, showing the basal lineages and the two major clades (Eurosid I and Eurosid II). Note that Stevens (2001 onward) does not include Saxifragales and Vitales in the Rosid clade, whereas Judd et al. (2002) take a more inclusive view of the group. The figure is adapted from Stevens (2001 onward) and Judd et al. (2002).

# BASAL ROSIDS

## 1. Order Saxifragales

**1.A  Saxifragaceae**—The Saxifrage Family. (Figure 14-19). 31–40 genera, 700 species. Cosmopolitan, especially in northern hemisphere.

§See page 292 for description and illustrations§

**1.B  Crassulaceae**—The Stonecrop Family. (Figure 14-20). 25–37 genera, 900 species. Cosmopolitan except Australia and Polynesia.

§See page 295 for description and illustrations§

**1.C  Grossulariaceae**—The Gooseberry or Currant Family (Figure 14-21). 1 genus, 150 species. North America, Eurasia, Andean region of South America.

§See page 297 for description and illustrations§

**1.D  Haloragaceae**—The Water Milfoil Family. 8–9 genera, 100 species. Cosmopolitan but most species in southern hemisphere.

Herbs, often aquatic (rarely shrubs or small trees). Leaves alternate, opposite or whorled, simple to pinnately dissected, estipulate. Plants synoecious, monoecious or polygamous. Inflorescence of solitary axillary flowers or terminal spikes, racemes or panicles. Flowers perfect or imperfect, regular. Sepals 3–4 (0), connate. Petals 4 (0), distinct. Stamens 2–8, distinct. Carpels 2–4, connate; ovary inferior with 2–4 locules with 1 apical-axile ovule per locule; styles 2–4, distinct. Fruit a nut, drupe or schizocarp.

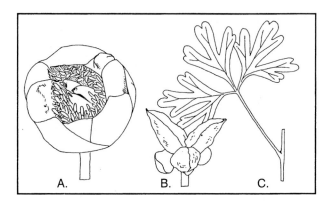

**Figure 14-2** Examples of Paeoniaceae. A–B. *Paeonia californica.* A. Flower. B. Cluster of follicles. C. Leaf.

**1.E    Paeoniaceae**—The Peony Family (Figure 14-2). 1 genus, 30 species. Mostly temperate Eurasia, few North American.

Herbs or shrubs. Leaves alternate, 1–2 pinnate, estipulate. Plants synoecious. Inflorescence of solitary flowers. Flowers perfect, ± regular. Sepals 3–5, distinct, imbricate, persistent. Petals 5–10 (–20), distinct, spiralled. Stamens many, distinct, centrifugal. Carpels 2–5, distinct, subtended by a lobed disk; ovaries superior, unilocular with several-many marginal ovules; styles 1 per carpel. Fruit a cluster of follicles.

**1.F    Peridiscaceae**—2 genera, 2 species. South America.

**1.G    Cercidiphyllaceae**—1 genus, 2 species. Southeast Asia.

**1.H    Hamamelidaceae**—The Witch-Hazel Family (Figure 14-22). 26–29 genera, 100 species. Discontinuously distributed in temperate and tropical regions of Old and New Worlds.

§See page 300 for description and illustrations§

**1.I    Daphniphyllaceae**—1 genus, 35 species. Eastern Asia southeast to Malaya.

# 2.  Order Vitales

**2.A    Vitaceae**—The Grape Family (Figure 14-3). 11–14 genera, 700 species. Mostly tropical and subtropical with a few widespread temperate representatives.

Woody vines with tendrils, sometimes succulent, (less commonly shrubs or small trees). Leaves alternate (rarely opposite), opposed to the tendrils, sim-

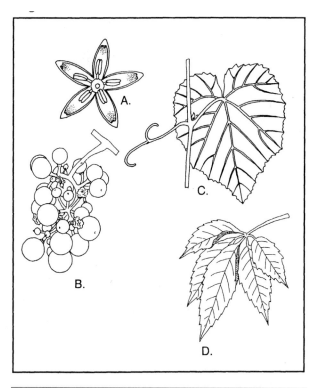

**Figure 14-3** Examples of Vitaceae. A–C. *Vitis vinifera.* A. Flower. B. Panicle of berries. C. Branchlet with tendril and leaf. D. Leaf of *Parthenocissus.*

ple and often palmately veined or lobed, or pinnately or palmately compound, stipulate or estipulate. Plants synoecious, monoecious, dioecious or polygamous. Inflorescence cymose, often paniculiform or corymbiform, sometimes spicate or racemose. Flowers small, perfect (sometimes imperfect), regular. Sepals 4–5 (3–7), connate, often very small, Petals 4–5 (3–7) (0), distinct (connate) or apically ± coherent. Stamens 4–5 (3–7), opposite the petals, distinct or anthers ± connate. Nectary disc present, Carpels 2 (3–6), connate; ovary superior with 2 (3–6) locules and 1–2 axile ovules per locule; style 1, undivided, or stigma sessile. Fruit a berry.

**2.B    Leeaceae**—1 genus, 70 species. Old World tropics, especially in Asia.

# 3.  Order Crossosomatales

**3.A    Crossosomataceae**—The Crossosoma Family. 3 genera, 10 species. Western United States and Mexico.

Shrubs. Leaves alternate or opposite, simple, stipulate or estipulate. Plants synoecious. Inflorescence of solitary flowers or axillary clusters. Flowers perfect, regular. Hypanthium sometimes developed. Sepals 4–5, distinct or connate. Petals 4–5 distinct. Stamens 4-many, distinct. Carpels 1–9, distinct; ovaries superior with 2-many marginal ovules; styles 1 per carpel. Fruit a cluster of follicles.

**3.B    Geissolomataceae**—1 genus, 1 species. South Africa.

**3.C    Staphyleaceae**—The Bladdernut Family. 5 genera, 50 species. North temperate and southeast Asia.

Shrubs or trees. Leaves opposite (alternate), trifoliolate or pinnate (unifoliolate), stipulate (estipulate). Plants synoecious (rarely polygamous or dioecious). Inflorescence of racemes or paniculiform clusters. Flowers perfect (rarely imperfect), regular. Sepals 5, distinct or basally connate. Petals 5, distinct. Stamens 5, distinct. Nectary disc present. Carpels 2–4, connate (rarely 3–4 distinct); ovary or ovaries superior with 2–4 locules and 1-few axile (rarely marginal) ovules; styles 2–4 to 1 and undivided. Fruit an inflated capsule, a cluster of follicles, a berry or a drupe.

**3.D    Stachyuraceae**—1 genus, 5–6 species. Southeast Asia.

# CORE ROSIDS

## 4. Order Geraniales

**4.A    Geraniaceae**—The Geranium Family (Figure 14-23). 11 genera, 700 species. Widespread in temperate regions, particularly in the Old World.

§See page 302 for description and illustrations§

**4.B    Melianthaceae**—The Melianthus Family. 2 genera, 8–35 species. Africa.

Shrubs or trees. Leaves alternate, pinnate, stipulate. Plants synoecious or polygamous. Inflorescence of racemes. Flowers perfect (rarely imperfect), regular or irregular. Sepals 5, distinct or basally connate. sometimes short-spurred or saccate. Petals 5, distinct. Stamens 4–5, distinct or filaments ± connate. Nectary disc present. Carpels 4–5, connate; ovary superior with 4–5 ovules and 1–5 axile ovules per locule; style 1, undivided. Fruit a capsule.

**4.C    Greyiaceae**—1 genus, 3 species. Southern Africa.

## 5. Order Myrtales

**5.A    Myrtaceae**—(Figure 14-24) 127–140 genera, 3000 species. Tropics and subtropics, especially in southern hemisphere

§See page 304 for description and illustrations§

**5.B    Lythraceae**—The Loosestrife Family. 24–28 genera, 500 species. Primarily tropical, few temperate.

Herbs, shrubs, or trees. Leaves opposite or whorled (rarely alternate), simple, stipulate or estipulate. Plants synoecious. Inflorescence of cymes, panicles, racemes or solitary flowers. Flowers perfect, regular (less commonly irregular). Hypanthium present. Sepals 4–8 (3–16) connate, sometimes subtended by bractlets. Petals 4–8 (3–16) (0) distinct. Stamens 8–16 (fewer) (many), distinct, often unequally inserted in hypanthium. Carpels 2–6, connate; ovary superior with 2–6 locules and several to many axile ovules (rarely with 1 locule and parietal ovules), sometimes axile below and parietal above; style 1, undivided or stigma sessile. Fruit a capsule.

**5.C    Sonneratiaceae**—2 genera, 10 species. Old World tropics.

**5.D    Trapaceae**—1 genus, 15 species. Tropical Africa and Asia.

**5.E    Punicaceae**—The Pomegranate Family. 1 genus, 2 species. Eurasia.

Shrubs or trees. Leaves opposite, simple and entire, estipulate. Plants synoecious. Inflorescence of solitary flowers or axillary clusters. Flowers perfect, regular. Hypanthium present. Sepals 5–8, basally connate. Petals 5–8, distinct. Stamens many, distinct, developing centripetally. Carpels 3–12, connate; ovary inferior with 2–12 locules, some of which are superimposed above the others. Ovules many with axile to parietal placentation; style 1. Fruit a many-seeded berry with a leathery exocarp; seeds with fleshy outer seed coats.

**5.F    Onagraceae**—(Figure 14-25) The Evening-primrose Family. 16–17 genera, 645 species. Temperate northern hemisphere especially western North America.

§See page 306 for description and illustrations§

**5.G    Combretaceae**—The Combretum Family. 20 genera, 400 species. Pantropical, especially well-represented in Africa.

Shrubs, woody vines and trees. Leaves alternate or opposite, simple and entire, estipulate. Plants synoecious (rarely dioecious or polygamous). Inflorescences racemes, spikes, heads or panicles. Flowers perfect (rarely imperfect), regular. Hypanthium present, often elongated, basally adnate to ovary. Sepals 4 or 5 (8), ± connate. Petals 4 or 5 (8) (0), distinct, adnate near top of hypanthium. Stamens 4–10 (2–3) (many), distinct, equally or unequally adnate to the hypanthium. Carpel apparently 1 (more than 1 and wholly connate); ovary partially to wholly inferior, with 1 locule and 2–6 apical ovules; style 1 and undivided. Fruit a 1-seeded drupe or several-winged samara (seed viviparous in mangrove forms).

**5.H    Melastomataceae**—The Melastome Family (Figure 14-4). 194–200 genera, 4000 species. Tropics and subtropics, especially in South America.

Herbs, shrubs, woody vines or trees. Leaves opposite (alternate or whorled), simple, estipulate; blades usually with 3–9 palmately spreading to nearly parallel major veins and numerous smaller cross veins. Plants synoecious. Inflorescence variously cymose, sometimes paniculiform racemiform or corymbiform, or flowers solitary or in axillary clusters. Flowers perfect, regular (rarely ± irregular). Hypanthium present, ± adnate to ovary. Sepals 4–5 (6) connate. Petals 4–5 (6), distinct. Stamens 8–10 (4-many); filaments distinct, jointed and usually bent. Carpels 4–5 (2-many), connate; ovary inferior (rarely superior to half inferior) with 4–5 (2-many) locules and many (few) axile ovules

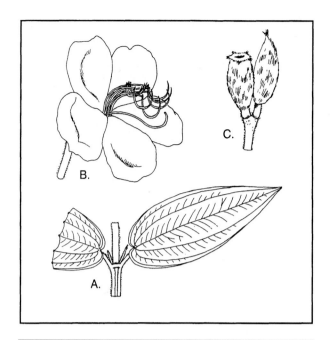

**Figure 14-4**    Examples of Melastomataceae. A–C. *Tibouchina* sp. A. Leaves and stem. B. Flower. C. Developing capsules.

(1 locule and ovules free-central); style 1, undivided. Fruit a capsule or berry.

**5.I    Penaeaceae**—7 genera, 20 species. South Africa.

**5.J    Crypteroniaceae**—1–3 genera, 4 species. India to the Philippines and the Malay Archipelago.

**5.K    Vochysiaceae**—7 genera, 200 species. Mostly tropical American, 1 African.

**5.L    Trigoniaceae**—3–4 genera, 26 species. Disjunct from American tropics to Madagascar and southeastern Asia.

# EUROSID I

## 6. Order Zygophyllales

**6.A    Zygophyllaceae**—The Caltrop Family (Figure 14-5). 27–30 genera, 250 species. Arid tropics and subtropics.

Herbs, shrubs or trees. Leaves opposite (alternate), simple to pinnate or deeply divided, stipulate. Plants synoecious (dioecious). Inflorescence variously cymose or flowers solitary. Flowers perfect (imperfect), regular. Sepals 4–5 distinct or basally connate. Petals (0) 4–5, distinct (sometimes connate near base). Stamens 10 (3–15), distinct. Carpels 5 (2–6), connate; ovary superior with (2–12) locules and 1–2 (–many) axile ovules per locule; style 1, undivided. Fruit a capsule, schizocarp, berry or drupe.

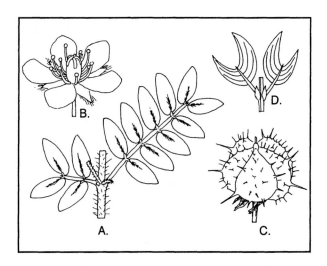

**Figure 14-5** Examples of Zygophyllaceae. A–C, *Tribulus terrestris* A. Stem with pinnately compound leaves. B. Flower. C. Schizocarp. D. Stem with bilobed leaves of *Larrea tridentata*.

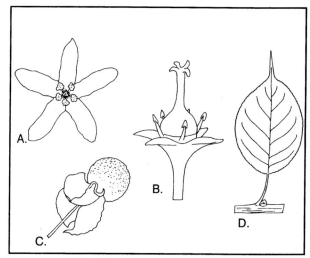

**Figure 14-6** Examples of Celastraceae. A–D. *Celastrus scandens*. A. Flower. B. Flower after petals have fallen. C. Capsule with arillate seeds. D. Leaf.

**6.B   Krameriaceae**—The Ratany Family. 1 genus, 15 species. Desert regions of New World.

Herbs or shrubs (small trees), hemiparasitic on the roots of other plants. Leaves alternate, simple (rarely trifoliolate), estipulate. Plants synoecious. Inflorescence of racemes or axillary flowers. Flowers perfect, irregular. Sepals 4–5, distinct, petaloid. Petals 5, the upper 3 of them often connate at base, the lower 2 thickened and nectar-secreting. Stamens 3–4, distinct, alternating with the upper petals, free or filaments adnate to petals; anthers with pores. Carpels 2 but only one fertile, the second reduced, empty and easily overlooked; ovary superior with 1 locule and 2 apparently marginal ovules; style 1, undivided. Fruit an indehiscent 1-seeded spiny pod.

# 7. Order Celastrales

**7.A   Celastraceae**—The Bittersweet Family (Figure 14-6). 50–85 genera, 800 species. Pantropical, less common in temperate regions.

Shrubs, woody vines, or trees. Leaves alternate or opposite, simple, stipulate (rarely estipulate). Plants synoecious (infrequently monoecious, dioecious, or polygamous). Inflorescence variously cymose or in axillary clusters (rarely of solitary flowers or racemes). Flowers perfect (rarely imperfect), regular. Hypanthium sometimes well developed. Sepals 3–5, connate (rarely distinct). Petals 3–5 (0), distinct. Stamens 3–5 (10), distinct, alternate with the petals. Nectary disc present. Carpels 2–5, connate; ovary superior to half inferior with 1–5 locules and 2 (1) (3–6) (many) axile ovules per locule; style 1, undivided or 2–5-lobed. Fruit a capsule, samara, drupe, berry or indehiscent dry fruit; seeds usually with brightly colored aril.

**7.B   Hippocrateaceae**—2 genera, 300 species. Pantropical.

**7.C   Stackhousiaceae**—3 genera, 20–25 species. Australia, New Zealand and islands of western Pacific.

# 8. Order Malpighiales

**8.A   Malpighiaceae**—The Malpighia Family. 60–65 genera, 1200 species. Pantropical, especially common in South America.

Shrubs, woody vines or trees. Leaves opposite (very rarely alternate) simple, stipulate. Plants synoecious. Inflorescences variously determinate or indeterminate, often racemes. Flowers perfect, regular, hypogynous. Sepals 5, distinct or connate. Petals 5, distinct, usually long-clawed. Stamens 10 (5 or 15), distinct or filaments basally connate, some often modified as staminodes. Carpels 3 (2–5), connate; ovary superior with 2–5 locules and 1 axile ovule per locule (rarely carpels distinct, ovaries with 1 marginal ovule); styles 3 (2–5), distinct (1 and undivided). Fruit a schizocarp with the mericarps often winged, or a nut or drupe.

**8.B   Euphorbiaceae**—The Spurge Family (Figure 14-26). 310–320 genera, 7500 species. Cosmopolitan, especially common in tropical regions.

§See page 308 for description and illustrations§

**8.C   Clusiaceae (Guttiferae)**—The Mangosteen Family (Figure 14-27). 47–50 genera, 1200 species. Mostly tropical with only 2 temperate genera.

§See page 311 for description and illustrations§

**8.D   Rhizophoraceae**—The Red Mangrove Family. 14 genera, 100 species. Pantropical.

Trees, shrubs and woody vines. Leaves opposite, simple and entire, stipulate. Plants synoecious (rarely polygamous). Inflorescence axillary, flowers solitary or in cymes or racemes. Flowers perfect (imperfect), regular. Hypanthium present or absent. Sepals 4–5 (3–16), distinct or nearly so. Petals 4–5 (3–16), distinct. Stamens 8–10 (–many); filaments distinct or basally connate or anthers sessile. Carpels 2–12; ovary superior or inferior with 2–12 locules and 2 (4 or more) apical-axile ovules or locule 1 and ovules apical-parietal; style 1 and undivided or lobed above (rarely styles 1 per carpel). Fruit a 1-few seeded berry (rarely a capsule) (seed viviparous in mangrove species).

**8.E   Violaceae**—The Violet Family (Figure 14–28). 16–20 genera, 800 species. Cosmopolitan.

§See page 313 for description and illustrations§

**8.F   Passifloraceae**—The Passion-flower Family. 16–17 genera, 650 species. Pantropical and subtropical, especially in Africa and the Americas.

Herbs, shrubs, tendril-bearing woody vines or trees. Leaves alternate, simple to palmate. Plants synoecious (rarely monoecious or dioecious). Inflorescence of solitary flowers, racemes or cymes, sometimes paniculiform. Flowers perfect or imperfect, regular. Hypanthium present. Sepals 5 (4–6), distinct or basally connate, often petaloid. Petals 5 (4–6), distinct or basally connate. Corona with numerous linear lobes developed between petals and stamens. Stamens 5 (3–10), distinct or filaments connate into a tube or in bundles, sometimes adnate to style. Carpels 3–5, connate; ovary superior with 1 locule and numerous parietal ovules; styles 3–5 or 1 and 3–5 lobed or stigmas nearly sessile. Fruit a berry or capsule.

**8.G   Salicaceae**—The Willow Family (Figure 14–29). 2 genera, 350 species. Nearly cosmopolitan but most species in north temperate zone.

§See page 315 for description and illustrations§

**8.H   Achariaceae**—3 genera, 3 species. South Africa.

**8.I   Chrysobalanaceae**—17 genera, 450 species. Pantropical, best developed in New World.

**8.J   Podostemaceae**—40–50 genera, 200 species. Mostly tropical.

**8.K   Dichapetalaceae**—3 genera, 325. Pantropical, especially Africa.

**8.L   Linaceae**—The Flax Family. 6 genera, 220 species (Figure 14–7). Widely distributed, most common in temperate and subtropical regions.

Herbs or shrubs. Leaves alternate (rarely opposite or whorled), simple and entire, stipulate (rarely estipulate). Plants synoecious. Inflorescence of cymes or appearing racemose. Flowers perfect, regular. Sepals 5 (4) distinct or ± connate. Petals 5 (4), distinct. Stamens 5, alternate with (rarely opposite) the petals, staminodes sometimes present, filaments basally connate. Carpels 2–5, connate; ovary superior with 2–5 locules, sometimes further divided by false septa, ovules 1–2 per locule; styles 2–5 or 1 and 2–5-lobed. Fruit a capsule (rarely a schizocarp).

**8.M   Ixonanthaceae**—5 genera, 30 species. Pantropical.

**8.N   Hugoniaceae**—7 genera, 60 species. Pantropical, especially Old World.

**8.O   Humiriaceae**—8 genera, 50 species. Mostly South American, 1 African.

**8.P   Erythroxylaceae**—4 genera, 200 species. Pantropical, especially New World.

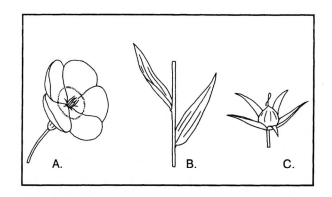

**Figure 14-7**   Examples of Linaceae. A–C. *Linum grandiflorum.* A. Flower. B. Stem with leaves. C. Capsule.

**8.Q    Pandaceae**—3–4 genera, 26 species. Tropical Africa, Asia, New Guinea.

**8.R    Ochnaceae**—The Ochna Family. 26–30 genera, 400 species. Pantropical, especially Brazil.

Shrubs or trees. Leaves alternate, simple (rarely pinnate), stipulate. Plants synoecious. Inflorescence variously cymose, racemose or paniculiform. Flowers perfect, regular (rarely irregular). Sepals 5, distinct or basally connate. Petals 5 (4) (6–12), distinct. Stamens 4–many, distinct or with filaments ± connate. Carpels 2–5 (10–15), connate or basally distinct and connate apically; ovary or ovaries superior, sometimes protruding from a fleshy receptacle, locules 2–5 (10–15) with 1–many ovules per locule; style 1. Fruit a berry, aggregate of drupes or capsule.

**8.S    Caryocaraceae**—2 genera, 23 species. Tropical America, especially Brazil.

**8.T    Quiinaceae**—4 genera, 40 species. Tropical America, especially Amazonia.

**8.U    Elatinaceae**—The Waterwort Family. 2 genera, 40 species. Nearly cosmopolitan.

Herbs or shrubs, sometimes aquatic. Leaves opposite or whorled, simple, stipulate. Plants synoecious. Inflorescence of solitary or cymosely clustered flowers. Flowers perfect, regular. Sepals 2–5, distinct or basally connate. Petals 2–5, distinct. Stamens 2–10, distinct. Carpels 2–5, connate; ovary superior with 2–5 locules and few-many axile ovules; styles 2–5. Fruit a capsule or schizocarp.

**8.V    Medusagynaceae**—1 genus, 1 species. Seychelles Islands.

**8.W    Flacourtiaceae**—79–85 genera, 800 species. Widespread in tropical and subtropical regions.

Shrubs or trees (rarely woody vines). Leaves alternate (rarely opposite or whorled), simple, stipulate (infrequently estipulate). Plants synoecious (rarely dioecious or monoecious). Inflorescence variously cymose, often paniculiform or corymbiform, sometimes of solitary or clustered axillary flowers. Flowers perfect or imperfect, regular. Sepals 2–15, distinct or basally connate. Petals 0–15 distinct or ± connate. Stamens (5–10), distinct or basally connate, sometimes in bundles opposite the petals. Carpels 2–10, connate; ovary superior (rarely half inferior to inferior) with 1 locule and many (few) parietal ovules (rarely with 2–10 locules and axile

ovules); styles 2–10, distinct or 1 and entire. Fruit a berry or capsule (rarely a drupe or achene).

**8.X    Lacistemataceae**—2 genera, 20 species. Tropical America.

**8.Y    Scyphostegiaceae**—1 genus, 1 species. Borneo.

**8.Z    Turneraceae**—8–10 genera, 20 species. Tropical-subtropical Africa and America.

**8.AA    Malesherbiaceae**—1–2 genera, 25 species. Western South America.

**8.AB    Balanopaceae**—1 genus, 9 species. Southwestern Pacific region.

# 9.  Order Oxalidales

**9.A    Oxalidaceae**—The Wood-sorrel Family (Figure 14-8). 6–8 genera, 900 species. Mostly tropical and subtropical, few temperate.

Herbs (shrubs or trees). Leaves alternate or all basal, pinnate or palmate, often trifoliolate, estipulate. Plants synoecious (polygamous). Inflorescence cymose, often umbelliform (sometimes flowers solitary or in axillary clusters). Flowers perfect (imperfect), regular. Sepals 5, distinct. Petals (0) 5, distinct or sometimes connate or coherent new base. Sta-

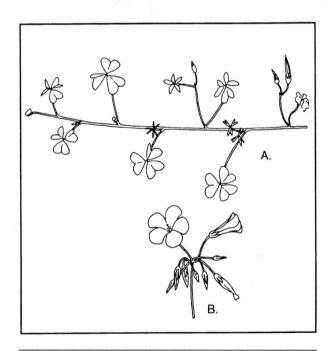

**Figure 14-8**    Examples of Oxalidaceae. A. Leafy stem of *Oxalis corniculata* with developing capsules. B. Inflorescence of *O. pes-caprae*.

mens 10 (15), distinct or filaments basally connate. Carpels (3–) 5, connate; ovary superior with (3–) 5 locules and 1–many axile ovules; styles 3–5. Fruit a capsule (berry).

**9.B  Cephalotaceae**—1 genus, 1 species. S.W. Australia.

**9.C  Cunoniaceae**—23–25 genera, 350 species. Mexico and South America, South Africa and Madagascar, southeast Asia to Australia.

**9.D  Brunelliaceae**—1 genus, 50 species. Tropical America.

**9.E  Connaraceae**—12–24 genera, 300–400 species. Pantropical, especially Old World.

**9.F  Elaeocarpaceae**—3 genera, 28 species. Australia and Tasmania.

## 10.  Order Fabales (Figs. 14-30–14-32)

**10.A  Fabaceae** (Leguminosae) = [Mimosaceae, Caesalpiniaceae, and Papilionaceae]—The Pea Family (Figure 14-30). 590–800 genera, 14,000–20,000 species. Cosmopolitan.

§See page 317 for description and illustrations§

**10.B  Polygalaceae**—The Milkwort Family. 12–19 genera, 750 species (Figure 14-9). Nearly cosmopolitan.

Herbs or shrubs (rarely trees). Leaves alternate (rarely opposite or whorled), simple, estipulate

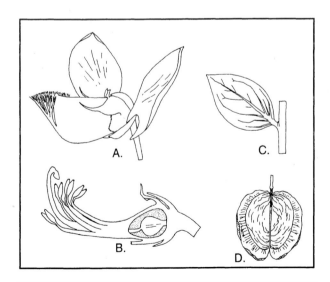

**Figure 14-9**  Examples of Polygalaceae. A–C. *Polygala dalmaisiana* A. Intact flower. B. Flower (l.s.). C. Leaf. D. Capsule of *P. californica*.

(rarely stipulate). Plants synoecious. Inflorescence of racemes, spikes, panicles or solitary flowers. Flowers perfect, irregular. Sepals 5, distinct (rarely connate), often petaloid, often very unequal. Petals 3 (5), distinct or basally connate, very unequal, adnate to stamen tube, the lowermost often fringed and keel-like. Stamens 8 (3–10), filaments connate into a tube open on one side (rarely all distinct). Carpels 2 (1–5), connate; ovary superior with 2 (1–5) locules and 1 (2–6) axile (marginal) ovule per locule; style 1, undivided. Fruit a capsule, drupe, samara or nut.

**10.C  Surianaceae**—3–4 genera, 7 species. Australia and maritime tropics.

**10.D  Xanthophyllaceae**—1 genus, 40 species. Indo-Malaysian region.

## 11.  Order Rosales

**11.A  Rosaceae**—The Rose Family (Figure 14-33 to 14-36). 97–100 genera, 3000 species. Cosmopolitan, especially in northern hemisphere.

§See page 318 for description and illustrations§

**11.B  Elaeagnaceae**—The Oleaster Family. 3 genera, 50 species. Temperate to subtropical regions.

Shrubs or trees, densely pubescent with stellate hairs or scales. Leaves alternate or opposite, simple and entire, estipulate. Plants synoecious, dioecious or polygamous. Inflorescence of solitary or clustered axillary flowers or racemes. Flowers perfect or imperfect, regular. Sepals 2 or 4, connate or nearly distinct. Stamens 4 or 8, distinct, free or filaments adnate to calyx tube. Carpel 1; ovary superior or seemingly inferior with 1 locule and 1 basal ovule; style 1. Fruit an achene surrounded by (but not adnate to) the fleshy base of the calyx tube, berry-like or drupelike in appearance.

**11.C  Rhamnaceae**—The Buckthorn Family (Figure 14-37). 50–55 genera, 900 species. Cosmopolitan, most common in warmer regions.

§See page 325 for description and illustrations§

**11.D  Ulmaceae**—The Elm Family (Figure 14-10). 15–18 genera, 150 species. Temperate to tropical, especially in the northern hemisphere.

Shrubs or trees. Leaves alternate, simple, stipulate. Plants synoecious, monoecious or polygamous. Inflorescence of solitary or clustered axillary

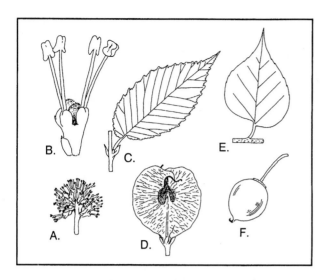

Figure 14-10 Examples of Ulmaceae. A–D. *Ulmus* sp. A. Cluster of flowers. B. Individual flower. C. Leaf. D. Samara. E–F. *Celtis reticulata.* E. Leaf. F. Drupe.

flowers. Flowers perfect or imperfect, regular. Sepals 4–8, distinct or basally connate. Petals absent. Stamens 4–8, opposite and sometimes adnate the sepals. Carpels 2, connate; ovary superior with 1 locule and 1 apical ovule; style 2, sometimes winged or apically forked. Fruit a samara or drupe.

**11.E Moraceae**—The Mulberry Family (Figure 14-38). 37–70 genera, 1000–2500 species. Widespread in tropical and subtropical latitudes of both Old and New Worlds. Sparingly distributed in temperate regions.

§See page 327 for description and illustrations§

**11.F Urticaceae**—The Nettle Family (Figure 14-39). 45–48 genera, 700–1000 species. Cosmopolitan with greatest diversity in the tropics.

§See page 330 for description and illustrations§

**11.G Cannabaceae**—The Hemp Family. 2 genera, 2 species, North temperate zone.

Herbs. Leaves alternate or opposite, simple or palmate, stipulate. Plants dioecious Inflorescence of axillary spikes or panicles. Flowers imperfect, regular. Sepals 5, distinct. Petals absent. Stamens 5, distract, opposite the sepals. Carpels 2, connate; ovary superior with 1 locule and 1 apical ovule; style 1 with 2 branches. Fruit an achene, sometimes enclosed by papery bracts.

**11.H Cecropiaceae.** 6 genera, 275 species. Tropics, especially Americas.

**11.I Barbeyaceae.** 1 genus, 1 species. Northeastern Africa to Arabia.

# 12. Order Cucurbitales

**12.A Cucurbitaceae**—The Cucumber Family (Figure 14-40). 90–120 genera, 700 species. Tropical and subtropical regions with a few temperate representatives.

§See page 332 for description and illustrations§

**12.B Begoniaceae**—The Begonia Family. 3–5 genera, 1000+ species. Pantropical, except Polynesia and Australia.

Herbs or shrubs. Leaves alternate, simple, stipulate. Plants monoecious. Inflorescence of solitary flowers or cymes (rarely in racemes). Flowers imperfect, regular or irregular. Sepals 2 in staminate flowers, 2–5 (10) in pistillate flowers, distinct. Petals 2 or 0 in staminate flowers, absent in pistillate flowers. Stamens 4–many, centripetal, distinct or basally connate. Carpels 3 (2) (5), connate; ovary inferior (rarely half-inferior) with 3 (2) locules and numerous axile ovules or with 1 locule and numerous parietal ovules; styles 3 (2) (5), distinct or basally connate. Fruit a capsule or berry.

**12.C Datiscaceae**—The Durango-root Family. 3 genera, 4 species. Disjunct in west-central Asia, Indo-Malaysia and the western United States.

Herbs. Leaves alternate, simple to pinnate, estipulate. Plants dioecious or polygamous. Inflorescence of axillary clusters or leafy spike-like racemes. Flowers imperfect (less commonly perfect), regular. Sepals 3–9, distinct in staminate flowers, 3–8, connate and adnate to ovary in pistillate flowers. Petals absent or 6–8, small, distinct. Stamens numerous in staminate flowers, 2–5 in perfect flowers. Carpels 3–8, connate; ovary inferior with 1 locule and many parietal ovules; styles 3–5, entire or bifid. Fruit a capsule.

**12.D Corynocarpaceae**—1 genus, 5 species. Australian region.

**12.E Anisophylleaceae**—4 genera, 40 species. Tropical or subtropical Africa to Asia.

# 13. Order Fagales

**13.A Fagaceae**—The Beech Family (Figure 14-41). 7 genera, 800–1000 species. Temperate and tropical regions of northern hemisphere plus

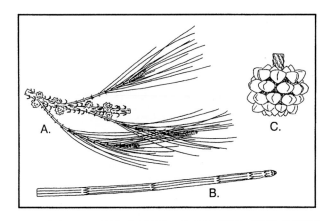

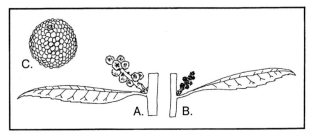

**Figure 14-12** Examples of Myricaceae. A–C. *Myrica californica.* A. Branchlet with leaf and spike of pistillate flowers. B. Branchlet with leaf and spike of staminate flowers. C. Drupe.

**Figure 14-11** Examples of Casuarinaceae. A–C. *Casuarina* sp. A. Branch with slender branchlets and pistillate inflorescences. B. Close-up of jointed branchlet with whorled scale leaves. C. Cone-like clusters of fruits.

one genus in temperate and tropical regions of southern hemisphere.

§See page 336 for description and illustrations§

**13.B  Betulaceae**—The Birch Family (Figure 14-42). 6 genera, 120–170 species. Widespread in Northern Hemisphere from Arctic to Tropics, extending south along mountains into Andes of South America

§See page 336 for description and illustrations§

**13.C  Casuarinaceae**—The She-oak Family (Figure 14-11). 1 genus, 50 species. Australia and Indo-Malayan region.

Trees. Leaves tiny and scale like, whorled and basally connate, estipulate. Plants dioecious. Staminate inflorescence an elongate pendulous verticillate spike; pistillate inflorescence a dense cone-like spike. Flowers imperfect, very reduced. Sepals 2, distinct in staminate flowers, absent in pistillate flowers. Petals absent. Stamen 1. Carpels 2, connate; ovary superior, 2-locular with one locule abortive, the other 2 ovulate; style 1 with 2 elongated stigmas. Fruit a small samara enclosed by a pair of woody bractlets, grouped together in cone-like multiple fruits.

**13.D  Myricaceae**—The Bayberry Family (Figure 14-12). 3 genera, 50 species. Widespread from cold-temperate to tropical regions of the northern hemisphere, south to temperate regions of South America and southern Africa. Shrubs or trees. Leaves alternate, simple, stipulate or estipulate, punctate. Plants monoecious (rarely polygamous).

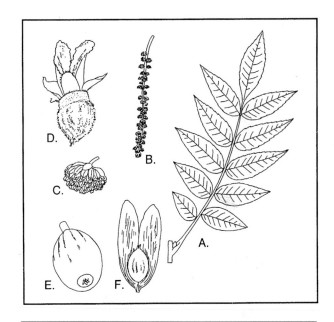

**Figure 14-13** Examples of Juglandaceae. A–E. *Juglans* sp. A. Leaf. B. Catkin of staminate flowers. C. Individual staminate flower. D. Pistillate flower. E. Drupe. F. Winged nut of *Pterocarya.*

Staminate and pistillate inflorescences both spikes. Flowers imperfect (rarely perfect) subtended by tiny bractlets. Sepals absent. Petal absent. Stamens 4 (2–20), distinct or filaments basally connate. Carpels 2, connate; ovary superior (inferior when flowers perfect) with 1 locule and 1 basal ovule; style 1 with 2 branches. Fruit a drupe.

**13.E  Juglandaceae**—The Walnut Family (Figure 14-13). 8 genera, 60 species. Primarily north-temperate with a few extending into South America.

Trees. Leaves alternate (opposite), pinnate, estipulate. Plants monoecious. Staminate inflorescence a catkin-like raceme or spike; pistillate inflorescence a solitary flower or spike (catkin). Flowers imperfect, regular. Sepals (0) 3–6 and ± connate in staminate flowers, 4 connate and adnate to ovary in pistillate flowers. Petals absent. Stamens 3–many, distinct. Carpels 2–3, connate; ovary inferior, with 1 locule and 1 basal ovule; styles 2–3. Fruit a drupe or nut, sometimes winged with adnate bracts.

**13.F  Nothofagaceae**—The Southern Beech Family. 1 genus, 35 species. Southern South America, New Zealand, Australia, East Indies.

# EUROSID II

## 14. Order Brassicales

**14.A  Brassicaceae (Cruciferae)**—The Mustard Family (Figure 14-43). 350–381 genera, 3000 species. Mostly temperate regions in both northern and southern hemispheres.

§See page 339 for description and illustrations§

**14.B  Tropaeolaceae**—The Nasturtium Family. 3 genera, 92 species. Primarily New World tropics and subtropics.

Herbs, sometimes climbing with twining petioles. Leaves alternate (opposite), simple to deeply divided, often peltate, stipulate (estipulate). Plants synoecious. Inflorescence of solitary axillary flowers or leafy racemes. Flowers perfect, irregular. Sepals 5, distinct (connate), the uppermost spurred. Petals (2) 5, distinct. Stamens 8, distinct. Carpels 3, connate; ovary superior with 3 locules and 1 axile ovule per locule; style 1, shallowly 3-lobed at apex. Fruit a schizocarp (rarely a berry or samaroid schizocarp).

**14.C  Capparaceae**—The Caper Family 42–45 genera, 800 species. Widespread in tropical and subtropical areas with a few temperate representatives.

Herbs, shrubs or trees. Leaves alternate (opposite), simple or palmate, stipulate. Plants synoecious (rarely dioecious). Inflorescence of racemes, corymbs, axillary clusters or solitary flowers. Flowers perfect (imperfect), regular or irregular. Sepals 3–8, distinct or connate. Petals 4–16 (0), distinct. Stamens 4-many, centrifugal, distinct. Carpels 2 (3–12); connate; ovary superior, often borne on an elongated stipe, with 1 locule and 2-many parietal ovules (rarely with 2–12 locules separated by false septa and 2-many parietal ovules); style 1. Fruit a capsule, silique or silicle, samara, schizocarp, nut, berry or drupe.

**14.D  Bataceae**—The Saltwort Family (Figure 14-14). 1 genus, 2 species. Warm coastal regions of North and South America.

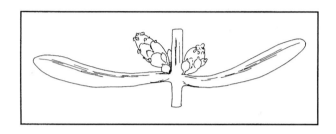

**Figure 14-14**  Examples of Bataceae, *Batis maritima.* Node with opposite fleshy leaves and axillary spikes.

Semi-succulent subshrubs. Leaves opposite simple and entire, estipulate. Plants dioecious. Both staminate and pistillate inflorescences spikes. Sepals 2, distinct in staminate flowers, absent in pistillate flowers. Petals [actually petaloid staminodes] 4, distinct in staminate flowers, absent in pistillate flowers. Stamens 4, distinct. Carpels 4, connate; ovary superior with 4 locules and 4 ovules; stigma sessile. Fruit a cluster of coalesced berries.

**14.E  Caricaceae**—4 genera, 30 species. New World tropics and tropical western Africa.

**14.F  Moringaceae**—1 genus, 10 species. Mediterranean region to India, disjunct in southwestern Africa and Madagascar.

**14.G  Resedaceae**—The Mignonette Family. 6–7 genera, 70 species. Europe to west-central Asia and northern Africa, disjunct in southern Africa and the western deserts of North America.

Herbs or shrubs. Leaves alternate, simple to pinnate, stipulate. Plants synoecious (rarely monoecious). Inflorescence of racemes or, spikes. Flowers perfect (rarely imperfect), irregular. Sepals 4–8, distinct or basally connate. Petals 4–8 (rarely 0 or 2), distinct (rarely connate) often unequal. Stamens 3–many, distinct, often clustered to 1 side of flower. Carpels 2–7, distinct or connate; ovary or ovaries superior, often open at apex at or shortly after anthesis, locules 1 with parietal or basal ovules or locule

1 per carpel with 1–2 ovules; styles 2–7 often very short. Fruit a capsule, berry or cluster of follicles.

**14.H   Akariaceae**—2 genera, 2 species. Australia, China.

**14.I   Limnanthaceae**—The Meadow-foam Family. 2 genera, 11 species. Temperate North America.

Herbs. Leaves alternate, pinnately dissected, estipulate. Plants synoecious. Inflorescence of solitary flowers. Flowers perfect, regular, hypogynous. Sepals 3–6, distinct. Petals 3–6, distinct. Stamens 6–10, distinct. Carpels 3–5, nearly distinct, united only by the styles; ovaries superior, one-locular with 1 basal ovule per carpel, style 1; gynobasic, with 3–5 branches. Fruit a schizocarp.

**14.J   Salvadoraceae**—3 genera, 12 species. Africa and Madagascar.

**14.K   Tovariaceae**—1 genus, 2 species. Tropical North and South America.

**14.L   Gyrostemonaceae**—5 genera, 17 species. Australia.

# 15.   Order Malvales

**15.A   Malvaceae**—The Mallow Family (Figure 14-44). 100–119 genera, 1500 species. Cosmopolitan.

§See page 341 for description and illustrations§

**15.B   Cistaceae**—The Rock-rose Family (Figure 14-15). 7–8 genera, 200 species. Temperate, especially in the Mediterranean region.

Herbs or shrubs. Leaves alternate or opposite, simple, stipulate or estipulate. Plants synoecious

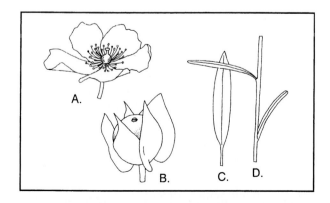

**Figure 14-15**   Examples of Cistaceae. A–C. *Cistus creticus.* A. Flower. B. Developing capsule with calyx. C. Leaf. D. Leaf of *Helianthemum scoparium.*

(rarely polygamous). Inflorescence variously cymose or of solitary flowers. Flowers perfect (imperfect), regular. Sepals 5, distinct, often unequal. Petals 5 (3) (0), distinct. Stamens many, centrifugal, distinct. Carpels 3 or 5–10, connate; ovary superior with 1 locule and 3-many parietal or basal-parietal ovules; styles 1–5. Fruit a capsule.

**15.C   Dipterocarpaceae**—The Dipterocarp Family. 16,600 species. Tropical Africa and Indo-Malaysia plus a single South American genus.

Trees. Leaves alternate simple and entire, stipulate. Inflorescence of racemes, panicles or cymes. Flowers perfect, regular. Sepals 5, distinct or connate. Petals 5, distinct or weakly connate. Stamens 10–15 (5-many), centrifugal, distinct or filaments ± connate. Carpels 3 (2–5), connate; ovary superior or inferior with 3 (2–5) locules and 2 (–4) axile ovules per locule. Fruit a nut surrounded by adherent samaroid fruiting calyx.

**15.D   Thymelaeaceae**—The Mezereum Family. 50–58 genera, 500 species. Cosmopolitan with greatest diversity in Africa.

Shrubs or trees (rarely herbs). Leaves alternate or opposite, simple and entire, estipulate. Plants synoecious (rarely dioecious). Inflorescence of racemes, umbels, heads or axillary clusters, sometimes involucrate. Flowers perfect (rarely imperfect), regular. Hypanthium present. Sepals 4–5, connate, petaloid. Petals 4–5 (8 or more) or absent, often much reduced, distinct. Stamens 2–10, distinct, sometimes unequally inserted in the hypanthium. Carpels 1–2 (3–5), connate; ovary superior with 1–2 (3–5) locules and 1 (2) apical ovule per locule; style 1, undivided. Fruit a nut, achene or drupe (rarely a capsule or berry).

**15.E   Neuradaceae**—3 genera, 10 species, Africa to India.

**15.F   Sphaerosepalaceae**—2 genera, 14 species. Madagascar.

**15.G   Sarcolaenaceae**—9–10 genera, 30 species. Madagascar.

**15.I   Tiliaceae**—The Linden Family (Figure 14-16) 50–53 genera, 450 species. Widespread in the tropics and subtropics, plus a few temperate representatives.

Herbs, shrubs or trees. Leaves alternate (rarely opposite), simple, stipulate. Plants synoecious (rarely monoecious). Inflorescence variously

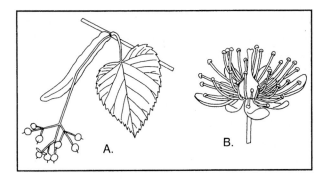

**Figure 14-16** Examples of Tiliaceae. A–B. *Tilia americana.* A. Branchlet with leaf and fruiting inflorescence. B. Flower.

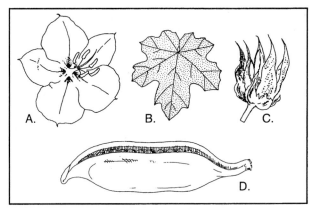

**Figure 14-17** Examples of Sterculiaceae. A–C. *Fremontodendron californicum.* A. Flower with large petaloid sepals. B. Leaf C. Capsule. D. Follicle of *Brachychiton* sp.

cymose. Flowers perfect (rarely imperfect), regular. Sepals 4–5, distinct or connate. Petals 4–5 (0), distinct. Stamens 10-many centrifugal, distinct or filaments basally connate in a ring or in groups, sometimes weakly adnate at base to petals. Carpels 2–10, connate; ovary superior with 2–10 (many) locules and 1–many axile ovules; style 1. Fruit a capsule or schizocarp.

**15.J Sterculiaceae**—The Cacao Family (Figure 14-17). 65–67 genera, 1000 species. Mostly tropical and subtropical.

Shrubs or trees (herbs). Leaves alternate, simple, stipulate. Plants synoecious, monoecious or polygamous. Inflorescences variously cymose (less commonly flowers solitary). Flowers perfect or imperfect, regular. Sepals 5, connate, sometimes petaloid. Petals 5 or 0, distinct, free or adnate to stamen column. Stamens 5 or 10-many in bundles opposite the petals, centrifugal, often 5 staminodes present, opposite the sepals; filaments connate into tube surrounding the ovary (rarely distinct). Carpels 2–12 (–many) connate, or sometimes the ovaries distinct and only the styles connate; ovary superior with axile placentation or basally distinct with marginal placentation; style 1 and 2–12 (-many) branched. Fruit a capsule, schizocarp, berry or duster of follicles.

**15.K Bombacaceae**—The Kapok Family. 20–30 genera, 200 species. Tropics, especially in New World.

Trees. Leaves alternate, simple or palmate, stipulate. Plants synoecious. Inflorescence of solitary flowers or short cymes (infrequently in racemiform or paniculiform clusters). Flowers perfect, regular (rarely slightly irregular), often subtended by an involucel. Sepals 5, distinct or basally connate.

Petals 5 (0), distinct, often individually adnate at base to stamen column. Stamens 5-many, centrifugal, filaments distinct or connate into a tube surrounding the style, anthers often one-locular. Carpels 2–5, connate; ovary superior with 2–5 locules and 2-many axile ovules per locule; style 1. Fruit a capsule or nut (rarely a berry or schizocarp).

**15.L Bixaceae**—1–3 genera, 15 species. Tropical America.

# 16. Order Sapindales

**16.A Sapindaceae**—The Soapberry Family. 134–140 genera, 1500 species. Mostly tropical or subtropical; only a few temperate.

Shrubs, tendril-bearing vines, or trees (rarely herbs). Leaves alternate (very rarely opposite), pinnate or bipinnate, sometimes trifoliolate (rarely simple), stipulate or estipulate. Plants monoecious, dioecious or polygamous. Inflorescence of racemiform or paniculiform cymes. Flowers imperfect, sometimes appearing perfect but functionally imperfect, regular or irregular. Sepals 5, distinct (rarely basally connate) or appearing 4 with 2 wholly connate. Petals 4–5 (0), distinct. Stamens 4–10 (many), distinct. Nectary disc usually present. Carpels 3 (2 or 4–6), connate; ovary superior with 3 (2–6) locules with 1 (2-several) axile ovules per locule (rarely 1 locule with parietal ovules); style 1 and 3-(2–6-)lobed (rarely styles 2–6 and distinct). Fruit a berry, drupe, capsule, nut, samara or schizocarp.

**16.B Rutaceae**—The Citrus Family (Figure 14-45), 150–158 genera, 900—1500 species. Widespread, especially common in tropics and subtopics.

§See page 343 for description and illustrations§

**16.C Meliaceae**—The Mahogany Family, 51 genera, 550 species. Widespread, mostly tropical and subtropical.

Shrubs or trees (rarely herbs). Leaves alternate, pinnate, (simple, bipinnate or more), estipulate. Plants synoecious (rarely dioecious or polygamous). Inflorescence of axillary clusters or paniculiform cymes (rarely of solitary flowers). Sepals 4–5 (2–7) basally connate (rarely distinct). Petals (3) 4–5 (–14) distinct or connate, free or adnate to stamen tube. Stamens (4–5) 8–10 (–many); filaments connate into a tube (rarely distinct). Nectary disc present. Carpels (1) 2–5 (6–many), connate; ovary superior with (1–) 2–5 (–many) locules and 1–2 (–many) ovules per locule; style 1, undivided or stigma sessile. Fruit a berry or capsule (drupe).

**16.D Simaroubaceae**—24–27 genera, 150 species. Pantropical, with a few subtropical and temperate representatives.

Shrubs or trees. Leaves alternate, simple or pinnate, estipulate (rarely stipulate). Plants dioecious (less commonly polygamous, monoecious or synoecious). Inflorescence paniculiform or variously cymose (rarely flowers solitary). Flowers imperfect (perfect), when imperfect, non-functional carpels or stamens often present, regular. Sepals 3–8, distinct or connate. Petals (0) 3–8 distinct. Stamens 6–16 (equal in number to petals) (many), distinct. Nectary disc usually present. Carpels (1) 2–10, distinct; ovaries superior, one-loculed with 1–2 (few) marginal ovules *or* carpels connate, the ovary superior with 2–10 locules and 1–2 (few) axile ovules per locule; styles 1–8, or stigma sessile. Fruit a schizocarp, capsule, or aggregate of samaras or drupes.

**16.E Anacardiaceae**—The Sumac Family (Figure 14-46). 60–80 genera, 600 species. Pantropical, with some well-distributed temperate representatives.

§See page 345 for description and illustrations§

**16.F Burseraceae**—The Frankincense Family. 16–20 genera, 600 species. Pantropical, especially well represented in the New World tropics and northeastern Africa.

Shrubs or trees with resin ducts in the bark and leaves; herbage aromatic. Leaves alternate (opposite), pinnate or trifoliolate (simple), usually estipulate. Plants monoecious, dioecious, polygamous or synoecious. Inflorescence paniculiform (flowers solitary or in racemes or heads). Flowers perfect or imperfect [if imperfect usually with abortive parts of the opposite sex present], regular. Sepals 4–5 (3), connate. Petals 4–5 (3), distinct. Stamens 5–10, distinct or filaments weakly connate at base. Nectary disc present Carpels 3–5 (2) connate; ovary superior with 3–5 (2) locules and 1–2 apical-axile ovules per locule; style 1; stigma lobed or capitate and unlobed. Fruit a 1–5-seeded drupe or a capsule.

**16.G Aceraceae**—The Maple Family (Figure 14-47). 2 genera, 120 species. Widely distributed in North temperate regions and tropical southeast Asia.

§See page 347 for description and illustrations§

**16.H Hippocastanaceae**—The Horse-chestnut Family (Figure 14-18). 2 genera, 16 species. Temperate regions of Asia, Europe, and North America to the tropics of southeast Asia and northern South America.

Shrubs or trees. Leaves opposite palmate, estipulate. Plants synoecious or polygamous. Inflorescence a raceme, paniculiform cyme or panicle. Flowers perfect or some functionally staminate, irregular. Sepals 4–5, distinct or basally connate. Petals 4–5, distinct. Stamens 5–9, distinct. Nectary disc present, small. Carpels 3 (2–4), connate; ovary superior with 3 (2–4) locules and 2 axile ovules per

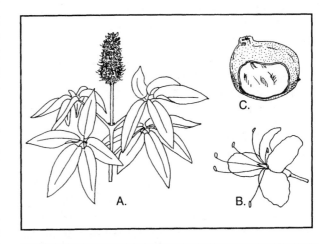

**Figure 14-18** Examples of Hippocastanaceae. A–C. *Aesculus californica.* A. Leafy branch of with panicle of flowers. B. Flower. C. Seed.

locule; style 1, undivided. Fruit a capsule with large seeds, often only 1–2 seeds maturing.

**16.I** **Leitneriaceae**—1 genus, 1 species. Southeastern United States.

# 17. Order Rafflesiales

**17.A** **Rafflesiaceae**—The Rafflesia Family. 7–8 genera, 50 species. Tropics and subtropics, especially in the Old World.

Herbaceous internal fungus-like parasites on stems or roots of host plants. Leaves opposite or alternate, scale-like or absent, non-chlorophyllous; estipulate. Plants monoecious or dioecious (rarely synoecious). Inflorescence of solitary flowers, spikes or racemes, emerging directly from host plant tissues. Flowers imperfect (rarely perfect), regular, very tiny to enormous. Sepals 3–10, connate. Petals absent. Stamens 8–many; filaments connate. Carpels 4-many, connate; ovary inferior (rarely superior to half-inferior) with 1 locule and many parietal ovules or with scattered locules, each lined with ovules; style 1 or stigma sessile. Fruit a capsule or berry.

# References

Allen, O. N., and E. T. Allen. 1981. *The Leguminosae: A Source Book of Characteristics, Uses, and Nodulation.* Univ. of Wisconsin Press, Madison.

Angiosperm Phylogeny Group II. 2003. An update of the Angiosperm Phylogeny Group classification for the orders and families of flowering plants: APG II. *Botanical Journal of the Linnean Society* 141:399–436.

Bensel, C. R., and B. Palser. 1975. Floral anatomy in the Saxifragaceae sensu lato. I. Introduction, Parnassioideae and Brexioideae. II. Saxifragoideae and Iteoideae. III. Kirengeshomoideae, Hydrangeoideae and Escallonioideae. IV. Baueroideae and conclusions. *American Journal of Botany* 62:176–185; 661–675; 676–687; 688–694.

Brizicky, G. K. 1962. The genera of Rutaceae in the southeastern United States. *Journal of the Arnold Arboretum* 43:1–22.

———. 1962. The genera of Anacardiaceae in the southeastern United States. *Journal of the Arnold Arboretum* 43:173–186.

———. 1963. The genera of Sapindales in the southeastern United States. *Journal of the Arnold Arboretum* 44:462–501.

———. 1964. The genera of Celastrales in the southeastern United States. *Journal of the Arnold Arboretum* 45:206–234.

———. 1964. The genera of Rhamnaceae in the southeastern United States. *Journal of the Arnold Arboretum* 45:439–463.

———. 1965. The genera of Vitaceae in the southeastern United States. *Journal of the Arnold Arboretum* 46:286–307.

Cronquist, A. 1981. *An Integrated System of Classification of Flowering Plants.* Columbia University Press, New York, pp. 521–857.

———. 1988. *The Evolution and Classification of Flowering Plants,* 2nd ed. New York Botanical Garden, Bronx, N.Y., pp. 359–413.

Dahlgren, R., & R. F. Thorne. 1984 (1985). The order Myrtales: circumscription, variation, and relationships. *Annals of the Missouri Botanical Garden* 71:633–699.

Elias, T. S. 1974. The genera of Mimosoideae (Leguminosae) in the southeastern United States. *Journal of the Arnold Arboretum* 55:67–118.

Eyde, R. H. 1966. The Nyssaceae in the southeastern United States. *Journal of the Arnold Arboretum* 47:117–125.

Ferguson, I. K. 1965. The Cornaceae in the southeastern United States. *Journal of the Arnold Arboretum* 47:106–116.

Gleason, H. A., and A. Cronquist. 1991. *Manual of Vascular Plants of Northeastern United States and Adjacent Canada,* 2nd ed. New York Botanical Garden, Bronx.

Graham, S. A. 1964. The genera of Lythraceae in the southeastern United States. *Journal of the Arnold Arboretum* 45:235–250.

———. 1964. The Elaeagnaceae in the southeastern United States. *Journal of the Arnold Arboretum* 45:274–278.

———. 1964. The genera of Rhizophoraceae and Combretaceae in the southeastern United States. *Journal of the Arnold Arboretum* 45:285–301.

Graham, S. A. 1966. The genera of Araliaceae in the southeastern United States. *Journal of the Arnold Arboretum* 47:126–136.

Harborne, J. B., D. Brothers, and B. L. Turner (eds.). 1971. *Chemotaxonomy of the Leguminosae.* Academic Press, London.

Herendeen, P. S., and D. L. Dilcher (eds.). 1992. *Advances in Legume Systematics*. Parts 4. The Fossil Record. Royal Botanic Gardens, Kew.

Heywood, V. H, (ed.). 1971. *The Biology and Chemistry of the Umbelliferae*. Publ. for Linnaean Society of London by Academic Press, London.

Heywood, V. H. (ed.). 1993. Flowering Plants of the World, updated edition. Oxford University Press, New York. pp. 137–221.

Johnson, L. A. S., and B. G. Briggs. 1984 (1985). Myrtales and Myrtaceae—a phylogenetic analysis. *Annals of the Missouri Botanical Garden* 71:700–756.

Judd W. S., C. S. Campbell, E. A. Kellogg, P. F. Stevens, and M. J. Donoghue. 2002. *Plant Systematics: A Phylogenetic Approach*. 2nd ed. Sunderland, MA: Sinauer Associates.

Kuijt, J. 1982. The Viscaceae in the southeastern United States. *Journal of the Arnold Arboretum* 63:401–410.

Koutnik, D. L. 1993. *Chamaesyce* and *Euphorbia*. Pp. 568–576 *in* J. C. Hickman (ed.), *The Jepson Manual*. University of California Press.

Miller, N. G. 1971. The Polygalaceae in the southeastern United States. *Journal of the Arnold Arboretum* 52:267–284.

Morgan, D. R., and D. E. Solits. 1993. Phylogenetic relationships among members of Saxifragaceae s.1. based on *rbc*L sequence data. *Annals of the Missouri Botanical Garden* 80:631–660.

Nevling, L. I. 1962. The Thymelaeaceae in the southeastern United States. *Journal of the Arnold Arboretum* 43:428–434.

Pimenov, M. G., and M. V. Leonov. 1993. *The Genera of the Umbelliferae*. Royal Botanic Gardens, Kew.

Pohill, R. M., and P. H. Raven (eds.). 1981. *Advances in Legume Systematics*. Parts 1 and 2. Royal Botanic Gardens, Kew Raven, P. H. 1964. The generic subdivision of Onagraceae, tribe Onagreae. *Brittonia* 16:276–288.

———. 1976. Generic and sectional delimitation in Onagraceae, tribe Epilobieae. *Annals of the Missouri Botanical Garden* 63:326–340.

———. 1979. A survey of reproductive biology in Onagraceae. *New Zealand Journal of Botany* 17:575–593.

Porter, D. M. 1972. The genera of Zygophyllaceae in the southeastern United States. *Journal of the Arnold Arboretum* 53:531–552.

Robertson, K. R. 1971. The Malpighiaceae in the southeastern United States. *Journal of the Arnold Arboretum* 53:101–112.

———. 1972. The genera of Geraniaceae in the southeastern United States. *Journal of the Arnold Arboretum* 53:182–201.

———. 1973. The Krameriaceae in the southeastern United States. *Journal of the Arnold Arboretum* 54:322–327.

———. 1974. The genera of Rosaceae in the southeastern United States. *Journal of the Arnold Arboretum* 55:303–332; 344–401; 611–662.

———. 1975. The Oxalidaceae in the southeastern United States. *Journal of the Arnold Arboretum* 56:223–239.

———, and Y.-T. Lee. 1976. The genera of Caesalpinioideae (Leguminosae) in the southeastern United States. *Journal of the Arnold Arboretum* 57:1–53.

Soltis, D. E., D. R. Morgan, A. Grable, P. S. Soltis, and R. Kuzoff. 1993. Molecular systematics of Saxifragaceae sensu stricto. *American Journal of Botany* 80:1056–1081.

Soltis, D. E., P. S. Soltis, M. W. Chase, M. E. Mort, D. C. Albach, M. Zanis, V. Savolainen, W. H. Hahn, S. B. Hoot, M. E. Fay, M. Axtell, S. M. Swenson, L. M. Prince, W. J. Kress, K. C. Nixon, and J. S. Farris. 2000. Angiosperm phylogeny inferred from 18S rDNA, *rbcL,* and *atpB* sequences. *Bot. J. Linn. Soc.* 133: 381–461.

Spongberg, S. A. 1971. The Staphyleaceae in the southeastern United States. *Journal of the Arnold Arboretum* 52:196–203.

Spongberg, S. A. 1972. The genera of Saxifragaceae in the southeastern United States. *Journal of the Arnold Arboretum* 53:409–498.

———. 1978. The genera of Crassulaceae in the southeastern United States. *Journal of the Arnold Arboretum* 59:197–248.

Stevens, P. F. (2001 onwards). Angiosperm Phylogeny Website, Version 5, May 2004 [and more or less continuously updated since]. http://www.mobot.org/MOBOT/research/APweb/.

Stirton, C. H. (ed.). 1987. *Advances in Legume Systematics*. Part 3. Royal Botanic Gardens, Kew.

Thorne, R. F. 1973. Inclusion of the Apiaceae (Umbelliferae) in the Araliaceae. *Notes from the Royal Botanical Garden of Edinburgh* 32:161–165.

Van Gelderen, D. M., P. C. De Jong, H. J. Oterdoom, and J. R. P. Van Hoey Smith. 1994. *Maples of the World*. Timber Press, Portland, Oregon.

Webster, G. L. 1967. The genera of Euphorbiaceae in the southeastern United States. *Journal of the Arnold Arboretum* 48:303–430.

———. 1987. The saga of the spurges: a review of classification and relationships in the Euphorbiales. *Botanical Journal of the Linaean Society* 94:3–46.

———. 1994. Classification of the Euphorbiaceae. *Annals of the Missouri Botanical Garden* 81:3–32.

———. 1994. Synopsis of the genera and suprageneric taxa of the Euphorbiaceae. *Annals of the Missouri Botanical Garden* 81:33–144.

Wilson, K. A. 1960. The genera of Myrtaceae in the southeastern United States. *Journal of the Arnold Arboretum* 41:270–278.

Wood, C. E. 1975. The Balsaminaceae in the southeastern United States. *Journal of the Arnold Arboretum* 56:413–426.

Wurdack, J. J., and R. Kral. 1982. The genera of Melastomataceae in the southeastern United States. *Journal of the Arnold Arboretum* 63:429–439.

# Saxifragaceae

## The Saxifrage Family
Saxifragales ■

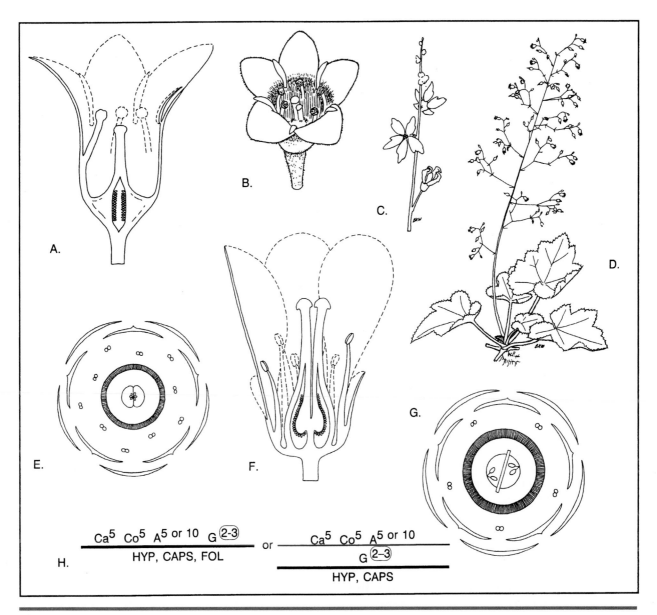

$$\overline{Ca^5\ Co^5\ A^{5\ or\ 10}\ G\ \textcircled{2-3}}$$
$$\text{HYP, CAPS, FOL}$$

or

$$\overline{\begin{array}{c} Ca^5\ Co^5\ A^{5\ or\ 10} \\ \overline{G\ \textcircled{2-3}} \end{array}}$$
$$\text{HYP, CAPS}$$

H.

**Figure 14-19** Features of Saxifragaceae. A. Flower of *Heuchera* (l.s.). B. Flower of *Heuchera*. C. Inflorescence of *Lithophragma*. D. Habit of *Tiarella*. E. Floral diagram of *Tiarella*. F. Flower of *Bergenia* (l.s.). G. Floral diagram of *Heuchera*. H. Generalized floral formulas.

# FAMILY DESCRIPTION

Herbs. Leaves alternate, opposite or all basal, simple (rarely bi- or triternate), estipulate (rarely stipulate). Plants synoecious (rarely polygamous). Inflorescences various, determinate or indeterminate, often paniculiform. Flowers perfect (rarely imperfect), regular (rarely irregular). Hypanthium present, often well developed (rarely absent). Sepals 5 (4–7), connate. Petals 5 (2–7) (0), distinct. Stamens 5–10 (4) (12, 14), distinct. Carpels 2 (3–5), distinct to connate, often fused only at the base; ovary or ovaries superior or inferior with 1–2 (3–5) locules and numerous marginal, parietal, or axile ovules; styles 2–5. Fruit a capsule or a cluster of follicles.

# FAMILY NOTES

The Saxifragaceae is a family of about 41 genera and 370 species. It has a worldwide distribution with the largest concentration of species in the northern hemisphere. It is well represented in temperate North America. Common genera include *Saxifraga* (saxifrage), *Mitella* (mitrewort), *Heuchera* (alum-root), *Tiarella* (foam-flower), and *Lithophragma* (woodland star).

The family as treated here is rather uniform in habit and vegetative features. Most are perennial herbs with usually alternate or basal leaves that lack stipules. The leaves are often palmately veined and lobed (Figure 14-19 D). There is considerable floral variation, however, particularly in the degree of connation and adnation of the carpels. The flowers are usually perfect, biseriate and regular. The perianth is clearly differentiated into sepals and petals but both the calyx and the corolla are sometimes petaloid. In some species the sepals are more conspicuous than the petals (Figure 14-19 B). The petals in several genera are toothed or lobed. The flowers often possess a conspicuous hypanthium (Figure 14-19 A, E) that in some members of the family resembles a fused corolla. The gynoecium is variable, ranging from apocarpous to fully syncarpous. The placentation may be marginal, axile or parietal, and sometimes the placentation varies from the top of the ovary to its base. The ovary position varies from superior (Figure 14-19 E) to wholly inferior; often the lower portion is adnate to the hypanthium and the apical portion is free (Figure 14-19 A).

The concept of the Saxifragaceae presented here is more restricted than that employed in many floras. In some works the families treated here as Escalloniaceae, Grossulariaceae, and Hydrangeaceae are lumped together with the Saxifragaceae *(sensu stricto)* as a broadly construed and ill-defined Saxifragaceae *(sensu lato)*. These groups all share an apparent close relationship with the Rosaceae but no consistent set of features unifies them. Some recent workers have recognized as many as 15 families in place of the "traditional" inclusive Saxifragaceae.

Table 14-1 compares 6 groups of "Saxifragaceae" that formerly were united under this name. Some botanists consider each to be a separate family. Cronquist (1981, 1988), recognized three families (Saxifragaceae, Hydrangeaceae, and Grossulariaceae) with habit being a conspicuous character. He combined the Grossularioideae and Escallonioideae in his concept of the Grossulariaceae. Evidence has been acumulating that the Grossularioideae and Escallonioideae are not closely related and we have chosen to treat them as separate families.

The saxifrages have limited economic importance. They are rarely common, yet they seem to always be present, especially in the spring floras. Various species are used as ornamentals, especially in rock-garden plantings. These include various species of *Astilbe, Saxifraga* (saxifrage), *Heuchera* (coral bells, alum-root), and *Bergenia* (bergenia).

## HERBS

### Saxifragaceae

**subf. Penthoroideae**

All stamens fertile, carpels 5–7, distinct fruits circumscissile, hypanthium absent, usually basal and leaves cauline, toothed

**subf. Saxifragoideae**

All stamens fertile carpels 2–3, distinct fruits splitting lengthwise, hypanthium usually present, leaves cauline, often toothed.

**subf. Parnassioideae**

Some stamens present as staminodia; carpels 3–4, connate; fruits splitting lengthwise; hypanthium absent; leaves basal and one cauline, all entire.

## SHRUBS

### Hydrangeaceae

(Saxifragaceae subf. Hydrangeoideae) Leaves opposite, pinnately veined, stamens 2 × petals or more, ovary at least partly inferior, fruit a capsule.

### Escalloniaceae

(Saxifragaceae subf. Escallonioideae) Leaves alternate, pinnately veined, stamens usually = petals, ovary superior or inferior, fruit a capsule.

### Grossulariaceae

(Saxifragaceae subf. Ribesoideae) Leaves alternate, palmately veined, stamens = petal, ovary inferior, fruit a berry.

**Table 14-1** Comparison of the six subfamilies formerly associated under the Saxifragaceae. We are treating the Penthoroideae, Saxifragoideae, and Parnassioideae together as Saxifragaceae. We recognize each of the shrubby subfamilies as a separate family.

# Crassulaceae

## The Stone Crop Family
### Saxifragales ■

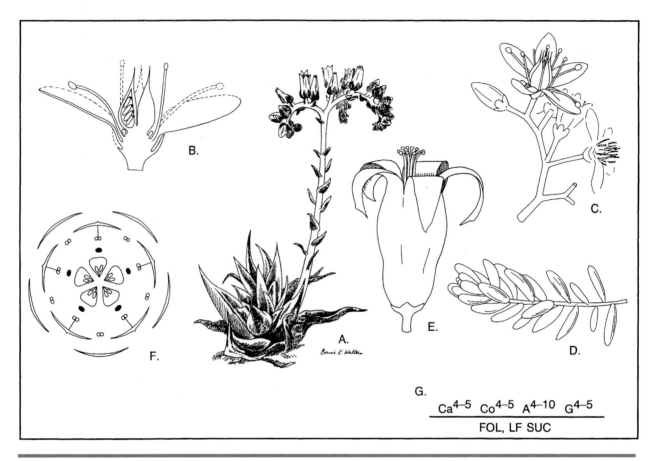

$$Ca^{4-5} \quad Co^{4-5} \quad A^{4-10} \quad G^{4-5}$$

FOL, LF SUC

**Figure 14-20** Features of Crassulaceae. A. Plant of *Dudleya lanceolata* with compound scorpioid cyme. B–D. *Sedum* sp. B. Flower (l.s.). C. Cymose inflorescence; note the distinct petals. D. Branchlet and succulent leaves. E. Flower of *Cotyledon* with a sympetalous corolla. F. Floral diagram of *Sedum*. G. Generalized floral formula.

## FAMILY DESCRIPTION

Herbs or shrubs, usually succulent. Leaves alternate or opposite (rarely whorled), simple, estipulate. Plants synoecious. Inflorescences mostly variously cymose. Flowers perfect, regular. Hypanthium absent or sometimes well developed. Sepals 4 or 5 (3–30), distinct. Petals 4 or 5 (3–30), distinct or connate. Stamens equal in number to or twice as many as the petals, distinct. Carpels 4 or 5 (3–30), distinct or weakly connate at base (rarely 4–5, connate); ovaries superior, unilocular with many (1-few) marginal ovules (rarely ovary with 4–5 locules and many axile ovules). Fruit a cluster of follicles (rarely a capsule).

# FAMILY NOTES

The Crassulaceae is a medium-sized family of 25–37 genera and 900–1200 species. The family has a nearly cosmopolitan distribution except for Australia and the islands of the Western Pacific. The family has its greatest diversity in South Africa where several genera are represented by numerous species. Members of the Crassulaceae are common in arid and semiarid regions, usually in stony soil or on rock outcrops. The largest genus, *Sedum* (stonecrop), is widespread in temperate regions and in tropical mountains. Some species occur at very high elevations or at high latitudes. One widespread annual species of *Crassula* is aquatic! *Sedum* is the most widespread genus in temperate North America. *Dudleya* is common in California and parts of Arizona. Several diminutive species of *Crassula* (pygmyweed) are widely distributed in the U.S.

The Crassulaceae is one the most easily recognized families. All have succulent leaves and stems (Figure 14-20 A, D). Some are herbs and others are shrubs, some with the habit of miniature trees. The flowers are generally arranged in cymes of one sort or another (Figure 14-20 A, B). The gynoecium is almost always apocarpous (Figure 14-20 A) with the number of carpels equalling the number of petals. In some genera the petals are distinct (Figure 14-20 B) and in others they are conspicuously connate (Figure 14-20 E). Even the diminutive annual species of *Crassula* are easily recognizable as members of this family.

Many members of the stonecrop family are grown as ornamentals for their unusual succulent foliage and their colorful flowers. They are particularly well-suited for rock-gardens. Genera with cultivated representatives include *Aeonium, Cotyledon, Crassula, Dudleya, Echeveria, Kalanchoe, Sedum,* and *Sempervivum*. Many of these can be grown outdoors in areas with mild winters, and almost all can be grown as houseplants.

Aside from their ornamental value, members of the Crassulaceae have little economic value. Extracts of some are used as folk medicines.

# Grossulariaceae

## The Gooseberry Family
### Saxifragales ■

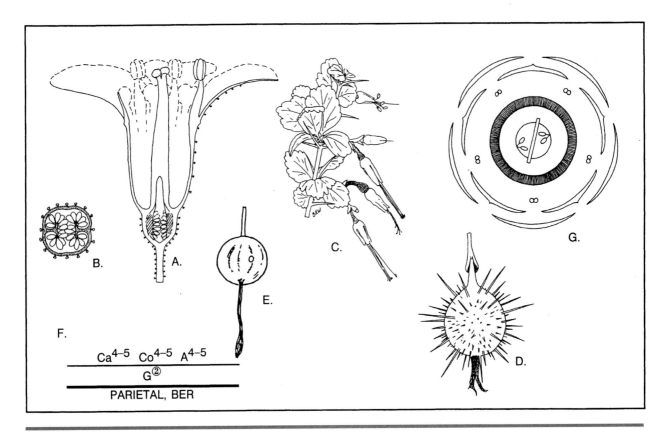

**Figure 14-21** Features of Grossulariaceae. A. Flower (l.s) of *Ribes*. B. Ovary of *Ribes* (x.s.) C. Habit of *Ribes speciosum*. C. Fruit of *Ribes roeslii*. E. Fruit of *Ribes aureum*. F. Generalized floral formula. G. Floral diagram of *Ribes*.

## FAMILY DESCRIPTION

Shrubs, sometimes armed with prickles and/or spines. Leaves alternate, simple, estipulate, petioled, palmately veined and often palmately lobed. Plants synoecious. Inflorescence of racemes, solitary flowers, or axillary clusters. Flowers perfect, regular. Hypanthium well developed, often petaloid. Sepals 4–5, distinct or basally connate. Petals 4–5, distinct. Stamens 4–5 distinct, alternate with petals. Carpels 2, connate; ovary inferior with 1 locule and many parietal ovules; styles 2 or 1 and 2-lobed. Fruit a berry.

# FAMILY NOTES

The Grossulariaceae is family of one genus and about 150 species. The family as we are treating it here consists of the genus *Ribes,* most of which occur in north temperate regions. A few species range into Central and the Andean region of South America.

As delimited by Cronquist (1981) the Grossulariaceae included the Escalloniaceae and several other smaller families that some authors treat separately. On the other hand, many floras include all of these in a very heterogeneous Saxifragaceae *(sensu lato).* We have elected to treat the Escalloniaceae as a family distinct from the Grossulariaceae.

Because of its size and distribution *Ribes* is commonly encountered by botanists in North America. Members of the genus generally have palmately veined and usually palmately lobed leaves (Figure 14-21 C). There are two principal groups of species within this genus. The currants are generally unarmed shrubs that usually bear flowers in racemes, Gooseberries are spiny and often prickly shrubs that most commonly bear solitary axillary flowers or small axillary clusters. Often even the berries are covered with prickles (Figure 14-21 D). Some authors separate the gooseberries from *Ribes* as the genus, *Grossularia.*

The edible fruits of some species of *Ribes* are grown commercially. Currants are often dried like raisins. In addition some species of *Ribes* have horticultural value, with colorful flowers and attractive foliage.

In some areas of North America this genus is economically important for a very different reason. Species of *Ribes* are the alternate hosts for the fungus that causes white pine blister rust. The fungus goes through part of its life cycle as a parasite on species of white pines and completes its life cycle as a parasite of *Ribes* spp. This Eurasian plant disease was accidentally introduced into North America where it has caused severe timber losses. In some areas foresters have attempted to extirpate species of *Ribes* to protect the valuable lumber trees. Laws have been passed forbidding the cultivation of gooseberries or currants in some regions.

# Hamamelidaceae

## Witch Hazel Family

### Saxifragales ∎

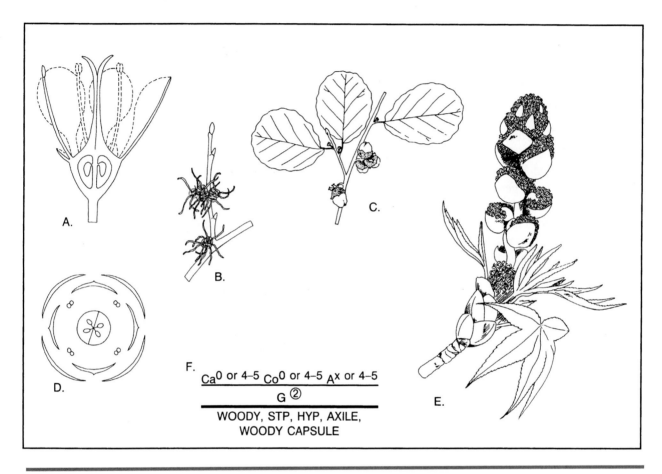

**Figure 14-22** Features of Hamamelidaceae. A. Flower of *Corylopsis* (l.s.). B. Leafless flowering twig of *Hamamelis*. C. Leafy twig of *Hamamelis* with ripening capsules. D. Floral diagram of *Hamamelis*. E. Twig of *Liquidambar* with leaves and heads of staminate and pistillate flowers. F. Generalized floral formula.

## FAMILY DESCRIPTION

Shrubs or trees. Leaves alternate, simple, stipulate. Plants synoecious, monoecious, or dioecious. Inflorescence of solitary or clustered axillary flowers, heads or spikes. Flowers perfect or imperfect, regular. Hypanthium present or absent, ± adnate to ovary. Sepals 4–5 (0), connate and usually ± adnate to ovary. Petals 0 or 4–5, distinct. Stamens 4–5 (2—many), distinct and alternate with petals. Carpels 2, basally connate; ovary superior to inferior, often half-inferior, with two locules and 1—many axile ovules; styles 2. Fruit a woody capsule (rarely a follicle when 1 carpel aborts).

# FAMILY NOTES

The Hamamelidaceae is one of the more primitive families in the Hamamelidae. It an ancient family with a long fossil record. The 26 genera and about 100 species have a highly disjunct distribution. Most of the genera are small, often monotypic, and apparently relictual. The largest concentrations of genera and species are in eastern Asia and eastern North America. Several genera have widely disjunct ranges. In both regions members of the family range from temperate regions south into the tropics. There are a few members of the family in southern and eastern Africa and in western Asia.

The flowers are quite variable, ranging from large and comparatively showy to small and inconspicuous. The family is transitional from an apparently petal-bearing ancestral condition to the largely apetalous condition predominating in the Hamamelidae. Some members of the Hamamelidacea have well-developed petals (Figures 14-22 A, B) and in others the perianth is reduced to very small sepals or is completely absent. Often the flowers are tightly grouped into dense heads (Figure 14-22 E) or spikes. The larger, more showy flowers have features in common with some members of the magnoliiobe and with the basal members of the Rosidae. The members of the Hamamelidaceae show a transition from ancestral animal pollination to the wind pollination common in the more advanced members of the Hamamelidae.

The gynoecium in the Hamamelidaceae is generally 2-carpellate. The bases of the carpels are connate and the tips are usually separate. The ovary is often half inferior. At maturity it ripens as a woody capsule (Figure 14-22 C) containing two to many seeds. In *Liquidambar* the pistillate flowers are aggregated into a globose head with the beaks of the capsules spreading in all directions.

The sweet gums *(Liquidambar)* produce a fragrant resin and are often grown as street trees. They are particularly handsome in autumn when the leaves change from green to brilliant red or golden. Members of various other genera are sometimes cultivated including species of *Corylopsis* and *Hamamelis,* both of which have the common name witch-hazel. Species of both genera have the unusual attribute of flowering in the autumn or winter months, sometimes while snow is on the ground. An extract of the dried leaves and twigs of *Hamamelis virginiana* is used to produce a soothing lotion. Other ornamentals include species of *Exbucklandia, Fothergilla* (witch-alder), *Loropetalum* and *Parrotia* (ironwood). The wood of several genera is used for lumber; these include *Altingia, Liquidambar, Parrotia* and *Symingtonia.*

# Geraniaceae

## The Geranium Family
### Geraniales ▪

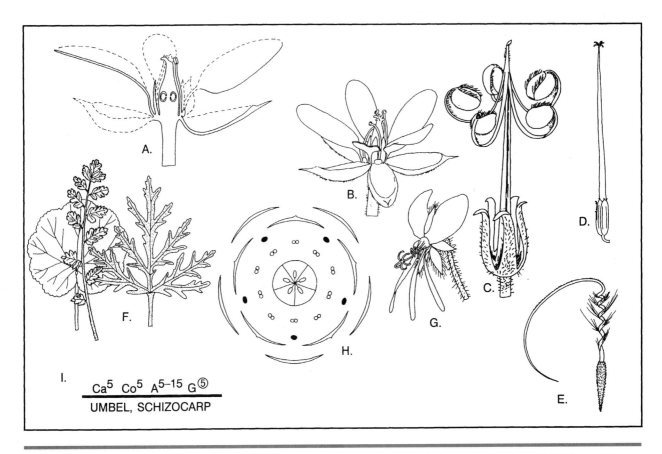

**Figure 14-23** Features of Geraniaceae. A. Flower of *Erodium* (l.s.). B. Intact radial flower of *Erodium*. C. Mature schizocarp of *Geranium* with mericarps separating from floral axis. D. Immature schizocarp of *Erodium* with elongated beak. E. Mericarp of *Erodium* with spirally coiled beak. F. Leaves of *Erodium* and *Pelargonium*. G. Bilateral flower of *Pelargonium* sp. H. Floral diagram of *Erodium*. I. Generalized floral formula.

## FAMILY DESCRIPTION

Herbs or shrubs, sometimes weak-stemmed and vine-like (rarely stem succulents or trees). Leaves alternate or opposite, simple to pinnately or palmately dissected or compound, stipulate (less commonly estipulate). Plants synoecious. Inflorescence of cymose umbels (rarely of solitary flowers or compound cymes). Flowers perfect, regular or irregular. Sepals 5 (4 or 8), distinct or basally connate, sometimes the uppermost spurred with the spur adnate to the pedicel. Petals 5 (0) (2, 4, or 8), distinct. Stamens 5–15 (2–4) distinct or filaments basally connate (rarely connate in 5 bundles of 3 each); staminodes sometimes present. Carpels 5 (3–8), connate; ovary superior with 5 (3–8) locules and 1–2 (many) axile ovules per locule; styles 5 (3–8) ± connate, often maturing as an elongated beak. Fruit a schizocarp, schizocarpic capsule or capsule.

# FAMILY NOTES

The Geraniaceae comprises 11 genera and about 700 species. It is widespread in temperate regions, particularly in the Old World.

Most members of the Geraniaceae have a very distinctive gynoecium. It is usually 5-carpellate and 5-lobed and is terminated by an elongated style with 5 distinct stigmas (Figures 14-23 B, G). The style matures as a stiff beak, sometimes several centimeters long. The ovary has one or two ovules in each locule and matures as a schizocarp. The mericarps separate initially at the base of the ovary and the split soon extends to the tip of the beak (Figure 14-23 C). The mericarps may retain the enclosed seeds or may split open at maturity. The segments of the beak become spirally coiled as they dry. They are hygroscopic and may coil and uncoil several times, in some cases forcing the tip of the mericarp into the soil, effectively planting the seeds. The beaked mericarps of *Erodium* spp. (filaree) may also cling to animals.

Several members of the Geraniaceae are cultivated as ornamentals for their flowers and foliage. These include numerous species and hybrids of *Pelargonium* (garden geranium) plus species of *Geranium* (wild geranium) and *Erodium*. Geranium oil, used in the perfume industry, is extracted from the herbage of *Pelargonium* species.

Various species of *Geranium* are native to North America. Some form colorful wildflower displays, especially in the mountainous regions of the western U.S. *Erodium cicutarium* (redstem filaree), a common weedy annual from Eurasia, has become widely naturalized in North America, especially in western range lands. Although it is palatable to cattle when it is young, its dried mericarps can cause injury to livestock. Several other species of *Erodium* are naturalized in California.

# Myrtaceae

## The Myrtle Family
### Myrtales ■

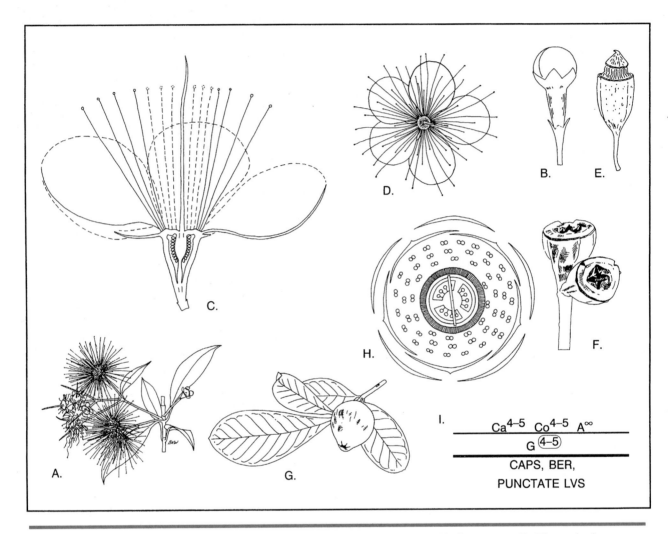

**Figure 14-24** Features of Myrtaceae. A–C. *Syzygium paniculatum*. A. Branchlet and inflorescence. B. Flower bud. C. Flower (l.s.). D. Flower of *Myrtus*. E–F. *Eucalyptus*. E. Bud with sepals and petals fused into a conical cap that is forced off as the stamens enlarge. F. Woody capsules. G. Branchlet and berry of *Psidium cattleianum*. H. Floral diagram of *Callistemon*. I. Generalized floral formula.

## FAMILY DESCRIPTION

Shrubs or trees. Leaves alternate, opposite or whorled, simple and entire, estipulate, punctate with oil glands and usually strongly scented. Plants synoecious. Inflorescences of simple or compound, sometimes paniculiform cymes, or of spikes, racemes, or solitary flowers. Flowers perfect, regu-lar. Hypanthium present, usually adnate to ovary. Sepals 4 or 5 (0) distinct or connate. Petals 4 or 5, distinct, or apparently absent (actually wholly connate and ± adnate to sepals forming a conical lid that falls at anthesis). Stamens (10) to many, distinct or filaments basally connate (sometimes connate in bundles opposite the petals). Carpels 2–6 (many), wholly connate; ovary inferior (very rarely superior) with 2–6 (many) locules and axile placentation or 1 locule and parietal placentation, ovules 1–many, style 1 undivided. Fruit a berry or capsule (rarely a drupe or nut).

# FAMILY NOTES

The Myrtaceae is a moderately large family of about 140 genera and 3000 species. The family is widely distributed, particularly in the tropics and subtropics of the southern hemisphere. They are especially well-represented in South America and the Australian region. The Myrtaceae is absent (except in cultivation) in temperate North America.

There are three subfamilies. The Leptospermoideae which are best represented in Australia usually have a 2–5 locular ovary that matures as a capsule (Figure 14-24 F). Leaves in this subfamily are often alternate. In tropical America and various other regions the subfamily Myrtoideae is common. These plants have mostly 2-carpellate ovaries that mature as berries (Figure 14-24 G) or drupes. Leaves in this subfamily are mostly opposite. The ovary in subf. Chamaelaucoideae matures as a dry, one-seeded nut.

The glands that dot the leaves and other plant parts in the Myrtaceae are embedded in the tissues of the leaf and are most readily visible by transmitted light. The oils are often strongly scented. In the areas where the plants are native there are often parasites and herbivores that have evolved the ability to attack the leaves, stems or flowers. However, in areas such as the United States which are outside the natural range of the Myrtaceae, the plants are generally pest-free. The oils deter most herbivores and sometimes also inhibit the growth of other plants as well.

The flowers usually have a well-developed hypanthium that is often lined with nectar-secreting tissue. The petals are relatively small and usually nearly circular in outline (Figure 14-24 A, D). In many species the numerous stamens are the most prominent feature of the flowers (Figure 14-24 A). In the large genus, *Eucalyptus,* the sepals and petals are wholly adnate into a cap that is pushed off as the stamens expand (Figure 14-24 E). Carpels in the Myrtaceae are wholly connate and the inferior ovary is tipped by a single style (Figure 14-24 C).

Most members of the Myrtaceae are frost-tender, but many are grown in areas with mild winters. Commonly cultivated ornamentals include *Myrtus communis* (myrtle) and species of *Eucalyptus* (gum), *Callistemon* and *Melaleuca* (bottle-brushes), *Syzygium* and *Leptospermum* (tea-tree). Species such as *Psidium guavaja* (guava), *P. cattleianum* (strawberry-guava, Figure 14-24 G), *Feijoa sellowiana* (pineapple guava), and species of *Eugenia* and *Syzygium* produce edible fruits. *Syzygium aromaticum* (cloves) and *Pimenta dioica* (allspice) are important spice-plants. Oils extracted from the foliage of various Myrtaceae, especially *Eucalyptus* and *Melaleuca,* have medicinal uses. Species of *Eucalyptus, Metrosideros,* and *Angophora* are important timber trees. Because of their rapid growth and disease resistance, *Eucalyptus* trees have been planted for lumber and wood pulp in many tropical and subtropical areas.

# Onagraceae

## The Evening Primrose Family
### Myrtales ■

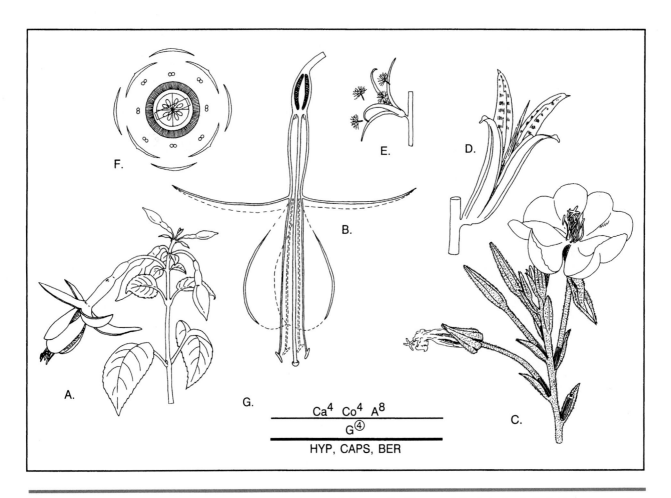

$$\frac{Ca^4 \ Co^4 \ A^8}{G^{\textcircled{4}}}$$

HYP, CAPS, BER

**Figure 14-25** Features of Onagraceae. A. Branch of *Fuchsia* with flower. B. Flower of *Fuchsia* (l.s.). C. Inflorescence of *Oenothera hookeri*. D. Capsule of *Oenothera hookeri*. E. Capsule of *Epilobium* sp. with comose seeds. F. Floral diagram of *Oenothera*. G. Generalized floral formula.

## FAMILY DESCRIPTION

Herbs or shrubs (trees). Leaves alternate or opposite, simple (rarely pinnate), stipulate or estipulate. Plants synoecious. Inflorescence of solitary axillary flowers, spikes, racemes or panicles. Flowers perfect, regular (sometimes irregular). Hypanthium present, adnate to ovary. Sepals 4–5 (2–3) (6), distinct or connate, sometimes petaloid. Petals 4–5 (0) (2–3) (6), distinct. Stamens 4–10 (1–2), distinct. Carpels 4 (2)(5) connate; ovary inferior with 4 (2) (5) locules with 1–many axile ovules per locule; style 1, stigmas branched or unbranched. Fruit a capsule, achene, nut or berry.

# FAMILY NOTES

The Onagraceae is a cosmopolitan family that is especially well represented in western North America where species occur from the coastal lowlands to the desert interior and high mountains. It is a moderate sized family with about 17 genera and 675 species. Members of the Onagraceae occur in a wide variety of terrestrial and aquatic habitats.

In North America most Onagraceae are herbs with a 4-merous perianth, 8 stamens, and an inferior ovary. Some have a well-developed hypanthium that much exceeds the ovary (Figure 14-25 B). In others the hypanthium is very short or absent. The family includes many showy wildflowers. Although most members of the family have 4-merous flowers, one genus, *Ludwigia* has 5-merous flowers and another, *Circaea,* has 2-merous flowers. The flowers of Onagraceae are mostly radially symmetric, but some species of *Clarkia* and *Lopezia* have bilateral flowers.

The common name, evening-primrose, describes the time of day when the flowers of *Oenothera* (Figure 14-25 C) open. These flowers are visited by night-flying moths. The flowers of most other members of the family are open in the daytime. Most are insect pollinated, but some species of *Epilobium* and *Fuchsia* are hummingbird-pollinated.

There is considerable variability in the fruits of the family. Most have capsules (Figure 14-25 D, E). In *Epilobium* (Figure 14-25 E) the capsule contains numerous tiny seeds, each with a tuft of hair. The hairs are like parachutes, carrying the seeds aloft in the winds. The wind-borne seeds of *Epilobium angustifolium* (fireweed) enable this species to invade areas burned over by forest fires. In the years after a fire, colorful masses of fireweed grow profusely over the burned landscape. The fruit of *Fuchsia* is a berry. Species of *Gaura* produce small nuts. The fruit of *Circaea* is a small achene or nutlet covered with hooked hairs that cling readily to fur or feathers.

With the exception of a few ornamental species, the Onagraceae has little economic importance. Common ornamentals include members of the genera *Oenothera* (evening-primrose), *Epilobium* (fireweed, California-fuchsia), *Clarkia,* and *Fuchsia.* A few species are rather weedy (e.g., some species of *Gaura* and *Oenothera*).

# Euphorbiaceae

## The Spurge Family

Malpighiales ■

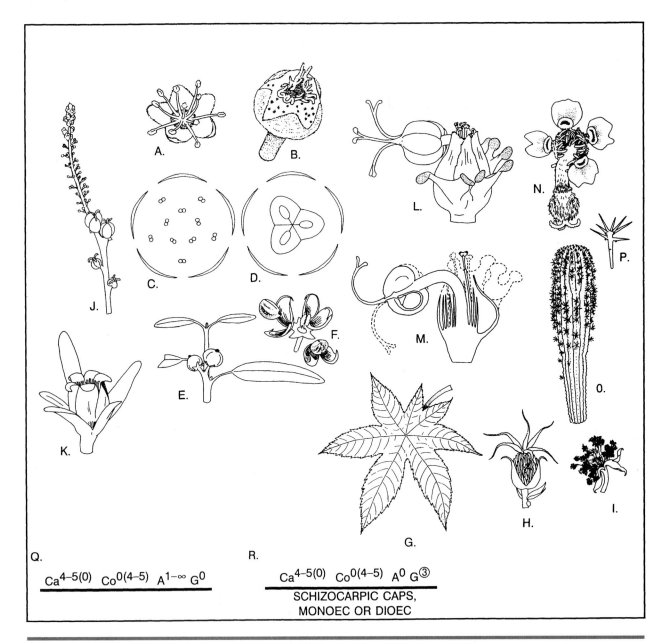

$$Ca^{4-5(0)} \quad Co^{0(4-5)} \quad A^{1-\infty} \, G^0$$

$$Ca^{4-5(0)} \quad Co^{0(4-5)} \quad A^0 \, G^{\textcircled{3}}$$

SCHIZOCARPIC CAPS,
MONOEC OR DIOEC

**Figure 14-26** Features of Euphorbiaceae. A–F. *Croton californicus*. A–B. Apetalous staminate and pistillate flowers. C–D. Floral diagrams of staminate and pistillate flowers. E. Branchlet with intact schizocarpic capsules. F. Dehiscing schizocarpic capsule. G. Peltate leaf of *Ricinus communis*. H–I. Apetalous staminate and pistillate flowers of *Ricinus*. Note fascicled stamens and branched stigmas. J. Inflorescence of *Stillingia paucidentata*. K. Sympetalous pistillate flower of *Jatropha cinerea*. L–M. Cyathium of *Euphorbia lathyris* (intact and l.s.). N. Cyathium of *Euphorbia marginata* with developing fruit exserted on pedicel; note petaloid appendages on glands. O. Succulent stem of *Euphorbia* sp. P. Close-up of branched thorn from succulent *Euphorbia*. Q–R. Generalized floral formulas for staminate and pistillate flowers.

# FAMILY DESCRIPTION

Herbs, shrubs, stem succulents, or trees, often with milky sap, sometimes climbing or twining. Leaves alternate or opposite (whorled), simple (rarely palmate), stipulate (less commonly estipulate). Plants monoecious or dioecious. Inflorescences variously determinate or indeterminate, sometimes very condensed forming a cyathium. Flowers imperfect, regular or naked. Sepals 0–8, mostly commonly 5, distinct or ± connate. Petals 0 (4–8), distinct or basally connate, free or adnate at base to stamens. Stamens 1–many, distinct or variously connate. Carpels 3 (1–20), connate; ovary superior, with 3 (1–20) locules and 1 or 2 apical-axile ovules per locule; styles 3 (1–20), often forked. Fruit a schizocarpic capsule (rarely a drupe, berry, legume-like pod, or samara).

# FAMILY NOTES

The Euphorbiaceae is a large (310–320 genera, 7500 species) pan-tropical family that is only sparingly represented in temperate areas. Two centers of distribution for the family are tropical America and tropical Africa. The family is one of extremely variable habit including annual and perennial herbs, vines, succulents, shrubs, and tall tropical trees. Most temperate species are herbaceous. Members of the Euphorbiaceae occur in many habitats from deserts to rain forests. The family is absent from regions with severely cold winters.

Genera of the Euphorbiaceae have flowers of varied forms. Relatively few members of the family have corollas. In those with a corolla, the petals may be small and inconspicuous (e.g., *Ditaxis*) or large and showy (e.g., some species of *Jatropha;* Figure 14-26 K); the petals may be distinct or connate. Most Euphorbiaceae have apetalous flowers and some lack sepals as well. Some (e.g., species of *Acalypha*) have showy stamens and stigmas and are insect pollinated. In many others the flowers are very inconspicuous and wind-pollinated.

The flowers are characteristically imperfect; some species are monoecious and others are dioecious. The staminate flowers may have as few as one stamen or as many as 400. The pistillate flowers most commonly have a superior, 3-carpellate and usually 3-lobed ovary (Figures 14-26 E, J, L). The styles or stigmas are often dichotomously forked one to several times (Figures 14-26 B, H, J). The ovary matures most commonly as a schizocarpic capsule (Figure 14-26 F); it breaks apart into mericarps that in turn split open releasing the one or two seeds inside. In many cases the fruit shatters elastically, throwing out the seeds.

Inflorescences are also extremely varied. Flowers in some species are solitary or in small to large cymose clusters. The primitive pattern of inflorescence development in the family is apparently cymose, but the inflorescences have been so highly modified in some genera that they no longer appear to develop in a cymose fashion. Various members of the family have spikes, racemes, or panicle-like clusters.

The large genus *Euphorbia* and its close relatives have a unique, highly modified inflorescence, called a **cyathium** (Figures 14-26 L–N). The outer portion of this inflorescence consists of a cuplike involucre of 4–5 connate bracts. Only the tips of the bracts are distinct. Each bract bears a pair of glands (highly modified stipules) that secrete nectar. Glands from adjacent bracts are usually fused together. In some species the glands bear flattened **petaloid appendages** (Figure 14-26 N) that strongly resemble small petals in both color and texture. Within each cyathium are several to many tiny staminate flowers and a solitary pistillate flower. The staminate flowers are aggregated together into 4–5 minute cymes, each subtended by a thin, membranous bractlet. Additional bractlets subtend the individual flowers.

The flowers within the cyathium are highly reduced. Each staminate flower consists of a solitary stamen on top of a short pedicel. The receptacle is represented by a constriction where the pedicel and filament join together. In most species the calyx is entirely absent. [A synsepalous calyx is present in two genera related to *Euphorbia*.] The pistillate flower is also generally naked, although in a few *Euphorbia* species and in several related genera a few small sepals may be present. Generally the flower consists of a 3-lobed, 3-carpellate ovary on a pedicel. The receptacle is generally a slight swelling at the apex of the pedicel. In many species the elongation of the pedicel of the pistillate flower gradually pushes the ovary out of the involucre (Figure 14-26 L–M).

The entire cyathium with its calyx-like involucre, its petal-like glands and highly reduced flowers may be only a few millimeters in diameter. It strongly resembles a small perfect flower and it is often described as a pseudanthium (false-flower). Those species with petaloid appendages on the cyathium glands are remarkably flower-like in appearance and function. However, a close examination of a cyathium reveals its tree nature. Cyathia in most genera are more or less radial, but in *Pedilanthus* the cyathia are strongly bilateral.

*Euphorbia* is an enormous genus (over 1000 species) and includes plants of extremely diverse habit. Many species are herbaceous, some annual and others perennial. Other species of *Euphorbia* are shrubs, including the poinsettia *(E. pulcherrima)* so familiar at the Christmas season. In the African deserts succulent species of *Euphorbia,* together with ice-plants (Aizoaceae), milkweeds (Asclepiadaceae) and stonecrops (Crassulaceae), occupy ecological niches similar to those occupied by the cacti (Cactaceae) in the New World tropics. Some of the succulent species are remarkable in their resemblance to the columnar cacti (Figure 13-15 O). The succulent Euphorbias are often armed with simple or branched thorns (Figure 14-26 P) or stipular spines. [Cacti have spines that originate from areoles (Figure 13-15 F).] Succulent *Euphorbia* species also occur in New World deserts, though these plants are not cactus-like in their appearance. Yet all of them have cyathia.

Recent floras differ in their treatment of *Euphorbia* and its close relatives. Some authors treat all or most of these plants as a broadly defined *Euphorbia* with several subgenera and sections. Gleason and Cronquist (1991), for instance, considered all the cyathium-bearing plants in the northeastern United States to be species of *Euphorbia.* Other botanists consider at least some of these groups to be distinct (though closely related) genera. Koutnik (1993), for instance, who treated *Euphorbia* for *The Jepson Manual* (the new flora of California) considered *Chamaesyce* to be a genus distinct from *Euphorbia.* Because all have cyathia, the subgenera or segregate genera are separated primarily on vegetative features and details of cyathium structure.

Members of the spurge family produce various chemicals. Members of the subfamilies Crotonoideae and Euphorbioideae produce latex, although clear sap predominates in the remaining three subfamilies. Rubber is produced from the latex of *Hevea brasiliensis.* The latex of some species is caustic and can cause skin irritation. Merely touching the trunk of a manchineel tree (*Hippomane* spp.) can cause severe contact dermatitis. Castor oil is extracted from the seeds of *Ricinus communis* (castor-bean). Tung-oil, used in various industries, is produced by species of *Aleurites.* Several genera produce waxes used in candle making. Most members of the family contain alkaloids of varying degrees of toxicity. Stinging hairs have evolved in two different lineages of the Euphorbiaceae. The fierce stings of *Cnidoscolus* are an equal to those of any true nettle (Urticaceae). *Manihot esculenta* (cassava, tapioca), cultivated as a source of starch in tropical regions, must have poisonous compounds removed before it is edible.

Various members of the Euphorbiaceae have ornamental value. Among them are species of *Acalypha, Codiaeum, Croton, Euphorbia, Jatropha, Phyllanthus, Ricinus,* and *Sapium.* Many African species of *Euphorbia* are grown in succulent gardens.

# Clusiaceae (Guttiferae)

## The Mangosteen Family
### Malpighiales ■

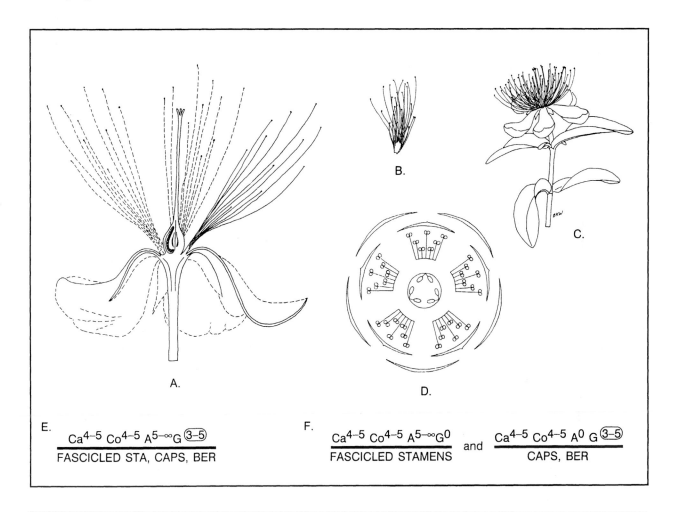

**Figure 14-27** Features of the Clusiaceae. A. Flower of *Hypericum* (l.s.). B. Stamen cluster from flower of *Hypericum*. C. Twig and flower of *Hypericum*. D. Floral diagram of *Hypericum*. E. Generalized floral formula for subf. Hypericoideae. F. Generalized floral formulas for subf. Clusioideae.

## FAMILY DESCRIPTION

Trees, shrubs, lianas or herbs. Leaves, opposite or whorled, simple and usually entire, estipulate and often glandular-punctate. Plants monoecious, dioecious, or synoecious. Inflorescence of solitary flowers or variously cymose. Flowers perfect or imperfect, regular. Sepals 4 or 5 distinct. Petals 4–6 (2–14) distinct or weakly connate at base. Stamens 9–many, centrifugal, distinct or connate into bundles opposite the petals. Carpels (1–) 3–5 (–20+) connate; ovary superior with one locule and many parietal ovules or 3–5 (–many) locules and many axile ovules; styles 3–5 (–many) or 1 and 3–5 (–many) lobed. Fruit a berry, drupe or capsule.

# FAMILY NOTES

The Clusiaceae, widespread in the tropics and in north temperate regions, comprises 47–50 genera and about 1200 species. This distinctive family can be divided into two rather easily recognized subfamilies. Members of the **Clusioideae** usually have imperfect flowers, non-punctate leaves, and arillate seeds, and are distributed in the tropics. The pistillate flowers often have stammodes. Members of the **Hypericoideae** have perfect flowers (Figure 14-27 A, D, E), punctate leaves, non-arillate seeds, and many are distributed in temperate regions. The latter is often treated as a separate family, the **Hypericaceae.** In both subfamilies stamens are often connate in bundles (Figure 14-27 B, E) that are inserted opposite the petals. In the United States one would expect to encounter only members of the Hypericoideae except in a large botanical garden.

Members of the Clusiaceae are economically important in various ways. The large genus *Hypericum* (St. John's-wort) includes a number of showy flowering herbs and shrubs. On the other hand, *Hypericum perforatum* (Klamath weed) and several other species are invasive weeds. Tropical members produce lumber *(Cratoxylum, Mesua),* drugs, dyes or gums *(Calophyllum, Garcinia, Clusia)* and edible fruits *Garcinia mangostana* (mangosteen) and *Mammea americana* (mammey apple).

# Violaceae

## The Violet Family
### Malpighiales ■

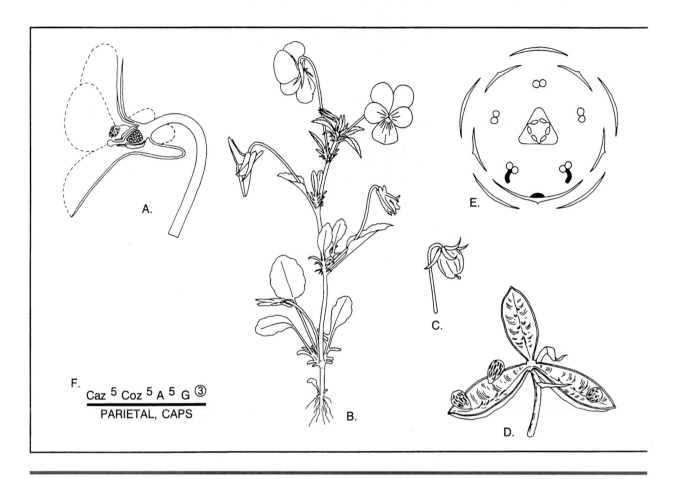

**Figure 14-28** Features of Violaceae. A. Flower of *Viola* (l.s.). B. Habit of *Viola*. C. Immature capsule of *Viola*. D. Mature capsule of *Viola*. E. Floral diagram of *Viola*. F. Generalized floral formula.

## FAMILY DESCRIPTION

Herbs or shrubs (rarely trees, woody vines). Leaves alternate, simple (sometimes divided), stipulate. Plants synoecious. Inflorescence of solitary flowers, racemes, panicles or cymes. Flowers perfect, regular or irregular, sometimes cleistogamous [never opening but setting seed]. Sepals 5, distinct. Petals 5, distinct, the lowermost often spurred. Stamens 5, distinct or connate around the ovary, sometimes the lower two with nectar spurs. Carpels 3, connate; ovary superior with 1 locule and 3–many parietal ovules; style 1. Fruit a capsule, berry, nut or samara.

# FAMILY NOTES

The Violaceae is a nearly cosmopolitan family of 16–20 genera and over 800 species. Botanists and students of temperate regions, however, are usually familiar with only the herbaceous genera, *Viola* (violets, pansies) and *Hybanthus* (green violet). The family has considerable diversity in the tropics, including shrubs, woody vines and even small trees. Although the flowers of *Viola* are strongly bilateral (Figures 14-28 A, B), those of some genera are regular or nearly so.

Stamens of the violet family have very short filaments and are tightly appressed against the ovary (but not fused to it). Some or all of the anthers bear gland-like or spur-like nectaries. In *Viola* only the two lower stamens bear nectary spurs and these are inserted into a spur or pouch in the lowest petal.

Some flowers of *Viola* are **chasmogamous** with well-developed petals; these flowers are usually formed in the spring months and are actively visited by pollinating insects. Others, often produced in summer, are **cleistogamous.** These flowers are small and apetalous; they never open and are self-pollinating.

Except for species of *Viola,* members of the violet family have little economic importance. Various species of *Viola* are cultivated as ornamentals, including the garden pansy (*V.* × *wittrockiana*) and the sweet violet Viola odorata *(V. odorata)* which is a source for scents used in perfumery. Extracts from the roots of several tropical genera are used to a limited extent as medicines.

# Salicaceae

## The Willow Family
## Malpighiales ■

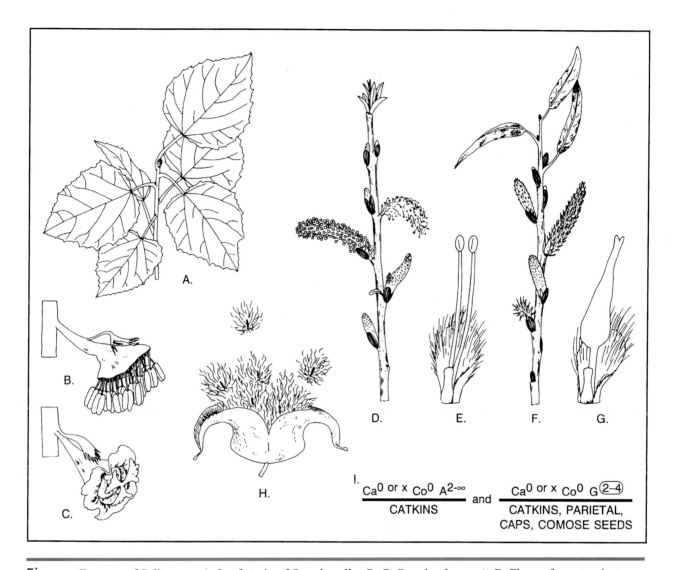

**Figure 14-29** Features of Salicaceae. A. Leafy twig of *Populus alba*. B–C. *Populus fremontii*. B. Flower from staminate catkin. C. Flower from pistillate catkin. D–G. *Salix lasiolepis*. D. Branch with staminate catkins. E. Staminate flower and subtending bractlet. F. Branch with pistillate catkins. G. Pistillate flower. H. Capsule of *Salix*. I. Generalized floral formulas.

## FAMILY DESCRIPTION

Shrubs or trees (sometimes dwarfed). Leaves alternate, simple, stipulate. Plants dioecious. Inflorescence of spicate or racemose catkins for both staminate and pistillate flowers. Flowers imperfect, subtended by bractlets. Sepals absent or wholly connate and reduced to a cupular disc. Petals absent. Stamens 2–many, distinct or filaments ± connate. Carpels 2–4, connate; ovary superior with numerous parietal ovules; styles distinct. Fruit a capsule with comose seeds.

# FAMILY NOTES

The Salicaceae consists of only two genera, *Salix* (willow) and *Populus* (aspen, poplar, and cottonwood), and approximately 350 species. The family is mostly north temperate in its distribution with a few species extending into the tropics and south temperate latitudes in South America and Africa. It is absent from the Australian region. All are dioecious trees or shrubs with imperfect flowers borne in catkins. Some of the willows of arctic and alpine areas are miniature shrubs only a few centimeters tall.

In some earlier classification systems (e.g., Engler and Prantl), the Salicaceae were placed into the Amentiferae, a group characterized primarily by catkins. In Cronquist's classification the catkin-bearers have been assigned to several different subclasses with the greatest number in the Hamamelidae. The catkins of the Salicaceae (Figures 14-29 D, F) have apparently evolved independently of those in members of the Hamamelidae. The structure of the gynoecium of the Salicaceae with parietal placentation and numerous ovules, along with several other characters, indicate that the family's affinities are with the Dilleniidae.

Flowers in the Salicaceae have a very reduced perianth. Staminate flowers of *Populus* have a disk-like calyx to which are attached numerous dangling stamens (Figure 14-29 B). The pistillate flowers have a similar calyx at the base of a 3–4 carpellate superior ovary (Figure 14-29 C). These plants are wind-pollinated. The calyx of both staminate and pistillate flowers of *Salix* (Figures 14-29 E, G) is reduced to 1–4 small nectary glands. Although they bear their flowers in catkins, many willows are insect-pollinated. In both *Salix* and *Populus* the ovary matures as a capsule (Figure 14-29 H). The seeds are wind dispersed; each bears a coma of soft hairs that can carry the tiny seeds for long distances.

The Salicaceae has limited economic importance. The wood is soft and is used primarily for pulp, boxes and match-stems. A few species, such as *Salix babylonica* (weeping willow), *Populus alba* (white poplar), *P. nigra* (black poplar, Lombardy poplar) and *P. tremuloides* (quaking aspen), are grown as shade trees or borders. Although they are fast growing, their value is lessened because they are short-lived, the wood is soft and easily broken, and branches and twigs often fall during even moderate winds. *Salix* and *Populus* are ecologically significant as components of plant communities throughout the northern hemisphere. Willows and cottonwoods are common along streams and in other wetlands where they provide habitats for a wide variety of wildlife.

# Fabaceae (Leguminosae)

## The Bean, Pea, Legume Family
### Fabales ▪

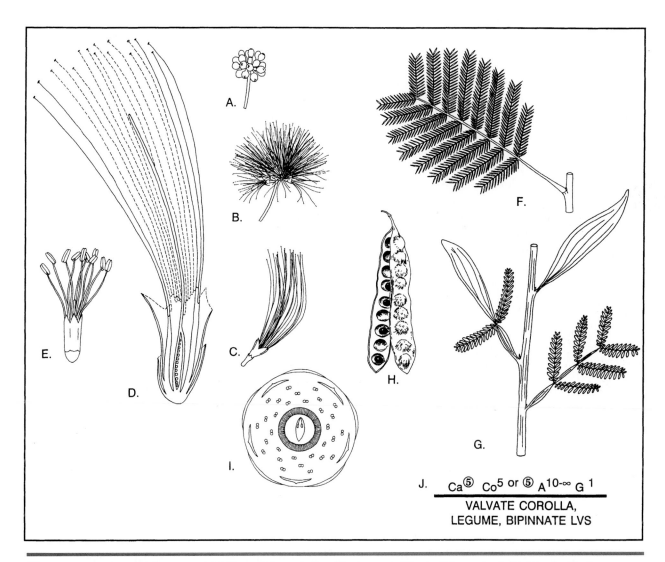

**Figure 14-30** Features of Fabaceae subf. Mimosoideae. A–D. *Albizia* sp. A. Head of flower buds. B. Head with flowers open. C. Intact flower. D. Flower (l.s.). E. Flower of *Prosopis*. F. Twig of *Acacia dealbata* with bipinnately compound leaf. G. Leafy twig of *Acacia melanoxylon* showing the transition from bipinnately compound leaves to phyllodes common in *Acacia*. H. Legume of *Albizia*. I. Floral diagram of *Acacia*. J. Generalized floral formula.

## FAMILY DESCRIPTION

Herbs, shrubs, woody vines or trees. Leaves alternate (rarely opposite), most commonly pinnate or bipinnate, less frequently simple or palmate, stipulate (or apparently estipulate). Plants synoecious (rarely monoecious or dioecious). Inflorescence of solitary or clustered axillary flowers, racemes, spikes, heads or umbels. Flowers perfect (rarely imperfect), regular to very irregular. Hypanthium present or absent. Sepals 5, connate. Petals 5 (0–1), distinct (connate in some genera with regular flowers) *or* lower 2 petals with bases distract and blades connate. Stamens 1–many, distinct or ± connate, often 10 and all connate or 0 connate and 1 distinct.

Carpel 1 (very rarely 2–16); ovary superior with 1 locule and 1–many marginal ovules (rarely 2-locular by intrusion of the placenta). Fruit a legume, loment or indehiscent pod (occasionally a samara).

# FAMILY NOTES

The legumes form the third largest family of flowering plants (after the Orchidaceae and Asteraceae) with 690–800 genera and 14,000–20,000 species. The Fabaceae is an ancient family with a fossil record that extends back into the Cretaceous Period. Members of the family occur in many habitats from deserts to rain forests, from the tropics to polar regions and from sea level to high mountains. The Fabaceae comprises three subfamilies (Table 14-2) that some workers (e.g., Cronquist, 1981) consider to be separate families. All three subfamilies are of extremely diverse habit, including herbs, shrubs and trees. Most members of subfamilies Mimosoideae (2000–3000 species) and Caesalpinioideae (2500–3000 species) occur in tropical or subtropical latitudes. Species of subf. Papilionoideae are common both in the tropics and in temperate areas. The Papilionoideae is by far the largest of the subfamilies with 10,000–15,000 species. Flowers in subf. Mimosoideae are usually densely clustered in spikes or heads (Figures 14-30 A, D). In most mimosoid flowers the stamens are much more conspicuous than are the petals (Figures 14-30 B, C). The ovary is often very small at the time of pollination. Commonly the ovaries of only a few of the flowers of an inflorescence mature as fruits.

Flowers in subf. Caesalpinioideae most commonly are moderately to strongly zygomorphic. The upper petal is typically different in shape and/or color from the two sets of lateral petals. The upper pair of lateral petals are outermost in bud (Figures 14-31 B, D). Their margins overlap those of the upper petal and the lower pair of lateral petals. The two lower petals are usually distinct, but in a few genera (e.g., *Cercis*, Figure 14-31 C) they form a keel.

Members of subf. Papilionoideae have a very characteristic floral structure (Figures 14-32 F–I). The term **papilionaceous** is often used to describe these distinctive flowers. The upper petal of the flower, the **banner** or **standard** is distinctly different in shape and size from the other petals. It is the outermost petal of the flower; both in bud and in open flowers, its margins overlap those of the spreading upper pair of lateral petals, the **wings**. The lower two lateral petals, the **keel petals**, are generally distinct at the base but are connate toward

| subf. Mimosoideae (Mimosaceae; Figure14-23) | subf. Caesalpinioideae (Caesalpiniaecae; Figure 14-24) | subf. Papilionoideae (Papilionaceae; Fabaceae sensu stricto; Figure 14-25) |
|---|---|---|
| Leaves usually bipinnate, sometimes reduced to simple phylloes; flowers regular; petals distinct or connate, valvate (inserted edge to edge in bud), stamens usually 10–many, distinct or connate; | Leaves pinnate or bipinnate, rarely simple; flowers regular or weakly to strongly irregular, petals distinct, overlapping in bud with the upper petal innermost in bud; all petals distinct; stamens 10 or fewer, distinct or connate, | Leaves pinnate or palmate, occasionally simple, never bipinnate; Flowers very irregular, petals distinct at base, overlapping in bud, with the upper petal outermost and the lower 2 petals connate above, forming a keel; stamens 10, 1 distinct and 9 with connate filaments or all 10 with connate filaments. |

Table
14-2  Comparison of the subfamilies of the Fabaceae.

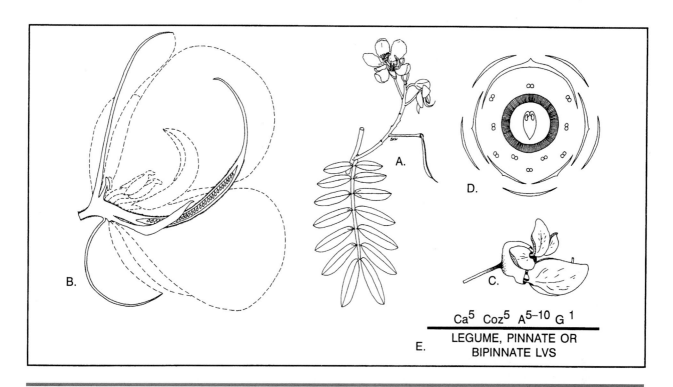

$$Ca^5 \ Coz^5 \ A^{5-10} \ G^1$$

LEGUME, PINNATE OR
BIPINNATE LVS

**Figure 14-31** Features of Fabaceae subf. Caesalpinioideae. A. Branch and inflorescence of *Senna tomentosa*. B. Flower of *Senna (l.s.)*. C. Flower of *Cercis occidentalis*. D. Floral diagram of *Senna*. E. Generalized floral formula.

the apex, forming a V-shaped **keel** that encloses the stamens and the carpel. The 10 stamens are usually either diadelphous (Figure 14-32 J) with 1 stamen distinct and 9 connate into a U-shaped tube around the ovary or monadelphous (Figure 14-32 K), with all 10 filaments connate.

The characteristic fruit of the pea family is the legume or pod. This one-carpellate fruit typically dehisces along two seams, splitting into two segments, releasing the enclosed seeds (Figure 14-32 M). Within the Fabaceae various modifications of the legume have evolved, often independently in different lineages. In some members of the family the pod is indehiscent and serves as the dispersal unit. Some indehiscent pods are single-seeded. In the peanut the indehiscent pod is forced underground by its stipitate base as it grows. In others, such as *Medicago polymorpha*, the pod bears prickles that cling to the fur or feathers of animals. The fruits of *Desmodium* (tickseed) and some other genera are loments (Figure 14-32 O); they break into one-seeded indehiscent pieces, each of which is an individual unit of dispersal. The legumes of other members of the family (e.g., *Tipuana tipu,* the tipu tree) are flattened and samaroid.

Economically the Fabaceae is one of the most important families of plants, second only to the Poaceae. The seeds or pods of some species are important sources of vegetable protein and starch. Important food crops include *Arachis hypogaea* (peanut), *Cicer arietinum* (garbanzo bean, chickpea), *Glycine max* (soybean), *Lens culinaris* (lentil), *Phaseolus* spp. (bean), *Pisum sativum* (pea), and *Vicia faba* (broadbean). Others are important forage crops, including *Medicago sativa* (alfalfa), *Trifolium* spp. (clover), *Vigna sinensis* (cowpea) and many others. Some are valuable, particularly in tropical regions as timber trees or as sources of cordwood. Many members of the family are important as ornamentals.

One of the most valuable contributions of members of this family is the restoration or maintenance of soil fertility. Most members of the family have symbiotic nitrogen-fixing bacteria in nodules on their roots. These bacteria are able to change molecular nitrogen which cannot be directly used by higher plants into nitrates. The amount of nitrogen compounds added to the soil is in excess of that used by the leguminous plant, and the breakdown of dead leaves, roots and other portions of the plant

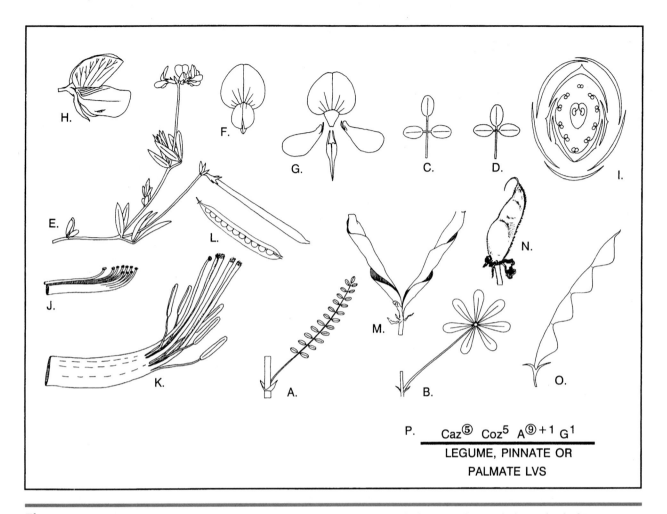

**Figure 14-32** Features of Fabaceae subf. Papilionoideae. A. Once pinnate leaf of *Astragalus*. B. Palmate leaf of *Lupinus*. C. Pinnately trifoliate leaf. D. Palmately trifoliate leaf. E. Branch of *Lotus corniculatus* bearing trifoliolate leaves with leaflike stipules. F. Intact corolla of *Lotus* (front view). G. Corolla of *Lotus* with petals separated. H. Flower of *Lupinus* (lateral view). I. Floral diagram of *Pisum*. J. Diadelphous stamens with nine filaments connate and one distinct. K. Monadelphous stamens of *Lupinus* with dimorphic anthers and all filaments connate. L. Legume of *Pisum* split apart showing rows of seeds inside the single locule. M. Legume of *Lupinus* dehisced with seed dispersal by twisting of the pericarp. N. Intact legume of *Lupinus*. O. Loment of *Desmodium*. P. Generalized floral formula.

further enriches the soil. The rotation of leguminous crops with crops that deplete the soil's nitrogen reserves is an alternative to expensive chemical fertilizers in many regions.

The seeds of some members of the legume family are very colorful and some have been used as beads. Unfortunately some of the most colorful are also very poisonous. Ingestion of the seeds of *Abrus precatorius, Ormosia* spp., *Sophora* spp. and some other genera can cause death, especially for children.

# Rosaceae

## The Rose Family
## Rosales ■

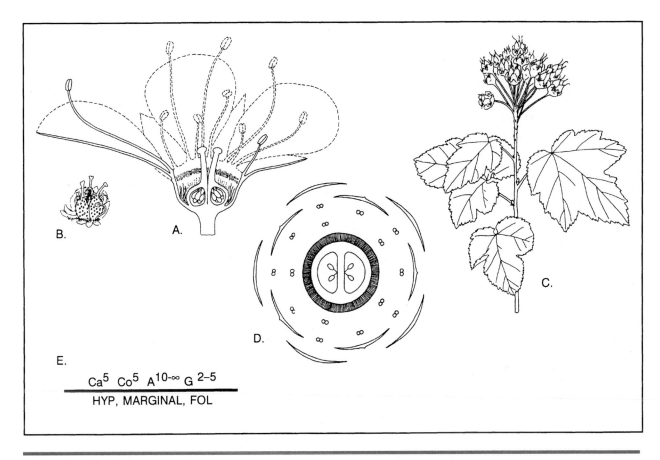

Ca⁵ Co⁵ A¹⁰⁻∞ G 2–5
HYP, MARGINAL, FOL

**Figure 14-33** Features of Rosaceae subf. Spiraeoideae; A. Flower of *Lyonothamnus floribundus* (l.s.). B. Aggregate of follicles of *Lyonothamnus*. C. Fruiting twig of *Physocarpus*. D. Floral diagram of *Lyonothamnus*. E. Generalized floral formula.

## FAMILY DESCRIPTION

Herbs, shrubs or trees. Leaves alternate (rarely opposite), simple to pinnate or palmate (rarely bipinnate or tripinnate), stipulate (rarely estipulate). Plants synoecious (rarely monoecious, dioecious or polygamous). Inflorescences various, determinate or indeterminate. Flowers perfect (rarely imperfect), regular (rarely irregular). Hypanthium present, usually well developed. Sepals 5 (4–many), connate, sometimes subtended by bractlets. Petals 5 (4–many or rarely absent), distinct. Stamens (1-) 10–many, distinct, when numerous developing centripetally. Carpels 1–many, distinct, or 2–5, connate; ovaries superior, unilocular with 1–2 (rarely many) marginal ovules (rarely with 2–5 locules and axile ovules) *or* ovary inferior (superior) with 2–5 locules and 1–2 (many) axile ovules per locule; styles 1 per carpel, distinct, undivided. Fruit a drupe, achene, follicle or pome (rarely a capsule) or an aggregate of achenes, drupelets or follicles, sometimes with fleshy accessory tissues derived from receptacle or hypanthium.

# FAMILY NOTES

The rose family is worldwide in distribution with a preponderance of species in the north temperate zone. It is a moderately large family with about 100 genera and 3000 species. The diversity of the family has been variously treated as four to six subfamilies. They are differentiated primarily by the type of fruit. The four subfamilies indigenous to North America are compared in Table 14-3 and illustrated in Figures 14-33–14-36.

The members of the Rosaceae are very diversified. Plants in the family range from herbs (most of them perennial) to small trees, The leaves are most commonly compound, although simple leaves are quite common; alternate leaves predominate and most are stipulate. The majority of the members of the family have perfect flowers with a biseriate perianth of 5 sepals and 5 petals. The flowers generally have a well-developed hypanthium that ranges from flat and disc-like to cup-shaped or tubular. The number of stamens is often numerous, and rarely is it fewer than twice the number of petals.

The gynoecium is highly variable. As can be seen in Table 14-3, it can be apocarpous, monocarpous, or syncarpous, with the ovary superior or inferior. Ovaries may mature as dry fruits including follicles, achenes or rarely capsules or as such fleshy fruits as drupes, drupelets, or pomes. In several genera with apocarpous gynoecia, fleshy accessory tissues ripen together with the ovaries (e.g., the receptacle of a strawberry or the hypanthium of a rose). In the pome-fruited members of the family the bulk of the flesh of the fruit is ripened hypanthium tissue. In several genera of the Rosoideae (e.g., *Cercocarpus, Fallugia,*) the styles are long and plumose and serve in wind-dispersal of the achenes. The Rosaceae is an important family in several respects. The combination of primitive features exhibited by members of the family place them in a near basal position within the Rosidae. Their evolutionary relationship to other primitive plants, both extinct and extant is a subject of ongoing research. In addition to their important evolutionary position, members of the rose family are extremely important economically, aesthetically and ecologically. The family is represented in cultivation by numerous ornamentals and an extraordinary diversity of fruit-crops. A few representative genera are: *Rosa* (rose); *Rubus* (blackberry, raspberry); *Fragaria* (strawberry); *Pyrus* (apple, pear); *Prunus* (almond, apricot, cherry, peach, plum); *Spiraea* (spiraea); *Potentilla* (cinquefoil); *Eriobotrya* (loquat); *Pyracantha* (firethorn); *Crataegus* (hawthorn); *Cydonia* (quince); and *Sorbus* (mountain-ash).

In introductory texts the pomes of subf. Maloideae are commonly described with reference to apples (*Malus* spp.). The pomes of *Malus* have fully syncarpous inferior ovaries in which the hypanthium greatly enlarges, forming the bulk of the fruit.

---

**Spiraeoideae** (Figure 14-33)

Woody plants with distinct or rarely connate carpels; ovules 2–many per carpel; ovaries superior, maturing as an aggregate of follicles or rarely as a capsule. Chromosome base no.: $x = 8, 9$

**Rosoideae** (Figure 14-34)

Herbaceous or shrubby plants with 1 carpel or 2-many distinct carpels; ovules usually 1–2 per carpel; ovaries superior, maturing as aggregates of achenes or drupelets, sometimes with accessory hypanthium or receptacle. Chromosome base no.: $x = 7, 8, 9$

**Prunoideae** [Drupoideae] (Figure 14-35)

Woody plants, usually with one carpel or rarely 2–5 distinct carpels; ovules 1–2 per carpel; ovary superior, maturing as a drupe. Chromosome base no.: $x = 8$

**Maloideae** [Pomoideae] (Figure 14-36)

Woody plants with 2–5 ± connate carpels; ovules 1–2 per carpel; ovary or ovaries inferior, with adnate fleshy hypanthium, maturing as a pome or a pomelike berry. Chromosome base no.: $x = 17$

Table 14-3    A comparison of the four subfamilies of the Rosaceae indigenous to North America.

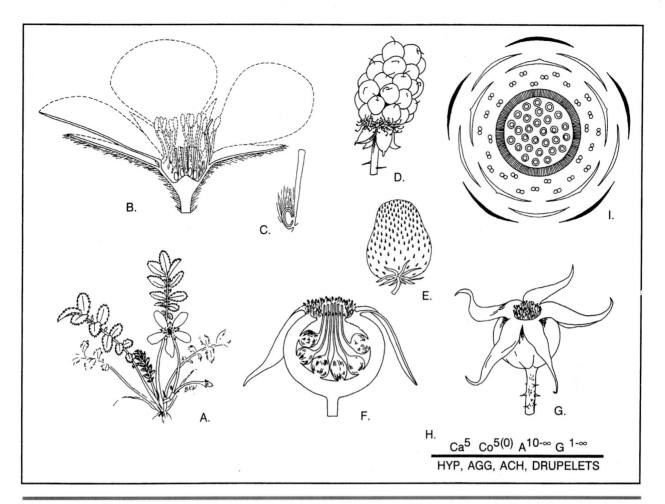

$$Ca^5 \ Co^{5(0)} \ A^{10-\infty} \ G^{1-\infty}$$

HYP, AGG, ACH, DRUPELETS

**Figure 14-34** Features of subf. Rosoideae. A. Habit of *Potentilla anserina*. B. Flower of *Potentilla* (l.s.). C. Carpel of *Potentilla*. D. Aggregate of drupelets of *Rubus*. E. Aggregate of achenes of *Fragaria* on fleshy receptacle. F–G. Sectioned and intact hips (fleshy hypanthia enclosing achenes) of *Rosa*. H. Generalized floral formula. I. Floral diagram of *Potentilla*.

In *Malus* the pericarp itself is differentiated into a fleshy mesocarp and cartilaginous endocarps that surround the seeds.

However, pomes are considerably more variable than this. In some genera (e.g., *Pyracantha*) the carpels are distinct, and in various other genera the carpels are only partially united. The degree of fusion between the ovary tissue and the surrounding hypanthium is also variable, ranging from ovaries that are nearly free from the hypanthium to those that are wholly adnate. Thus the ovary is often only partially inferior. In some genera the pomes are drupelike with hard, stonelike ovaries surrounded by fleshy hypanthium tissue.

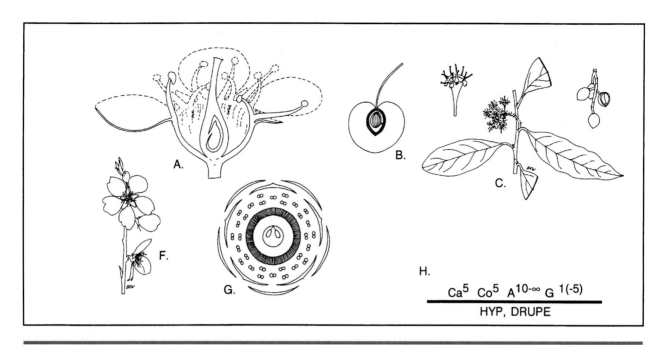

**Figure 14-35** Features of Rosaceae subf. Prunoideae. A. Flower of *Prunus* (l.s.). B. Drupe of *Prunus avium* (l.s.). C. Flowering twig of *Prunus caroliniana* D. Individual flower of *Prunus caroliniana*. E. Raceme of drupes of *Prunus caroliniana*. F. Flowering twig of *Prunus amygdalus*. G. Floral diagram of *Prunus*. H. Generalized floral formula.

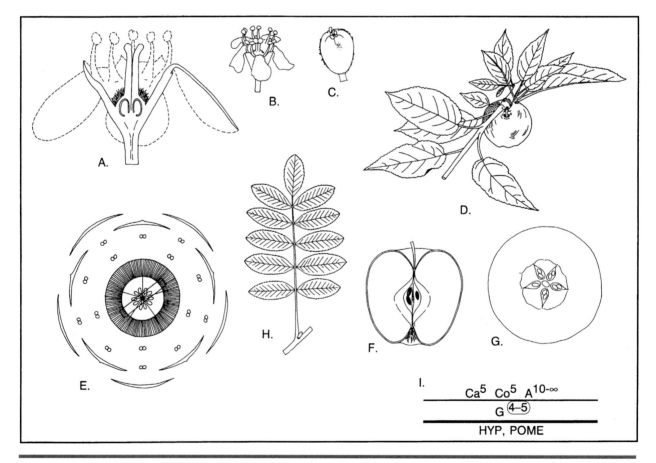

**Figure 14-36** Features of Rosaceae subf. Maloideae. A. Flower of *Pyracantha* (l.s.). B. Flower of *Heteromeles arbutifolia*. C. Fruit of *Heteromeles*. D. Fruiting twig of *Malus*. E. Floral diagram of *Malus*. F. Pome of *Malus* (l.s.) illustrating inferior ovary. G. Pome of *Malus* (x.s.) illustrating fleshy hypanthium adnate to five loculed compound ovary. H. Leaf of *Sorbus*. I. Generalized floral formula.

# Rhamnaceae

## The Coffeeberry Family
### Rosales ■

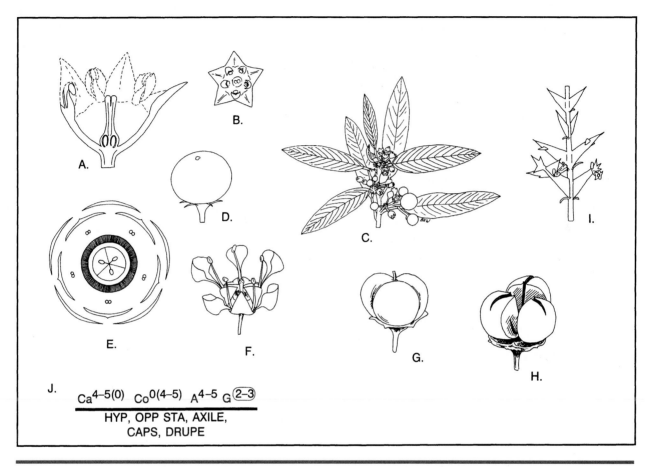

**Figure 14-37** Features of Rhamnaceae. A. Flower of *Rhamnus* (l.s). B. Flower of *Rhamnus* viewed from top. C. Flowering and fruiting twig of *Rhamnus californica*. D. Drupe of *Rhamnus*. E. Floral diagram of *Ceanothus*. F. Flower of *Ceanothus*. G–H. Intact and dehiscing capsule of *Ceanothus*. I. Branch of *Colletia cruciata* with flattened thorns. J. Generalized floral formula.

# FAMILY DESCRIPTION

Shrubs or trees (rarely woody vines, subshrubs). Leaves alternate or opposite, stipulate, sometimes very reduced. Plants synoecious (rarely monoecious). Inflorescence of cymes, umbels, spikes, heads, or panicles. Flowers perfect (imperfect), regular. Hypanthium present but often small. Sepals 4–5 (6–7), basally connate. Petals 4–5 (6–7) (0), often concave and hood-like, often long-clawed, distinct. Stamens 4–5 (6–7), opposite the petals, distinct. Nectary disc sometimes present. Carpels 2–3 (4) connate; ovary superior to inferior with 2–3 (1) (4) locules and 1 (2) basal-axile ovule per locule; style 1 with 2–3 (4) lobes. Fruit a drupe, berry or schizocarpic capsule (rarely a schizocarp, nut, or samara).

# FAMILY NOTES

The Rhamnaceae, a family of 51–55 genera and about 900 species, has a very widespread distribution, especially in tropical and subtropical regions. They have a center of diversity in California and adjacent portions of the western United States where the genus *Ceanothus* has diversified. Most are shrubs or small trees, usually with alternate, simple leaves. Species of arid regions are often stiffly branched and very thorny. A South American shrub, *Colletia cruciata* (Figure 14-37 I), has its stems transformed into flattened, thorn-tipped cladodes.

The flowers are small, usually with a biseriate, regular perianth and a ± well-developed hypanthium (Figure 14-37 A). The petals are frequently concave and hoodlike and often are clawed (Figure 14-37 F).

They are rudimentary or absent in some members of the family. The stamens are characteristically opposite the petals (Figure 14-37 B, E) and alternate with the sepals. The syncarpous gynoecium consists of a superior or half-inferior ovary with basal-axile placentation and a single ovule in each locule. The ovary matures either as a drupe (Figure 14-37 D) or a woody capsule (Figures 14-37 G, H).

Most Rhamnaceae have little direct economic value to man. Species of *Rhamnus*, particularly *R. purshiana* (cascara sagrada) are the source of a powerful laxative. Dyes are extracted from several *Rhamnus* species. There are a few minor fruits such as jujube *(Zizyphus jujuba)* and ornamental shrubs in several genera (e.g., *Ceanothus* and *Rhamnus*). Species of *Rhamnus* and *Ceanothus* are major components of the chaparral of southern California.

# Moraceae

## Mulberry or Fig Family
Moraceae ■

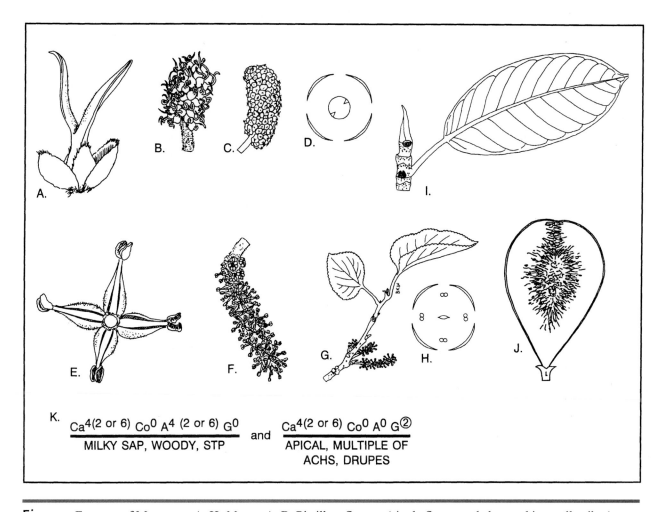

**Figure 14-38** Features of Moraceae. A–H. *Morus*. A–B. Pistillate flowers (single flower and clustered in small spikes). C. Multiple fruit (fruiting spike). D. Floral diagram of pistillate flower. E–F. Staminate flowers (individual flower and clustered in short spikes). G. Leafy twig with spikes of staminate flowers. H. Floral diagram of staminate flower. I–J. *Ficus*. I. Leaf and bud. J. Syconium (l.s.). K. Generalized floral formula.

# FAMILY DESCRIPTION

Shrubs or trees (rarely herbs), usually with milky sap. Leaves alternate, simple, stipulate. Plants monoecious or dioecious. Inflorescence variable, usually reduced or condensed, often with flowers ± embedded in a fleshy receptacle. Sepals (2) 4 (6), distinct, inconspicuous, in pistillate flowers free or adnate to ovary. Petals absent. Stamens (2) 4 (6) distinct, opposite the sepals. Carpels 2, connate; ovary superior or inferior with 1 locule and 1 apical ovule (rarely locules 2–3 and 1 apical-axile ovule per locule or ovule in only 1 locule); styles 2. Fruit an achene or drupelet, often coalescent into a dense multiple fruit with accessory tissues.

# FAMILY NOTES

The Moraceae is a primarily tropical family of about 40–70 genera and 1000–2500 species. Only two genera are indigenous to temperate North America—*Morus* (mulberry) and *Maclura* (Osage-orange). Both of these genera have the multiple fruits and milky sap characteristic of the family.

Flowers in the Moraceae are small and generally are densely aggregated into spikes, heads, or other tight inflorescences (Figures 14-38 B, F). Some members of the family are monoecious and others are dioecious. The perianth usually consists of 4 small, inconspicuous sepals (Figures 14-38 A, D, E, H). Most genera are wind-pollinated. The gynoecium is bicarpellate and usually has a one loculed ovary with a single apical ovule. The ovary matures either as an achene or a drupe and is often surrounded by fleshy accessory tissues derived from sepals or from the axis of the inflorescence. In most genera the densely clustered flowers and the accessory tissues ripen together as a multiple fruit (Figure 14-38 C). The multiple fruits in some members of the family are quite large. The spherical fruits of Osage-orange grow to more than 10 cm diameter and those of the breadfruit *(Artocarpus altilis)* are sometimes as much as 20 cm diameter. The champions though are the ovoid fruits of the jack fruit *(Artocarpus heterophyllus)* that can grow to 60 cm long or more and weigh as much as 18 kilograms (40 pounds).

*Ficus* (fig) is a very large tropical genus (more than 800 species) with an extremely varied ecology. It includes trees, shrubs, epiphytes, vines, and stranglers. A strangler begins its life as an epiphyte anchored to the branches of a tropical tree. From its branches aerial roots drop to the ground and begin to draw nutrients and water from the soil. Secondary growth enlarges these roots into trunks and the expanding canopy of the strangler fig envelops that of the host tree. The host tree eventually dies, its leaves in the shade and its trunk encircled by the expanding woody roots of the fig. The strangler then stands alone on its woody roots.

The inflorescence and fruiting structures of fig species is a unique structure called a **syconium** (Figure 14-38 J). It is a hollow inflorescence with flowers opening to the inside. There is a small pore or ostiole opposite the peduncle. Pistillate flowers mature first. They become pollen-receptive while the syconium is quite small. The flowers are polli-

nated by small wasps that lay their eggs in ovaries of some of the flowers. A female wasp, already dusted with pollen, enters the young syconium through the ostiole. She moves around within the syconium, laying eggs within the ovaries of some of the flowers. Having completed her egg-laying activities, the female wasp dies. Usually only a part of the ovaries are parasitized by wasp larvae. The remainder, pollinated by pollen from the wasp's body, mature normally. In each parasitized ovary a wasp larva consumes the developing ovule and eventually pupates.

By the time that the ovaries have matured as small achenes, the syconium has enlarged and become fleshy. At this stage the staminate flowers mature within the syconium and young wasps emerge from their pupal cases. The wasps mate within the syconium and the males soon die. The female wasps force their way to the surface of the syconium, becoming dusted with pollen in the process, and fly away. They are attracted by chemicals emitted from young syconia, which they enter and lay their eggs. The mature, ripened syconia are fleshy and berrylike. Animals of many kinds eat the berries and disseminate the seeds in their feces.

*Ficus carica* (edible fig) has two different sexual forms. In one form (Capri figs) both staminate and pistillate flowers are produced, as described above. The other form (Smyrna figs) produces only pistillate flowers. The shapes of the pistils of the two forms are different. The pollinating wasps are unable to distinguish immature Capri figs from immature Smyrna figs. If a female wasp enters a Capri fig, she proceeds to lay her eggs. She grasps the pistil and inserts her ovipositor and lays a single egg. She then moves to another pistil and does the same. If the wasp enters a Smyrna fig, however, the events are different. Because of the different shape of the pistil, the wasp is unable to lay her eggs. As she moves from flower to flower attempting to lay eggs, she dusts pollen onto the stigmas. Eventually she dies and the pollinated flowers go on to produce fruits. Seeds form but there are no wasp larvae. Smyrna figs are harvested for human consumption. Capri figs are necessary for pollination but are not harvested.

Members of the mulberry family are grown for food, fiber and ornament. Edible fruits are produced by *Ficus* (figs), *Morus* (mulberries), *Brosimum* (breadnut), and *Artocarpus* (breadfruit and jack fruit). *Brosimum alicastrum* was an important food tree in the Maya civilization of Mexico and Central America. The foliage of *Brosimum* is sometimes harvested as food for livestock. *Maclura* (Osage orange or hedge-apple) is commonly grown as a windbreak. It commonly persists along the edges of cultivated fields and in open woods in the eastern U.S. *Masclura* has a hard, fine-grained wood. High quality lumber is also produced by various species of *Ficus* and *Brosimum*. Numerous species of *Ficus* are grown as ornamentals, particularly in warm climates; some are grown indoors as potted plants. *Morus* is also cultivated as an ornamental, but its wind-disseminated pollen can cause hay-fever for sensitive individuals. The leaves of *Morus alba* are the food for silkworms.

# Urticaceae

## Stinging Nettle Family

### Rosales ■

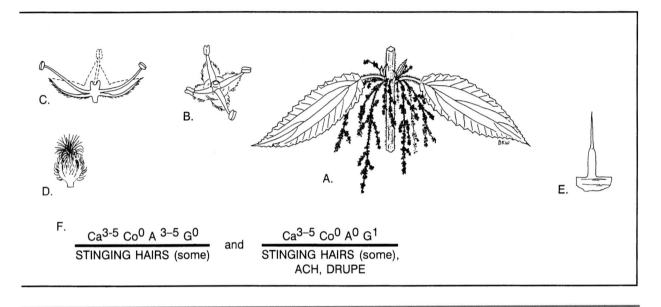

Ca$^{3-5}$ Co$^0$ A $^{3-5}$ G$^0$ / STINGING HAIRS (some)    and    Ca$^{3-5}$ Co$^0$ A$^0$ G$^1$ / STINGING HAIRS (some), ACH, DRUPE

**Figure 14-39** Features of Urticaceae. A–E. *Urtica.* A. Stem with leaves and flowers. B. Intact staminate flower. C. Staminate flower (l.s.). D. Pistillate flower. E. Stinging hair with glandular base and stiff, sharp tip. F. Generalized floral formulas for staminate and pistillate flowers.

## FAMILY DESCRIPTION

Herbs, shrubs or trees, sometimes armed with stinging hairs. Leaves alternate or opposite, simple, stipulate (sometimes estipulate). Plants monoecious or dioecious (polygamous). Inflorescence of solitary or variously cymose axillary clusters, sometimes paniculiform or headlike. Flowers imperfect (rarely perfect), regular. Sepals (0) 3–5, distinct or connate. Petals absent. Stamens (1–) 3–5, distinct, opposite the sepals. Carpel 1; ovary superior with 1 locule and 1 basal ovule; style 1 or stigma sessile. Fruit an achene or drupe.

## FAMILY NOTES

The Urticaceae is a cosmopolitan family with about 45 genera and 700–100 species. Although the family has a worldwide distribution, most species are tropical where they are common components of herbaceous layers of vegetation. All of the temperate members of the family are herbaceous, but in the tropics some species are shrubby or small trees.

The stinging nettle family is known in the north temperate zone primarily by the widespread genus, *Urtica* (stinging nettle; Figure 14-39 D). Stems and leaves of this genus and its close relatives are covered with stout hollow hairs that arise from bulbous liquid-filled bases (Figure 14-39 E). Each hair is like a miniature hypodermic needle. The sharp tip of the hair easily penetrates skin and breaks off. Slight pressure on the base of the hair causes irritating fluid to be injected beneath the skin. Pain is immediate and sometimes intense. In some cases the irritation lasts for only a short time, but the irritating fluids of some species may cause severe inflammation or allergic reactions. In addition to their presence in species of *Urtica,* stinging hairs are characteristic of *Hesperocnide, Laportea* and *Urera.* In the eastern United States *Laportea* is represented by an herbaceous species, but Australian species are commonly called stinging trees. Even though the nettles are nasty to handle, species including the giant nettle, *Urtica dioica,* can be prepared as cooked greens and eaten like spinach. Cooking softens the hairs and destroys the irritating qualities of the fluid. The stinging hairs are characteristic of only a few genera in the family.

Flowers in the Urticaceae are small and inconspicuous. In most genera they are imperfect (Figures 14-39 A–C), but in *Parietaria* (pellitory) the flowers are perfect. Members of the Urticaceae are generally wind-pollinated. Stamens often are explosively dehiscent, ejecting a puff of pollen grains into the air. *Pilea microphylla* is known as artillery plant because of its exploding anthers. The gynoecium appears to be monocarpous, but some authors consider it to be secondarily derived from the two-carpellate condition common in other families of the Urticales. In most genera the ovary matures as an achene which is often tightly enwrapped by persistent sepals.

The largest genera of the Urticaceae are *Elatostema* and *Pilea,* both of which lack stinging hairs. Several species of *Pilea* (aluminum plant), and *Soleirolia* (baby's tears) are grown as foliage plants. *Boehmeria nivea* (ramie) produces a commercial fiber of nearly pure cellulose that can be woven into a silk-like fabric. This plant has not lived up to its full potential, however, because of the difficulty of removing gums from the ensheathing bundle cells. Otherwise the family is of little economic importance.

# Cucurbitaceae

## The Cucumber Family
### Cocorbitales ■

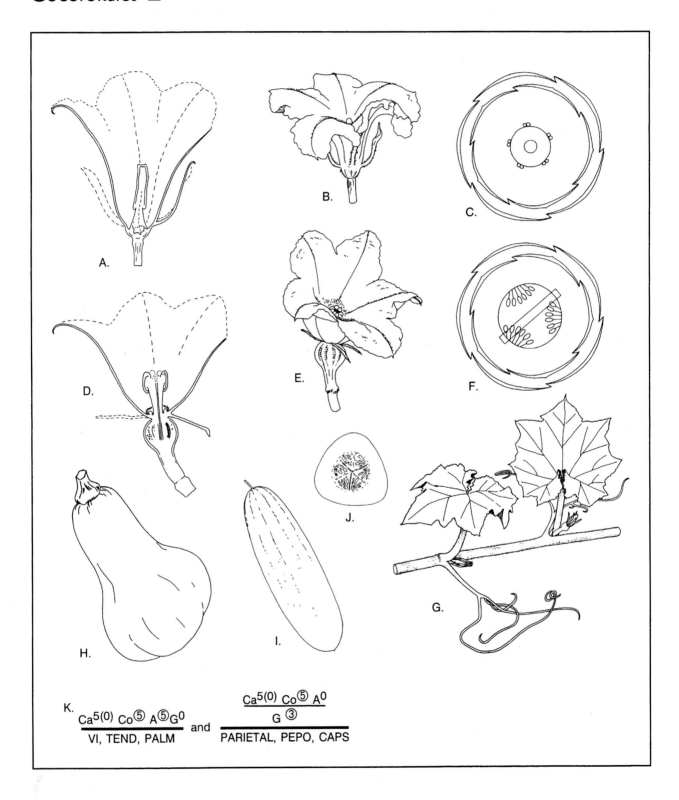

A.

B.

C.

D.

E.

F.

G.

H.

I.

J.

K. $\dfrac{Ca^{5(0)}\ Co^{\circledS}\ A^{\circledS}G^{0}}{VI,\ TEND,\ PALM}$ and $\dfrac{Ca^{5(0)}\ Co^{\circledS}\ A^{0}}{G\ \circledsmall{3}}$

PARIETAL, PEPO, CAPS

# FAMILY DESCRIPTION

Herbs, usually climbing by tendrils (rarely shrubs or trees). Leaves alternate, mostly palmately veined, simple to deeply lobed or divided, estipulate. Plants monoecious (rarely polygamous). Inflorescence of solitary flowers or variously cymose, sometimes paniculate or racemose. Flowers imperfect (rarely perfect), regular. Hypanthium sometimes present. Sepals 5 distinct or variously connate or very reduced and apparently absent. Petals 5, connate (rarely distinct). Stamens 5, distinct or variously connate. Nectary disc usually present. Carpels 2–5 (1), connate; ovary inferior with 1 locule and several–many (1) parietal (rarely marginal) ovules or with 2–5 locules and 1–many axile ovules; style 1, simple or ± lobed above (rarely 3 and distinct). Fruit a berry, pepo, capsule or achene.

# FAMILY NOTES

The Cucurbitaceae, a family of 90–120 genera and over 700 species, is distributed for the most part in tropical and subtropical regions. Temperate representatives are comparatively few. Members of the family are mostly perennial herbaceous vines, usually bearing helically coiled tendrils.

The imperfect flowers are rather variable. Both staminate and pistillate flowers sometimes have such a reduced calyx that the flowers often appear to have only one whorl of perianth parts. The corolla may be obviously sympetalous (Figure 14-40 A, B, D, E) or so deeply lobed that the petals appear to be distinct. The stamens are often connate (Figure 14-40 A, B), sometimes with the anthers strongly contorted and appearing to be reduced to three. The inferior ovary (Figure 14-40 D) is 3-carpellate in most genera, and most commonly has parietal placentation (Figure 14-40 F, J). Sometimes the placentae intrude into the interior of the ovary resulting in three separate locules or completely filling the interior of the ovary with fleshy tissue (Figure 14-40 J).

The ovary may mature as a large, water-filled, thick-walled fruit called a pepo (e.g., melons, squashes, gourds; (Figure 14-40 I, J), as a berry or as a dry capsule. The capsular fruits of some genera eject their seeds in a jet of fluid (e.g., *Ecballium elaterium,* the squirting cucumber, and *Marah* spp., manroot). The capsules in various genera are covered with stiff prickles. Some genera have a monocarpous gynoecium that matures as an achene (e.g., *Brandegea, Echinocystis*).

The family is economically important primarily for the diverse products of its fruits. Common edible fruits include *Cucumis* spp. (cucumber, melons), *Citrullus lanatus* (watermelon), *Cucurbita* spp. (squash. pumpkin) and *Sechium edule* (chayote). The dried pepos of *Lagenaria siceraria* (bottle gourd) have been used since ancient times in many areas of the world as containers. The common name, vegetable sponge, describes the use of the dried skeleton of the fruit of *Luffa cylindrica*. The seeds of some cucurbits are rich in oils, and in some arid regions wild melons are a valuable water source for both humans and wild animals.

**Figure 14-40** Features of Cucurbitaceae. A–B. Staminate flowers of *Cucurbita* (l.s. and intact). C. Floral diagram of staminate flower. D–E. Pistillate flowers of *Cucurbita* (l.s. and intact). F. Floral diagram of pistillate flower. G. Leafy stem of *Cucurbita* with tendrils and axillary Pistillate flower. H. Pepo of *Cucurbita pepo.* I–J. Pepo of *Cucumis* (intact and x.s.) illustrating parietal placentation. K. Generalized floral formulas.

# Fagaceae

## The Beech Family
### Fagales ■

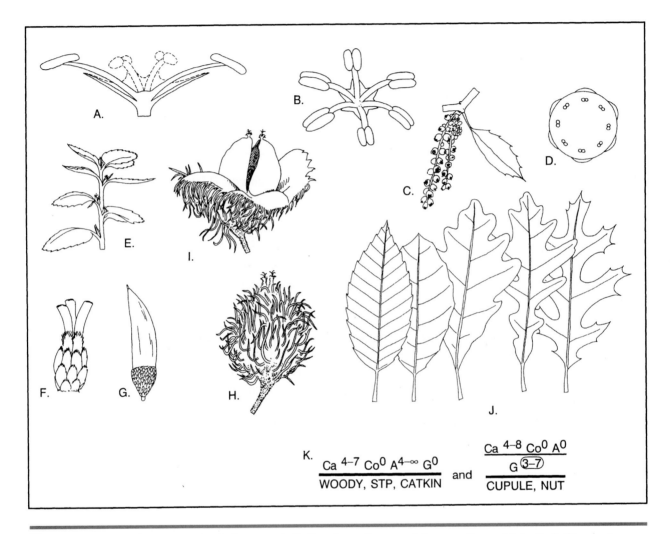

$$\frac{Ca^{4-7}\ Co^0\ A^{4-\infty}\ G^0}{\text{WOODY, STP, CATKIN}} \quad \text{and} \quad \frac{Ca^{4-8}\ Co^0\ A^0}{G\ \boxed{3-7}}$$

K.                                    CUPULE, NUT

**Figure 14-41** Features of Fagaceae. A–G. *Quercus.* A–B. Staminate flowers of *Quercus* (l.s. and intact). C. Staminate catkins of *Quercus.* D. Floral diagram of staminate flower of *Quercus.* E. Leafy twig of *Quercus agrifolia* with axillary pistillate flowers. F. Pistillate flower of *Quercus* enclosed by bractlets of involucral cupule. G. Nut of *Quercus* with cupule at base. H–I. *Fagus sylvatica.* H. Immature cupule and pistillate flowers. I. Two triangular nuts within the dehiscent woody cupule. J. Leaves of *Fagus* and various species of *Quercus.* K. Generalized floral formulas for staminate and pistillate flowers.

## FAMILY DESCRIPTION

Shrubs or trees. Leaves alternate, simple, stipulate. Plants monoecious. Staminate inflorescence an erect or pendulous spike or head (rarely staminate flowers solitary); pistillate inflorescence of 1–3 (–7) sessile flowers subtended or surrounded by an involucre of numerous ± connate bractlets. Flowers imperfect, regular. Sepals 4–7, connate in staminate flowers; 4–8, very reduced, connate and adnate to ovary in pistillate flowers. Stamens 4–many, distinct. Carpels 3–7, connate; ovary inferior with 3–7 locules and 2 ovules per locule, all but one aborting; styles 3–7.

Fruit a nut subtended or ± surrounded by a woody cup-like or bur-like involucre.

# FAMILY NOTES

The Fagaceae comprises 7 genera and about 800–1000 species. The family is very widespread in the north temperate zone and extends into the tropics in Central America, southeast Asia and the East Indies. The family is absent from all but the northern-most parts of Africa.

The genus *Nothofagus* (southern beech) has long been assigned to the Fagaceae as the sole southern hemisphere member of the family. It occurs in south temperate latitudes of the Australian region and southern South America and extends to equatorial regions in New Guinea. Recently some taxonomists have presented evidence that *Nothofagus* is probably not closely related to the northern hemisphere genera and should be segregated as its own family. Cronquist (1988) accepted this position and recognized the Nothofagaceae as a family distinct from the Fagaceae. We are following Cronquist's treatment here.

The Fagaceae are readily recognizable by their characteristic inflorescences. The staminate inflorescence may be a weak and dangling catkin (Figure 14-41 C), a headlike structure or a stiff erect spike. The bracts of the catkin are small and early deciduous. Some members of the family are wind-pollinated and others are insect-pollinated. The pistillate flowers are solitary or in groups of 2–7 in involucral cupules (Figures 14-41 F, H). These are actually very condensed cymes. The bracts of the cupules may mature as flattened scales in the acorn cup of *Quercus* (oak, Figure 14-41 G), as elongate spines as in the burs of *Castanea* (chestnut), *Castanopsis,* and *Chrysolepis* (chinquapin), or as elongate, tapering structures in the cupule of *Fagus* (beech: Figure 14-41 I).

Oaks and their kin are very important plants. They often dominate vegetation in temperate and subtropical regions. *Quercus* and *Fagus* are important components of the forests and woodlands of North America and Eurasia. Until it was nearly wiped out by the chestnut blight, *Castanea dentata* was one of the dominant forest trees in the deciduous forests of the eastern U.S. *Lithocarpus* and *Castanopsis* are mostly tropical genera of eastern Asia and the East Indies. One species of *Lithocarpus* occurs in forests of California and southern Oregon. Shrubby oaks are important components of the chaparral of western North America and in similar vegetation of the Mediterranean region. *Chrysolepis* (chinquapin) is a small genus of trees and shrubs of western North America. Members of the Fagaceae support many species of game animals through their bast (twigs and buds) and their large edible nuts.

The Fagaceae are economically important in various ways. Many members of the family produce valuable lumber that is used in a wide variety of applications. Some members of the family are large trees and their wood is strong and relatively resistant to decay. Oak wood is used extensively for cordwood and manufacturing charcoal. The bark of *Lithocarpus densiflora* (tanbark oak) and several species of *Quercus* are commercial sources of tannins. Cork is harvested from the bark of *Quercus suber* and several other species of *Quercus.*

Acorns were an important staple for some of the pre-Columbian Indian tribes in North America. However, because of their high tannin content, these nuts must be specially treated to make them edible. Species of *Castanea* have been used for food in various countries, and *Lithocarpus edulis,* a Japanese species, also produces edible fruits. Many members of the family are cultivated as ornamentals.

# Betulaceae

## Birch or Alder Family
### Fagales ■

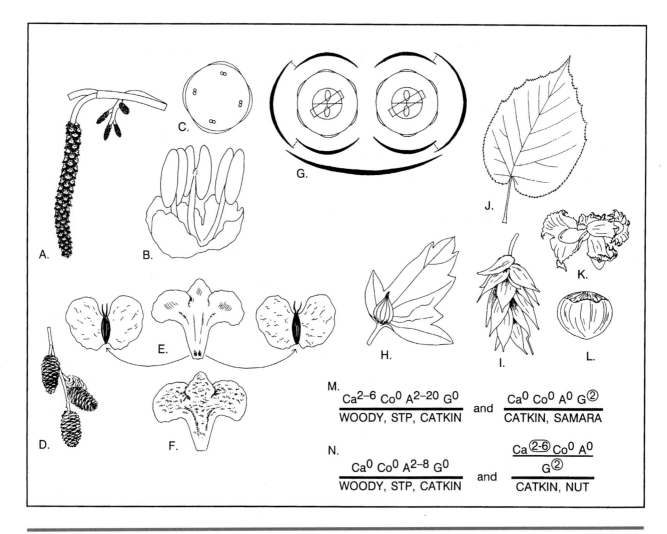

M. 
$$\frac{Ca^{2-6}\ Co^0\ A^{2-20}\ G^0}{\text{WOODY, STP, CATKIN}}$$ and $$\frac{Ca^0\ Co^0\ A^0\ G^{\textcircled{2}}}{\text{CATKIN, SAMARA}}$$

N. 
$$\frac{Ca^0\ Co^0\ A^{2-8}\ G^0}{\text{WOODY, STP, CATKIN}}$$ and $$\frac{Ca^{\textcircled{2-6}}\ Co^0\ A^0}{\dfrac{G^{\textcircled{2}}}{\text{CATKIN, NUT}}}$$

**Figure 14-42** Features of Betulaceae. A–D. *Alnus*. A. Staminate and pistillate catkins at time of pollination. B. Staminate flower. C. Floral diagram of staminate flower. D. Conelike ripened pistillate catkins. E–F. *Betula*. E. Pair of samaras with three fused subtending catkin bractlets (adaxial surface). F. Group of three fused bractlets from pistillate catkin (abaxial surface). G–H. *Carpinus*. G. Floral diagrams of pistillate flowers with subtending bractlets. H. Nut with three fused subtending bracts. I. Fruiting catkin of *Ostrya* with inflated bracts surrounding nuts. J–L. *Corylus*. J. Leaf. K. Immature nuts of *Corylus* surrounded by involucre of fused papery bracts. L. Mature nut of *Corylus* with bracts removed. M. Floral formulas for subf. Betuloideae. N. Floral formulas for subf. Coryloideae.

# FAMILY DESCRIPTION

Shrubs or trees. Leaves alternate, simple, stipulate. Plants monoecious. Staminate inflorescence a pendulous spike-like or raceme-like catkin with 2–3 flowered cymules subtended by scale-like bractlets; pistillate inflorescence a spike-like or head-like catkin of very reduced flowers in 1–3 flowered cymules subtended by scale-like bractlets. Flowers imperfect, regular. Sepals 2–4 (–6) and distinct (sometimes absent) in staminate flowers; 2–4, very reduced, connate and adnate to ovary or absent in pistillate flowers. Petals absent, Stamens 2–6 (–8), distinct or filaments basally connote. Carpels 2, connate; ovary inferior or superior, 1-loculed above, 2 loculed below with 1 axile ovule per locule; styles 2. Fruit an achene, nut or samara.

# FAMILY NOTES

The birch family comprises 6 genera and about 120–170 species. Most occur in cool temperate portions of the northern hemisphere. They extend southward along the mountains to the Andes of South America. The family includes important trees and shrubs of the cold temperate to subartic regions throughout the northern hemisphere. They are often associated with streams, lakes, or other poorly drained areas such as those that occur in permafrost areas.

Both the staminate and pistillate flowers are borne in catkins with persistent bractlets (Figure 14-42 A). Wind is the pollinating agent throughout the family. The catkins represent very condensed cymose clusters of flowers arranged in the axils of bractlets along short to elongated axes (Figure 14-42 G). Perianth parts are very inconspicuous (Figure 14-42 B) or completely absent. The pistillate flowers mature as samaras in *Betula* (birch; Figure 14-42 E) and *Alnus* (alder), and as small to large nuts in the remaining genera. In *Betula* and *Alnus* the mature pistillate catkins are dry and conelike (Figure 14-42 D). At maturity the bractlets and samaras of birch catkins (Figure 14-42 E, F) break free from the axis of the catkin and the catkin disintegrates. The bractlets of alder catkins remain attached to the catkin axis, and the samaras are released as the bractlets dry out. The mature bracts of the pistillate catkins become enlarged and leafy in *Carpinus* (ironwood; Figure 14-42 H) and inflated and sac-like in *Ostrya* (hop-hornbeam; Figure 14-42 I). The bracts fuse to form a cup-like to tubular involucre around the ovary of *Corylus* (hazelnut, filbert; Figure 14-42 K).

The Betulaceaes comprise two subfamilies and three tribes (Table 14-4). Some authors consider each to be sufficiently distinct to warrant recognition as a separate family.

Members of the Betulaceae are grown for lumber, edible nuts, and as ornamentals. The fine-grained wood of some birch species is particularly valuable for cabinet work, flooring and furniture. The wood of *Ostrya* and *Carpinus* is very dense and hard. The large edible nuts of *Corylus* (filbert, hazel nut) are harvested commercially. Oil of birch is used as a flavoring, and the sap can be fermented to make birch beer. Tannins are extracted from the bark of some member of the family. The bark of *Betula papyrifera* was used by some tribes of North American Indians to manufacture canoes and other implements. Various members of the family are used for cordwood and charcoal.

| subf. Betuloideae | subf. Coryloideae (Corylaceae) | |
| --- | --- | --- |
| **tribe Betuleae** (Figure 14-42 A–F) | **tribe Carpineae** (Fig. 14-42 G–I) | **tribe Coryleae** (Fig. 14-42 J–L) |
| Staminate flowers with a calyx; stamens 4–6 per flower; cymules of pistillate catkins 2–3 flowered; pistillate flowers naked; pistillate catkins cone-like at maturity, the bractlets of the catkin remaining small; fruit a samara. | Staminate flowers without a calyx, the 3 flowers of the cymule more or less coalescent; stamens 4–6 per flower; cymules of pistillate catkins 2-flowered; pistillate flowers with a tiny calyx; pistillate catkins elongate, more or less open at maturity; fruit a small nut, subtended by a deeply lobed leaf-like bract or enclosed in an inflated hollow bract. | Staminate flowers without a calyx, the 3 flowers of cymule more or less; cymules of coalescent; stamens 1–2-flowered; pistillate flowers with a tiny calyx; cymules of pistillate catkins 2-flowered; pistillate catkins short, few-fruited at maturity; fruit a nut, enclosed in a cuplike or tubular involucre. |

**Table 14-4**   Comparison of subfamilies and tribes of the Betalaceae.

# Brassicaceae (Cruciferae)

## Mustard Family

### Brassicales ■

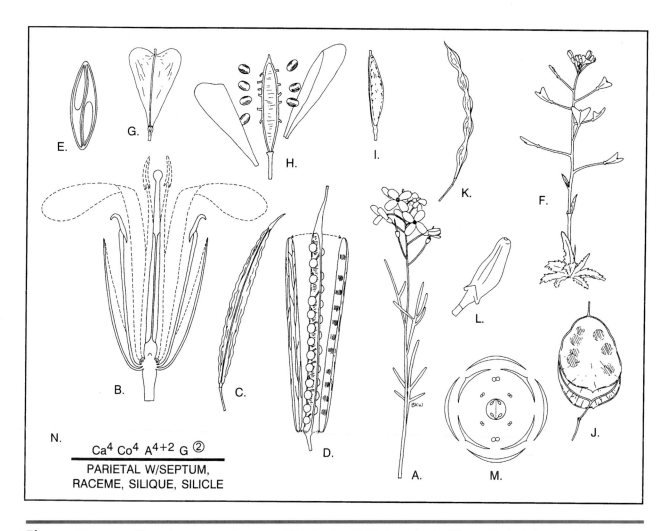

$$Ca^4 \ Co^4 \ A^{4+2} \ G \ ②$$

PARIETAL W/SEPTUM,
RACEME, SILIQUE, SILICLE

**Figure 14-43** Features of Brassicaceae. A. Raceme of *Cardamine*. B–D. *Brassica*. B. Flower (l.s.). C. Intact silique D. Silique dehiscing. E. Cross section of ovary of *Lobularia maritima,* compressed parallel to septum. F–H. *Capsella bursa-pastoris.* F. Habit with raceme of flowers and fruits. G. Intact silicle compressed at right angle to septum. H. Silicle dehiscing. I. Replum (septum) of *Rorippa* after valves and seeds have fallen. J. Silicle of *Lunaria* dehiscing. K. Indehiscent silique of *Raphanus.* L. Indehiscent silique of *Cakile maritima.* M. Floral diagram. N. Floral formula.

## FAMILY DESCRIPTION

Herbs or shrubs. Leaves alternate (opposite or whorled), estipulate, simple to bipinnate. Plants synoecious. Inflorescence of racemes (very rarely flowers solitary). Flowers perfect, regular. Sepals 4, distinct. Petals 4, distinct. Stamens 6 (4 or 2) (–16), 4 with long filaments and 2 with short filaments, distinct. Carpels 2, connate; ovary superior with 2 locules and 1–many parietal ovules (rarely ovary 1-locular); style 1. Fruit a silicle or silique, sometimes indehiscent, sometimes breaking transversely into seed-bearing joints.

# FAMILY NOTES

The Brassicaceae is a cosmopolitan family of 350–381 genera and about 3000 species with a preponderance of species in the north temperate zone. It is a family of mostly herbaceous plants that are often scented with compounds called mustard oils.

The mustard family is one of the easiest to recognize. The flowers are borne in bractless racemes (Figure 14-43 A, F) that often are corymbose when young. The perianth is 4-merous (Figure 14-43 A, M) with the petals arranged in the form of a cross. The alternate name for the family, Cruciferae, means cross-bearer. The stamens are usually **tetradynamous,** four of them with long filaments and two with short filaments (Figure 14-43 B, M). Although the 2-carpellate superior ovary has parietal placentation, there is a membranous septum, often called the **replum,** that divides the ovary into two locules.

The fruit is a peculiar capsule called a **silique** (if it is long and slender; Figure 14-43 C, D) or a **silicle** (if it is twice as long as wide or less; Figure 14-43 E, G, H, J). When the fruit dries, the sides of the fruit **(valves)** break free from the replum (Figure 14-43 E, H, J) and fall to the ground. The seeds fall out and the membranous replum remains attached to the receptacle (Figure 14-43 H, I). Some genera have a terminal indehiscent segment called a **beak** (Figure 14-43 C, D) that contains one or more seeds. In *Raphanus* (Figure 14-43 K) and *Cakile* (Figure 14-43 L) the fruits are indehiscent and break transversely into seed-bearing sections. In a few genera (e.g., *Stanleya* and *Lunaria;* Figure 14-43 J) the basal portion of the ovary is slender and stalk-like, forming a **stipe** that is inserted between the receptacle and the seed-bearing portion of the ovary. Such flowers have two stalks, the pedicel and the stipe.

The mustard family has considerable economic importance. Garden vegetables include *Armoracea rusticana* (horseradish), *Brassica oleracea* (cabbage, cauliflower, broccoli, kohlrabi, kale, brussels sprouts), *B. rapa* (turnips, Chinese cabbage), and *Raphanus sativus* (radish). Mustard is prepared from the ground seeds of *Brassica nigra, B. juncea* and *Sinapsis alba.*

Mustards are also important as oil crops. Rapeseed oil, produced from *Brassica napus* has numerous industrial uses. Ordinary rapeseed oil is unsuitable for human consumption. Canola oil is an edible form of rapeseed oil that, because of its reduced quantities of erucic acid, is palatable.

Ornamentals include *Matthiola* spp. (stock), *Lobularia maritima* (sweet alyssum), *Lunaria annua* (honesty), *Iberis* (candy-tuft) and *Erysimum* spp. (wallflower). Red-leafed cabbages are sometimes cultivated for the attractive foliage.

Some annual and biennial forms are aggressive and weedy. Chemicals in the foliage and seeds of some species are poisonous to livestock.

# Malvaceae

## The Mallow Family

Malvales ■

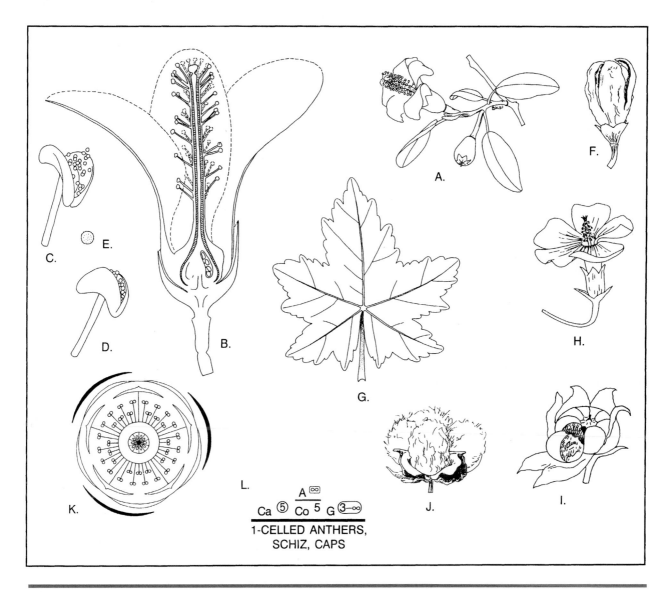

**Figure 14-44** Features of Malvaceae. A–B. *Lagunaria patersonii*. A. Flowering branch. B. Flower (l.s.). C–E. One-loculed anthers and pollen grain. F. Capsule of *Lagunaria*. G. Palmately veined leaf of *Lavatera assurgentiflora*. H. Flower of *Lavatera*. I. Schizocarp of *Lavatera*. J. Capsule of *Gossypium*. K. Floral diagram of *Malva*. L. Generalized floral formula.

# FAMILY DESCRIPTION

Herbs, shrubs or trees. Leaves alternate, simple, stipulate. Plants synoecious. Inflorescences various, usually determinate. Flowers perfect, regular. Sepals 5, distinct or connate, often subtended by bractlets. Petals 5, distinct, individually adnate to stamen column. Stamens many, centrifugal, filaments connate into a tube surrounding the pistil; anthers unilocular. Carpels 4–many, connate; ovary superior with 4–many locules and 1–many axile ovules; style 1 ± divided. Fruit a schizocarp or capsule (rarely a berry).

# FAMILY NOTES

The Malvaceae is a cosmopolitan family of 100–119 genera and about 1500 species. It has a center of diversity in the American tropics. Temperate species are mostly plants of open habitats ranging from marshlands to grasslands and deserts.

Stamens in the Malvaceae are monadelphous and have unilocular anthers. The filaments are connate forming a hollow tube (Figure 14-44 B, K). The tips of the filaments are distinct from the tube and each bears a crescent-shaped anther with a single continuous pollen sac (Figure 14-44 C, D). The petals in most Malvaceae are separate from each other but are **epistemonous,** individually adnate to the stamen tube (Figure 14-44 B). The combination of stamens and corolla falls as a single unit. The stamen tube surrounds the pistil but is not fused to it. The style extends upward through stamen tube (Figure 14-44 B) and the stigmas are exserted.

The syncarpous gynoecium matures either as a capsule (Figure 14-44 F, J) or as a schizocarp (Figure 14-44 I). In mallows with schizocarpic fruits, the carpels are often numerous and are attached in a ring to a conical central axis (Figure 14-44 I). The mericarps separate as wedge-shaped units that usually contain one or two seeds. In some cases the mericarps are dehiscent, releasing the enclosed seeds after separation. In many the mericarps are indehiscent and serve as the unit of dispersal. In *Malvaviscus* the mericarps are fleshy and berrylike.

Members of the Malvaceae are economically important in various ways. Important ornamentals include *Alcea rosea* (hollyhock), numerous species of *Hibiscus, Abutilon,* and *Malvastrum.* The only significant food-crop in the family is *Hibiscus esculentus* (okra). At one time one of the principal ingredients of marshmallows was the root-sap of *Alcea officinalis.* By far the most important crop is cotton (*Gossypium* spp.; Figure 14-44 J). Cotton fiber is pubescence on the seeds. The seeds of cotton are a source of oil and feed for livestock. Several members of the family are important field weeds, particularly *Abutilon theophrasti* (velvet-leaf) and species of *Malva* (mallow).

# Rutaceae

## The Citrus Family
### Sapindales ■

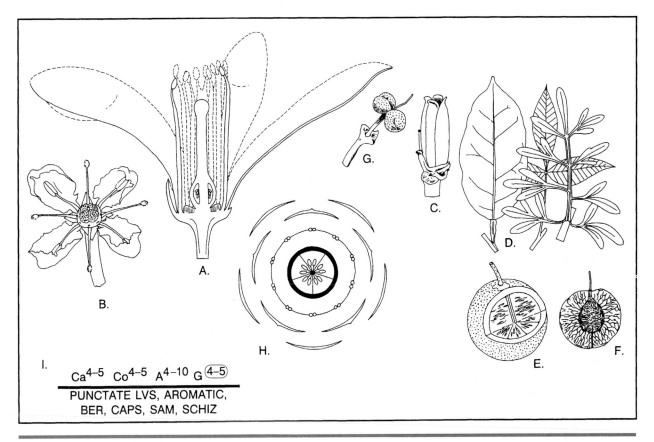

**Figure 14-45** Features of Rutaceae. A. Flower of *Citrus* (l.s.). B. Flower of *Ruta graveolens*. C. Flower of *Thamnosma montana*. D. Leaves of *Citrus, Ptelea,* and *Ruta* (note the winged, jointed petiole of *Citrus*. E. Hesperidium of *Citrus*. F. Samara of *Ptelea*. G. Leathery capsule of *Thamnosma*. H. Floral diagram of *Citrus*. I. Generalized floral formula.

## FAMILY DESCRIPTION

Herbs, shrubs or trees. Leaves alternate (rarely opposite), simple to pinnate, usually punctate with oil glands and often strongly scented, estipulate. Plants synoecious (rarely polygamous or dioecious). Inflorescences various, often corymbiform or paniculiform. Flowers perfect (rarely imperfect), regular (rarely irregular). Sepals 4–5 (2–3) ± connate. Petals 4–5, distinct (connate). Stamens 4–10 (rarely 15–many), distinct. Nectary disc present. Carpels 2–5 (1)(–many), connate (distinct or nearly so); ovary superior with 2–5 (1) (–many) locules and 1–many axile ovules (ovaries with 1–many marginal ovules); style 1. Fruit a capsule, drupe, berry, hesperidium, samara, schizocarp, or cluster of follicles.

# FAMILY NOTES

The Rutaceae is widely distributed in tropical and temperate regions. It is a moderate sized family of about 15–158 genera and 900–1500 species. The family consists mostly of tropical and subtropical species, many of which are unfamiliar to botanists of temperate regions.

*Citrus* (grapefruits, lemons, limes, oranges) is probably the most familiar genus in the Rutaceae, and it is often used to represent the family. However, the citrus family is quite variable in habit, foliage, floral features and fruits. Figure 14-45 D illustrates the contrasting foliage of *Citrus, Ptelea,* and *Ruta* (rue), three very dissimilar members of the family. *Citrus* is unusual in having apparently simple leaves that are actually compound leaves in which all but the terminal leaflet have been lost; there is a distinct joint where the leaflet attaches to the petiole. The closely related genus, *Poncirus* (trifoliate-orange) has compound leaves with three well-developed leaflets. Some members of the family do have truly simple leaves. *Citrus, Ruta* and most other members of the family have oil glands in the leaves and often in other parts of the plant. These glands are one of the most noticeable characteristics of the family. The oils are often very strongly scented.

Flowers in the Rutaceae are rather variable. Most have 4 or 5 distinct petals, but in *Correa* the corolla is sympetalous. Most are radially symmetric although a few are somewhat bilateral (e.g., *Dictamnus*). There is usually a prominent nectary disk (Figures 14-45 A, B), below which the stamens are inserted. In most genera there are twice as many stamens as petals. In *Calodendrum* five of the stamens are modified into petaloid staminodes. The stamens may be wholly distinct or basally connate. The ovary is superior in almost all genera. In many members of the family (e.g., *Ruta*) it is deeply 4–5-lobed with a gynobasic style and numerous ovules. In *Citrus,* on the other hand, the ovary is unlobed and has a terminal style and only 1 or 2 ovules per locule.

There is much variety in the fruits of the Rutaceae. Both *Ruta* and *Citrus* have berries. The large leathery-skinned berry of *Citrus* and its close relatives is called a hesperidium. *Ptelea* produces flat, round samaras (Figure 14-45 F). *Thamnosma* has a leathery capsule (Figure 14-45 G). The fruit of *Calodendrum* is a woody capsule. The distinct carpels of *Zanthoxylum* mature as follicles. Other members of the family produce schizocarps or drupes.

Economically, the most important members of the Rutaceae are the various species of *Citrus*. Edible fruits are also produced by *Fortunella* spp. (kumquat). Many other members of the Rutaceae are cultivated as ornamentals or as sources of perfumes. These include *Calodendrum capense* (Capechestnut), *Phellodendron* spp. (cork-tree), *Correa* spp. (Australian-fuchsia), *Dictamnus albus* (dittany, burning bush), and *Ptelea trifoliata* (hop-tree).

*Ruta* (rue) is often cultivated as an ornamental and has been used in the past as a culinary herb or medicinal plant. However, its foliage is extremely poisonous and should not be taken internally. In susceptible individuals contact with the herbage of this plant can cause severe contact dermatitis similar to that caused by poison-oak or poison-ivy.

# Anacardiaceae

## The Sumac Family
## Sapindales ■

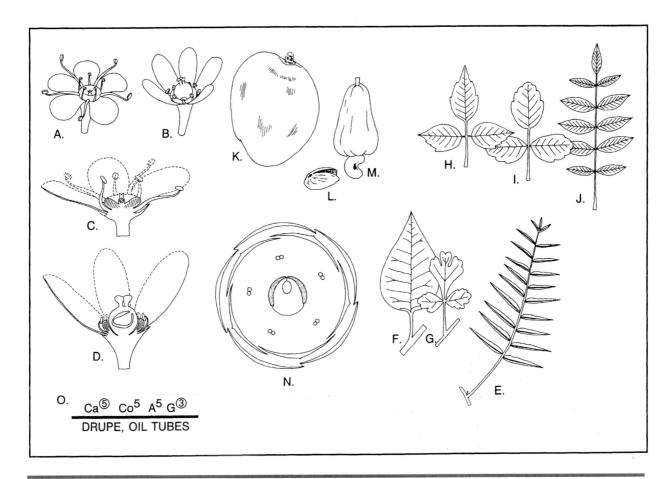

**Figure 14-46** Features of Anacardiaceae. A–D. *Schinus.* A–B. Intact staminate and pistillate flowers. C–D. Staminate and pistillate flowers of (l.s.). E–J. Leaves of representative Anacardiaceae. E. *Schinus molle.* F. *Rhus ovata.* G. *Rhus trilobata.* H. *Toxicodendron radicans.* I. *Toxicodendron diversilobum.* J. *Toxicodendron vernix.* K–M. Fruits of representative *Anacardiaceae.* K. Drupe of *Mangifera indica* (mango). L. Dry drupe of *Pistacia vera* (pistachio) with dehiscent endocarp. M. Kidney-shaped drupe of *Anacardium occidentale* (cashew) with swollen pedicel. N. Floral diagram of *Rhus.* O. Floral formula of *Rhus.*

## FAMILY DESCRIPTION

Shrubs, woody vines or trees. Leaves alternate, simple to pinnate or trifoliolate, estipulate. Plants monoecious, dioecious, polygamous or synoecious. Inflorescence paniculiform. Flowers perfect or imperfect [if imperfect usually with abortive parts of the opposite sex present], regular. Hypanthium sometimes present, but small. Sepals 5 (3–7), connate. Petals 5 (3–7), distinct. Stamens 5–10 (1) (many), distinct or filaments weakly connate at base. Nectary disc present. Carpels 3 (2–12) connate (rarely 1 or 5 and distinct); ovary (ovaries) superior (inferior) with 1 locule and 1 apparently apical or basal [actually axile but only 1 locule developed] or marginal ovule per locule (rarely with 4–5 locules and 1 axile ovule per locule); soles 3 (1 or 4–5), distinct or nearly so. Fruit a drupe or berry, sometimes dry at maturity.

## FAMILY NOTES

The Anacardiaceae is a predominately tropical family of about 75 genera and 600 species that extends into the temperate zones of North America and Eurasia. The genera most familiar in temperate regions are *Rhus* (sumac) and *Toxicodendron* (poison-ivy, poison-oak, and poison-sumac). Some botanists consider the two to constitute a single genus.

Flowers in the Anacardiaceae are typically small and 5-merous and arranged in few- to many-flowered paniculiform clusters. The petals and stamens are inserted around the rim of a well-developed nectary disk. Flowers in the family are often imperfect. Usually in the staminate flowers there is an abortive ovary (Figures 14-46 A, C), and staminodes are usually present in the pistillate flowers (Figures 14-46 B, D). The flowers of most genera have 10 stamens, but in *Rhus* five stamens alternate with the petals. The gynoecium is rather variable. In most members of the family it is 3-carpellate, and the most common type of ovary has one locule with one apical or basal ovule. In some genera the ovary is 2–5 locular with one apical-axile ovule per locule. In a few genera the carpels are distinct, and in *Anacardium* and *Mangifera* the gynoecium is one-carpellate. In six tropical genera the ovary is inferior; in all the others it is superior.

The stems and leaves often contain resinous compounds or volatile substances with a strong aroma. Some species, especially in the genera *Toxicodendron* and *Metopium* (poisonwood), produce a resinous substance called urushiol, a skin irritant that causes swelling, rashes, itching and blisters. Some other members of the family also produce small amounts of urushiol or similar compounds. The degree of sensitivity to this substance varies greatly from individual to individual. People highly sensitive to urushiol may be affected by mangos, pistachios or cashews. Lacquer produced from an extract of *Toxicodendron vernicifluum* also contains urushiol and can cause skin irritation.

The Anacardiaceae contains several economically important plants. Edible fruits or seeds are produced by *Anacardium occidentale* (cashew), *Mangifera indica* (mango) and *Pistacia vera* (pistachio). The gums and resins extracted from members of various genera have industrial applications. A few members of the family produce usable lumber. As mentioned above, lacquer is produced from an extract of *Toxicodendron vernicifluum*. In addition there are a number of ornamentals including species of *Schinus* (pepper-tree), *Cotinus* (smoke tree), *Rhus* (sumac) and *Pistacia* (pistachio).

# Aceraceae

## The Maple Family
### Sapindales ■

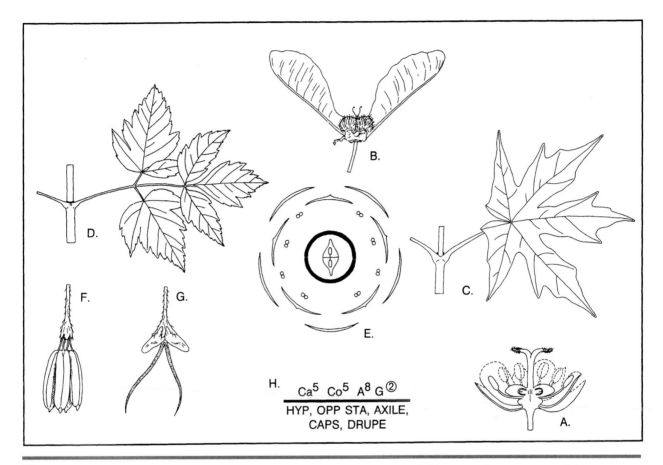

Ca$^5$ Co$^5$ A$^8$ G$\circled{2}$
HYP, OPP STA, AXILE,
CAPS, DRUPE

**Figure 14-47** Features of Aceraceae. A. Flower of *Acer macrophyllum* (l.s.). B. Schizocarp of two samaras of *Acer*. C. Opposite palmate leaves of *Acer macrophyllum*. D. Leaf of *Acer negundo*. E. Floral diagram of *Acer macrophyllum*. F–G. Staminate and pistillate flowers of *Acer negundo*. H. Generalized floral formula.

## FAMILY DESCRIPTION

Shrubs or trees. Leaves opposite, simple and palmately veined or palmately or pinnately compound, estipulate. Plants synoecious, monoecious, dioecious or polygamous. Inflorescence a panicle, raceme, corymb or umbel-like axillary cluster. Flowers perfect or imperfect, regular. Sepals (4)5 (6–9), distinct or basally connate, sometimes very reduced. Petals (4) 5 (6–9), distinct or absent. Stamens 8 (4–12), distinct. Nectary disc often present. Carpels 2 (3+), connate, winged; ovary superior with 2 locules and 2 axile ovules per locule; styles 2 or 1 and deeply divided. Fruit a samaroid schizocarp with 1 seed maturing per locule.

# FAMILY NOTES

The Aceraceae is a small but ecologically important family that comprises only two genera and about 120 species. Species of *Acer* (maple) are important and often dominant components of the temperate deciduous forests in both the Old and New Worlds. They extend into the tropics of southeast Asia. The other genus in the family, *Dipteronia,* occurs only in China.

Flowers of maples range from comparatively showy and insect-pollinated to small, inconspicuous and wind-pollinated. Some have a biseriate perianth (Figure 14-47 A) and in others (e.g., *Acer negundo*) the perianth is reduced to a minute rim (Figure 14-47 F, G). Maple flowers are often imperfect. In some species staminate, pistillate and perfect flowers occur together in the same inflorescence. The gynoecium is usually flattened and even in flower has two projecting wing-like protuberances (Figure 14-47 A). These expand into airfoils as the fruit matures as a samaroid schizocarp (Figure 14-47 B). In some maples only one of the mericarps typically develops; the second aborts. At maturity the fruits flutter to the ground like tiny helicopters.

Several species of *Acer* are valuable as timber trees, producing a fine-grained wood useful for furniture. The sugar-rich sap of *Acer saccharum* is the source of maple sugar products. Maples are widely used for shade and street trees. They dominate large areas of the temperate deciduous forests of eastern North America. When their leaves change color in autumn, they are among the most colorful trees of the forest.

# Exercises

## ■ Study Questions

### Saxifragaceae

1. What kinds of gynoecium occur in the Saxifragaceae?

   Ovary position?

   Placentation?

   Flower insertion?

2. Most members of the Saxifragaceae have __-merous flowers.

3. What fruit types are found in the Saxifragaceae?

### Crassulaceae

4. What features are characteristic of the leaves in the Crassulaceae?

5. What kind(s) of corollas can be found among members of the Crassulaceae?

   Corolla symmetry?

6. How many stamens would you expect to find in a flower of the Crassulaceae?

7. What kind of gynoecium characterizes the Crassulaceae?

   What is the ovary position?

   Placentation?

8. What is the characteristic fruit type in the Crassulaceae?

### Grossulariaceae

9. What kind of gynoecium characterizes the Grosulariaceae?

   Ovary position?

   Placentation?

10. Most members of the Grossulariaceae have __-merous flowers.

11. What is the type fruit in the Grossulariaceae?

12. On a lab exam how would you be able to decide if a plant is in the Grossulariaceae or in the Onagraceae?

### Hamamelidaceae

13. What growth habits are represented among members of the Hamamelidaceae?

    Leaf arrangement?

    Leaf complexity?

14. What is unusual about the flowering time of *Hamamelis?*

15. Flowers of the Hamamelidaceae are unusual among members of the Hamamelidaceae in having _____.

16. How many carpels comprise the gynoecium in the Hamamelidaceae?

    Placentation?

    Ovary position?

    Fruit type?

## Geraniaceae

17. Flowers in the Geraniaceae are mostly ___-merous.

18. How many stamens do flowers in the Geraniaceae usually have?

    What is unusual about one of the sepals of *Pelargonium?*

19. Describe the gynoecium of a *Geranium* or *Erodium* flower.

    How many carpels?

    Placentation?

    Ovary position?

    What is the beak?

20. What is the fruit type in most Geraniaceae?

## Myrtaceae

21. What kind of gynoecium is characteristic of the Myrtaceae?

    What is the ovary position?

    What is the flower insertion?

22. How many stamens would you expect to find in a flower of a member of the Myrtaceae?

    What are these stamens attached to?

23. What kind of leaf margin do we generally find in the Myrtaceae?

    What is another characteristic feature of these leaves?

24. To which subfamily in the Rosaceae are members of the Myrtaceae most similar?

    What features do their flowers have in common?

    In what ways are they different?

    How would you decide which one you were looking at on a lab exam?

25. What do a bottlebrush (Myrtaceae) and a member of Fabaceae subf. Mimosoideae have in common?

    How could you distinguish them?

26. What are the fruit types in the Myrtaceae?

27. In what way(s) do the flowers of *Eucalyptus* differ from those of other Myrtaceae?

## Onagraceae

28. Most members of the Onagraceae have __-merous flowers.

29. What kind of gynoecium characterizes the Onagraceae?

    What is the ovary position?

    Placentation?

    Flower insertion?

30. On a lab exam how would you be able to decide if a plant is in the Onagraceae or in the Papaveraceae?

## Euphorbiaceae

31. Do all members of the Euphorbiaceae have milky sap?

32. What is the sexual condition of flowers in the Euphorbiaceae?

    What about the sexual condition of the plants?

33. Are all flowers in the Euphorbiaceae apetalous?

34. What kind of gynoecium characterizes the Euphorbiaceae?

    How many carpels do we usually find per flower?

    Placentation?

    Ovary position?

35. The fruits of many Euphorbiaceae are described as schizocarpic capsules.

    How can a fruit be both a schizocarp and a capsule?

36. What is a cyathium?

    Describe the individual flowers of the genus *Euphorbia*.

    Do all members of the Euphorbiaceae have cyathia? Explain.

37. Some members of the genus *Euphorbia* look remarkably like cacti.

    What would you look for to decide if a particular plant is a succulent *Euphorbia* or a cactus?

    Can you explain how such distantly related plants could come to look so similar?

## Clusiaceae (Guttiferae)

38. What type of corolla do members of the Clusiaceae have?

    How many petals?

    Symmetry?

39. What is unusual about the stamens in most Clusiaceae?

40. What kind of gynoecium do we find in this family?

    Ovary position?

    Placentation?

41. What are the fruit types?

42. How is subf. Hypericoideae differentiated from subf. Clusioideae?

# Violaceae

43. What is the habit of the temperate members of the Violaceae?

44. What kind of corolla does *Viola* have?

    Symmetry?

    Unusual features?

45. How many stamens do Violaceae flowers have?

    What features are unusual?

46. What kind of gynoecium characterizes the Violaceae?

    Ovary position?

    Placentation?

47. What is a cleistogamous flower?

    How does it differ from an ordinary violet flower?

# Salicaceae

48. What kind of inflorescence characterizes the Salicaceae?

49. What is the sexual condition of the flowers?

    What is the sexual condition of the plants?

50. What kind of perianth do Salicaceae flowers have?

51. What is the fruit type in the Salicaceae?

    What is unusual about the seeds?

# Fabaceae

52. What kind of gynoecium characterizes the Fabaceae?

    Ovary position?

    Placentation?

53. Fabaceae flowers are __-merous.

54. What is the fruit in most Fabaceae?

    Describe some of the modifications of this fruit type?

55. What is the traditional family name for the Fabaceae?

56. Which subfamily of the Fabaceae has radial symmetry?

57. Which has a banner petal, two wing petals, and two fused keel petals?

58. How does the banner petal of the Papilionoideae differ from that of the Caesalpinioideae?

59. Which subfamily usually has diadelphous stamens?

60. In which are the stamens generally more showy than the petals?

61. In which are the stamens often numerous?

62. In which are the leaves generally bipinnate?

In which are they usually once pinnate but never bipinnate?

In which are they sometimes bipinnate and sometimes once pinnate?

# Rosaceae

63. What habit classes are represented in the Rosaceae?

What leaf complexity types?

Are vegetative features going to be very useful in helping you to recognize the Rosaceae? Explain.

64. Are inflorescence features going to be useful in characterizing the Rosaceae?

65. Flowers in the Rosaceae are generally ___-merous.

What type of corolla is characteristic of Rosaceae flowers?

What are the petals attached to?

66. How many stamens do we find in most Rosaceae flowers?

What are the stamens attached to?

67. What do all Rosaceae flowers have that all Ranunculaceae flowers lack?

68. What is a subfamily?

69. What is the ovary position in Rosaceae subf. Spiraeoideae?

in subf. Rosoideae?

in subf. Prunoideae?

in subf. Maloideae?

70. What is/are the gynoecium type(s) in Rosaceae subf. Spiraeoideae?

in subf. Rosoideae?

in subf. Prunoideae?

in subf. Maloideae?

71. What is/are the fruit type(s) in Rosaceae subf. Spiraeoideae?

in subf. Rosoideae?

in subf. Prunoideae?

in subf. Maloideae?

72. What are the important fruit crops of the Rosaceae subf. Rosoideae?

of Rosaceae subf. Prunoideae?

of Rosaceae subf. Maloideae?

# Rhamnaceae

73. What kind of gynoecium characterizes the Rhamnaceae?

Placentation?

Ovary position?

74. Rhamnaceae flowers are usually ___-merous.

    What features characterize their petals?

75. What feature of the stamens helps to distinguish Rhamnaceae from other families?

    What are the stamens attached to?

76. What are the fruit types in the Rhamnaceae?

77. Many members of the Rhamnaceae produce thorns.

    What is a thorn?

    How is it different from a spine?

    How does it differ from a prickle?

## Moraceae

78. What kind of sap do members of the Moraceae generally have?

79. What is the sexual condition of Moraceae flowers?

80. What kind of perianth characterizes flowers in the Moraceae?

81. What kind of gynoecium characterizes the Moraceae?

    How many carpels do we usually find per flower?

    What is the ovary position?

82. Individual ovaries in the Moraceae generally ripen as _____.

83. What is a mulberry?

    What is the fleshy tissue?

    What are the little hard things?

84. What kind of inflorescence do we find in the genus *Ficus?*

    What are the characteristic features of these inflorescences?

    What is the pollinator in *Ficus?*

    How do these pollinators bring about pollination?

85. What is the crunch in your fig-newton?

## Urticaceae

86. What causes nettles to sting?

    Do all members of the Urticaceae sting?

87. Describe the perianth of a flower in the Urticaceae.

    How many sepals?

    How many petals?

88. What is the sexual condition of Urticaceae flowers?

89. What kind of gynoecium do Urticaceae flowers have?

    How many locules?

    Placentation?

    Fruit type?

## Cucurbitaceae

90. What unusual vegetative feature occurs in the Cucurbitaceae?

91. What is the leaf venation in most Cucurbitaceae?

92. What is the sexual condition of Cucurbitaceae flowers?

    What is the sexual condition of the plants?

93. What kind of gynoecium characterizes the Cucurbitaceae?

    How many carpels do we usually find per flower?

    What is the ovary position?

94. What feature of the stamens helps to distinguish Cucurbitaceae from other families?

    What are these stamens attached to?

95. What kind of fruits do we find in the Cucurbitaceae?

    What are the characteristic features of these fruits?

    What are some common examples of these fruits?

96. Some members of the Cucurbitaceae lack sepals.

    How do we know that it is the sepals that are missing and not the petals?

## Fagaceae

97. What is the habit of members of the Fagaceae?

    Leaf complexity?

    Venation?

98. What is the sexual condition of plants in the Fagaceae?

99. What kind of inflorescence characterizes the staminate flowers of the Fagaceae?

    What about the pistillate flowers?

100. What are the parts of a staminate flower in the Fagaceae?

    Of a pistillate flower?

101. What kind of gynoecium characterizes the Fagaceae?

    Placentation?

    Ovary position?

102. What is the fruit type in the Fagaceae?

    What surrounds the developing fruit?

# Betulaceae

103. What kinds of habit occur among members of the Betulaceae?

    Leaf complexity?

    Leaf venation?

104. What is the sexual condition of plants in the Betulaceae?

105. What kind of staminate inflorescence characterizes the Betulaceae?

    How do these differ from similar inflorescences in the Fagaceae?

    Why are these inflorescences considered to be spike-like instead of true spikes?

106. What variations in the perianth do we find in staminate flowers in the Betulaceae?

107. What kind of pistillate inflorescence characterizes the Betulaceae?

    When we view one of these inflorescences externally, what structures are visible to us?

108. What variations in the perianth do we find in pistillate flowers in the Betulaceae?

    How many carpels do we find per flower?

    Ovary position?

109. What is the pollination mechanism in the Betulaceae?

110. What are the fruit types in the Betulaceae?

# Brassicaceae (Cruciferae)

111. What kind of inflorescence characterizes the Brassicaceae?

112. Brassicaceae flowers are ___-merous.

    What type of corolla do they have?

    How does the appearance of these flowers relate to the alternate name of the family?

113. What feature of the stamens helps to distinguish Brassicaceae from other families?

    What term describes this condition?

    What are the stamens attached to?

114. What kind of gynoecium characterizes the Brassicaceae?

    How many carpels do we find per flower?

    What is unusual about the placentation?

    What is the ovary position?

115. What are the main fruit types of the Brassicaceae?

    What is the difference between them?

    How do these differ from other kinds of capsules?

## Malvaceae

116. What is the most common pattern of leaf venation in the Malvaceae?

117. Are the sepals of Malvaceae distinct or connate?

    What structures commonly surround the calyx?

118. Malvaceae flowers are usually _____-merous.

    What type of corolla do they usually have?

    What are the petals attached to?

119. What feature of the stamens helps to distinguish Malvaceae from other families?

    What are the stamens attached to?

120. What kind of gynoecium characterizes the Malvaceae?

    Placentation?

    Ovary position?

121. What are the fruit types in the Malvaceae?

## Rutaceae

122. What are the characteristic features of the foliage of most Rutaceae.

    Leaf arrangement?

    Variations in leaf complexity?

    Surface features?

123. Flowers in most Rutaceae are ____-merous or ____-merous.

    Are the petals of most genera distinct or connate?

124. What is the ovary position of most Rutaceae?

125. What kind of fruit is a grapefruit?

    What does each segment represent?

    What other kinds of fruits are found in the Rutaceae?

## Anacardiaceae

126. What is the leaf arrangement in most Anacardiaceae.

    Describe the variations in leaf complexity.

127. The trifoliolate leaves of immature box-elder trees *(Acer negundo)* often strongly resemble those of poison-ivy *(Toxicodendron radicans)*. What is a sure way of telling the two apart?

128. Most members of the Anacardiaceae have ____-merous flowers.

    How many stamens are there in the flowers of most genera?

    How many are there in the flowers of *Rhus* and *Toxicodendron?*

129. Describe the variations in the gynoecium of the Anacardiaceae.

    Which is the most common type?

# Aceraceae

130. How are the leaves arranged in the Aceraceae?

Describe a typical maple leaf.

How are the leaves of *Acer negundo* (box-elder) different?

131. Describe the gynoecium of a maple flower.

132. Some texts describe the fruit of a maple as a samara. Others describe it as a schizocarp. Which is correct? Explain.

133. List the families and species of rosids that you identified in the lab.

_____    _____

_____    _____

_____    _____

_____    _____

_____    _____

_____    _____

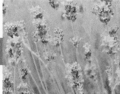

# Asterids

chapter 15

The Asteridae is a large group of dicots, comprising 10 orders, 49 families, and about 60,000 species. One third of these are placed into one order, the Asterales.

The Asterids, is a highly unified group with many features shared by its orders and families. Although all habit classes are represented in the Asteridae, herbs predominate in many families. Inflorescences are diverse, sometimes uniform within a family and in other cases highly varied. In several lineages there is a strong tendency for the inflorescence to become so condensed that it, rather than the individual flower, serves as the pollination unit. Flowers in the subclass usually have 4-merous or 5-merous sympetalous corollas with epipetalous stamens. The stamens are equal in number to the petals or fewer and occupy positions alternate with the corolla lobes. The gynoecium comprises 2–5 connate carpels (most often 2). The ovary position varies from superior to inferior, but flowers lack a free hypanthium. A nectar-bearing disk is often present. Ovules have a single massive integument and a thin nucellus. Several families have distinctive fruit types although such generalized fruits as capsules, berries and achenes are common.

When the features of the Asterids are compared with those of other dicots, a general set of advancements can be noted. Unlike the situation in some portions of the Magnoliidae, Rosidae, and Caryophyllidae, the number of perianth parts is almost always fixed at 4 or 5. Although certain members of other subclasses have sympetalous corollas, the feature is nearly uniform in the Asteridae. The number of stamens is determinate, almost always equal in number to the corolla lobes or fewer. No members of the Asteridae have a truly apocarpous gynoecium. A large portion of the Asterids has been recognized (under different names) as comprising a coherent group in earlier classification systems.

Figure 15-1 shows the relationships within the group. There are two major clades in the group, and these are recognized here as Euasterid I and Euasterid II. The Cornales and Ericales appear to be sister taxa to the Asterid clade.

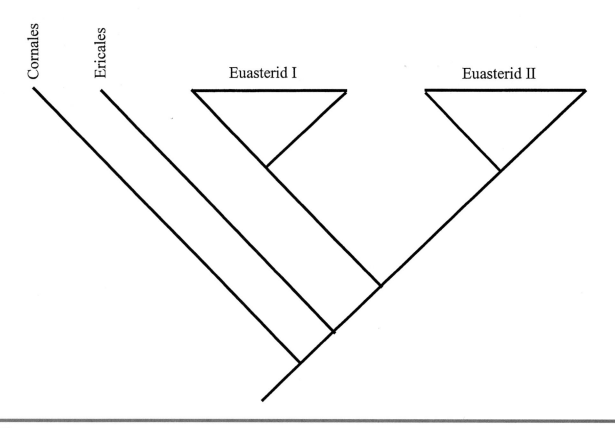

**Figure 15-1** Phylogeny of the Asterid clade, showing the basal lineages and two major clades (Euasterid I and Euasterid II). The figure is adapted from Stevens (2001 onward).

# BASAL TO THE ASTERIDS LINEAGES

## 1. Order Cornales

**1.A   Cornaceae**—The Dogwood Family (Figure 15-2). 6–11 genera, 100 species. Widespread in north temperate regions, less common in tropics and southern hemisphere.

Shrubs or trees (rarely herbs). Leaves alternate or opposite, simple, estipulate. Plants synoecious or dioecious. Inflorescence of various compound cymes, sometimes paniculiform or corymbiform, or of involucrate heads. Flowers perfect or imperfect, regular. Sepals 4 or 5 (0), distinct or basally connate. Petals 4 or 5 (0), distinct. Stamens 4 or 5, distinct. Carpels 2 (1–4), connate; ovary inferior with 1–4 locules and 1 apical-axile (infrequently apical parietal) ovule per locule; style 1 and undivided or 2–4 branched or styles 2–4, distinct. Fruit a drupe or berry, sometimes coalesced forming a multiple fruit.

**1.B   Alangiaceae**—1 genus, 20 species. Eastern Asia to Australia.

**1.C   Nyssaceae**—3 genera, 8 species. Eastern North America and eastern Asia.

**1.D   Hydrangeaceae**—The Mock-orange Family (Figure 15-3). 16–17 genera, 170 species. Temperate Northern Hemisphere. Shrubs (herbs, trees). Leaves alternate or opposite, simple, estipulate. Plants synoecious (rarely polygamous). Inflorescence of cymes, corymbiform clusters, heads or racemes (rarely flowers solitary). Flowers perfect (occasionally some flowers imperfect) regular. Hypanthium present, ± adnate to ovary. Sepals 4–5 (6–10), distinct or basally connate. Petals 4–5 (6–10), distinct. Stamens 4–many, distinct or filaments basally connate, when many developing centripetally. Carpels 2–7, connate; ovary superior to inferior, often half-inferior, with 2–7 locules and axile placentation or 1 locule and parietal placentation, sometimes parietal above and axile below, ovules many (1) per locule; styles 1–7, distinct to wholly connate. Fruit a capsule (rarely a berry).

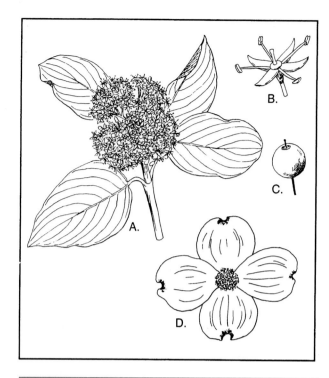

Figure Examples of Cornaceae. A–C. *Cornus sericea.*
15-2 A. Leafy branch with umbelliform cyme.
B. Flower. C. Drupe. D. Head of *Cornus canadensis* with subtending petaloid bracts.

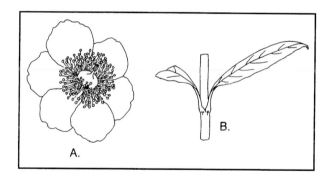

Figure Examples of Hydrangeaceae. A–B. Flower and
15-3 node with opposite leaves of *Carpenteria-californica.*

**1.E Hydrostachyaceae**—1 genus, 20 species. Africa and Madagascar.

**1.F Loasaceae**—The Blazing Star Family (Figure 15-4). 14–15 genera, 200 species. Temperate to tropical Americas, southwestern Africa and the Arabian peninsula.

Herbs or shrubs, sometimes twining, usually very harshly pubescent. Leaves alternate or opposite,

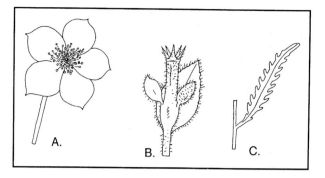

Figure Examples of Loasaceae. A–C. *Mentzelia* sp.
15-4 A. Flower. B. Immature capsule. C. Leaf.

simple to deeply divided, estipulate (stipulate). Plants synoecious. Inflorescence of cymes or racemes (heads). Flowers perfect, regular. Sepals 4–7, distinct or basally connate. Petals 4–7 (many), distinct (rarely basally connate). Stamens 5–many (2), centripetal or centrifugal, distinct or filaments connate in bundles opposite the petals. Carpels 3–7 (1), connate; ovary inferior with 1 locule and 1–many parietal (rarely marginal) ovules (rarely with 2–3 locules and 1–3 axile ovules); style 1. Fruit a capsule or achene.

**1.G Grubbiaceae**—1 genus, 3 species. South Africa.

# 2. Order Ericales

**2.A Ericaceae**—The Heath Family (Figure 15-28). 116–125, 3500 species. Widespread in temperate regions and tropical mountains.

§See page 385 for description and illustrations§

**2.B Balsaminaceae**—The Touch-me-not Family. 2 genera, 450 species. Mostly tropical Asia and Africa, a few widespread in temperate regions.

Herbs or shrubs. Leaves alternate, opposite or whorled, simple, estipulate (rarely stipulate). Plants synoecious. Inflorescence of solitary flowers, axillary clusters, racemes or panicles. Flowers perfect, irregular. Sepals 3–5, distinct, the lowermost spurred or saccate. Petals 5, distinct or the lateral petals connate in pairs. Stamens 5, coherent to connate around the ovary. Carpels 5, connate; ovary superior with 5 locules and 1–many axile ovules; style 1, short or stigmas sessile. Fruit an explosively dehiscent capsule (drupe).

**2.C     Pellicieraceae**—1 genus, 1 species. Mangrove habitats of the western coast of South America.

**2.D     Polemoniaceae**—The Phlox Family. 18–20 genera, 300 species (Figure 15-29). Temperate areas of northern hemisphere south into the mountains of South America. Particularly well represented in the arid regions of the American southwest with over half in California.

§See page 388 for description and illustrations§

**2.E     Fouquieriaceae**—The Ocotillo Family (Figure 15-5). 1 genus, 11 species. Deserts and arid tropical forests of North America.

Shrubs or trees. Leaves alternate, simple, estipulate, the petioles of some leaves persistent as spines. Plants synoecious. Inflorescence a raceme or paniculiform cluster. Flowers perfect, regular. Sepals 5, distinct. Petals 5, connate into a tube. Stamens 10–many, distinct, free. Carpels 3, connate; ovary superior with 1 locule and 4–6 parietal ovules per placenta; style 1, 3-lobed or entire. Fruit a capsule.

**2.F     Pentaphylacaceae**—1 genus, 1 species. Southern China to Malaya and Sumatra.

**2.G     Theaceae**—The Tea Family (Figure 15-6). 25–40 genera, 600. Pantropical with species extend-

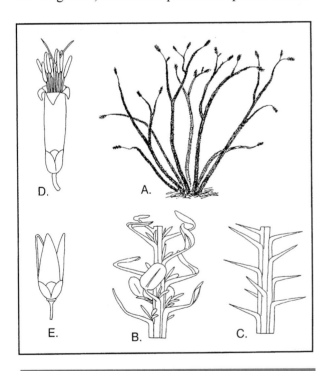

**Figure 15-5**   Examples of Fouquieriaceae. A–E. *Fouquieria splendens*. A. Habit. B. Stem with young primary leaves and axillary fascicles. C. Stem after leaf blades have separated from spiny petioles and fallen. D. Flower. E. Capsule.

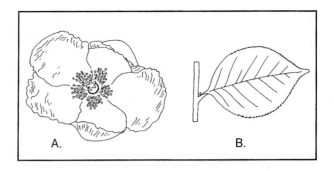

**Figure 15-6**   Examples of Theaceae. A–B. *Camellia japonica*. A. Flower. B. Leaf.

ing into the temperate regions of eastern Asia and eastern North America.

Shrubs or trees (rarely woody vines). Leaves alternate, simple, estipulate. Plants synoecious (rarely dioecious). Inflorescence of solitary flowers (less commonly cymes, racemes or panicles). Flowers perfect (rarely imperfect), regular. Sepals 4–7, distinct or basally connate. Petals 4–5 (–many), distinct or basally connate. Stamens many, centrifugal, distinct or filaments basally connate in a ring or in groups, sometimes basally adnate to the petals. Carpels 2–5 (6–many), connate; ovary superior (rarely inferior) with 2–5 (6–many) locules and 2–many axile ovules per locule; styles 2–5 (6–many), distinct or ± connate or 1 and undivided. Fruit a capsule, drupe, berry or achene.

**2.H     Ebenaceae**—The Ebony Family. 3–5 genera, 450 species. Widespread in the tropics and subtropics with a few temperate representatives.

Shrubs or trees (rarely woody vines). Leaves alternate, simple and entire, estipulate. Plants dioecious (rarely monoecious or synoecious). Inflorescence of solitary axillary flowers or cymes (rarely of racemes or paniculiform clusters). Flowers imperfect (rarely perfect), regular. Sepals 3–8, connate. Petals 3–8, connate. Stamens 3–many, distinct, free or filaments adnate to petals. Carpels 2–16, connate; ovary superior (rarely inferior) with 2–16 locules and 1–2 axile ovules per locule; styles 2–8 or 1 and simple or lobed. Fruit a berry.

**2.I     Theophrastaceae**—4 genera, 100 species. New World tropics.

**2.J     Primulaceae**—The Primrose Family. 23–30 genera, 1000 species (Figure 15-30). Boreal and north temperate regions and tropical mountains.

§See page 390 for description and illustrations§

**2.K Myrsinaceae**—30–39 genera, 1000 species. Widespread in warm temperate and tropical regions.

Shrubs or trees. Leaves alternate (rarely opposite), simple, estipulate, usually punctate. Plants synoecious or dioecious. Inflorescence of axillary clusters, cymes, racemes, corymbs or panicles. Flowers perfect (rarely imperfect), regular. Sepals 4–6, distinct or connate. Petals 4–6, connate (rarely distinct). Stamens 4–6, distinct or filaments connate, adnate to and opposite the petals. Carpels 4–6, connate; ovary superior (rarely inferior) with 4–6 locules and 1-several axile ovules per locule or with 1 locule and ovules free-central or basal; style 1 and undivided. Fruit a drupe, berry or nut.

**2.L Styracaceae**—The Storax Family. 10–11 genera, 150 species. In three widely disjunct areas: The Mediterranean region, southeast Asia and warm-temperate to tropical North and South America.

Shrubs or trees. Leaves alternate, simple, estipulate. Plants synoecious (rarely polygamous). Inflorescence of axillary clusters, racemiform or paniculiform cymes. Flowers perfect (imperfect), regular. Sepals 4–8, connate. Petals 4–8, distinct or connate. Stamens 4–many; filaments basally connote and adnate to petals (rarely free). Carpels 3–5, connate; ovary superior or inferior with 3–5 locules at base and 1 locule toward apex, placentation axile below and parietal above, ovules 2–8 per placenta; style 1, undivided. Fruit a drupe or capsule.

**2.M Lecythidaceae**—The Brazil-Nut Family. 20 genera, 400 species. Pantropical, especially well represented in South America.

Trees or shrubs. Leaves alternate, simple, stipulate or estipulate. Inflorescence of racemes, panicles, solitary flowers or clusters borne on the trunk. Flowers perfect, regular or irregular. Sepals 4–6 (2–12), distinct or connate. Petals 4–6, (0), distinct or connate. Stamens many, centrifugal, filaments connate at base and adnate to corolla. Carpels 2–6, connate; ovary inferior with 2–6 locules and 1–many axile ovules per locule. Fruit a capsule (drupe or berry).

**2.N Scytopetalaceae**—5 genera, 20 species. Tropical west Africa.

**2.O Sapotaceae**—The Sapodilla Family. 53–70 genera, 800 species. Pantropical.

Shrubs or trees with milky sap. Leaves alternate (rarely opposite), simple, stipulate or estipulate.

Plants synoecious. Inflorescence of cymes, solitary flowers or axillary clusters. Flowers perfect, regular (rarely irregular). Sepals 4–12, distinct or basally connate, sometimes in 2 or more series. Petals 4–many, connate, in 1 or 2 series. Stamens 3–many, distinct, filaments adnate to corolla. Carpels 4 or 5 (1–14), connate; ovary superior with 4 or 5 (1–14) locules and 1 axile ovule per locule; style 1, undivided or apically lobed. Fruit a berry.

**2.P Mitrastemonaceae**—1 genus, 2 species. Central America and western Pacific.

**2.Q Actinidiaceae**—The Kiwi-fruit Family. 3 genera, 300 species. Tropics and subtropics, especially Asia and west Pacific.

Shrubs or trees, sometimes climbing. Leaves alternate, simple, estipulate (rarely stipulate). Plants synoecious, dioecious or polygamous. Inflorescence of axillary clusters or variously cymose (rarely of heads). Flowers perfect or imperfect, regular. Sepals 5, distinct. Petals 5 (4) (0), distinct (rarely connate). Stamens 10–many, centrifugal, distinct, sometimes basally adnate to the petals. Carpels 5–many, connate; ovary superior with 5–many locules and axile ovules; styles 5–many, distinct or apically connate. Fruit a berry or capsule.

**2.R Sarraceniaceae**—The Pitcher-plant Family (Figure 15-7). 3 genera, 8 species. Acid bogs, Eastern and western North America and northern South America.

Herbs. Leaves basal and alternate, simple, estipulate, the basal leaves with elongate hollow, often

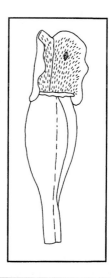

**Figure 15-7** Example of Sarraceniaceae.

water-filled petioles and reduced blades, the stem leaves reduced and bractlike. Plants synoecious. Inflorescence of solitary scapose flowers or of scapose racemes. Flowers perfect, regular. Sepals 4–6, distinct. Petals 4–6 (0), distinct. Stamens 12–many, distinct. Carpels 3–6, connate; ovary superior with 3–6 locules and axile ovules; style 1, entire or 4–6 lobed. Fruit a capsule.

**2.S  Clethraceae**—The Clethra Family. 1 genus, 65 species. Warm-temperate and tropical regions of eastern Asia to the East Indies, and of North and South America.

Shrubs or trees. Leaves alternate, simple, estipulate. Plants synoecious. Inflorescence of racemes or panicles. Flowers perfect, regular. Sepals 5, distinct. Petals 5, distinct or basally connate. Stamens 10, distinct, anthers opening by pores. Carpels 3, connate; ovary superior with 3 locules and many axile ovules per locule; style 1, apically 3–many lobed. Fruit a capsule.

**2.T  Cyrillaceae**—3 genera, 14 species. Tropical and subtropical regions of North and South America.

**2.U  Tetrameristaceae**—2 genera, 2 species. Malaysia and South America.

**2.V  Marcgraviaceae**—5 genera, 100 species. Tropical America.

**2.W  Empetraceae**—The Crowberry Family. 3 genera, 5 species. Widespread in boreal regions, disjunct in southern South America.

Shrubs. Leaves alternate, simple and entire, estipulate. Plants dioecious (rarely polygamous). Inflorescence of axillary flowers, racemes or heads. Flowers imperfect (rarely perfect), regular. Sepals 2–3, distinct, sometimes petaloid. Petals 3, distinct. Stamens 2–4, distinct. Carpels 2–9, connate; ovary superior with 2–9 locules and 1 axile ovule per locule; style 1 with 2–9 stigmas. Fruit a dry or fleshy berry or drupe.

**2.X  Diapensiaceae**—7–8 genera, 18 species. Circumboreal.

**2.Y  Lissocarpaceae**—1 genus, 2 species. Tropical South America.

**2.Z  Symplocaceae**—The Sweetleaf Family. 1 genus, 300–400 species. Moist tropics and subtropics of the New World, southeast Asia Australia and the East Indies

Shrubs or trees. Leaves alternate, simple, estipulate. Plants synoecious (rarely polygamous). Inflorescence a raceme or panicle. Flowers perfect (rarely imperfect), regular. Sepals 5, distinct or basally connate. Petals 5 or 10, connate at base. Stamens 4–many, distinct or filaments basally connate; filaments adnate to corolla. Carpels 2–5, connate; ovary inferior or half-inferior with 2–5 locales and 2–4 axile ovules per locule; style 1, undivided. Fruit a berry or drupe.

# Euasterids I

## 3. Order Garryales

**3.A  Garryaceae**—The Silk-Tassel Family. 1 genus, 13 species. Western North America to southern Mexico and Cuba.

Shrubs or small trees. Leaves opposite, simple, estipulate. Plants dioecious. Staminate inflorescence a pendulous raceme; pistillate inflorescence a stiff or pendulous spike. Flowers imperfect. Sepals 4, distinct in staminate flowers, 0, 2 or 4, connate and adnate to ovary in pistillate flowers. Petals absent. Stamens 4, distinct. Carpels 2 (3), connate; ovary inferior with 1 locule and 2 apical-parietal ovules; styles 2 (3), persistent. Fruit a 1–2 seeded berry.

**3.B  Icacinaceae**—50–53 genera, 400 species. Pantropical.

**3.C  Oncothecaceae**—1 genus, 1 species. New Caledonia.

**3.D  Eucommiaceae**—1 genus, 1 species. China.

## 4. Order Gentianales (APG)

**4.A  Gentianaceae**—Gentian Family (Figure 15-8). 75–76 genera, 1000 species. Cosmopolitan.

Herbs (rarely shrubs). Leaves opposite or whorled (rarely alternate), simple and usually entire, estipulate. Plants synoecious. Inflorescence variously cymose or flowers solitary. Flowers perfect, regular or slightly irregular. Sepals 4 or 5, connate. Petals 4 or 5 (–12), connate often with scale-like appendages within the tube. Stamens 4 or 5, distinct, alternate with petals and filaments adnate to corolla, Nectary disc or distinct nectar glands present Carpels 2, connate; ovary superior with 1 locule and many parietal ovules (2 locules and many axile ovules; rarely with free-central ovules); style 1, undivided. Fruit a capsule (rarely a berry).

**4.B   Rubiaceae**—The Madder Family. 450–606 genera, 6500 species (Figure 15-31). Pantropical with some temperate representatives.

§See page 392 for description and illustrations§

**4.C   Dialypetalanthaceae**—1 genus, 1 species. Brazil.

**4.D   Saccifoliaceae**—1 genus, 1 species. South America.

**4.E   Loganiaceae**—The Logania Family. 20 genera, 500 species. Pantropical plus a few representatives in temperate regions.

Herbs, shrubs, woody vines and trees. Leaves opposite (rarely alternate or whorled), simple, stipulate (sometimes one of a pair modified into a tendril). Plants synoecious. Inflorescence of cymes or flowers solitary. Flowers perfect, regular. Sepals 4 or 5, distinct or connate. Petals 4 or 5 (8–16), connate. Stamens 5 (1–16), distinct, alternate with petals; filaments adnate to corolla. Carpels 2, connate; ovary superior (rarely half-inferior) with 2 locules and

many (few) axile ovules (rarely 4-locular with placental septa) or with 1 locule and many (few) parietal ovules; style 1, usually undivided. Fruit a capsule (rarely a drupe or berry).

**4.F   Apocynaceae**—The Dogbane Family (Figure 15-9). 168–200 genera, 2000 species. Widespread, mostly tropical.

Herbs, shrubs, woody vines or trees, often with milky sap. Leaves opposite or whorled, simple and entire, estipulate (rarely stipulate). Plants synoecious. Inflorescence paniculiform or cymose or flowers solitary. Flowers perfect, regular. Sepals 5, distinct or connate. Petals 5, connate, often bearing scale-like or petal-like appendages within the tube. Stamens 5, distinct, alternate with petals; filaments adnate to corolla; anthers adnate or adherent to stigma. Nectary disc or distinct nectar-glands sometimes present. Carpels 2, distinct at base and connate by stigma or wholly connate; ovaries superior or half-inferior with 1–many marginal ovules *or* ovary superior with 1 or 2 locules and 1–many parietal or axile ovules; styles 2, distinct at base or connate at stigma, or style 1, undivided. Fruit a berry, drupe, schizocarp or pair of follicles.

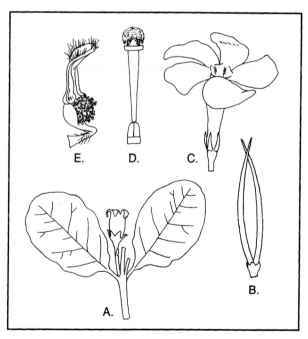

**Figure 15-9**   Examples of Apocynaceae. A–B. *Apocynum androsaemifolium*. A. Stem with leaves and flower. B. pair of immature follicles. C–E. *Vinca minor*. C. Intact flower. B. Pistil. E. Stamen.

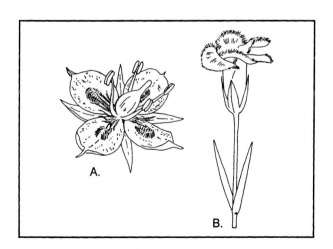

**Figure 15-8**   Examples of Gentianaceae. A. Flower of *Swertia* sp. B. Leaves and flower of *Gentiana detonsa*.

**4.G Asclepiadaceae**—The Milkweed Family. 315 genera, 2000 species (Figure 15-32). Widespread in tropical, subtropical and warm temperate regions with greatest diversity in the tropics of South America.

§See page 394 for description and illustrations§

**4.H Theligonaceae**—1 genus, 3 species. Eurasia.

**4.I Boraginaceae**—The Borage Family. 110–131 genera, 2400 species (Figure 15-33). Cosmopolitan with centers of diversity in the Mediterranean region, western North America and the American tropics.

§See page 396 for description and illustrations§

**4.J Lennoaceae**—The Lennoa Family. 2–3 genera, 5 species. Western North America.

Fleshy-stemmed herbs without chlorophyll, parasitic on roots of host plants. Leaves alternate, simple and scale-like, estipulate. Plants synoecious. Inflorescence a head or paniculiform cluster. Flowers perfect, regular or irregular. Sepals 5–10, distinct or basally connote. Petals 5–8, connate. Stamens 5–10, distinct; filaments adnate to corolla tube. Carpels 6–14, connate; ovary superior with 12–28 locules [because of false septa] and 1 axile ovule per locule; style 1, simple. Fruit a capsule.

# 5. Order Lamiales (APG notes Lamiales tree is highly provisional)

**5.A Lamiaceae (Labiatae)**—The Mint Family. 200–212 genera, 3200 species (Figure 15-34). Cosmopolitan, especially common in the Mediterranean region.

§See page 398 for description and illustrations§

**5.B Oleaceae**—The Olive Family (Figure 15-10). 24–30 genera, 600 species. Cosmopolitan.

Herbs, shrubs, woody vines or trees. Leaves opposite (alternate or whorled), simple or pinnate (rarely trifoliolate), estipulate. Plants synoecious (rarely dioecious or polygamous). Inflorescence variously cymose, often paniculiform or racemiform (rarely flowers solitary or in axillary clusters). Flowers perfect or imperfect, regular. Sepals 4 (0) (5–15), connate. Petals 4 (0) (2–12), distinct or connate. Stamens 2 (3–5), distinct; filaments adnate to corolla. Carpels 2, connate; ovary superior with 2 locules and 2 (1–10) axile ovules; style 1 with

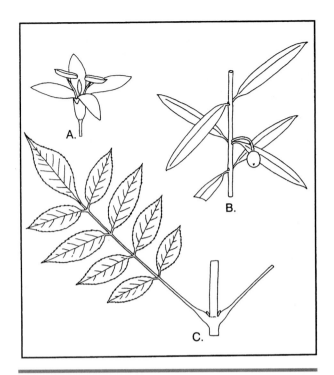

**Figure 15-10** Examples of Oleaceae. A–B. *Olea europea.* A. Flower. B. Leafy branchlet with drupe. C. Node of *Fraxinus* sp. with pinnately compound leaves.

1–2 stigmas or stigmas sessile. Fruit a berry, drupe, capsule, schizocarp, nut or samara.

**5.C Gesneriaceae**—The Gesneria Family (Figure 15-11). 120–133 genera, 2500 species. Pantropical.

Herbs or shrubs (rarely woody vines or trees), often epiphytic. Leaves opposite or all basal (infrequently alternate), simple (sometimes divided), estipulate. Plants synoecious. Inflorescence of various types of cymes, racemes or solitary flowers. Flowers perfect, irregular. Sepals 5, connate. Petals 5, connate; corolla often bilabiate. Stamens 2 or 4, distinct; filaments adnate to corolla. Carpels 2, connate; ovary superior to inferior with 1 locule and

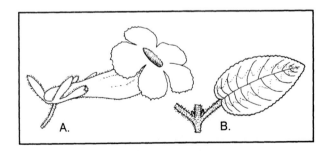

**Figure 15-11** Example of Gesneriaceae. A–B. *Episcea* sp. A. Flower. B. Node with leaves.

many parietal ovules *or* 2–4 locular by placental fusion. Style 1, undivided. Fruit a capsule or berry.

**5.D    Acanthaceae**—The Acanthus Family (Figure 15-12). 228–250 genera, 2500 species. Pantropical with a few temperate species.

Herbs or shrubs (less commonly woody vines or trees). Leaves opposite or all basal, simple, estipulate. Plants synoecious. Inflorescence extremely variable, of cymes, axillary verticils, racemes, spikes, heads, panicles or solitary flowers. Flowers perfect, irregular, usually subtended by a pair of bractlets. Sepals 4–5, distinct or connate. Petals 5 (4), connate; corolla often bilabiate. Stamens 4 (2 or 5), distinct or ± connate in pairs; filaments adnate to corolla. Carpels 2, connate; ovary superior with 2 locules and 2 (1 or 3–many) axile ovules per locule; style 1, undivided. Fruit a capsule, often explosively dehiscent, or a drupe.

**5.E    Medoniaceae**—2–4 genera, 60 species. South America, Africa, Madagascar.

**5.F    Buddlejaceae**—8–10 genera, 150 species. Pantropical with a few temperate species.

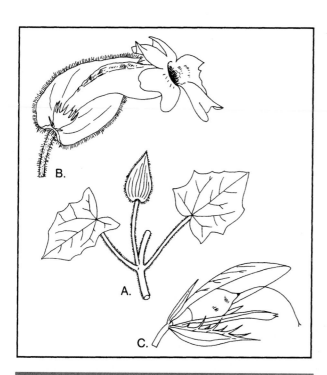

**Figure 15-12**    Examples of Acanthaceae. A–B. *Thunbergia alata*. A. Node with pair of leaves and fruit concealed by cordate bracts. B. Flower subtended by bract (front bract removed. C. Immature capsule of *Acanthus mollis* with bract and calyx.

**5.G    Bignoniaceae**—The Bignonia Family (Figure 15-13). 100–111 genera, 800 species. Pantropical with a few temperate species.

Shrubs, woody vines or trees (rarely herbs). Leaves opposite (less commonly whorled), simple, pinnate or bipinnate, estipulate; some leaflets often modified into tendrils. Plants synoecious. Inflorescence variously cymose, sometimes paniculiform, or flowers solitary. Flowers perfect, irregular or nearly regular. Sepals 5, connate. Petals 5, connate, corolla sometimes bilabiate. Stamens 4 (2 or 5), distinct; filaments adnate to corolla. Carpels 2, connate; ovary superior with 2 locules and several to many axile ovules or with 1 locule and several to many parietal ovules; style 1, undivided. Fruit a capsule with the seeds often winged or a berry.

**5.H    Byblidaceae**—1–2 genera, 4 species. Australia and South America.

**5.I    Lentibulariaceae**—The Bladderwort Family. 3–5 genera, 200 species. Cosmopolitan.

Herbs, frequently carnivorous, sometimes aquatic or semiaquatic, sometimes rootless. Leaves alternate or all basal or without clear distinction between stems and leaves, simple to much-dissected, estipulate; animal-trapping leaves bearing sticky glands or hollow, hinged bladder-like traps, or modified into hollow pitchers. Plants synoecious. Inflorescence of solitary flowers or racemes, rising above water surface in aquatic species. Flowers perfect, irregular. Sepals 2–5, distinct or basally connate. Petals 5, connate; corolla bilabiate with lower lip usually spurred. Stamens 2, distinct; filaments adnate to corolla. Carpels 2, connate; ovary superior with 1 locule and numerous free-central ovules; style 1, short, or stigma sessile. Fruit a capsule.

**5.J    Martyniaceae**—The Unicorn Plant Family. 3 genera, 10 species. New World, mostly tropical.

Herbs. Leaves alternate or opposite, simple, estipulate. Plants synoecious. Inflorescence of solitary flowers or racemes. Flowers perfect, irregular. Sepals 5, distinct or connate. Petals 5, connate. Stamens 2 or 4; anthers coherent in pairs; filaments distinct, adnate to corolla. Carpels 2, connate; ovary superior with 1 locule and many parietal ovules or sometimes with 2 locules separated by a placental septum; style 1, undivided. Fruit a large woody capsule, usually with a long, curved beak. (Included by Cronquist in Pedaliaceae.)

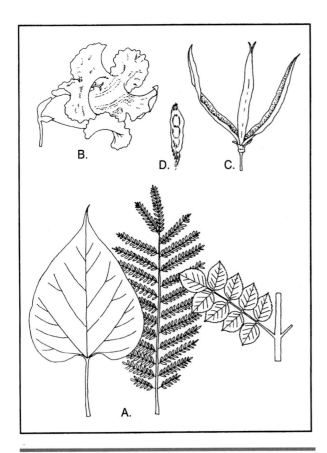

**Figure 15-13** Examples of Bignoniaceae. A. Leaves of *Catalpa bignonioides, Jacaranda acutiloba,* and *Tecomaria capensis.* B–D. *Catalpa bignonioides.* B. Flower. C. Capsule. D. Winged seed.

**5.K    Myoporaceae**—The Myoporum Family. 3–4 genera, 125 species. Mostly Australasia, 1 South America, 2 Africa.

Herbs, shrubs or trees. Leaves alternate or opposite, simple, estipulate, usually punctate. Plants synoecious. Inflorescence of axillary clusters, cymes or solitary flowers. Flowers perfect, regular or irregular. Sepals 5, connate. Petals 5, connate; corolla sometimes bilabiate. Stamens 4 (5), distinct; filaments adnate to corolla. Carpels 2, connate; ovary superior with 2 locules and 2–10 axile ovules per locule (rarely with 3–10 locules by intrusion of placental septa); style 1, undivided. Fruit a berry or drupe.

**5.L    Orobanchaceae**—The Broomrape Family (Figure 15-14). 17 genera, 150 species. Northern hemisphere.

Fleshy-stemmed herbs without chlorophyll, parasitic on roots of host plants. Leaves alternate, simple and scale-like, estipulate. Plants synoecious. Inflo-

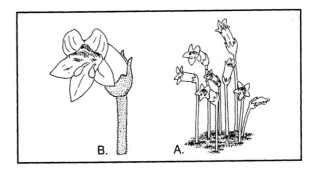

**Figure 15-14** Example of Orobanchaceae. A–B. *Orobanche uniflora.* A. Habit. B. closeup of flower.

rescence of solitary flowers, racemes, corymbs or spikes. Flowers perfect, irregular. Sepals 2–5, distinct or connate. Petals 5, connate. Stamens 4, distinct; filaments adnate to corolla. Carpels 2 (3), connate; ovary superior with 1 locule and many parietal ovules; placentae sometimes 2 per carpel; style 1, undivided. Fruit a capsule.

**5.M    Pedaliaceae**—14–17 genera, 70 species. Old World.

**5.N    Plantaginaceae**—The Plantain Family (Figure 15-15). 3 genera, 250 species. Cosmopolitan.

Herbs or shrubs. Leaves all basal (rarely alternate or opposite), simple and parallel-veined or deeply pinnately divided, estipulate; leaf bases or petioles sheathing. Plants synoecious. Inflorescence a spike (in one genus in 3-flowered clusters). Flowers perfect (imperfect), regular or weakly isobilateral. Sepals 4, distinct. Petals 4, connate, scarious. Stamens 4 (1–2), distinct, alternate with petals and filaments adnate to corolla. Carpels 2, connate; ovary superior with 2 locules (sometimes 3–4 locules because of false septa) and 1–many axile ovules per locule, or with q locule and 1–many free-central or basal ovules; style 1, 2-lobed. Fruit a circumscissile capsule or a small nut.

**5.O    Globulariaceae**—The Globularia Family. 2–10 genera, 300 species. Africa, Madagascar, Europe and western Asia.

Herbs or shrubs. Leaves alternate, simple and entire, estipulate. Plants synoecious. Inflorescence of involucrate heads, sometimes secondarily grouped into spicate or paniculiform clusters. Flowers perfect, irregular. Sepals 5, connate. Petals 5, connate; corolla bilabiate. Stamens 4; filaments adnate to corolla. Carpels 2 or seemingly 1; ovary

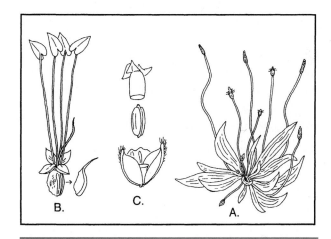

Figure 15-15 Examples of Plantaginaceae. A–C. *Plantago lanceolata.* A. Habit. B. Flower with subtending bract. C. Calyx with circumscissile capsule and seed.

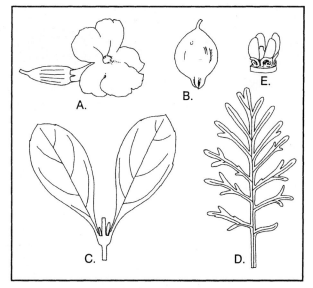

Figure 15-17 Examples of Verbenaceae. A–C. *Duranta erecta.* A. Flower. B. Drupe enclosed by persistent calyx. C. Node with pair of leaves. D–E. *Glandularia bipinnatifida.* D. Leaf. E. Nutlets on receptacle.

superior with 1 locule and 1 apical ovule; style 1, undivided or slightly bilobed. Fruit an achene.

**5.P Hippuridaceae**—1 genus, 1 species. Widespread in temperate regions.

**5.Q Callitrichaceae**—The Water Starwort Family (Figure 15-16). 1 genus, 35 species. Cosmopolitan.

Herbs, usually aquatic. Leaves opposite, simple and entire, estipulate. Plants monoecious. Inflorescence of solitary axillary flowers. Flowers imperfect, naked, subtended by a pair of tiny bractlets. Sepals absent. Petals absent. Stamen 1 (–3). Carpels 2, connate; ovary superior with 2 or 4 locules and 1 apical-axile ovule per locule; styles 2. Fruit of 2 or 4 nutlets.

**5.R Scrophulariaceae**—The Figwort Family. 190–275 genera, 4000 species (Figure 15-35). Cosmopolitan.

§See page 400 for description and illustrations§

**5.U Retziaceae.**—1 genus, 1 species. South Africa.

**5.S Verbenaceae**—The Vervain Family (Figure 15-17). 86–100 genera, 2600 species. Pantropical with a few temperate species.

Herbs, shrubs, woody vines or trees. Leaves opposite or whorled (rarely alternate), simple to pinnately or palmately lobed or compound, estipulate. Plants synoecious (rarely polygamous). Flowers perfect (rarely imperfect), regular or irregular. Sepals 5 (4–8), connate (rarely distinct). Petals 5 (4–many), connate; corolla sometimes ± bilabiate. Stamens 4 (2 or 5), distinct; filaments adnate to corolla. Carpels 2 (4 or 5), connate; ovary superior, often ± 4-lobed (–10-lobed) with 4 (–10) locules separated by true and false septa, with 1 axile ovule per locule; style 1, terminal (rarely ± gynobasic), undivided. Fruit a drupe with a lobed endocarp or breaking apart into nutlets (rarely a capsule).

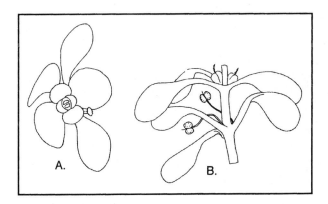

Figure 15-16 Examples of Callitrichaceae. A–B. *Callitriche marginata.* A. Stem tip with floating leaves with flowers. B. Stem with axillary fruits.

# 6. Order Solanales (APG)

**6.A    Solanaceae**—The Potato Family. 85–96 genera, 2800 species (Figure 15-36). Cosmopolitan.

§See page 402 for description and illustrations§

**6.B    Sphenocleaceae**—1 genus, 2 species. Pantropical.

**6.C    Convolvulaceae**—The Morning-glory Family (Figure 15-18). 50–54 genera, 1500 species. Cosmopolitan.

Herbs, often twining, shrubs or trees, with sap usually milky. Leaves alternate, simple to deeply divided or compound, estipulate (stipulate). Plants synoecious (rarely dioecious). Inflorescence of solitary flowers or variously cymose, the flowers generally subtended by a pair of bracts. Flowers perfect (rarely imperfect), regular (occasionally slightly irregular). Sepals 5, distinct (sometimes basally connate). Petals 5, connate. Stamens 5, distinct, alternate with petals; filaments adnate to corolla. Nectarydisc usually present. Carpels 2 (3–5), connate (rarely nearly distinct and only united by style); ovary superior with 2 (3–5) locules and 1–4 basal-axile ovules per locule; style 1 (2), simple or 2-lobed, terminal or gynobasic. Fruit a capsule (rarely a schizocarp, nut or berry).

**6.D    Cuscutaceae**—The Dodder Family. 1 genus, 150 species. Cosmopolitan.

Twining parasitic herbs without chlorophyll. Leaves absent. Plants synoecious. Inflorescence of umbelliform, corymbiform or headlike clusters. Flowers perfect, regular. Sepals 4 or 5, distinct or connate. Petals 4 or 5, connate, often with scale-like appendages within the tube. Stamens 4 or 5, distinct, alternate with petals; filaments adnate to the corolla. Carpels 2, connate; ovary superior with 2 locules and 2 axile ovules per locule; styles 2 or 1 and 2-lobed. Fruit a capsule. (Often included in Convolvulaceae.)

**6.E    Hydrophyllaceae**—The Waterleaf Family. 19–20 genera, 230 species (Figure 15-37). Widespread in New World, scattered in widely separated regions of Old World. Most diverse in semiarid regions of American southwest with over half in California.

§See page 404 for description and illustrations§

**6.F    Duckeodendraceae**—1 genus, 1 species. South America.

**6.G    Nolanaceae**—The Nolana Family. 2 genera, 66 species. South America.

Herbs or shrubs. Leaves alternate or opposite, simple, estipulate, often succulent. Plants synoecious. Inflorescence of solitary axillary flowers. Flowers perfect, regular. Sepals 5, connate. Petals 5, connate. Stamens 5, distinct, alternate with petals and filaments adnate to corolla. Nectary disc usually present. Carpels 5, connate or nearly distinct; ovary or ovaries superior with 5 locules and 5–many basal-axile ovules, sometimes lobed and vertically segmented into one-ovulate units; style 1, undivided, terminal or gynobasic. Fruit a schizocarp or divided into 5–many nutlets. (Some authorities include the Nolanaceae in the Solanaceae.)

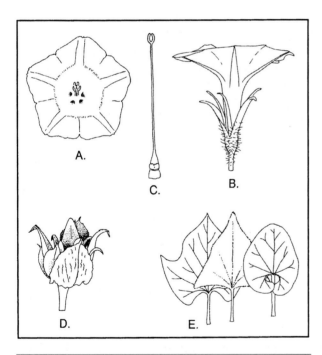

**Figure 15-18**    Examples of Convolvulaceae. A–C. *Ipomoea* sp. A. Corolla. B. Flower. C. Pistil. D. Capsule of *Calystegia macrostegia* with calyx and subtending bracts. E. Leaves of *Ipomoea* and *Calystegia* spp.

# EUASTERIDS II

## 7. Order Aquifoliales

**7.A**    **Aquifoliaceae**—The Holly Family. 4 genera, 300–400 species. Cosmopolitan.

Shrubs or trees. Leaves alternate (rarely opposite or apparently whorled), simple, stipulate (rarely estipulate). Plants synoecious or dioecious (rarely polygamous). Inflorescence of solitary or clustered axillary flowers, cymes, racemes or panicles. Flowers perfect or imperfect, regular. Sepals 4 (-7) (0), basally connate (distinct). Petals 4 (-7), distinct or basally connate. Stamens 4–9 (-12), distinct, free or filaments adnate to corolla. Carpels (2–3) 4–6 (8–many), connate; ovary superior with 2–many locules and 1 (2) apical-axile ovules per locule; style 1, undivided or stigma sessile. Fruit a drupe or berry.

**7.B**    **Cardiopteridaceae**—1 genus, 3 species. Southeast Asia to New Guinea and northern Australia.

## 8. Order Apiales

**8.A**    **Apiaceae**—The Carrot Family. 428 genera, 3000 species (Figure 15-38). Cosmopolitan, especially in temperate areas.

§See page 406 for description and illustrations§

**8.B**    **Araliaceae**—The Ginseng Family (Figure 15-19). 47–70 genera, 700 species. Widely distributed in temperate and tropical regions.

Herbs, shrubs, woody vines, trees. Leaves alternate (rarely opposite), often very large, simple to several times compound, stipulate (rarely estipulate). Plants synoecious, polygamous or dioecious. Inflorescence of heads or umbels, these often secondarily clustered into panicles, compound umbels, or corymbiform clusters. Flowers perfect or imperfect, regular. Sepals 5, distinct, usually much reduced or almost absent. Petals 5 (3–10), distinct or united toward apex. Stamens 5 (3–10) (many), distinct or united toward apex. Carpels 2–15 (1) connate; ovary inferior (very rarely superior) with 2–15 (1) locules and 1 apical/axile ovule per locule; styles 2–5, distinct or more or less connate or sometimes stigmas sessile, sometimes sub-

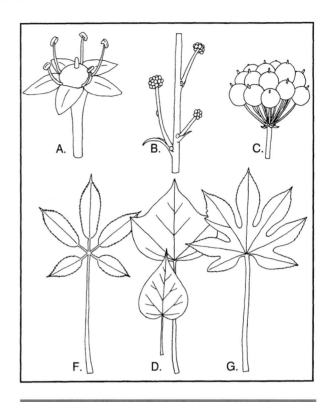

**Figure 15-19**    Examples of Araliaceae. A–E. *Hedera canariensis*. A. Flower. B. Portion of immature inflorescence. C. Umbel of drupes. D–E. Leaves from flowering and vegetative branchs. F. Leaf of *Panax quinquefolia*. G. Leaf of *Fatsia japonica.*

tended by enlarged stylopodia. Fruit a berry, drupe or schizocarp.

**8.C**    **Pittosporaceae**—The Pittosporum Family (Figure 15-20). 9 genera, 200 species. Africa and southeast Asia to Australia.

Shrubs, woody vines or trees. Leaves alternate, simple, and entire, estipulate. Plants synoecious. Inflorescence of solitary flowers or cymes, sometimes corymbiform or paniculiform. Flowers perfect, regular (rarely irregular). Sepals 5, distinct or basally connate. Petals 5, distinct or basally connate. Stamens 5, distinct (rarely filaments connate). Carpels 2–5, connate; ovary superior with 1 locule and many parietal ovules or locules 2–5 with many axile ovules; style 1, undivided. Fruit a capsule or berry.

**8.D**    **Paracryphiaceae**—1 genus, 1 species. New Caledonia.

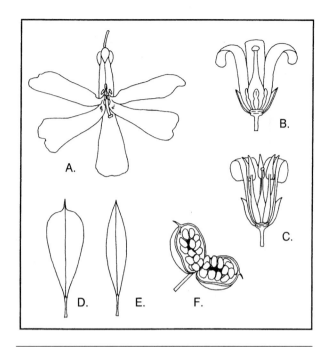

**Figure 15-20** Examples of Pittosporaceae. A. Flower of *Hymenosporum flavum*. B–C. Pistillate and staminate flowers of *Pittosporum* sp. D–E. Leaves of *Pittosporum* spp. F. Fleshy capsule of *Pittosporum tobira*.

**8.E    Escalloniaceae**—The Escallonia family. 20–24 genera, 200 species.

Shrubs or trees. Leaves alternate, simple, stipulate or estipulate. Plants synoecious (rarely dioecious or polygamous). Inflorescence of racemes, cymes, solitary flowers or axillary clusters, sometimes paniculiform. Flowers perfect (rarely imperfect), regular. Hypanthium often well-developed. Sepals 4–5, distinct or connate. Petals 4–5, distinct or basally connate. Stamens 4–5 (rarely 8–10) distinct (sometimes with 4–5 staminodes).

## 9. Order Asterales

**9.A    Asteraceae**—The Sunflower Family. 1300–1540 genera, 20,000–23,000 species. Cosmopolitan.
**9.B    Campanulaceae**—The Bellflower Family. 70–84 genera, 2000 species. (Figure 15-39). Cosmopolitan.

§See page 408 for description and illustrations§

**9.C    Pentaphragmataceae**—1 genus, 30 species. Southeast Asia.

**9.D    Alseuosmiaceae**—3–4 genera, 40 species. New Zealand and New Caledonia.

**9.E    Stylidiaceae**—5 genera, 155 species. Australasia and southern South America.

**9.F    Donatiaceae**—1 genus, 2 species. New Zealand and southern South America.

**9.G    Menyanthaceae**—5 genera, 35 species. Cosmopolitan.

**9.H    Goodeniaceae**—12–14 genera, 300 species. Mostly Australia with a few species in other tropical regions.

**9.I    Brunoniaceae**—1 genus, 1 species. Australia.

**9.J    Calyceraceae**—6 genera, 60 species. South and Central America.

**9.K    Bruniaceae**—12 genera, 75 species. Southern Africa.

## 10. Order Dipsacales

**10.A    Dipsacaceae**—The Teasel Family. 7–10 genera, 270 species. Africa and Eurasia.

Herbs (less commonly shrubs). Leaves opposite (sometimes whorled), simple to pinnately divided or compound, estipulate. Plants synoecious. Inflorescence an involucrate spike or head. Flowers perfect, irregular or nearly regular, surrounded by an adnate involucel of bractlets. Sepals 5, wholly connate and cuplike or divided into 5–20 pappus-like awns or bristles. Petals 5 or apparently 4, connate; corolla sometimes bilabiate. Stamens 2–4, distinct; filaments adnate to corolla. Carpels 2, connate; ovary inferior with 1 locule and 1 apical ovule; style 1, undivided. Fruit an achene.

**10.B    Adoxaceae**—1–3 genera, 1 species. Circumboreal.

**10.C    Caprifoliaceae**—The Honeysuckle Family. 13–15 genera, 400 species (Figure 15-40). Mostly north temperate and with some species in tropical mountains. Greatest diversity is in eastern Asia and eastern North America.

§See page 410 for description and illustrations§

**10.D    Valerianaceae**—The Valerian Family. 13–15 genera, 300 species. Nearly cosmopolitan but most common in north temperate regions.

Herbs or shrubs, often ill-scented. Leaves opposite, simple to pinnately divided or compound, estipulate. Plants synoecious (rarely dioecious). Inflorescence variously cymose, often much branched and paniculiform. Flowers perfect (rarely imperfect), irregular (rarely nearly regular). Sepals 5, usually reduced to an unlobed rim or sometimes absent, after flowering sometimes growing out into a pappus of 5–many long bristles. Petals 4 or 5, connate; corolla sometimes bilabiate, often saccate or spurred at base, Stamens 1–4, distinct; filaments adnate to corolla. Carpels 3, connate; ovary inferior with 1–3 locules, one of which contains 1 apical ovule; style 1, undivided. Fruit an achene.

**10.E  Columelliaceae**—1 genus, 4 species. Andean region, Columbia and Bolivia.

# ASTERID II—A Focus on the Asteraceae

Because of the large size and great diversity of the Asteraceae (Compositae), the family is treated here separate from the remainder of the Asterales. The Asteraceae is the largest family of dicots and the second largest family of flowering plants [only the Orchidaceae has more species]. It contains 1400–1540 genera and about 20,000–23,000 species and has a cosmopolitan distribution.

Members of the family occur from polar regions to the tropics, from sea level to the tops of high mountains and in an extraordinary diversity of habitats ranging from extremely arid deserts to aquatic environments. In many areas of the world members of the Asteraceae comprise 10 to 20 percent of the flora. The family includes many common wild and cultivated plants. It is a very important family for students to learn.

The Asteraceae have a reputation as a "difficult" family. In part this comes from the size of the family and the superficial resemblance many Asteraceae have for each other. In addition the beginner must learn a set of unfamiliar terms that describe some unfamiliar features of the flowers, fruits, and inflorescences. The reputation is not deserved, though. Once you get past the terminology hurdle, you should be able to key members of the Asteraceae with no greater difficulty than any other family. This section is designed to get you to that point.

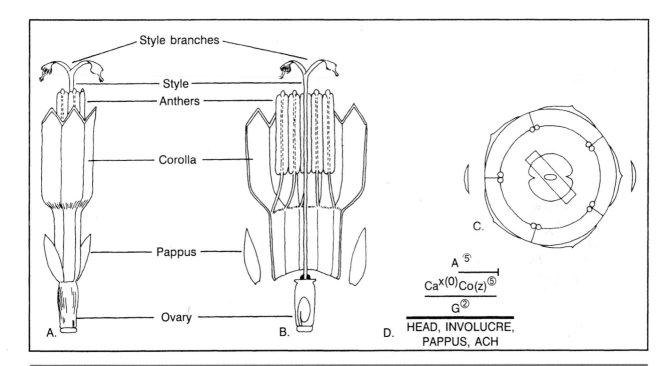

**Figure 15-21**  Disk florets. A. Intact floret. B. Floret dissected to show parts: Note that anthers are connate but filaments are distinct and adnate to corolla. The ovary is inferior with one locule and one basal ovule. There is one style with two style branches tipped by brush-like hairs. C. Floral diagram. Note that although the calyx is illustrated here as two pappus scales, the number of parts is extremely variable. D. Floral formula for disk flower or ligulate flower.

# Asteraceae (Compositae)

## The Sunflower Family
### Asterales ■

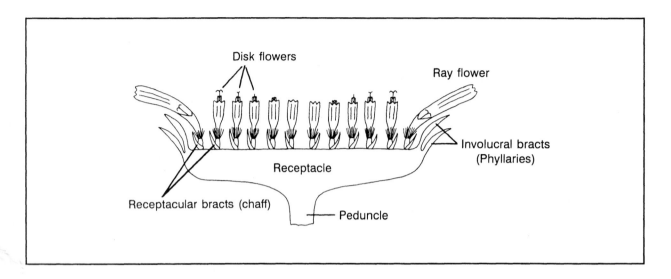

**Figure 15-22** Parts of a radiate head. Note that receptacular bracts are often absent.

## FAMILY DESCRIPTION

Herbs or shrubs (woody vines, stem-succulents, or trees). Leaves alternate or opposite (rarely whorled), simple to bipinnate, estipulate. Plants synoecious or polygamous, less commonly monoecious (rarely dioecious). Inflorescence of involucrate heads that are often secondarily aggregated in various cymose clusters, less commonly heads racemosely clustered (rarely heads grouped into secondary heads). Flowers perfect or imperfect, regular or irregular, of several kinds, each with a characteristic morphology (see descriptions below). Sepals absent or modified into a pappus of scales, awns or bristles, never green. Petals 5 (4) or apparently 3 (rarely completely absent in pistillate flowers), connate; corolla sometimes bilabiate or with a single lip. Stamens 5 (4); filaments distinct (rarely connate); anthers connate into a tube around the style (rarely distinct). Carpels 2, connate; ovary inferior with 1 locule and one basal ovule; style 1 with 2 branches, these sometimes very short or essentially absent. Fruit an achene (very rarely a berry or a drupe).

## DISCUSSION OF COMPOSITE MORPHOLOGY

In a discussion of the Asteraceae it is necessary to introduce a set of specialized terms that botanists use to describe the highly modified flowers and inflorescences. An understanding of the features that these terms describe is essential to a basic working knowledge of the family. Individual flowers (often called **florets;** Figures 15-21, 15-22, 15-26) in the Asteraceae are generally rather small and inconspicuous. They are characteristically clustered in heads (Figures 15-22 and 15-27) and

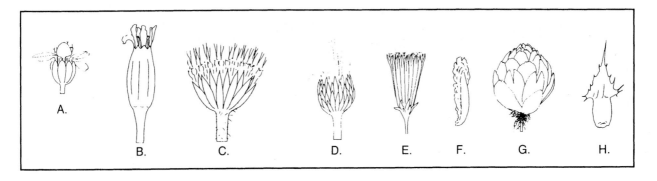

**Figure 15-23** Variation in the involucre. A. Uniseriate involucre of distinct phyllaries. B. Uniseriate involucre of connate phyllaries. C. Involucre of 2 series of subequal phyllaries. D. Involucre of strongly imbricated phyllaries in several unequal series. E. Involucre with principal phyllaries in one series and smaller bractlets (calyculum) at base. F. Individual phyllary with scarious margin. G. Involucre of imbricated phyllaries that are wholly scarious. H. Individual phyllary with fringed, spine-tipped terminal appendage.

are surrounded by an involucre of one to several series of bractlets (**phyllaries;** Figures 15-22 and 15-23). The individual flowers are attached to a common receptacle (the swollen end of the peduncle) and in some genera are individually subtended by receptacular bractlets (often called **chaff scales,** pales, or **paleae;** Figure 15-22). Each head is an indeterminate inflorescence with the most mature flowers toward the periphery and the least mature in the center. In many members of the Asteraceae the heads are aggregated in secondary clusters. Unlike the indeterminate arrangement of flowers in the head, the clustering of heads on the plant is generally determinate.

Flowers in the Asteraceae are quite small and individually play a relatively minor role in the attraction of animal pollinators. Instead, the functional pollination unit in the family is usually the head of several to many flowers. One visit by a pollinator can simultaneously pollinate a large number of individual flowers. The family's alternate name, Compositae, comes from the superficial resemblance of a head to a single large individual flower. The structure that appears both to the casual observer and to the pollinator to be an ordinary flower is actually a "composite" of small flowers.

Floral structure in the Asteraceae is very characteristic (Figure 15-21). The calyx (sometimes absent) is modified into whitish or brownish bristles, scales, or awns, collectively called the **pappus.** The units that comprise the pappus are *never* green and sepal-like. The number, size, and form of the structures comprising the pappus is remarkably variable (Figure 15-24). The features of the pappus are very useful taxonomically. When you key out mem-

bers of the Asteraceae you can expect to have to choose among different kinds of pappus structures.

The corolla is sympetalous and varies from actinomorphic to strongly zygomorphic. The corolla in radially symmetric flowers is generally 5-merous. In bilaterally symmetric flowers it may be 5-merous or apparently 3-merous (Figure 15-26). An individual head may contain more than one kind of flower.

The stamens are epipetalous, alternating with the corolla lobes. The filaments are inserted into the corolla at the junction of the corolla tube and throat. The filaments are usually distinct, but the anthers are connate into a tube (Figure 15-21) that surrounds the style. Dehiscence of the anthers is introrse (toward the axis). As a result the anthers release pollen to the interior of the anther tube. Each anther usually bears a terminal flattened sterile appendage derived from the tip of the connective (Figure 15-25 A–D). In some members of the family the bases of the anther sacs are sagittate or bear tail-like appendages (Figure 15-25 B, C).

Each flower has a 2-carpellate inferior ovary with one locule and a solitary basal ovule. The ovary matures as an achene [also called a cypsella]. There is one style that generally branches near its apex (Figure 15-25 E–M). As a flower opens, its style grows in length. The style or its branches generally have brush-like sweeping hairs that push pollen grains from the anther tube as the style elongates. These may be located in a ring around the style, at the tips of the branches, or on sterile appendages at the ends of the style branches. The stigmatic surfaces may be restricted to the inner faces of the style branches, to the margins of the branches, or localized near the tip of the style.

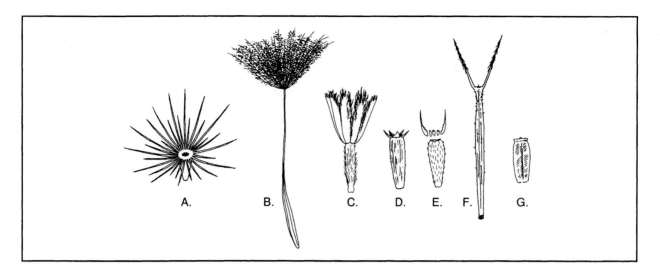

**Figure 15-24** Variations of achenes and pappus structures. A. Pappus of simple bristles. B. Beaked achene with pappus of plumose (feather-like) bristles. C. Pappus of fringed scales. D. Pappus of a crown of low scales. E. Pappus of deciduous scales. F. Pappus of barbed awns. G. Epappose achene (without a pappus).

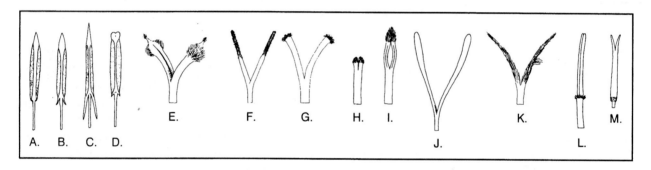

**Figure 15-25** Variations in anthers (A–D) and style branches (E–M). A. Anther of *Rudbeckia* (Heliantheae). B. Sagittate anther of *Gaillardia* (Heliantheae). C. Long-tailed anther of *Acourtia* (Mutisieae). D. Anther of *Pectis* (Heliantheae). E. Style branches of *Dahlia* (Heliantheae). F. Style branches of *Rudbeckia* (Heliantheae). G. Truncate style branches of *Achillea* (Anthemideae). H. Very short style branches of functionally staminate disk flower of *Osteospermum* (Calenduleae). I. Style with adherent branch-tips of *Aster* (Astereae). J. Clavate style branches of *Ageratum* (Eupatorieae). K. Elongated, tapering style branches of *Vernonia* (Vernonieae). L. Style with ring of hairs of *Centaurea* (Cardueae). M. Style of *Arctotis* (Arctoteae) with short branches and ring of hairs.

# Variation in Floret and Head Structure

In conjunction with the diversification of the Aster-aceae, several kinds of flowers have evolved, differing in corolla morphology and/or sexual condition (Figure 15-26). The most common type is the **disk flower** (Figure 15-21 A–B, Figure 15-26 E). It has a radially symmetric tubular corolla. In most cases disk flowers are perfect, forming both pollen and seeds. Disk flowers are sometimes borne in heads by themselves (**dis-coid heads;** Figure 15-27 B) or are borne in the center of heads that also contain another kind of flower. In some composites the disk flowers are bilabiate (Figure 15-26 D) with a 1- to 4-lobed posterior (adaxial) lip and a 4- to 1-lobed anterior (abaxial) lip.

Some authors differentiate two or more kinds of bilabiate corollas. Those with a 3-lobed anterior lip and a 2-lobed posterior lip are described as true bil-abiate corollas. Those with other arrangements are described as pseudobilabiate. The Mutisia tribe typically has true bilabiate corollas. In the Barnedesia tribe bilateral corollas are typically pseudobilabiate with 1-lobed posterior lips and 4-lobed anterior lips.

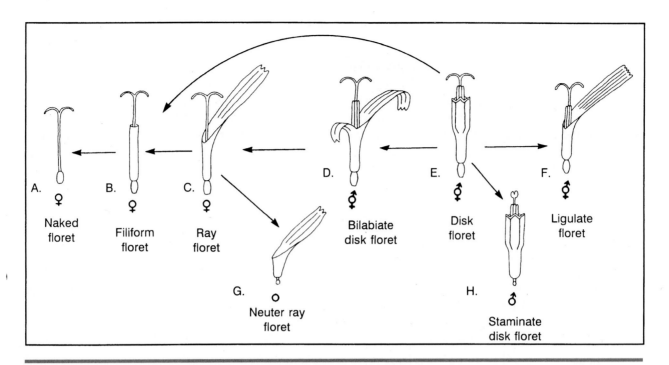

**Figure 15-26** Putative evolutionary derivation of flower types in the Asteraceae. Radially symmetric disk flowers are assumed to be ultimately ancestral to all other types.

Heads of many composites have strongly zygomorphic peripheral flowers (**ray flowers;** Figures 15-22, 15-26 C, G). These usually have a strap-shaped 3-lobed anterior lip (**ligule**) and essentially no posterior lip (rarely a short posterior lip is present). Ray flowers generally have no stamens (rarely staminodes are present) and sometimes lack a functional gynoecium. (An ovary is always present, but it may be much reduced; the style may be present or absent.) A ray flower is generally either pistillate or neuter. Ray flowers are almost always borne in heads together with disk flowers (**radiate heads;** Figure 15-27 A). The

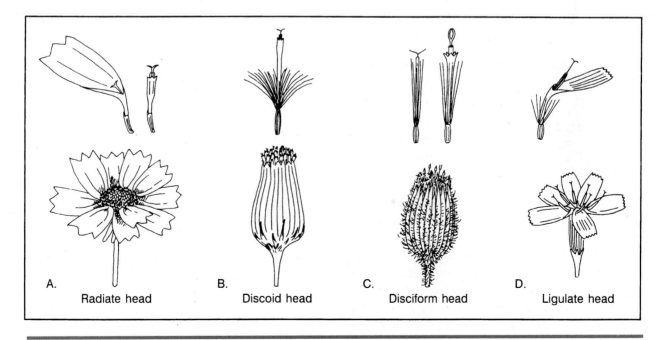

**Figure 15-27** Types of heads and the flowers that they contain. A. Radiate head with ray and disk flowers. B. Discoid head with only disk flowers. C. Disciform heads with filiform flowers and disk flowers. D. Ligulate head with only ligulate flowers.

showy ligules of ray flowers are important in attracting pollinators.

Ray flowers have evidently evolved from disk flowers (Figure 15-26), probably more than once in the diversification of the Asteraceae. This apparently involved a change in the symmetry of the peripheral flowers of the head, followed by suppression of the posterior lip and the androecium. In some cases the change has not been complete. In genera such as *Tridax* and *Monolopia* the 2-lobed posterior lip is present but very reduced. Occasional production of staminodes in ray flowers is evidence of incomplete suppression of the stamens. In certain cultivated composites, horticulturists have applied strong selection for extra rays, sometimes resulting in heads in which most or all the flowers are zygomorphic.

Some composites have pistillate flowers with very slender tubular corollas (**filiform flowers).** These may have evolved from ray flowers by suppression of the ligule or from disk flowers by loss of the anthers (Figure 15-26 B). Some have a vestigial ligule or highly reduced corolla lobes. Many filiform flowers can be crowded together in a small head. Species that produce filiform flowers also produce disk flowers, either in the same heads or in different heads. (Some are monoecious and some dioecious.) Because these heads resemble discoid heads (neither has spreading rays), they are known as **disciform heads** (Figure 15-27 C). The final stage in this reduction series is relatively uncommon; it is the complete loss of the corolla in the pistillate flowers. These are **naked flowers** (Figure 15-26 A).

Still another kind of flower occurs in the tribe Lactuceae. **Ligulate flowers** have corollas that resemble disk flowers sliced open along one side and spread to the side forming a 5-lobed ligule (Figure 15-26 F). Ligulate flowers are perfect and generally do not occur together in a head with other kinds of flowers. Plants with **ligulate heads** (Figure 15-27 D) usually also have milky sap, a feature absent in most other composites.

## Economic Importance

Although the Asteraceae is an enormous family, comparatively few have major economic importance. Vegetables include *Lactuca sativa* (lettuce), *Cichorium endiva* (endive), *Cynara scolymus* (artichoke), and *Helianthus tuberosus* (Jerusalem artichoke). A few, including *Helianthus annuus* (sunflower), *Carthamus baeticus* (safflower), and *Guizotia abyssinica* (niger seed), are important

oilseed crops. The seeds of *Helianthus annuus* are edible and used in various ways. The roots of *Cichorium intybus* (chicory) are used as a coffee substitute or additive. The foliage of *Artemisia dracunculus* (tarragon) is used as a culinary herb. *Parthenium argentatum* (guayule) and its relatives have been investigated as rubber source. An insecticide is extracted from the dried heads of *Tanacetum cineriifolium* (pyrethrum). *Brachylaena huillensis* and *Vernonia arborea* are timber trees.

Many Asteraceae are grown as ornamentals. These include *Dendranthema* spp. (florist's chrysanthemum), *Callistephus, Cosmos, Dahlia, Ageratum, Gazania, Felicia, Helianthus, Arctotis, Arctotheca, Tagetes* (marigold), *Calendula* (pot-marigold), *Zinnia,* and many others. Some are cultivated mainly for their foliage and others for their showy flower heads.

Some genera include both ornamental species such as *Centaurea cyanus* (bachelor's buttons) and weedy species such as *C. solstitialis* (yellow star thistle) and *C. maculosa* (spotted knapweed). There are many other weedy species such as *Taraxacum officinale* (common dandelion), *Sonchus oleraceus* (sow-thistle), *Chondrilla juncea* (skeleton weed), *Conyza canadensis* (horseweed), *Cirsium arvense* (Canada thistle), *C. vulgare* (bull thistle), *Carduus nutans* (nodding thistle), *Senecio jacobea* (tansyragwort), *Xanthium strumarium* (cocklebur), *X. spinosum* (clot bur), and many more. *Ambrosia* spp. (ragweeds) are notorious as causes of hay fever.

# Synopsis of the Subfamilies and Tribes of the Asteraceae

Members of the Asteraceae are very easy to recognize to family. However, because of the great diversity of taxa within the family, it has been divided into subfamilies and tribes [a tribe is a taxonomic rank between subfamily and genus] that are thought to represent evolutionary lineages. There is some dispute among taxonomists as to the number of tribes that should be recognized. Most experts in composite taxonomy recognize 12 to 17. Tribes are very useful categories for advanced students and professional taxonomists but are most useful to introductory students as intermediate steps in keying.

Recent studies involving both morphology and chloroplast DNA investigations, coupled with cladistic analyses have brought about some major rearrangements of the tribes. The arrangement of tribes that follows is based on the recent work by Bremer (1994) and his colleagues.

## Subfamily Barnedesioideae

The Barnedesioideae comprises one small tribe. The features of the subfamily are described below in the discussion of the tribe. The Barnedesioideae were formerly included in the Mutisieae. However, studies of chloroplast DNA by Jansen and his colleagues (e.g., Jansen and Palmer 1987, 1988; Bremer and Jansen, 1992) indicate that early in the evolution of the Asteraceae a split occurred between the Barnedesioid lineage and the remainder of the Asteraceae. There is a large inverted segment in the chloroplast DNA of all of the Asteraceae except the Barnedesioideae. The DNA sequence of the Barnedesioid lineage is shared with other angiosperm families.

1. **Barnedesieae**—The Barnedesia Tribe. 9 genera, 92 species. South America.

Members of this tribe are morphologically diverse, ranging from perennial herbs to shrubs and trees. Leaves are entire and usually alternate. In the majority of species there are axillary spines. The heads vary from discoid to disciform or radiate. The involucral bracts are borne in several series and often are spine-tipped. The receptacle lacks chaff scales.

There is much variation in the flowers of the Barnedesieae. Flower color ranges from white to pink, purple, yellow, orange, and red. Disk corollas vary from radially symmetric to variously bilateral and sometimes ligulate. The bilateral corollas are usually pseudobilabiate and typically have a 4-lobed posterior lip and a 1-lobed anterior lip. The apparent ray flowers are actually pseudobilabiate with a reduced posterior lip. The anthers have short to elongated basal tails. Style branches are generally short and glabrous to papillate. The pappus consists of softly plumose bristles or sometimes scales. The achenes of the Barnedesieae lack the 2-celled double hairs that are present in the remainder of the Asteraceae.

## Subfamily Cichorioideae

The Cichorioideae is composed of six tribes. It is a diverse group that includes mostly herbs and shrubs, though there are a few large forest trees in the tribe Vernonieae. Members of some of the tribes produce latex. Heads are quite variable in form and structure. Discoid heads are the most common type, but ligulate heads are characteristic of the tribe Lactuceae and most members of the Arctoteae and Liabeae have radiate heads. Head types are diverse in the Mutisieae. Disciform heads are scattered among the tribes. Flower color varies widely, but comparatively few members of the subfamily are yellow-flowered. Corollas are typically larger and more deeply lobed than those in the subfamily Asteroideae. All types are present in one or another members of the subfamily. Ligulate flowers characterize the Lactuceae. Most Mutisieae have bilabiate corollas, some of which are often ray-like with well developed, radially spreading anterior lips and much reduced posterior lips. Anthers commonly have the bases prolonged into tail-like appendages (Figure 15-25 C). Style branches are very diverse, varying from long and slender (Figure 15-25 K) to very short (15-25 L–M). The ventral surfaces of the style branches are stigmatic throughout. The achenes typically bear two-celled double hairs.

The Cichorioideae are diverse and are united primarily by the lack of the more derived features that characterize the subfamily Asteroideae. The six tribes apparently do not form a monophyletic group. However, the relationships among some of the included taxa are poorly understood and not well resolved. Future studies may result in a restructuring of the tribal and subfamilial boundaries in this group of plants.

1. **Mutisieae**—The Mutisia Tribe. 76 genera, 970 species. Mostly tropical and subtropical North and South America, with some in Africa, Asia, Australia and Hawaii. Only a few occur in the continental United States.

The tribe is morphologically diverse, composed of herbs, shrubs, trees and vines, usually with alternate leaves. Heads usually have several series of phyllaries and a naked receptacle. Corolla forms are diverse in the Mutisieae. Many have bilabiate disk flowers. Some members of this tribe have a mixture of bilabiate and ray flowers, and some have a mixture of actinomorphic disk flowers, bilabiate flowers and ray flowers! The ray flowers have been derived from bilabiate forms and commonly have reduced posterior lips. Typical radiate heads are uncommon. Anthers in this tribe are usually sagittate at the base with long basal appendages (Figure 15-25 C). Style branches are either very short or moderately long, truncate, and tipped with brush-like hairs. The pappus usually consists of capillary bristles.

2. **Cardueae** [formerly called Cynareae]—The Thistle Tribe. 83 genera, 2500 species. Some in North and South America but mostly Eurasian.

Leaves in the thistles are often spiny-margined. The numerous imbricated phyllaries are usually tipped with a spine or a fringed appendage (Figure 15-23 H). In the majority of genera the surface of the receptacle is covered with long bristles. Most have discoid heads with long slender flowers. In most genera the heads are many-flowered, but in *Echinops* the heads are one-flowered and clustered into secondary heads. The corolla is usually bent at the junction of tube and throat and in most genera the lobes are long and slender. Some thistles have ray-like, pistillate or neuter outer flowers but none have true rays. The stamens have slender tail-like basal appendages. The style (Figure 15-24 L) is elongated and bears a node-like swelling below the flat, elliptical, often externally pubescent branches. The pappus is usually bristly.

3. **Lactuceae** [formerly called Cichorieae]—The Lettuce Tribe. 98 genera, 1550 species. Widespread but most common in North Temperate zone.

Members of this most distinctive of the tribes are mostly herbaceous plants (rarely shrubs or trees) with alternate leaves, and most have milky sap, a feature rarely encountered elsewhere in the Asteraceae. The heads are characteristically ligulate (Figure 15-27 D) and most have a naked receptacle. Flower color ranges from white or yellow to blue or purple and all the flowers of the head are generally the same color. The anthers vary from sagittate to short-tailed. The pollen grains are often elaborately sculptured. Style branches range from short to elongate and are sometimes blunt and sometimes taper; sweeping hairs vary from long to short. The pappus usually consists of capillary bristles or sometimes scales, or is occasionally absent.

4. **Vernonieae**—The Ironweed Tribe. 98 genera, 1300 species. Mostly tropical with centers in South America and Africa.

Members of the Vernonieae range from herbs to large trees and have alternate leaves. The heads are characteristically discoid, usually with many imbricated phyllaries. The deeply lobed disk flowers are usually reddish or bluish (very seldom yellow or orange). The anthers are long and basally sagittate. The style branches are long and slender, uniformly short-pubescent and taper to acute tips (Figure 15-25 K). The pappus usually consists of long slender bristles.

The tribe has generally been considered in the past to consist of *Vernonia*, a very large and polymorphic genus and a group of smaller, more tightly defined genera. Some recent studies have proposed that *Vernonia* should be broken into numerous smaller, more well-defined genera.

5. **Liabeae**—The Liabum Tribe (14 genera, 160 species), New World tropics.

Members of this small tribe are herbs, shrubs, vines or trees and most have opposite tomentose leaves. The yellow or orange, radiate or discoid heads have 3-several series of phyllaries. The lobes of the disk flowers are long and slender. The anthers are sagittate. The style branches vary from long and slender to short and cylindrical and are short pubescent throughout. The pappus (when present) consists of capillary bristles. The Liabeae formerly was included within the Vernonieae.

6. **Arctoteae** (often called Arctotideae)—The Arctotis Tribe. 16 genera, 200 species. Almost all African.

The Arctoteae are closely related to the Cardueae. The plants are non-spiny herbs and shrubs that bear radiate heads with numerous imbricated phyllaries. The receptacular surface varies from naked to bristly. Anthers are usually not tailed. The style (Figure 15-25 M) usually bears a node-like hairy swelling below the ± flattened branches. The pappus consists of scales or bristles or is absent.

## Subfamily Asteroideae

The Asteroideae is composed of ten tribes. It is a diverse group that includes mostly herbs and shrubs. There are few trees in the subfamily. Very few members of the Asteroideae produce latex. Discoid and radiate heads are the most common types, and disciform heads are common in some tribes. Ligulate and bilabiate flowers are seldom encountered. Flower color varies widely, but many members of the subfamily are yellow-flowered. Corollas are typically smaller and less deeply lobed than those in the subfamily Cichorioideae. Anther bases are generally not tailed. The style branches are usually more or less elongated and usually have stigmatic areas separated in marginal bands. The style branches often have terminal sterile appendages with sweeping hairs. Achenes typically have two-celled double hairs.

1. **Inuleae**—The Inula Tribe. 38 genera, 480 species. An Old World tribe with most representa-

tives in Eurasia and North Africa and a few extending to tropical and southern Africa.

Until recently the Inuleae was treated as including the genera here segregated into the tribes Plucheeae and Gnaphalieae. In its more limited circumscription, the Inuleae comprises mostly herbs, subshrubs, and shrubs with alternate or less commonly opposite leaves. The heads range from large and radiate to very small and discoid or disciform. The involucral bracts are borne in several series and range in texture from green and herbaceous to dry and parchment-like. The receptacle may be chaffy but more commonly is naked. The disk flowers have short lobes and vary in color from yellow to white or purple. The anthers characteristically have tailed or sagittate bases. The style branches have 2 marginal stigmatic lines; the stylar appendages are generally rounded. Sweeping hairs extend down the style branches to about the branching point. The achenes lack a carbonized layer. The pappus usually consists of slender capillary bristles or flattened scales; it is absent in a few genera.

2. **Plucheeae**—The Pluchea Tribe. 28 genera, 220 species. Widespread but most diverse in the tropics and arid regions of Africa, Asia, Australia, North America, and South America.

The Plucheeae comprises herbs, shrubs, and trees with alternate or less commonly subopposite opposite leaves. The heads range from large and radiate to very small and disciform. In most representatives of the tribe the heads are disciform. Many have scarious or hyaline involucral bracts (Figure 15-23 G). The receptacles are usually naked; in a few they are chaffy. The disk flowers have short lobes and vary in color from yellow to white or purple. The anthers may have tailed or tailless bases. The style branches have 2 marginal stigmatic lines; the stylar appendages are generally rounded. Sweeping hairs extend down the style to well below the branching point. The achenes lack a carbonized layer. The pappus usually consists of slender capillary bristles, but is composed of elongated scales or absent in some genera.

3. **Gnaphalieae**—The Everlasting Tribe. 162–181 genera, 2000 species. Widespread but most diverse in southern Africa, Australia, South America, and the Mediterranean region.

This tribe comprises mostly herbs and shrubs with alternate or opposite, entire leaves. Most North American representatives of the Inuleae are covered to some extent with felt-like pubescence. The heads are mostly small and disciform or discoid; only

rarely are they radiate. In some small-headed groups the heads are clustered into secondary heads. Many have scarious or hyaline involucral bracts (Figure 15-23 G). The receptacle is usually naked and only rarely chaffy. In some of the small-headed species the corollas of the pistillate flowers are extremely reduced. Disk flowers have short lobes and vary in color from yellow to white or purple. The anthers characteristically have tailed or sagittate anther bases, but this is difficult to observe in the plants with very small flowers. The style branches have 2 marginal stigmatic lines; the stylar appendages vary from abruptly truncate to rounded or conical and have several different arrangements of sweeping hairs. The achenes lack a carbonized layer. The pappus usually consists of slender capillary bristles, but is composed of scales, reduced, or absent in some genera.

4. **Calenduleae**—The Calendula Tribe. 7-8 genera, 110 species. Mostly Africa with some in the Mediterranean region.

This small tribe consists of herbs and shrubs (rarely small trees) with alternate or less commonly, opposite leaves. The heads are radiate with a campanulate to hemispherical involucre of distinct scarious-margined phyllaries in 1-several series The receptacle is naked. The rays are variously colored, from white to yellow or purple. The disk flowers, which may be the same color as the rays or differently colored, have short corolla lobes. In most genera the disk flowers are functionally staminate and only the ray flowers produce fruits. In one genus the rays are sterile and the disk flowers are fertile, and in one both ray and disk flowers are fertile. Anther bases range from sagittate to tailed. The style branches of those with fertile disk flowers are very short and bear with short sweeping hairs near the apices. The styles of functionally staminate flowers bear a ring of sweeping hairs just below the rudimentary style branches (Figure 15-25 H). The achenes are epappose throughout the tribe. In one genus the ovaries mature as drupes.

5. **Astereae**—The Aster Tribe. 135–170 genera, 3000 species. Worldwide in distribution, but most common in temperate areas of the New World, Australia, New Zealand, and southern Africa.

Most members of the Astereae are herbs or shrubs, and the leaves are usually alternate. The heads are often comparatively small and may be radiate, discoid or disciform. Most lack a chaffy receptacle. Rays may be the same color as the usually yellow disk or differently colored. Disk flowers have comparatively short lobes. The anthers bear a triangular terminal appendage and are obtuse and not at all tailed at the base. The style branches of the disk flowers are subterete and are tipped by subulate to triangular appendages covered with collecting hairs. In many genera the style branches remain together at the apex in open flowers but separate in the center, thereby forming a loop (Figure 15-25 I). The achenes lack a carbonized layer. The pappus in most consists of capillary bristles and is sometimes reduced or absent. There is much controversy regarding delimitation of certain genera, particularly in North America.

6. **Anthemideae**—The Mayweed Tribe. 109 genera, 1740 species. Mostly temperate regions, most abundant in Eurasia.

The tribe consists of herbs and shrubs with usually alternate, often very deeply dissected or compound leaves. Many members of the tribe have glands containing strongly scented volatile oils embedded in their leaves. The heads may be radiate, disciform or discoid. The involucre consists of one to several series of scarious or scarious-margined phyllaries (Figure 15-23 F). The receptacle can be naked or chaffy. Ray flowers vary greatly in color and may be pistillate or neuter. Disk flowers are usually yellow and functionally perfect. The anthers are truncate or rarely short-tailed. The style branches are usually truncate with very short apical hairs (Figure 15-25 G) and marginal stigmatic areas. Achenes vary from terete to strongly flattened. The pappus consists of distinct or connate scales or is absent. This tribe contains several culinary herbs and numerous ornamentals. In some semiarid areas of the world species of *Artemisia* (sagebrush) are dominant. This large genus is sometimes broken apart into several smaller genera.

7. **Senecioneae**—The Senecio Tribe. 100–120 genera, 3000 species. Cosmopolitan.

All habit classes are represented in this morphologically diverse tribe. Leaves are usually alternate. Heads may be radiate, disciform or discoid. The involucre usually consists of 1 or 2 series of ± equal phyllaries, often subtended by several very short outer phyllaries. The receptacle is naked. Flowers are variously colored with yellow the most common. Rays are pistillate and fertile. Disk flowers have short corolla lobes and are usually perfect. The anthers range from truncate at base to sagittate or tailed. Style branches are most commonly truncate with terminal brush-like hairs but sometimes are variously appendaged. The achenes are usually ± terete and lack a carbonized layer. The pappus usually consists of numerous slender capillary bristles. *Senecio,* the largest genus of the Asteraceae makes up about half the tribe. This enormous cosmopolitan genus includes herbs, shrubs, vines, succulents and even trees. Various attempts have been made to divide *Senecio* into smaller genera but these attempts have usually resulted only in the removal of a few ± distinctive peripheral groups, leaving unchanged the bulk of *Senecio.*

8. **Helenieae** [including Tageteae]—The Sneezeweed Tribe. 110 genera, 800 species. Mostly New World, especially in North America.

In some classifications of the Asteraceae, the genera here treated as Helenieae have been included in the Heliantheae. Various workers have suggested mergers between these tribes (e.g., Cronquist, 1955, 1977; Robinson 1981). Karis and Ryding (Chapter 21 *in* Bremer 1994), however, chose to separate these admittedly closely related groups as separate tribes.

Herbs, subshrubs, and shrubs predominate. In most at least the lower leaves are alternate. Both radiate and discoid heads are common. The involucral bracts are usually in one or two series and may be distinct or connate. The receptacle is naked in most genera. Yellow flowers are very common but some have white, pink, red or other hues. Rays may be the same color as or differently colored than the disk. The disk corollas have rather short lobes in most Helenieae. The anthers usually have well-developed apical appendages (Figure 15-25 A, B, D) but lack elongated basal tails (they may be sagittate, however). Style branches are usually flattened and each style branch has two distinct stigmatic lines (Figure 15-25 E, F). Stylar appendages are rather variable. The achenes, like those of the Eupatorieae and Heliantheae, generally have a carbonized layer darkened by phytomelanin pigments. Pappus types are extremely diverse although flattened scales are the most common.

9. **Heliantheae**—The Sunflower Tribe. 189 genera, 2500 species. Cosmopolitan but most in New World.

In the most widely followed classifications of the Asteraceae, the Heliantheae has been restricted to plants with a chaffy receptacle. Related plants lacking chaff were placed into the Helenieae. Various recent treatments have suggested mergers between these tribes. As mentioned above, we are following the treatment of Karis and Ryding (Chapter 22 *in* Bremer, 1994) that separates the two.

The Heliantheae is the most morphologically diverse of the tribes. All habit types are represented though herbs and shrubs predominate. Both radiate and discoid heads are common. Involucres vary from one- to many-seriate. Receptacles are chaffy in most genera. Yellow flowers are very common but some have white, pink, red or other hues. Rays may be the same color as or differently colored than the disk. The disk corollas have rather short lobes in most Heliantheae. The anthers usually have well-developed apical appendages (Figure 15-25 A, B, D) but lack elongated basal tails (they may be sagittate, however). Style branches are usually flattened and each style branch has two distinct stigmatic lines (Figure 15-25 E, F). Stylar appendages are rather variable. The achenes, like those of the Helenieae and Eupatorieae have a carbonized layer darkened by phytomelanins. Pappus types are extremely diverse although flattened scales are perhaps most common.

10. **Eupatorieae**—The Eupatorium Tribe. 160–170 genera, 2400 species. Mostly New World tropics with a few pantropical genera.

Most are herbs or shrubs with opposite leaves. The heads are mostly small and all are discoid with 1-several series of phyllaries. Most have a naked receptacle. The flowers range in color from white to purple, but never are bright yellow or orange. All the flowers are perfect and fertile and the corollas have short lobes. The anthers are usually small and are never tailed. The style branches are long and clavate and often very conspicuous (Figure 15-25 J). The achenes have a blackish layer containing phytomelanin pigments, a feature shared only with the Heliantheae and Helenieae. In most Eupatorieae the pappus consists of capillary bristles but in some it is reduced to scales or is absent. Traditionally most species in the tribe have been placed in *Eupatorium,* a large polymorphic genus. Since the late 1960's studies of the Eupatorieae by Robert M. King and Harold Robinson have used many previously overlooked microscopic features to divide *Eupatorium* into a multitude of smaller genera and have reorganized most other genera as well. These changes have been very controversial.

# References

Anderberg, A. A. 1991. Taxonomy and phylogeny of the tribe Gnaphalieae (Asteraceae). *Opera Botanica* 104:1–195.

————. 1991. Taxonomy and phylogeny of the tribe Inuleae (Asteraceae). *Plant Systematics and Evolution* 176:75–123.

————.1991. Taxonomy and phylogeny of the tribe Plucheeae (Asteraceae). *Plant Systematics and Evolution* 176:145–177.

Angiosperm Phylogeny Group II. 2003. An update of the Angiosperm Phylogeny Group classification for the orders and families of flowering plants: APG II. *Botanical Journal of the Linnean Society* 141:399–436.

Bentham, G. 1873. Ordo 88: Compositae. *in* G. Bentham and J. D. Hooker. *Genera Plantarum* 2(1):163–533, 536, 537. L. Reeve Co., London.

Bremer, K. 1987. Tribal relationships in the Asteraceae. Cladistics 3:210–253.

————. 1994. *Asteraceae, Cladistics and Classification.* Timber Press, Portland, Oregon.

————, and R. K. Jansen. 1992. A new subfamily of the Asteraceae. Annals of the Missouri Botanical Garden 79:414–415.

Carlquist, S. 1976. Tribal interrelationships and phylogeny of the Asteraceae. *Aliso* 8:465–492.

Cronquist, A. 1955. Phylogeny and taxonomy of the Compositae. *American Midland Naturalist* 53:478–511.

Cronquist, A. 1977. The Compositae revisited. *Brittonia* 29:137–153.

————. 1981. *An Integrated System of Classification of Flowering Plants.* Columbia University Press, New York, pp. 1020–1028.

————. 1988. *The Evolution and Classification of Flowering Plants,* 2nd ed. New York Botanical Garden, Bronx, N.Y., pp. 446–448.

Donoghue, M. J. 1983. The phylogenetic relationships of *Viburnum,* pp. 143–166 *in* N. I. Platnick and V. A. Funk (eds.). *Advances in Cladistics, vol. 2. Proceedings of the Second Meeting of the Willi Hennig Society.* Columbia Univ. Press, New York.

Ferguson, I. K. 1966. The genera of Caprifoliaceae in the southeastern United States. *Journal of the Arnold Arboretum* 47:33–59.

Grant, V. 1959. *Natural History of the Phlox Family.* M. Nijhoff, The Hague.

————, and K. A. Grant. 1965. Flower Pollination in the Phlox Family. Columbia Univ. Press, New York.

Hawkes, J. G., R. N. Lester and A. D. Skelding (eds.) 1979. *The Biology and Taxonomy of the Solanaceae.* Linnaean Society of London Symposium Series No. 7. Academic Press, London.

Heywood, V. H., J. B. Harborne and B. L. Turner (eds.). 1977. *The Biology and Chemistry of the*

*Compositae.* 2 Vol. Academic Pr., London, New York and San Francisco. 1189 pp.

Heywood, V. H. (ed.). 1993. *Flowering Plants of the World,* updated edition. Oxford University Press, New York, pp. 222–268.

Jones, S. B. 1982. The genera of the Vernonieae (Compositae) in the southeastern United States. *Journal of the Arnold Arboretum* 63:489–507.

Judd W. S., C. S. Campbell, E. A. Kellogg, P. F. Stevens, and M. J. Donoghue. 2002. *Plant Systematics: A Phylogenetic Approach.* 2nd ed. Sunderland, MA: Sinauer Associates.

Long, R. W. 1970. The genera of Acanthaceae in the southeastern United States. *Journal of the Arnold Arboretum* 51:257–309.

Robinson, H. 1981. A revision of the tribal and subtribal limits of the Heliantheae (Asteraceae). *Smithsonian Contributions to Botany* 51:1–102.

Rogers, G. K. 1987. The genera of Cinchonoideae (Rubiaceae) in the southeastern United States. *Journal of the Arnold Arboretum* 68:137–183.

Rosatti, T. J. 1984. The Plantaginaceae in the southeastern United States. *Journal of the Arnold Arboretum* 65:533–562.

———. 1986. The genera of Sphenocleaceae and Campanulaceae in the southeastern United States. *Journal of the Arnold Arboretum* 67:1–64.

Solbrig, O. T. 1963. The tribes of Compositae in the southeastern United States. *Journal of the Arnold Arboretum* 44:436–461.

Soltis, D. E., P. S. Soltis, M. W. Chase, M. E. Mort, D. C. Albach, M. Zanis, V. Savolainen, W. H. Hahn, S. B. Hoot, M. E. Fay, M. Axtell, S. M. Swenson, L. M. Prince, W. J. Kress, K. C. Nixon, and J. S. Farris. 2000. Angiosperm phylogeny inferred from 18S rDNA, *rbcL,* and *atpB* sequences. *Bot. J. Linn. Soc.* 133:381–461.

Stevens, P. F. (2001 onwards). Angiosperm Phylogeny Website, Version 5, May 2004 [and more or less continuously updated since]. http://www.mobot.org/MOBOT/research/APweb/.

Thieret, J. W. 1967. Supraspecific classification of the Scrophulariaceae: a review. *Sida* 3:87–106.

———. 1971. The genera of Orobanchaceae in the southeastern United States. *Journal of the Arnold Arboretum* 52:404–434.

———. 1977. The Martyniaceae in the southeastern United States. *Journal of the Arnold Arboretum* 58:25–39.

Vuilleumier, B. S. 1969. The genera of Senecioneae in the southeastern United States. *Journal of the Arnold Arboretum* 50:104–123.

———. 1969. The tribe Mutisieae (Compositae) in the southeastern United States. *Journal of the Arnold Arboretum* 50:620–625.

———. 1973. The genera of Lactuceae (Compositae) in the southeastern United States. *Journal of the Arnold Arboretum* 54:42–93.

Wagenitz, G. 1975. Systematics and phylogeny of the Compositae (Asteraceae). *Plant Systematics and Evolution* 125:29–46.

Wagenitz, G. 1977. New aspects of the systematics of Asteridae, pp. 375–395 *in* K. Kubitzski (ed.), *Flowering Plants. Evolution and Classification of Higher Categories.* Plant Systematics and Evolution, suppl. 1. Springer-Verlag, Vienna.

Wilson, K. A. 1960. The genera of Hydrophyllaceae and Polemoniaceae in the southeastern United States. *Journal of the Arnold Arboretum* 41:197–212.

———. 1960. The genera of Convolvulaceae in the southeastern United States. *Journal of the Arnold Arboretum* 41:298–317.

———, and C. E. Wood. 1959. The genera of Oleaceae in the southeastern United States. *Journal of the Arnold Arboretum* 40:270–278.

Wood, C. E. 1983. The genera of Menyanthaceae in the southeastern United States. *Journal of the Arnold Arboretum* 64:435–449.

———. 1982. The genera of Gentianaceae in the southeastern United States. *Journal of the Arnold Arboretum* 63:441–487.

# Ericaceae

## Heather or Blueberry Family

### Ericales ■

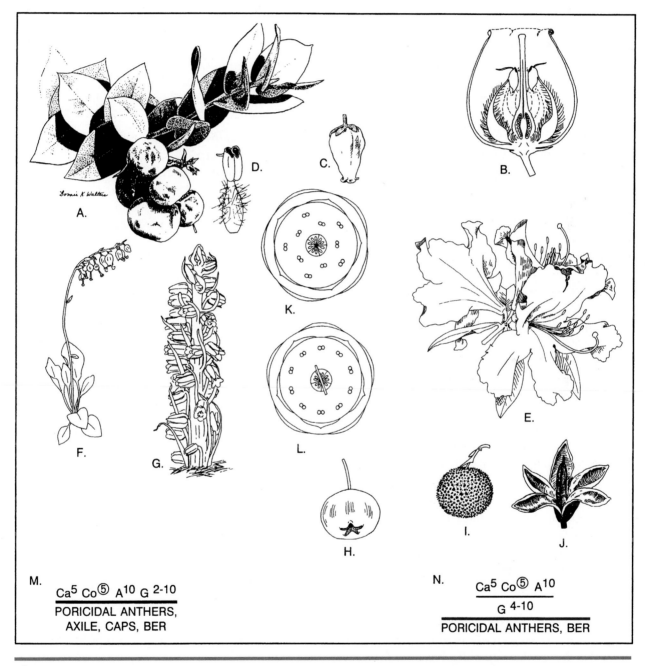

M. $Ca^5 Co^\circledS A^{10} G^{2\text{-}10}$
PORICIDAL ANTHERS, AXILE, CAPS, BER

N. $Ca^5 Co^\circledS A^{10}$ $G^{4\text{-}10}$
PORICIDAL ANTHERS, BER

**Figure 15-28** Features of Ericaceae. A. Branch of *Arctostaphylos luciana* with mature drupes. B–C. Flowers of *Arctostaphylos* (intact and l.s.). D. Stamen of *Arctostaphylos* with dorsal appendage and apparently terminal pores of the anthers. E. Flower of *Rhododendron*. F. Plant of *Pyrola secunda* with terminal raceme. G. Plant of *Sarcodes sanguinea*. H. Berry of *Vaccinium* with persistent calyx lobes on inferior ovary. I. Berry of *Arbutus* with rough surface. J. Capsule of *Rhododendron*. K. Floral diagram of *Arctostaphylos*. L. Floral diagram of *Vaccinium*. M. Generalized floral formula for subf. Arbutoideae. N. Generalized floral formula for subf. Vaccinioideae.

# FAMILY DESCRIPTION

Herbs (sometimes fleshy and mycotrophic and lacking chlorophyll), shrubs (sometimes climbing), or trees. Leaves alternate (rarely opposite or whorled), simple, estipulate (reduced to bladeless scales in mycotrophic species). Plants synoecious. Inflorescences of solitary flowers, axillary clusters, racemes, corymbs, umbels and panicles. Flowers perfect, regular or irregular. Sepals 5 (4–7) distinct or connate (absent). Petals 5 (4–7) connate (distinct); corolla often urceolate. Stamens 10 (4–many), distinct, free or only basally adnate to corolla tube, anthers often appendaged, usually opening by pores. Nectary disk usually present. Carpels 2–10, connate; ovary superior or inferior with 2–10 locules and 1–many axile (parietal) ovules; style 1, undivided. Fruit a capsule, berry or drupe.

# FAMILY NOTES

The Ericaceae is a very widespread family of 116–125 genera and about 3500 species. Members of the family occur from polar regions to the tropics in both hemispheres. Most commonly they grow in nutrient-poor acid soils. The roots almost always have mycorrhizal fungi that are essential to the survival of the plants.

Floral form in the Ericaceae has considerable variability. The corollas are generally strongly sympetalous. Many members of the family have urceolate corollas (Figures 15-28 B, C), but a wide variety of other shapes ranging from narrowly tubular to funnelform (Figure 15-28 E) or broadly bowl-shaped are also present. Most genera have actinomorphic corollas but in the large genus, *Rhododendron,* flowers are often strongly zygomorphic.

Although the sympetalous corollas of the Ericaceae resemble those of families of the subclass Asteridae, the stamens are quite different. In the Asteridae, stamens are generally equal in number to the petals or fewer and are usually epipetalous. In the Ericaceae, stamens are often more numerous than the corolla lobes (Figures 15-28 K, L) and the filaments are free from the corolla or nearly so. Anthers in the Ericaceae are generally dehiscent by pores and often bear bristle-like appendages (Figure 15-28 D). In *Vaccinium* and its close relatives the anthers have long, tubular extensions that are terminated by pores.

The ovary in most genera is superior (Figure 15-28 B, K), but in the subfamily **Vaccinioideae** it is inferior (Figure 15-28 H, L). Fruits vary from dry capsules (Figure 15-28 J) to berries (Figure 15–28 H, I) and drupes (Figure 15-28 A). In the drupes of some *Arctostaphylos* species, each seed is enclosed by its own endocarp.

The classification of the Ericaceae and its close relatives varies considerably (Table 15-1). Most taxonomists recognize several subfamilies, but not all botanists are in agreement regarding the composition or taxonomic rank of these taxa. Some combine the Arbutoideae and Vaccinioideae as a single subfamily with variable ovary position. Others consider the Vaccinioideae to be a separate family, the Vacciniaceae, characterized by inferior ovaries and certain other characters. In the classification accepted here we include the Vaccinioideae within the Ericaceae but maintain it as a subfamily distinct from the Arbutoideae.

Many botanists segregate subfamilies Monotropoideae as families distinct from but closely related to the Ericaceae. We are treating them as subfamilies of the Ericaceae. Some botanists merge the Monotropaceae and Pyrolaceae under the latter name. Regardless of the taxonomic rank given to these groups, botanists are unanimous in considering them all to be closely related.

There are numerous characters shared by the taxa listed above. All have unusual stamens. The anthers are structurally inverted in a mature flower and dehisce by terminal (actually basal) pores (Figure 15-28 D). Pollen grains are released from the anthers as tetrads (clusters of 4). Certain cellular details of the anther tissues are also characteristic; the anthers lack fibrous tissues and have a multinucleate tapetum. The style is hollow and internally lobed. The ovules have a single integument and a thin nucellus; the seed coat is one cell layer thick.

| subf. Ericoideae | subf. Rhododendroideae | subf. Arbutoideae |
|---|---|---|
| Shrubs or subshrubs; leaves green, often scale-like or needle-like; inflorescences axillary or on short lateral branches; corolla urceolate to tubular, sometimes campanulate, persistent in fruit; ovary superior. | Shrubs or trees; leaves green, flat, well developed; inflorescences terminal on main branches; corolla tubular to bowl-shaped or funnelform (Figure 15-28 E), less often urceolate, deciduous in fruit; ovary superior. | Shrubs or trees; leaves green, flat well developed; inflorescence usually axillary or on short lateral branches; corolla usually urceolate (Figure 15-28 B, C), deciduous in fruit; ovary superior. |
| **subf. Vaccinioideae** [Vacciniaceae] | **subf. Pyroloideae** [Pyrolaceae] | **subf. Monotropoideae** [Monotropaceae] |
| Shrubs or woody vines; leaves green, flat, well developed; inflorescence usually axillary or on short lateral branches; corolla usually urceolate, deciduous in fruit; ovary inferior (Figure 15-28 H). | Herbs (Figure 15-28 F) or subshrubs, sometimes without chlorophyll; leaves flat, well developed or reduced to bladeless scales; inflorescence terminal; corolla of distinct or only basally connate petals, tardily deciduous in fruit; ovary superior. | Fleshy herbs without chlorophyll (Figure 15-28 G); leaves scale-like; inflorescence terminal; corolla usually urceolate or petals distinct, deciduous or persistent in fruit; ovary superior. |

**Table 15-1**   Subfamilies of the Ericaceae.

Within the ovule the endosperm forms cell-walls as it develops and absorbs nutrients from the nucellus at both ends of the ovule.

The Ericaceae includes many handsome ornamentals and a few fruit crops. *Rhododendron* is a giant genus of over 800 species ranging from polar regions to the tropics. Many of its species are grown as ornamentals and many species and artificial hybrids are prized for their beautiful flowers. *Erica* (heather) with about 600 species, also has numerous cultivated forms and hybrids. Many other genera are also in cultivation. In California species of *Arctostaphylos* (manzanita) are important components of the chaparral plant communities and some are cultivated as ornamentals as well. The most important fruit crops of the Ericaceae are products of the genus *Vaccinium* (blueberry, cranberry).

# Polemoniaceae

## The Phlox Family

### Ericales ▪

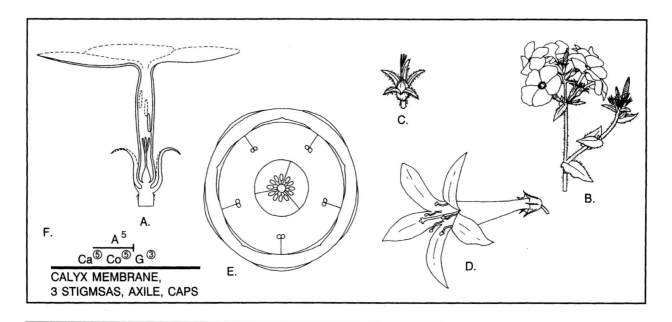

**Figure 15-29** Features of Polemoniaceae. A–C. *Phlox* sp. A. Flower (l.s.). B. Inflorescence. C. Immature capsule. D. Flower of *Ipomopsis*. E. Floral diagram of *Phlox*. F. Generalized floral formula.

## FAMILY DESCRIPTION

Herbs or shrubs (rarely woody vines or trees). Leaves alternate or opposite, simple to variously divided or compound, estipulate. Plants synoecious. Inflorescence variously cymose, sometimes corymbiform or headlike, or flowers solitary. Flowers perfect, regular (occasionally irregular). Sepals 5, connate, the membrane connecting adjacent sepals often transparent or translucent. Petals 5, connate. Stamens 5, alternate with petals; filaments adnate to corolla. Carpels 3 (2–4), connate; ovary superior with 3 (2–4) locules and 1–many axile ovules or with 1 locule and 3–many parietal ovules; style 1, entire or 3 (2–4) branched. Fruit a capsule.

## FAMILY NOTES

The Polemoniaceae is a comparatively small family (18–20 genera, 300 species), restricted for the most part to temperate regions. Most Polemoniaceae occur in the New World and the family is particularly well represented in the arid regions of the American southwest. More than half the species in the family occur in California. The range of the family extends southward into the mountains of South America. Most temperate members of the family are herbaceous plants, and many of these are annuals. The tropical American members of the Polemoniaceae include trees, shrubs and woody vines.

Flowers in the Polemoniaceae are almost always 5-merous and radially symmetric (Figure 15-29 B). In most genera the lobes of the sympetalous calyx have green centers and hyaline (nearly transparent) margins. (The calyx lobes are wholly green in *Cobaea* and *Polemonium*.) The corolla tube is often very slender. The epipetalous stamens are sometimes inserted at different levels within the corolla (Figure 15-29 A). The gynoecium is almost always 3-carpellate. The superior ovary is tipped by a slender style and a 3-lobed stigma. Placentation is axile and the fruit is a small capsule (Figure 15-29 C).

The biology of the Polemoniaceae has been thoroughly investigated by Grant (1959). Grant and Grant (1965) studied the interaction between flowers and pollinators in the family. This study has provided pollination biologists with much information about floral evolution. Pollinators of members of the Polemoniaceae include various kinds of insects, hummingbirds, and in the tropical vine, *Cobaea*, bats.

A few members of the Polemoniaceae, particularly species and hybrids of *Phlox,* are cultivated as ornamentals. Also cultivated are species of *Cantua, Ipomopsis, Gilia,* and *Polemonium.* Otherwise the family is of little economic importance. The family includes numerous conspicuous wildflowers.

# Primulaceae

## The Primrose Family
Ericales

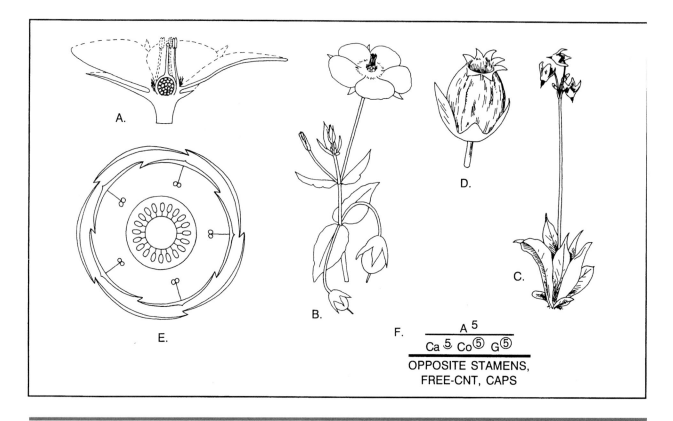

**Figure 15-30** Features of Primulaceae. A. Flower of *Anagallis arvensis* (l.s.). B. Flowering and fruiting branch of *Anagallis*. C. Habit of *Dodecatheon*. D. Denticidal capsule of *Dodecatheon*. E. Floral diagram of *Anagallis*. F. Generalized floral formula.

## FAMILY DESCRIPTION

Herbs. Leaves alternate, opposite or whorled, sometimes all basal, simple (rarely pinnately dissected), estipulate (rarely stipulate). Plants synoecious. Inflorescence a raceme, spike, umbel or panicle or flowers solitary or in axillary clusters. Flowers perfect, regular (irregular). Sepals 5 (4–9), connate. Petals 5 (4–9) (0), connate at least at base (distinct).

Stamens 5 (4–9), opposite the petals; filaments distinct or connate, adnate to corolla. Carpels 4–8 [number difficult to determine], connate; ovary superior (rarely half inferior) with 1 locule and many free-central ovules; style 1, undivided. Fruit a capsule, sometimes circumscissile.

# FAMILY NOTES

The Primulaceae comprises 23–30 genera and about 1000 species, very widespread in temperate regions, but with only a few tropical representatives. The majority of species occur in cool north-temperate regions, often in high mountain areas.

The primrose family is unusual in the Dilleniidae in having a sympetalous corolla and few stamens together with an ovary with free-central placentation. The corolla tube is often so short that the petals appear to be distinct (Figure 15-30 A). The stamens are inserted opposite the corolla lobes, and in a few genera alternate with staminodes. The number of carpels (usually 5) is difficult to determine without anatomical evidence. The carpels are wholly connate and the stigma is unlobed. The free-central placenta usually provides no clues. In some genera the tips of the carpels separate when the capsule dehisces (Figure 15-30 D), but in others the capsule is circumscissile.

The Primulaceae is of limited economic importance. It contains a number of showy wildflowers and several ornamentals. Conspicuous wildflowers include *Dodecatheon* spp. (shooting stars; Figure 15-30 C), *Primula* spp. (primroses), *Anagallis* spp. (pimpernel; Figure 15-30 B), *Lysimachia* spp. (loosestrife) and a few others. Garden flowers include species and hybrids in the genera *Cyclamen*, *Soldanella, Cortusa, Primula,* and *Dodecatheon.*

# Rubiaceae

## The Madder Family
### Gentianales ■

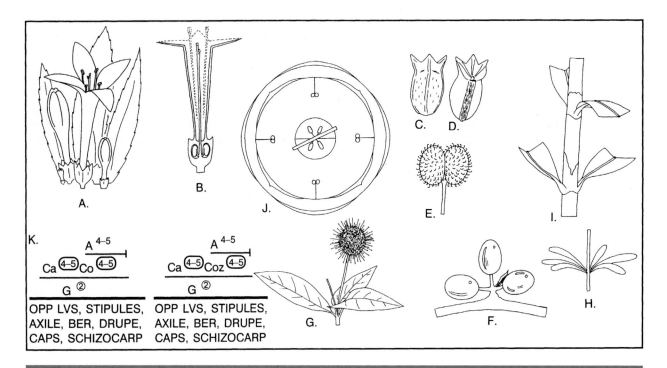

**Figure 15-31** Features of Rubiaceae. A. Inflorescence and flower of *Sherardia arvensis*. B. Flower of *Sherardia*. C–D. Abaxial and adaxial views of a mericarp of *Sherardia*. E. Schizocarp of *Galium*. F. Stem of *Coffea* with drupes. G. Branch of *Cephalanthus occidentalis* with whorled leaves and head flowers. H. Leafy node of *Galium* showing apparently whorled leaves and absence of stipules. I. Leafy nodes of *Gardenia* showing opposite leaves and prominent stipules. J. Floral diagram of *Hedyotis*. K. Generalized floral formula.

## FAMILY DESCRIPTION

Herbs, shrubs, woody vines or trees. Leaves opposite or whorled, simple, stipulate; stipules sometimes indistinguishable from leaves in plants with apparently whorled leaves. Plants synoecious (rarely dioecious). Inflorescence variously cymose, sometimes flowers solitary or in heads. Flowers perfect, regular (rarely irregular). Sepals 4 or 5, connate or distinct (sometimes apparently absent by reduction).

Petals 4 or 5 (8–10), connate. Stamens 4 or 5, distinct; filaments adnate to corolla. Carpels 2 (1–many), connate; ovary inferior (very rarely superior) with 2 (3–many) locules and 1–many axile ovules per locule (rarely with 1 locule and many parietal ovules); style 1, undivided. Fruit a berry, drupe, capsule, or schizocarp.

# FAMILY NOTES

The Rubiaceae is a large, mostly tropical family with 450–606 genera and 6500 species. Many of the tropical members of the family are trees or shrubs. However, most temperate members of the family are herbaceous. Students and botanists who are acquainted with only the temperate members of the Rubiaceae receive a very biased view of the diversity of the family. The only large woody species in temperate North America is *Cephalanthus occidentalis* (buttonbush, buttonwillow), which attains the stature of a large shrub or small tree. Many of the tropical members of the family resemble each other in so many features that they are very difficult to identify.

Plants in the Rubiaceae generally have entire opposite or whorled leaves with well-developed stipules (Figure 15-31 I), a feature absent from most families of the Asteridae. In *Galium* (bedstraw), a common herbaceous genus of temperate regions, the stipules are as large as the foliage leaves and are nearly indistinguishable from them. As a result, *Galium* appears to have whorled leaves.

Flowers in the Rubiaceae are usually borne in cymes (Figure 15-31 A). The sepals are often rather short (Figure 15-31 B) and in *Galium* usually appear to be absent. The sympetalous 4–5 merous corolla is often showy. The interior ovary is usually 2-carpellate. It may mature as a schizocarp (Figure 15-31 C–E), a capsule, a berry, or a drupe (Figure 15-31F).

Economically the most important member of the Rubiaceae is *Coffea* (coffee), native to Africa and now widely cultivated in the tropics. *Cinchona* species produce quinine. The most commonly cultivated ornamental is *Gardenia*. Several other genera are of minor horticultural significance.

# Asclepiadaceae

## The Milkweed Family
### Gentianales ■

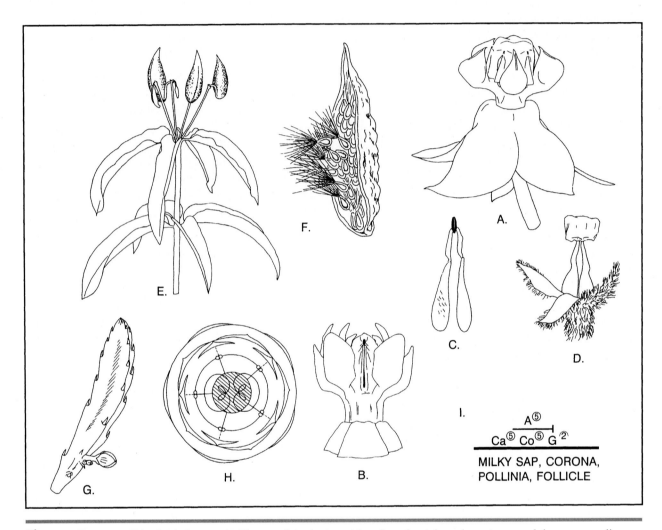

**Figure 15-32** Features of Asclepiadaceae. A. Flower of *Asclepias* with reflexed petals and erect corona lobes surrounding gynostegium. B. Corona and gynostegium of *Asclepias*. Note vertical stigmatic slit on surface of gynostegium and gland (corpusculum) above the slit. C. Pollinia of *Asclepias* attached by translator arms to corpusculum. D. Calyx and gynoecium of *Asclepias* with corolla and androecium removed. Note that styles and ovaries are distinct but stigmas are connate. E. Habit of *Asclepias* sp. with follicles. F. Follicle of *Asclepias syriaca*. G. Branch of *Stapelia* sp. with flower bud. H. Floral diagram of *Asclepias*. I. Generalized floral formula.

## FAMILY DESCRIPTION

Herbs, shrubs, woody vines, stem-succulents, or trees, usually with milky sap. Leaves opposite (less frequently alternate or whorled), simple and entire (in succulent species sometimes vestigial), estipulate. Plants synoecious. Inflorescence a cyme or umbel (raceme), or in succulent species flowers often solitary. Flowers perfect, regular, hypogynous to partially epigynous. Sepals 5, connate or nearly distinct. Petals 5, connate. Stamens 5, adnate to stigma, the filaments connate (distinct) and bearing appendages that form a prominent petaloid corona; lobes of corona alternate with petals and basally adnate to corolla, sometimes bearing horn-like appendages; anthers connate and adnate to drum-shaped stigma; pollen sacs of adjacent stamens united to a glandular appendage; pollen united into waxy pollinia. Carpels 2, distinct at base, connate apically; ovaries 2, superior or basally inferior with many marginal ovules; styles 2, united to 1 massive stigma with 5 vertical stigmatic slits. Fruit a pair of follicles, one of which often aborts; seeds often bearing a tuft of hairs.

# FAMILY NOTES

The Asclepiadaceae comprises about 315 genera and 2000 species. It is widespread in tropical, subtropical and warm temperate regions with the greatest diversity in the tropics of South America. Members of the family occur in many different habitats ranging from marshes and swamps to deserts and rain forests. There is an unusual diversity of growth forms ranging from perennial herbs and vines to shrubs, trees and cactus-like succulents.

The milkweeds the most specialized members of the order Gentianales. The flowers are very unusual. The sepals and petals are comparatively normal in appearance. However, the stamens bear little resemblance to ordinary stamens. The filaments bear inflated or otherwise modified appendages that form a conspicuous corona (Figure 15-32 A, B). In some species each lobe of the corona bears a central horn-like appendage. In addition, the filaments are connate into a tube surrounding the ovaries and styles. The anthers are connate into a ring and are adnate to the sides of the drum-shaped stigma forming a compound structure called a **gynostegium** (Figure 15-32 B). The pollen grains are united together into waxy **pollinia** (Figure 15-32 C). Each anther consists of two pollinia that are concealed behind flaps of tissue on the sides of the gynostegium. The gynostegium has 5 vertical stigmatic slits that alternate with the 5 anthers. A gland (**corpusculum**) connected to a pollinium from each of two adjacent anthers by a pair of slender **translator arms** is located above the apex of each stigmatic slit. The pollination mechanism involves the adherence of one of the glands to the leg of a pollinating insect which then by chance inserts one of the connected pollinia into one of the stigmatic slits. Although there is a low probability that such a pollination event will be successful, a single successful pollination results in the production of many seeds.

Members of the Asclepiadaceae have a very peculiar gynoecium in other respects. The carpels are completely separate at the base and each has its own style (Figure 15-32 D). They are united only by their peculiar stigma. The fruit is a pair of follicles, usually with long-comose seeds that are wind-dispersed (Figure 15-32 E). Usually only one of the two carpels develops into a follicle, and often only one or two follicles develop per inflorescence. The Asclepiadaceae shares these features with many members of the closely related Apocynaceae.

Some African Asclepiadaceae are cactus-like succulents. The flowers of *Stapelia,* one of these succulents, have a flaring corolla that has the color and aroma of decaying flesh. These are attractive to flies that feed on carrion and are pollinated by these insects. Flies often lay their eggs on the flowers, but the maggots die because of a lack of food.

The milky sap of the milkweeds is accompanied by alkaloids and other bitter-tasting and poisonous substances. Most animals avoid eating the leaves and stems. An exception is the larva of the monarch butterfly which eats the leaves and stores the distasteful substances in its own tissues. Adult monarch butterflies retain the bitter taste and are avoided by insect-eating birds.

Some members of the Asclepiadaceae are cultivated as ornamentals or as greenhouse curiosities. Some species of *Asclepias* are particularly handsome in flower. Hoya *carnosa* (wax plant), *Ceropegia woodii* (rosary vine), *Stephanotis floribunda,* and several other members of the family are often grown as house plants. Stem succulents such as *Caralluma* spp., *Huernia* spp., and *Stapelia* spp. are often grown in succulent gardens in warm climates or as greenhouse curiosities. They lose their appeal indoors, however, when their carrion-scented flowers open.

# Boraginaceae

## The Borage Family

### Gentionales ■

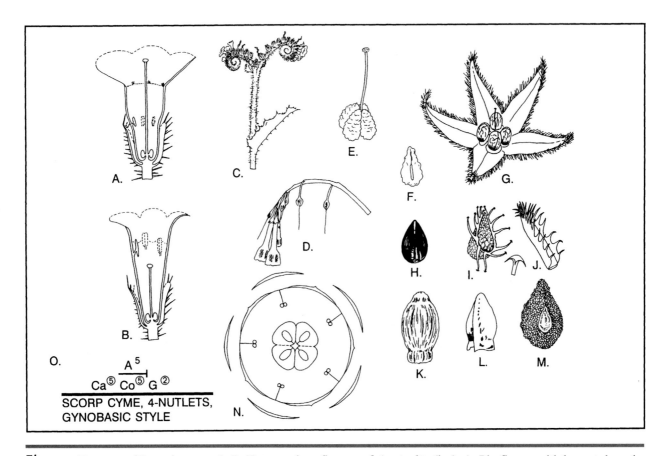

**Figure 15-33** Features of Boraginaceae. A–B. Heterostylous flowers of *Amsinckia* (l.s.). A. Pin flower with long style and anthers set low in corolla. B. Thrum flower with short style and anthers inserted near top of corolla. C. Scorpioid cyme of sessile flowers of *Amsinckia*. D. Scorpioid cyme of pedicelled flowers of *Mertensia*. E–F. 4-lobed ovary and single detached nutlet of *Amsinckia*. G. Calyx and maturing nutlets of *Borago officinale*. H–M. Nutlets. H. *Myosotis arvensis*. I. *Lappula redowski*. J. *Pectocarya linearis*. K. *Borago officinale*. L. *Anchusa* sp. M. *Amsinckia tesselata*. N. Generalized floral diagram. O. Generalized floral formula.

## FAMILY DESCRIPTION

Herbs or shrubs (rarely woody vines or trees). Leaves alternate (rarely opposite) simple and usually entire, estipulate. Plants synoecious (rarely polygamous). Inflorescence cymose, usually scorpioid, sometimes secondarily clustered and paniculiform, exceptionally solitary. Flowers perfect (rarely imperfect), regular or irregular. Sepals 5, distinct or connate, sometimes fused in groups. Petals 5, connate, often bearing appendages that sometimes conceal the stamens. Stamens 5, distinct, alternate with petals; filaments adnate to corolla. Carpels 2 (–4), connate; ovary superior, shallowly 4-lobed or deeply divided into 4 one-ovulate segments; style 1, terminal or gynobasic, entire or 2- (4) lobed. Fruit of 1–4 nutlets (a drupe with a 2–4-lobed endocarp).

# FAMILY NOTES

The Borage family has a cosmopolitan distribution. It comprises 110–131 genera and 2400 species. Members of the family occur in habitats ranging from extremely arid deserts to temperate woodlands and tropical forests. Centers of diversity include the Mediterranean region, western North America, and the American tropics.

The Boraginaceae shares several features with the Hydrophyllaceae. In both families the most common inflorescence is a scorpioid cyme (Figure 15-33 C). Both have regular, sympetalous 5-merous corollas with epipetalous stamens (Figure 15-33 A, B). Leaves in the Boraginaceae are almost always entire. Although some members of the Hydrophyllaceae have simple entire leaves, the leaves of many members of that family are toothed, lobed, or compound.

The most reliable feature for distinguishing the two families is the gynoecium. Most members of the Boraginaceae have a peculiar gynoecium (Figure 15-33 E, G). The ovary is 2-carpellate and each carpel contains a pair of ovules. The pericarp closely enwraps the four ovules. The result is a deeply 4-lobed ovary with the style arising from the base of the ovary (**gynobasic style**) or from the top of the ovary axis. At maturity the ovary breaks apart into four 1-seeded indehiscent nutlets (Figure 15-33 E, F,

G). These structures are often ornamented with prickles, wrinkles, bumps, ridges, etc. (Figure 15-33 H–M). The ornamentation is often used in taxonomic keys to separate genera and species.

The subfamily **Ehretioideae** is often segregated as a separate family, the Ehretiaceae. These plants are mostly woody and mostly tropical. The inflorescence, though cymose, is often non-scorpioid. The ovary is not as deeply lobed as in the subfamily **Boraginoideae,** and the style is usually terminal. In most genera the fruit is a drupe with 2–4 endocarps or with a 2–4 lobed endocarp. The gynoecium in this group apparently represents the primitive condition for the Boraginaceae. Subfamily **Heliotropioideae** is intermediate in its characters between the Boraginoideae and the Ehretioideae. The style is terminal and in some the ovary matures as a drupe. In others it breaks apart into nutlets.

Economically important members of the Boraginaceae include ornamental herbs and shrubs, most from subfamily Boraginoideae and timber trees from subfamily Ehretioideae. Ornamentals include *Borago officinale* (borage) and species of *Echium* (pride-of-Madeira), *Heliotropium* (heliotrope), *Mertensia* (bluebells), *Myosotis* (forget-me-not), and various others. Timber trees include species of *Cordia, Ehretia,* and *Patagonula.*

# Lamiaceae (Labiatae)

## The Mint Family

Lamiales ■

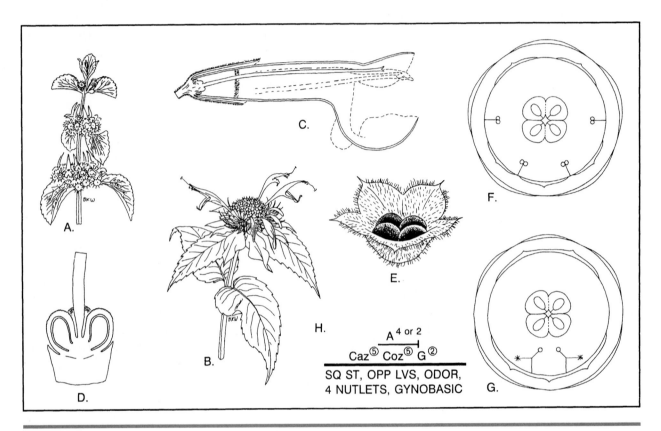

**Figure 15-34** Features of Lamiaceae. A. Stem and inflorescences of *Marrubium vulgare* illustrating opposite leaves and axillary verticils of flowers. B. Branch of *Monarda* with verticils aggregated into a terminal head-like cluster. C. Bilabiate flower of *Stachys* (l.s.). D. Gynoecium with deeply lobed ovary and gynobasic style (l.s.). E. Calyx and nutlets of *Stachys bullata*. F. Floral diagram of *Stachys*. G. Floral diagram of *Salvia* showing the two fertile stamens each with only one fertile anther-sac and a hinged connective. H. Generalized floral formula.

## FAMILY DESCRIPTION

Herbs or shrubs (rarely woody vines or trees) with young stems usually 4-angled. Leaves opposite (rarely whorled), simple (deeply divided or pinnate), estipulate, usually punctate and aromatic. Plants synoecious. Inflorescence of axillary cymes or verticils, sometimes racemiform, paniculiform or head-like, or flowers solitary. Flowers perfect, irregular. Sepals 5, variously connate, sometimes appearing fewer because of connation. Petals 5, connate; corolla often bilabiate (lobes sometimes appearing as 4 because of connation). Stamens 2 or 4, distinct (rarely filaments connate), filaments adnate to corolla. Carpels 2, connate; ovary superior, 4-lobed or divided into 4 locules with 1 basal-axile ovule per locule; style 1, gynobasic (terminal), 2-lobed near apex, deciduous with corolla. Fruit of 4 (1–3) nutlets.

# FAMILY NOTES

The mints comprise a large and cosmopolitan family of 200–212 genera and 3200 species. They are especially common in the Mediterranean region. They occur in a wide diversity of habitats and at almost all latitudes.

The combination of vegetative and floral features characteristic of most mints makes the Lamiaceae one of the easiest families to recognize. The leaves are opposite and the stem is square (at least when young). Most have tiny glands containing scented compounds on the leaves, stems and flowers. The inflorescence is often **verticillate** (Figure 15-34 A). The flowers are crowded into whorls or head-like cymes in the axils of the opposite leaves or bracts. Sometimes the verticils are congested at the ends of the stem into raceme-like or spike-like clusters, or into a terminal headlike cluster (Figure 15-34 B). The flowers are usually strongly bilabiate with a 2-lobed upper corolla lip and a 3-lobed lower lip (Figure 15-34 C). In some cases the corolla lobes are fused to the tip and the actual number of lobes is difficult to determine. There are either 4 or 2 stamens that are inserted alternately with the lateral corolla lobes (Figure 15-34 F, G). Members of the Lamiaceae have a gynoecium similar to that of the Boraginaceae (Figure 15-34 D). Each carpel of the 2-carpellate ovary contains two ovules. Pericarp tissue tightly surrounds each ovule resulting in a deeply 4-lobed ovary. The style is gynobasic, arising from the base of the ovary lobes. The style is usually deciduous from the ovary at the time the corolla falls. At maturity the ovary breaks apart into one-seeded nutlets (Figure 15-34 E).

In *Mentha* the two lobes of the upper corolla lip are fused nearly or all the way to the tip. The upper lip is similar in size and shape to the three lobes of the lower lip. The corolla thus appears to be 4-lobed and is nearly radial.

In flowers of *Salvia* there are only two stamens (Figure 15-34 G). The connective between the two anther sacs is elongated and jointed to the filament. The connective is cantilevered on the tip of the filament. In some species one of the two anther sacs is highly reduced or absent. The sterile ends of the connectives project into the corolla throat whereas the ends of the connective that are tipped by fertile anthers are concealed in the upper corolla lip. When a pollinator probes into the flower, it pushes against the sterile end of the connective. This causes the anthers to rotate out of the upper corolla lip and dust pollen against the body of the pollinator.

The mint family includes many economically important plants. The aromatic oils of some mints are used commercially as flavorings, in pharmacology, or in the perfume industry. These include *Lavendula* spp. (lavender), *Mentha* spp. (mint), *Ocimum basilicum* (basil), *Origanum* spp. (marjoram, oregano), *Rosmarinus officinalis* (rosemary), *Salvia officinalis* (sage), and *Thymus officinalis* (thyme).

Many other members of the Lamiaceae are cultivated. *Coleus* and *Plectranthus* are commonly cultivated as house plants. There are many ornamental herbs and shrubs including species of *Lavendula, Leonurus, Monarda, Phlomis, Rosmarinus, Salvia, Stachy,* and many others. *Nepeta cataria* (catnip) has long been known for the peculiar effect it has on the behavior of cats.

Common wildflowers include *Monarda didyma* (Oswego-tea), *Prunella vulgaris* (self-heal), *Salvia columbariae* (chia), and *Stachys* spp. (hedge-nettle). Shrubby members of the family are common in California, including *Lepechinia calycina* (pitcher sage), *Salvia* spp. (sage), and *Trichostemma lanatum* (woolly blue curls).

# Scrophulariaceae

## The Figwort Family
### Lamiales ■

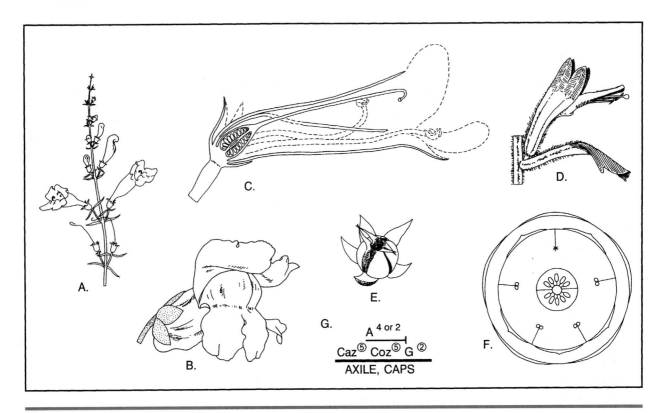

**Figure 15-35** Features of Scrophulariaceae. A. Inflorescence of *Penstemon*. B. Bilabiate flower of *Antirrhinum majus*. C. Bilabiate flower of *Penstemon* (l.s.). D. Flower and subtending bract of *Castilleja*. E. Capsule of *Keckiella cordifolia*. F. Floral diagram of *Penstemon*. G. Generalized floral formula.

## FAMILY DESCRIPTION

Herbs or shrubs. Leaves alternate, opposite or whorled, simple to deeply divided or compound, estipulate. Plants synoecious. Inflorescence variously cymose or racemose, sometimes spicate or paniculate. Flowers perfect, irregular (rarely nearly regular). Sepals 5, distinct or variously connate. Petals 5, connate; corolla often bilabiate. Stamens 4 or 2 (5), distinct; filaments adnate to corolla. Carpels 2, connate; ovary superior with 2 locules and 2–many ovules per locule; sole 1, undivided. Fruit a capsule (very rarely a berry).

# FAMILY NOTES

The Scrophulariaceae is a large (190–275 genera, 4000 species), diverse family with a cosmopolitan distribution. Members of the family occur in many habitats from sea level to the tops of high mountains and from wetlands to deserts.

The flowers of the figwort family resemble those of the mints (Lamiaceae) in many respects. Both families have bilaterally symmetric sympetalous corollas and, except in one genus, flowers of the Scrophulariaceae have either four or two stamens (Figure 15-35 F). The flowers are often strongly bilabiate (Fig 15-35 A, D). The gynoecium is very different, however. It is 2-carpellate with a terminal style and numerous axile ovules, and it matures as a capsule (Figure 15-35 E).

Some members of the Scrophulariaceae establish a root connection with other plants and are parasitic in varying degrees. Most of these are green and photosynthetic. Root parasites include most or all species of *Castilleja* (Figure15-35 D), *Orthocarpus, Cordylanthus, Pedicularis* and several other genera. Species of *Striga* (witchweed) are strongly parasitic and can cause significant crop damage. The Orobanchaceae, a small family of non-photosynthetic root parasites is closely related to the Scrophulariaceae and included in this family by some authorities.

Despite its large size the Scrophulariaceae is of limited economic importance. An important drug, digitalis, is extracted from *Digitalis purpurea.* There are numerous genera with ornamental species including *Antirrhinum* (snapdragon), *Calceolaria* (slipperflower), *Hebe,* and *Linaria* (toadflax). Many members of the family are conspicuous wildflowers. These include species of *Mimulus* (monkeyflower), *Castilleja* (Indian paintbrush), *Orthocarpus* (owl's clover), *Penstemon* (beardtongue), *Chelone* (turtleheads), and many others.

# Solanaceae

## The Potato Family
### Solanales ■

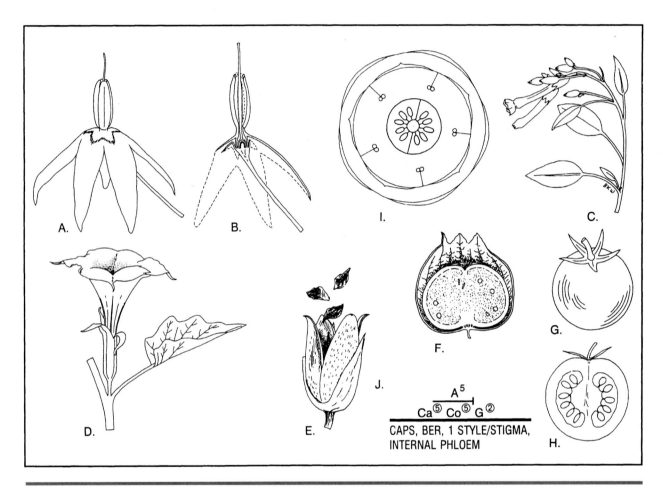

**Figure 15-36** Features of Solanaceae. A–B. *Solanum*. A. Intact flower with rotate corolla. B. Flower (l.s.). C. Leafy twig of *Nicotiana glauca* with inflorescence of tubular flowers. D. Leafy twig and flower of *Datura wrightii*. E. Capsule of *Nicotiana*. F. Berry of *Physalis* with enlarged calyx (l.s.). G–H. Berries of *Lycopersicon* (intact and l.s.). I. Floral diagram. J. Generalized floral formula.

# FAMILY DESCRIPTION

Herbs, shrubs, woody vines or trees. Leaves alternate or subopposite but sometimes arrangement obscured by adnation of leaf bases to stems, simple to deeply divided or once or twice pinnate, estipulate. Plants synoecious. Inflorescence variously cymose or flowers solitary. Flowers perfect, regular (rarely irregular). Sepals 5 (4–7) distinct or connate. Petals 5 (4–7), connate. Stamens 5 (2) (4–7), alternate with petals; filaments adnate to corolla; anthers sometimes opening by pores. Carpels 2 (3–5), connate; ovary superior (rarely half inferior) with 2 (3–5) locules, sometimes with placental septa, and many (few) axile ovules *or* with 1 locule and many parietal ovules; style 1, undivided. Fruit a capsule or berry.

# FAMILY NOTES

The Solanaceae is a cosmopolitan family of 85–96 genera and 2800 species. Its greatest diversity is in the American tropics and Australia.

Most members of the Solanaceae have 5-merous regular flowers. Corolla shapes are extremely variable. Some genera, including *Solanum* (Figure 15-36 A), *Physalis* and *Lycopersicon,* have rotate corollas with very short corolla tubes. *Datura, Petunia* and various other genera have funnelform flowers (Figure 15-36 D). Others including species of *Nicotiana* and *Lycium* have narrowly tubular corollas (Figure 15-36 C). The five stamens are epipetalous (Figure 15-36 B) and alternate with the corolla lobes. The ovary is most commonly superior and 2-carpellate with axile placentation. In some species that have been bred for large fruits (tomatoes, bell peppers) the number of carpels is often 3–5. Peppers often have axile placentation at the base and parietal above. The ovary of *Datura* is 4-locular by fusion of the axile placentae to the carpel walls. Fruits in the family are about equally divided between berries (Figure 15-36 F, G) and capsules (Figure 15-36 E). Capsular fruits often dehisce by 4 valves.

The vascular bundles of the Solanaceae are unusual in having internal phloem. Each bundle has two zones of phloem, one next to the pith and one external to the xylem. In some Solanaceae the internal phloem forms a continuous cylinder between the pith and the primary xylem.

Economically the Solanaceae is one of the most important families. It is a source of several major food crops including *Lycopersicon esculentum* (tomato), *Solanum tuberosum* (potato), *Solanum melongena* (eggplants), and *Capsicum* spp. (bell peppers, chili peppers). *Physalis philadelphica* (tomatillo) is used in Mexican cooking.

Many members of the Solanaceae contain alkaloids, some of which are extremely poisonous and some that are used as drugs. The plant most widely used for its drug properties is *Nicotiana tabacum* (tobacco) which produces nicotine, which is both moderately poisonous and addictive. In addition to its presence in cigarettes, cigars, snuff, etc., nicotine is an active ingredient in some insecticides. Other drugs in the Solanaceae are produced by *Atropa belladonna* (atropine, belladonna), *Hyoscyamus niger* (hyoscyamine, scopolamine), and *Datura* spp. (stramonium). Species of *Datura* and *Brugmansia* (angel's trumpet) were used by some aboriginal societies to induce hallucinations, but most abandoned the practice because of the dangers involved. There is little difference between a hallucinogenic dose and a fatal dose.

Many members of the family have showy flowers. Ornamentals in the family include species of genera such as *Petunia* (petunia), *Nicotiana* (tobacco), *Schizanthus* (poor man's orchid), *Cestrum* (night-blooming jasmine), *Solandra* (cup of gold), *Brugmansia* (angel's trumpet), and *Physalis* (Chinese lanterns). Some species of *Solanum* are aggressive weeds.

# Hydrophyllaceae

## The Waterleaf Family

Solanales ■

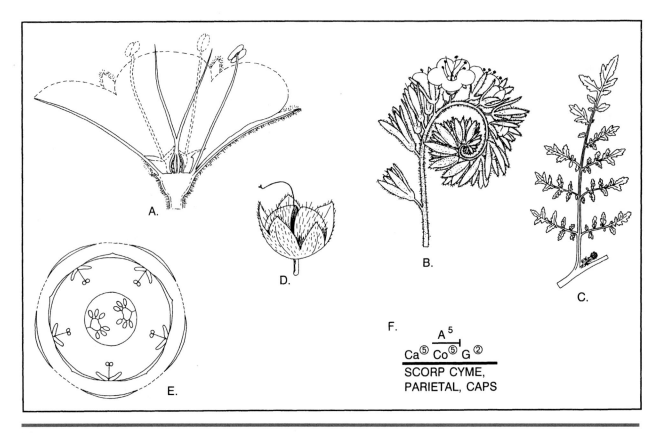

$$\underline{Ca^{\textcircled{5}} \; \overline{Co^{\textcircled{5}}}^{\; A^5} \; G^{\textcircled{2}}}$$

SCORP CYME,
PARIETAL, CAPS

**Figure 15-37** Features of Hydrophyllaceae. A. Flower of *Phacelia* (l.s.) B. Scorpioid cyme of *Phacelia*. C. Compound leaves of *Phacelia distans*. D. Loculicidal capsule of *Eucrypta chrysanthemifolia*. E. Floral diagram of *Phacelia*. F. Generalized floral formula.

# FAMILY DESCRIPTION

Herbs, shrubs, and small trees. Leaves alternate or opposite, simple to deeply lobed or pinnate, estipulate. Plants synoecious. Inflorescence cymose, usually scorpioid, or flowers solitary. Flowers perfect, regular. Sepals 5 (10–12), distinct or basally connate, sometimes alternating with auricles. Petals 5 (10–12), connate, often appendaged within. Stamens 5 (10–12), distinct, alternate with petals; filaments adnate to corolla. Carpels 2, connate; ovary superior (partially inferior) with 1 locule and few to many parietal ovules or with 2 locules and few to many axile ovules; styles 2 or 1 and 2-lobed. Fruit a capsule.

# FAMILY NOTES

The Hydrophyllaceae, a family of 19–20 genera and 250 species, is widespread in New World and scattered in widely separated regions of the Old World. The greatest diversity occurs in arid and semiarid regions of western North America. Over half the species in the family occur in California. Many of these are members of the genus *Phacelia*. Many of the temperate members of the Hydrophyllaceae are annual or perennial herbs. Species of *Eriodictyon* (yerba santa) from California and Arizona are woody and some occasionally attain the stature of small trees. Some species of *Wigandia*, a tropical American genus, also grow to tree size.

Members of the Hydrophyllaceae are usually rather easy to recognize. Although there is considerable variation in vegetative features, floral features are rather uniform. The flowers are borne in cymes, and these are most commonly scorpioid (Figure 15-37 B). In most species the flowers are blue, white or purple. The corollas are radial and often campanulate (Figure 15-37 B). Within the corolla tube there are often appendages that subtend or alternate with the stamens (Figure 15-37 A, E). The filaments are inserted near the base of the corolla tube. The style bears two short to elongated style branches. The ovary is one- or two-locular with the ovules attached to parietal (or less commonly axile) placentae. The fruit is a loculicidal capsule.

The waterleaf family is important primarily because of the beauty of its flowers. There are numerous showy wildflowers in the family including species of *Hydrophyllum* (waterleaf), *Nemophila* (baby blue-eyes, five-spot), *Pholistoma* (fiesta flower), *Nama*, and *Emmenanthe* (whispering bells). *Phacelia*, the largest genus in the family has many showy species. Species of *Phacelia*, *Nemophila*, *Wigandia* and a few other genera are cultivated as ornamentals. Otherwise the Hydrophyllaceae is of comparatively little economic significance.

# Apiaceae (Umbelliferae)

## The Carrot Family

Apiales ■

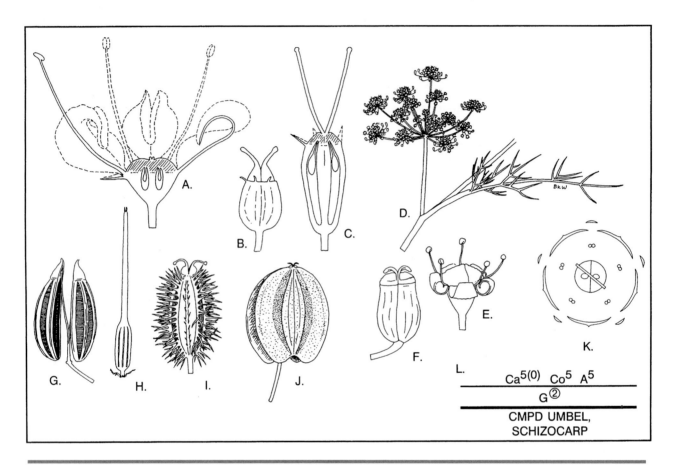

**Figure 15-38** Features of Apiaceae. A–C. *Oenanthe sarmentosa* A. Flower (l.s.). B. Immature schizocarp. C. Schizocarp (l.s.). D–G. *Foeniculum vulgare*. D. Leaf and compound umbel. E–F. Staminate and pistillate phase of protandrous flower; stamens and petals fall before styles elongate and stigmas become receptive to pollen. G. Mature mericarps splitting away from floral axis. H. Schizocarp of *Scandix pecten-veneris* with carpopodium prolonged into a beak. I. Schizocarp of *Daucus carota* armed with prickles. J. Winged schizocarp of *Lomatium*. K. Floral diagram of *Oenanthe*. L. Generalized floral formula.

# FAMILY DESCRIPTION

Herbs (rarely shrubs or trees). Leaves alternate (occasionally opposite) or all basal, simple to variously lobed or compound, estipulate (rarely stipulate) with petioles sheathing, with internal oil tubules and often strongly scented. Plants synoecious (occasionally polygamous, rarely dioecious). Inflorescence of compound umbels (rarely of simple umbels, heads, or axillary flowers). Flowers perfect (rarely imperfect) regular or irregular. Sepals 5, small, often very reduced and apparently absent, distinct. Petals 5 (0), distinct. Stamens 5, distinct, alternate with petals. Carpels 2, connate; ovary inferior with 2 locules and 1 apical-axile ovule per locule; styles 2, often subtended by bulging stylopodia (rarely elongated into a beak in fruit). Fruit a schizocarp with 2 mericarps, often strongly ribbed, sometimes winged and samaroid or covered with tubercles or prickles.

# FAMILY NOTES

The Apiaceae is a cosmopolitan family of about 428 genera and 3000 species with its greatest diversity in the north-temperate zone. Some are weedy and have greatly expanded their range along with man. They tend to be rare in the tropics.

The family is a highly natural one that can be recognized by a combination of distinctive features. The leaves are often large and compound with expanded, sheathing, many-veined petioles (Figure 15-38 D). The flowers are individually small, and almost always are borne in compound (or rarely simple) umbels (Figures 11-13 G, 15-38 D). The alternate name for the family, Umbelliferae, means "bearer of umbels."

The flowers are small and characteristically 5-merous (Figure 15-38 A). The sepals are usually very reduced (Figure 15-38 A–C), and often are so short that they do not exceed the inferior ovary to which they are fused (Figure 15-38 E–F). It is easy to determine that it is sepals and not petals that are [apparently] absent in such a flower. In plants that have both sepals and petals, the stamens are inserted alternately with the petals and opposite the sepals. In Apiaceae in which there is only one whorl of perianth, the missing parts are located opposite the stamens in the sepal position (Figure 15-38 E). The petals are usually early deciduous. They are usually radial, but in some genera the outer flowers of the umbels are strongly bilateral.

The inferior ovary is bicarpellate with one seed per carpel and often is ornamented with spines and wings (Figs. 15-38 I, J). Oil tubules typically extend through the pericarp of the resulting schizocarpic fruit. Often there are a pair of dome-shaped or conical **stylopodia** on the top of the ovary (Figures 15-38 F, H). The mature schizocarp breaks into a pair of mericarps attached to a pair of carpophores (Figure 15-38 G), the remnants of the ovary's axis.

Members of the Apiaceae have considerable economic importance. Vegetable crops include *Daucus carota* (carrot), *Foeniculum vulgare* (finochio), *Pastinaca sativa* (parsnip), and *Apium graveolens* (celery). Culinary herbs include the foliage of such species as *Anethum graveolens* (dill), *Anthriscus cerefolium* (chervil), *Coriandrum sativum* (cilantro), and *Petroselinum crispum* (parsley) and the mericarps ("seeds") of *Anethum graveolens* (dill), *Apium graveolens* (celery), *Carum carui* (caraway), *Coriandrum sativum* (coriander), *Cuminum cyminum* (cumin), *Foeniculum vulgare* (fennel), and *Pimpinella anisum* (anise).

Not all members of the family are edible, though. The family also includes a number of deadly poisonous plants including *Aethusa cynapium* (fool's parsley), *Conium maculatum* (poison hemlock) and *Cicuta* spp. (waterhemlock).

Ornamentals include species of *Angelica, Eryngium, Heracleum* and *Trachymene. Eryngium* is unusual in that its species have dense, spiny-bracted heads rather than umbels.

# Campanulaceae

## The Bellflower Family
### Asterales ■

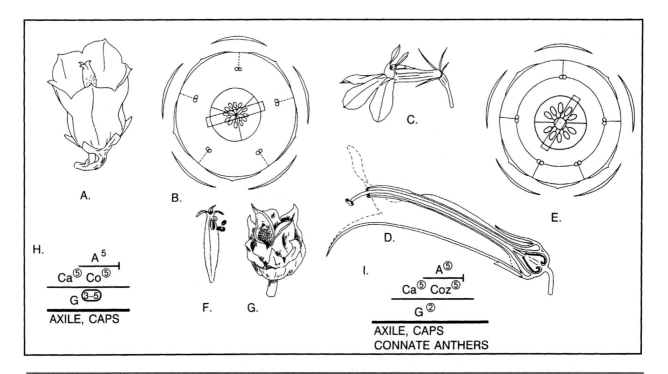

**Figure 15-39** Features of Campanulaceae. A. Radially symmetric flower of *Campanula*. B. Floral diagram of *Campanula*. C. Bilateral flower of *Lobelia*. D. Flower of *Lobelia* (l.s.). E. Floral diagram of *Lobelia*. F. Capsule of *Triodanis*. G. Capsule of *Lobelia*. H. Floral formula for subf. Campanuloideae, I. Floral formula for subf. Lobelioideae.

## FAMILY DESCRIPTION

Herbs (less frequently shrubs or trees), usually with milky sap. Leaves alternate (rarely opposite or whorled), simple, estipulate. Plants synoecious (rarely dioecious). Inflorescence variously cymose or racemose, often spicate or paniculiform or flowers solitary. Flowers perfect (rarely imperfect), regular or irregular. Sepals 5 (3–4 or 6–10), distinct or basally connate. Petals (0) 5 (3, 4 or 6), connate (rarely distinct); corolla sometimes bilabiate. Stamens 5, alternate with petals; filaments distinct or connate, free or adnate to base of corolla; anthers distinct or connate around the style. Carpels 2–5, connate; ovary inferior (superior) with 2–5 (10) locules and numerous (few) axile ovules or with 1 locule and numerous parietal ovules; style 1, entire or 2–5 lobed. Fruit a capsule or berry.

# FAMILY NOTES

The Campanulaceae is a cosmopolitan family of 70–84 genera and 1800–2000 species. Members of the Campanulaceae in temperate regions are mostly small herbs. However in certain tropical areas peculiar shrub and tree forms of the family have evolved. *Lobelia* and several related genera have diversified in the Hawaiian Islands. One Hawaiian genus, *Cyanea,* is composed of palm-like trees. Giant rosette tree forms of *Lobelia* occur at high elevations in the mountains of central Africa.

Flowers in the bellflower family vary considerably. The corolla varies from fully sympetalous to apparently apopetalous in a few genera. The stamens are often only slightly adnate to the corolla above the usually inferior ovary. The Campanulaceae comprises two subfamilies that are often treated as separate families.

Plants in subf. **Campanuloideae** have flowers actinomorphic or nearly so (Figure 15-39 A). In most members of the subfamily the stamens are distinct. The gynoecium comprises three to five carpels, with three the most common number. Common genera in North America are *Campanula* (bell-flower) and *Triodanis* (Venus looking glass).

Members of subf. **Lobelioideae** (Lobeliaceae) have zygomorphic flowers (Figure 15-39 C) that are often bilabiate and that have a pollen presentation method similar to that of the Asteraceae. The anthers (and sometimes the filaments as well) are connate into a tube that surrounds the style (Figure 15-39 D). Pollen grains are released into the interior of the tube and are pushed out by elongation of the style. There has been much controversy as to whether this feature has evolved independently in the two families or whether it was inherited from a common ancestor. Available evidence seems to support the first hypothesis. The flowers in Lobelioideae are resupinate (turned upside down). Because the pedicel has a half twist, what appears to be a 3-lobed lower corolla lip is actually the upper lip, and the 2-lobed upper lip originated as a lower lip. In North America the most common lobelioid genera are *Lobelia, Downingia,* and *Nemacladus.*

The fruit of the Campanulaceae is generally a capsule. These dehisce in various ways (Figure 15-39 F, G). A few genera produce berries.

Members of the Campanulaceae are important chiefly as flowering ornamentals. The flowers of many species of *Campanula* and *Lobelia* are particularly attractive. Species of various other genera are also commonly cultivated, including *Edrianthus, Jasione, Phyteuma, Platycodon,* and *Symphyandra.*

# Caprifoliaceae

## The Honeysuckle Family
### Dipsacales ■

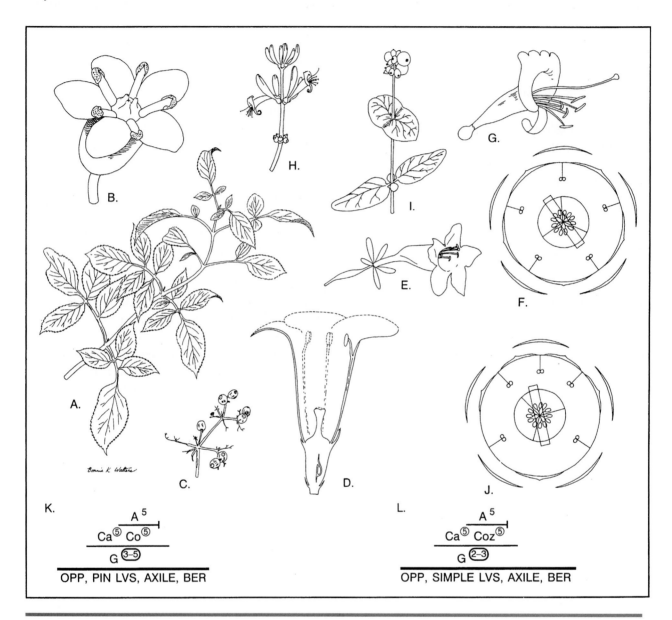

**Figure 15-40** Features of Caprifoliaceae. A. Leafy branch of *Sambucus mexicana* B. Flower of *Sambucus*. C. Cluster of berries of *Sambucus*. D. Flower of *Viburnum* (l.s.). E. Flower of *Abelia*. F. Floral diagram of *Sambucus*. G. Flower of *Lonicera*. H. Inflorescence of *Lonicera hispidula* with flowers in sessile cymes (appearing whorled). I. Twig of *Lonicera* with developing berries. J. Floral diagram of *Lonicera*. K. Floral formula of *Sambucus*. L. Floral formula of *Lonicera*.

# FAMILY DESCRIPTION

Herbs, shrubs, or woody vines. Leaves opposite (rarely whorled), simple or pinnate, estipulate (stipulate). Plants synoecious. Inflorescence variously cymose, sometimes corymbiform, paniculiform, or spicate, or flowers 1–few in leaf axils. Sepals 5, distinct or variously connate. Petals 5, connate; corolla regular to strongly irregular, sometimes bilabiate. Stamens 4 or 5, distinct; filaments adnate to corolla. Carpels 2–5, connate; ovary inferior with 2–5 locules and 1–few (many) apical-axile ovules per locule (rarely some locules sterile) or with 1 locule and 1-few parietal or apical ovules; style 1 or stigma sessile. Fruit a berry, drupe, capsule, or achene.

# FAMILY NOTES

The Caprifoliaceae is a comparatively small family of about 13–15 genera and 400 species. It is distributed mostly in north temperate and boreal regions with some species in tropical mountains. The greatest diversity is in eastern Asia and eastern North America, regions that share several genera and closely related species. Among these are *Linnaea, Lonicera, Sambucus, Symphoricarpos, Triosteum,* and *Viburnum.*

The Caprifoliaceae includes two genera with radially symmetric corollas, *Viburnum* and *Sambucus,* (Figure 15-38 B, D). The remaining genera all have more or less bilaterally symmetric flowers (Figure 15-38 E, G, H). Both groups have opposite leaves, cymose inflorescences and flowers with 5-merous corollas, epipetalous stamens, and an inferior ovary. Recent studies by Donoghue (1983) have suggested that the genera with radially symmetric corollas may not be closely related to the genera that have bilateral flowers. These studies indicate that the radial-flowered members of the Caprifoliaceae may be most closely related to the monotypic family, Adoxaceae. Other workers have suggested a relationship of the Caprifoliaceae to the Rubiaceae. Although similar in major floral features, the Caprifoliaceae and Rubiaceae differ in a number of minor, but important characters (e.g., nature of stipules, type of endosperm).

Features of the gynoecium vary considerably in the honeysuckle family. The number of carpels ranges from two to five. Most commonly the placentation type is axile. In *Abelia* the ovary has three locules, but only one contains ovules. The other two are empty. In *Viburnum* the ovary is unilocular and contains a single apical ovule. The ovary matures as a berry in most genera, as a capsule in a few, as a drupe in *Viburnum,* and an achene in *Abelia* and *Linnaea.*

The Caprifoliaceae includes several genera of woody ornamentals. These include members of such genera as *Lonicera* (honeysuckles), *Viburnum, Abelia,* and *Weigela.* Some species of *Sambucus* (elderberry), the only genus in the family with compound leaves, are used in preparation of wines and jellies. *Lonicera japonica,* widely cultivated as an ornamental, is an invasive weed in some forested regions of North America where it crowds out the native understory vegetation. Otherwise the family is of little economic importance.

# Exercises

## ■ Study Questions

## Ericaceae

1. What kind of corolla do we typically find in the Ericaceae?

   What is the most common corolla shape in this family?

2. To what structure are the stamens in the Ericaceae attached?

   How does this differ from the attachment in the families you have studied in the Asteridae?

   How do anthers in most Ericaceae dehisce?

   What other features of the anthers are unusual?

3. What kind of gynoecium characterizes the Ericaceae?

   Placentation?

   What ovary position do we find in most of the subfamilies?

   What is the ovary position in the Vaccinioideae?

4. What kinds of fruits do we find in the Ericaceae?

5. Members of the Ericaceae are often found on soils of [acidic or basic pH]?

6. What features of subf. Monotropoideae are unusual?

   How do these plants obtain their nutrients?

7. What features of subf. Pyroloideae are unusual?

## Polemoniaceae

8. What feature of the calyx is usually distinctive for the Polemoniaceae?

9. What kind of corolla do members of the Polemoniaceae have?

   Their flowers are usually ___-merous.

   What is the floral symmetry in most members of the family?

10. How many stamens are there in Polemoniaceae flowers?

    What are they attached to?

    What is their relation to the corolla lobes?

11. What kind of gynoecium characterizes the families of the Polemoniaceae?

    Placentation?

    Number of carpels?

    Ovary position?

12. What is the fruit type in the Polemoniaceae?

## Primulaceae

13. What is the habit of members of the Primulaceae?

14. What kind of corolla do members of the Primulaceae have?

    What is the relationship between the stamens and the petals?

15. What kind of gynoecium do we find in Primulaeae flowers?

    Placentation?

    Ovary position?

    Fruit type?

16. How could you determine that the Primulaceae is *not* a member of subclass Asteridae?

## Rubiaceae

17. What is the leaf arrangement in most Rubiaceae?

    What feature of the leaves of most Rubiaceae is unusual in the Asteridae?

18. What is unusual about the leaves of *Galium?*

19. Flowers in Rubiaceae are generally ____-merous or ____-merous.

    What is the usual flower symmetry?

20. What is the ovary position in Rubiaceae?

21. What are the fruit types in Rubiaceae?

## Asclepiadaceae

22. Why is the Asclepiadaceae called the milkweed family?

23. What is the inflorescence of most Asclepiadaceae?

24. Most Asclepiadaceae flowers have a corona. What is this structure?

25. What are the unusual characteristics of the stamens of the Asclepiadaceae?

26. What is unusual about the gynoecium?

27. Explain the structure and function of the gynostegium.

28. What is the characteristic fruit of the Asclepiadaceae?

## Boraginaceae

29. What is the characteristic inflorescence of most Boraginaceae?

30. What kind of corolla do members of the Boraginaceae have?

    Their flowers are usually ____-merous.

    Flower symmetry?

31. How many stamens are there in Boraginaceae flowers?

What are they attached to?

What is their relation to the corolla lobes?

32. What kind of gynoecium characterizes the Boraginaceae?

How many carpels do we find per flower?

Ovary position?

33. What is the fruit type in the Boraginaceae?

What other family has this type of fruit?

# Lamiaceae (Labiatae)

34. What vegetative features characterize the Lamiaceae?

What is the usual inflorescence in this family?

35. What kind of corolla do members of the Lamiaceae have?

Their flowers are usually ___-merous.

Flower symmetry?

How does this flower symmetry relate to the traditional name for the family?

36. How many stamens are there in Lamiaceae flowers?

What are they attached to?

What is their relation to the corolla lobes?

37. What kind of gynoecium characterizes the Lamiaceae?

How many carpels do we find per flower?

What is the ovary position?

What other family has this type of gynoecium?

How would you distinguish these families?

38. What is the fruit type in the Lamiaceae?

39. Some members of the Lamiaceae are used as culinary herbs. What characteristic of the Lamiaceae gives these plants such interesting tastes and aromas?

# Scrophulariaceae

40. What kind of corolla do members of the Scrophulariaceae have?

Their flowers are usually ___-merous.

What is the flower symmetry in most members of the family?

41. How many stamens are there in most Scrophulariaceae flowers?

What are they attached to?

What is their relation to the corolla lobes?

If the flower has a staminode, in what position would you find it?

42. What kind of gynoecium characterizes the Scrophulariaceae?

    How many carpels do we find per flower?

    What is the usual kind of placentation?

    Ovary position?

43. What is the fruit type in the Scrophulariaceae?

44. What features do Scrophulariaceae and Lamiaceae have in common?

    How would you distinguish them?

## Solanaceae

45. What kind of corolla do members of the Solanaceae have?

    Their flowers are usually ___-merous.

    Flower symmetry?

46. How many stamens are there in most Solanaceae flowers?

    What are they attached to?

    What is their relation to the corolla lobes?

47. What kind of gynoecium characterizes the Solanaceae?

    What is the most common number of carpels?

    What kind of placentation?

    Ovary position?

48. What are the fruit types in the Solanaceae?

49. What features do Solanaceae and Scrophulariaceae share?

    How would you distinguish them?

## Hydrophyllaceae

50. What is the characteristic inflorescence of most Hydrophyllaceae?

51. What kind of corolla do members of the Hydrophyllaceae have?

    Their flowers are usually ___-merous.

    Flower symmetry?

52. How many stamens are there in Hydrophyllaceae flowers?

    What are they attached to?

    What is their relation to the corolla lobes?

53. What kind of gynoecium characterizes the Hydrophyllaceae?

    How many carpels?

    Ovary position?

54. What is the fruit type in the Hydrophyllaceae?

55. What features do Hydrophyllaceae and Boraginaceae share?

    How would you distinguish them?

# Apiaceae

56. What kind of inflorescence is characteristic of the Apiaceae?

    How does this inflorescence relate to the traditional name for the family?

57. What kind of gynoecium characterizes the Apiaceae?

    How many carpels do we find per flower?

    What is the ovary position?

58. How many stamens would you expect to find in a flower of a member of the ?Apiaceae?

    What are these stamens attached to?

    Are they alternate with or opposite the petals?

59. Flowerse in the Apiaceae are __-merous.

60. What kind of fruit do we find in the Apiaceae?

61. Some members of the Apiaceae lack sepals.

    How do we know that it is the sepals that are missing and not the petals?

62. Some members of the Apiaceae are used as culinary herbs.

    What characteristic of the Apiaeae gives these plants such interesting tastes and aromas?

# Campanulaceae

63. What kind of corolla do members of the Campanulaceae have?

    Their flowers are ___-merous.

64. How many stamens are there in Campanulaceae flowers?

    What are they attached to?

    What is their relation to the corolla lobes?

65. What kind of gynoecium characterizes the Campanulaceae?

    How many carpels do we find per flower?

    What is the usual kind of placentation?

    Ovary position?

66. What is the fruit type in the Campanulaceae?

67. What features distinguish subf. Campanuloideae from subf. Lobelioideae?

# Caprifoliaceae

68. What vegetative features characterize the Caprifoliaceae?

69. What kind of corolla do members of the Caprifoliaceae have?

    Their flowers are usually ___-merous.

    These flowers may be either _____ symmetric or _____ symmetric.

70. How many stamens are there in most Caprifoliaceae flowers?

What are they attached to?

What is their relation to the corolla lobes?

71. What kind of gynoecium characterizes the Caprifoliaceae?

How many carpels may we expect to find per flower?

What kinds of placentation?

What is the ovary position?

72. What are the fruit types in the Caprifoliaceae?

## General Study

73. What features do all of the families above share?

Which ones have radial symmetry?

Which have bilateral symmetry?

Which have superior ovaries?

Which have inferior ovaries?

Which have 4 nutlets?

Which have capsules?

Which have berries?

Which have scorpioid cymes?

74. Based upon your answers to question 73, choose six of these families and write a key to them: Asclepiadaceae, Solanaceae, Polemoniaceae, Hydrophyllaceae, Boraginaceae, Lamiaceae. Scrophulariaceae, Campanulaceae, Rubiaceae, and Caprifoliaceae.

75. List the families and species of Asterids that you identified in the lab.

_____     _____

_____     _____

_____     _____

_____     _____

_____     _____

_____     _____

76. What is the alternate name for the Asteraceae?

How does this name relate to the features of this family?

77. What are the parts of the head of a member of the Asteraceae?

Is it a determinate or indeterminate inflorescence?

Is the arrangement of heads on the plant determinate or indeterminate?

78. What are the individual bracts of the involucre of the Asteraceae called?

What are the bracts of the receptacle of the head called?

79. What kind of corolla do members of the Asteraceae have?

80. What is the symmetry of most disk flowers?

Their flowers are basically all _____-merous.

How many stamens are there in a disk flower?

To what are the stamens attached?

81. Botanists who specialize in the Asteraceae are often called "synantherologists."

To what feature of the Asteraceae flower does this refer?

82. What is the calyx of an Asteraceae flower called?

In what ways does it differ from the calyx of most other families?

What are some of the common modifications of these structures?

83. What kind of gynoecium characterizes the Asteraceae?

How many carpels?

How can we tell?

Number of ovules per ovary?

Placentation type?

Ovary position?

84. What is the fruit type in the Asteraceae?

What do we typically find attached to the top of the fruit when it falls from the plant?

85. What is the difference between chaff and pappus?

Where is chaff attached?

Where is the pappus attached?

86. A head composed only of disk flowers is a _____ head.

87. What is the symmetry of a ray flower?

Their flowers generally appear to be _____-merous.

What is the flat, strap-shaped lip of a ray flower called?

What is the sexual condition of a typical ray flower?

88. What is a head with ray flowers around the outside and disk flowers in the center called?

89. What is the difference between a filiform flower and a ray flower?

What do a filiform flower and a typical ray flower have in common?

What is the sexual condition of a filiform flower?

90. What is a disciform head?

What is the difference between a disciform head and a discoid head?

91. What is the symmetry of a ligulate flower?

    Ligulate flowers are _____-merous.

    What is the flat, strap-shaped lip of a ligulate flower called?

    What is the sexual condition of a typical ligulate flower?

92. What do ligulate flowers and ray flowers have in common?

    How do they differ?

    What do ligulate flowers and disk flowers have in common?

    How do they differ?

93. What is a ligulate head?

    How does it differ from a radiate head?

    What is the sap like in plants with ligulate heads?

94. Which members of the Asteraceae are most important economically?

95. List the species of Asteraceae that you identified in the lab.

_____    _____

_____    _____

_____    _____

_____    _____

_____    _____

_____    _____

# Monocots

chapter 16

■ Subclasses Alismatidae, Arecidae, Zingiberidae, and Lillidae

The monocots (Liliopsida) are divided into 11 orders, 65 families and about 50,000 species. The monocots have been traditionally been divided into five subclasses. The new classification presented here represents a significant departure from this grouping of five subclasses. The Alismatales are comparable to Cronquist's Alismatidae and are largely aquatic with apocarpous gynoecia. Subclass Zingiberidae is retained as a clade of Zingiberales with colored petals and sepaloid sepals, but the Bromeliales have been transferred to the Poales. The subclassses Arecidae and Liliidae are not monophyletic and have been dismantled. The Commellinanae includes Cronquist's Commelinidae, with flowers reduced and perianth green, brown or wanting, but the addition of the Commelinales and Zingiberales to this group unites two groups with showy flowers with this large group of "grass-like" plants.

Monocots and dicots differ in various respects. There are fundamental differences in the internal anatomy of monocot and dicot vegetative organs. Monocots lack typical cambial secondary growth. Monocots lack the taproots typical of most dicots and instead have a well-developed system of fibrous roots. Monocot leaves often have parallel rather than pinnate or palmate venation. The flowers of most monocots are 3-merous, unlike the 4-merous and 5-merous flowers that predominate in the dicots.

The internal anatomy of a monocot stem is quite different from that of most dicots. Young dicot stems typically have a ring of vascular bundles that separates the stem's parenchymatous ground tissue into an outer cortex and a central pith area. As the stem ages, a cylindrical vascular cambium forms from a combination of undifferentiated procambium cells of the vascular bundles and parenchyma cells of the adjacent ground tissue. As the cambium produces secondary xylem and secondary phloem, the stem increases in diameter. The trunk of a dicotyledonous tree expands in diameter as the canopy enlarges.

Monocot stems have scattered vascular bundles embedded in ground parenchyma but do not have clearly defined cortex and pith zones. A typical vascular cambium never forms in monocot stems. As a result the stem diameter in most monocots does not change after primary growth has taken place. The trunk of a treelike monocot such as a palm tree or a giant bamboo reaches its full diameter as a result of primary growth and retains that diameter once primary growth ceases. A few monocots (e.g., *Yucca brevifolia,* the Joshua-tree) are exceptional in having a peculiar form of secondary growth in which a cambium-like cylinder of cells produces additional vascular bundles in a parenchyma matrix. These plants do show increase in diameter with age.

Monocot leaves most commonly have parallel venation. There are several principal veins that extend the full length of the leaf. Often additional small veinlets branch away from the main veins, sometimes extending from one principal vein to another like rungs of a ladder. Certain monocots do have pinnate or palmate venation, however. In the subclasses Arecidae and Zingiberidae pinnate or palmate venation predominate, and such venation occurs in some members of the Alismatidae and Liliidae as well. In many cases, however, the internal anatomy of the main veins is much different from that of most dicots. In dicot leaves with pinnate venation, the midrib of the leaf most commonly contains a single crescent-shaped vascular bundle. In pinnately veined monocot leaves, however, the anatomy of the midrib often resembles that of a monocot stem with several to many scattered vascular bundles in a parenchyma matrix. Often the major lateral veins in a pinnately veined monocot leaf are parallel. The bases of most monocot leaves enwrap the stem forming a cylindrical sheath. The leaf veins arise in a cylindrical pattern from the stem.

Monocot flowers are typically 3-merous. In some monocot flowers the sepals and petals are clearly differentiated from each other and in others both whorls are petaloid or sepaloid. In the latter situation, the petals and sepals may look so much alike that they can be differentiated only by their position in the bud. Perianth parts are reduced or absent in some monocots. Monocot flowers most commonly have 6 stamens. The stamens are numerous in only a few genera. In several lineages the number is reduced to 3 or fewer and sometimes staminodes are present. A gynoecium of 3 carpels is the most common condition. In the Alismatidae the flowers in

certain families have several to many distinct carpels. In some cases the number of carpels is reduced to 1 or 2. The degree of carpellary fusion varies in several subclasses from apocarpous or nearly so to fully syncarpous.

Monocots and dicots apparently separated in the lower Cretaceous period not long after the first dicots made their appearance in the fossil record. The morphology of the earliest monocots and of their immediate ancestors is not known. The lack of cambial activity and the predominately parallel venation of monocot leaves are probably primitive within the group and have been used to suggest that monocots are descended from a dicotyledonous ancestor that was aquatic or semiaquatic. (Cambial activity is lost in some aquatic dicots; e.g., members of the Nymphaeaceae.) Various botanists have suggested that the monocot leaf may have originated as a bladeless sheathing petiole with parallel veins. The leaf-blades of modern monocots would be secondarily expanded if this hypothesis is true. It should be noted here that not all botanists are in

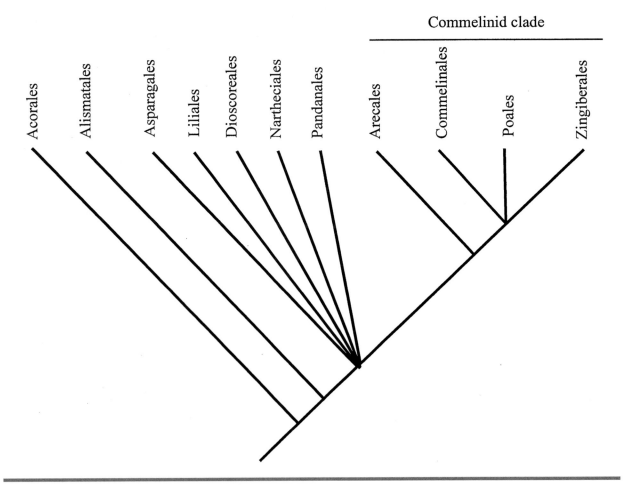

**Figure 16-1** Phylogeny of the monocots. The Figure is adapted from Stevens (2001 onward) and Judd et al. (2001).

agreement regarding a petiolar origin of the monocotyledonous leaf. Dahlgren et al. (1985) have proposed that the original monocots had net-veined, slender-petioled leaves like those of most dicots. The original monocots probably had a 3-merous flower with an apocarpous gynoecium.

Although vegetative and floral characters are of importance in monocot classification, many of the major taxonomic characters within the monocots are less readily observable physiological, chemical, anatomical, or ecological features. The unusual stem and leaf anatomy of monocots apparently restricts the amount of variation possible. Floral features are important in the monocots, but they are less so than in the dicots because there is less variation in monocot flowers. The relationship and generic composition of many monocot families are still the basis of much debate. The monocots are currently recognized in eleven orders. The major clade the Lilianae has been separated into five unresolved orders: Asparagales, Liliales, Dioscoreales, Nartheciales, Pandanales, and the Commelinid is comprised of the Arecales, Poales, Commelinales, and the Zingiberales (Figure 16-1).

# BASAL TO THE MONOCOTS LINEAGES

## 1. Order Acorales

**1.A    Acoraceae**—The Sweet-Flag Family. 1 genus, 2 species. Circumboreal, southeast Asia, East Indies.

## 2. Order Alismatales

**2.A    Araceae**—The Aruhm Family (Figure 16-7). 105–110 genera, 1800–2450 species. Cosmopolitan with greatest diversity of species in tropical and subtropical areas.

§See page 432 for description and illustrations§

**2.B    Hydrocharitaceae**—The Wateweed Family. 15–19 genera, 100 species. Cosmopolitan.

Submersed or floating aquatic herbs. Leaves alternate, opposite, whorled or all basal, simple, sheathing or sheathless, of diverse forms. Plants monoecious or dioecious (or rarely synoecious). Inflorescence of solitary flowers or umbels, subtended by spathes. Flowers perfect or imperfect, regular or weakly irregular. Hypanthium sometimes well developed. Sepals 3 (2), distinct, green. Petals 3 (0 or 2), distinct. Stamens 1–many, distinct or filaments ± connate. Carpels 2–15, connate; ovary inferior with 1 locule and many laminar-parietal ovules or with 6–9 locules and many laminar-axile ovules per locule; styles 2–15, often bifid or trifid. Fruit a capsule or berry.

**2.C    Butomaceae**—The Flowering-rush Family. 1 genus, 1 species. Eurasia.

Aquatic or semiaquatic herbs. Leaves basal and alternate, simple. Without distinct petiole, stipulate, basally sheathing. Plants synoecious. Inflorescence an umbelliform cyme. Flowers perfect, regular. Sepals 3, distinct. Petals 3, distinct. Stamens 9, distinct. Carpels 6, basally weakly connate; ovaries superior with 1 locule and many laminar ovules; style 1 per carpel, undivided. Fruit a follicle.

**2.D    Alismataceae**—The Water Plantain Family (Figure 16-5). 11–14 genera, 75–100 species. Nearly cosmopolitan but with greatest diversity in tropical and subtropical regions in the northern hemisphere.

§See page 434 for description and illustrations§

**2.E    Limnocharitacae**—The Water-poppy Family. 3–4 genera, 7–12 species. Tropics to subtropics.

Aquatic or semiaquatic herbs. Leaves basal and alternate, simple, usually with exposed blade distinct from petiole, stipulate, basally sheathing. Plants synoecious. Inflorescence an umbelliform cyme. Flowers perfect, regular. Sepals 3, distinct. Petals 3, distinct. Stamens 6 or 9 or many, distinct. Carpels 3–9, distinct or basally weakly connate; ovaries superior with 1 locule and many laminar ovules; style 1 per carpel, undivided. Fruit a follicle.

**2.F    Scheuchzeriaceae**—The Scheuchzeria Family. 1 genus, 1 species. Cooler regions of northern hemisphere.

Semiaquatic herbs. Leaves alternate and basal, simple and entire, linear, basally sheathing and ligulate. Plants synoecious. Inflorescence a raceme.

Flowers perfect, regular. Sepals 3, distinct, green. Petals 3, distinct, similar to sepals. Stamens 6, distinct. Carpels 3 or 6, nearly distinct, connate only at base; ovaries superior with one locule and 2–many basal ovules per carpel; styles absent, stigmas sessile. Fruit a cluster of follicles.

**2.G Aponogetonaceae**—The Capepondweed Family. 1 genus, 40–50 species. Old World tropics.

Aquatic herbs. Leaves all basal, simple, estipulate, some strap-shaped and linear, others with floating blades or erect and bladeless. Plants synoecious or monoecious. Inflorescence of 1–10 spikes on a long peduncle, subtended by ephemeral spathes. Flowers perfect or imperfect, regular, irregular or naked. Tepals 1–3, distinct or absent. Stamen 5 to many, distinct. Carpels 3–8, distinct; ovaries superior with 1 locule and 2–8 basal or marginal ovules per carpel; style 1 per carpel, undivided. Fruit a cluster of follicles.

**2.H Juncaginaceae**—The Arrow-grass Family. 3–5 genera, 20 species. Cold and temperate regions of both hemispheres.

Herbs, sometimes aquatic. Leaves basal, simple and entire, linear, basally sheathing, ligulate. Plants synoecious, monoecious, dioecious or polygamous. Inflorescence a raceme or spike. Flowers perfect or imperfect, regular. Sepals 1 or 3, distinct, green or red. Petals 0 or 3, distinct, similar to sepals. Stamens 1 or 4 or 6, distinct; filaments very short. Carpels 1 or 4 or 6, distinct or connate; ovary superior with 4 or 6 locules or ovaries superior and each 1-locular, ovule 1 per locule, basal; style 1 per carpel, very short or stigmas sessile. Fruit a cluster of follicles or achenes.

**2.I Posidoniaceae**—1 genus, 3 species. Mediterranean and Australia.

**2.J Ruppiaceae**—The Ditch-grass Family. 1 genus, 1–7 species. Nearly cosmopolitan.

Submersed aquatic herbs. Leaves alternate or opposite, simple, linear; basal sheath with stipule-like lobes, ligulate or eligulate. Plants synoecious. Inflorescence of 2-flowered axillary spikes, umbelliform in fruit because of long-stipitate carpels; peduncle helically coiled in fruit. Petals absent. perfect, naked. Sepals absent. Petals absent. Stamens 2, distinct; anthers sessile. Carpels 4, distinct; ovaries superior with 1 locule and 1 apical ovule, sessile in flower but long stipitate in fruit; styles absent, stigmas sessile. Fruit an achene.

**2.K Cymodoceaceae**—5 genera, 16–18 species. Tropical sea coasts.

**2.L Zosteraceae**—The Eel-grass Family. 3 genera, 18 species. Subtropical to subarctic sea coasts.

Submersed marine aquatic herbs. Leaves alternate, 2-ranked, simple, linear, sheathing with stipule-like outgrowths. Plants monoecious or dioecious. Inflorescence a flattened spadix enclosed by a spathe-like leaf sheath. Flowers imperfect, naked or with a single perianth segment. Sepal 1 or 0, attached near margin of spadix. Petals absent. Stamen 1, with a single anther sac. Carpel 1; ovary superior with 1 locule and 1 apical ovule; style 1, two-branched. Fruit an achene.

**2.M Potamogetonaceae**—The Pondweed Family (Figure 16-2). 1 genus, 100 species. Cosmopolitan.

Aquatic herbs, submersed or with floating leaves. Leaves alternate (or rarely opposite or whorled), 2-ranked, simple and entire, sheathing with stipule-like ligules. Plants synoecious. Inflorescence of spikes. Flowers perfect, regular. Tepals 4, distinct, inconspicuous. Stamens 4, distinct, opposite the tepals and basally adnate to them; anthers sessile. Carpels 4, distinct; ovaries superior with 1 locule and 1 marginal ovule; styles 1 per carpel, short or stigmas sessile. Fruit an achene or a drupe.

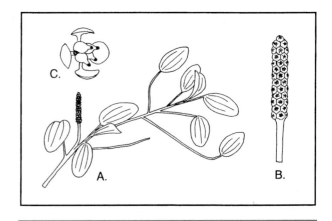

**Figure 16-2** Examples of Potamogetonaceae. A–C. *Potamogeton* sp. A. Stem with floating leaves and fruiting spike. B. Fruiting spike. C. Calyx with maturing achenes.

## 3. Order Petrosaviales

**3.A   Petrosaviaceae**—1 genus, 2 species. Southeast Asia.

## 4. Order Dioscoreales

**4.A   Burmanniaceae**—15–20 genera, 130 species. Pantropical, few temperate.

**4.B   Dioscoreaceae**—The Yam Family. 5–9 genera, 620 species. Widespread in the tropics and subtropics, few temperate.

Herbs, shrubs or woody vines. Leaves alternate (or rarely opposite), simple and palmately veined, estipulate, non-sheathing. Plants synoecious or dioecious. Inflorescence of spikes, racemes, panicles or solitary flowers. Flowers perfect or imperfect, regular, small. Hypanthium usually present but often short. Sepals 3, distinct, green or petaloid. Petals 3, distinct, similar to sepals. Stamens 3 or 6, distinct or filaments basally connate, ± adnate to hypanthium. Carpels 3, connate; ovary inferior with 3 locules and 2–many axile ovules per locule; styles 3 or 1 and 3-lobed. Fruit a capsule, berry or samara.

## 5. Order Pandanales

**5.A   Velloziaceae**—5–10 genera, 250–360 species. South America, Africa, Madagascar and the Arabian peninsula.

**5.B   Triuridaceae**—6–7 genera, 70—80 species. Tropics to subtropics.

**5.C   Stemonaceae**—3–4 genera 30–40 species. Southeast Asia to Australia and disjunct in the southeastern United States.

**5.D   Pandanaceae**—3 genera, 680–900 species. Old World tropics.

**5.E   Cyclanthaceae**—11–12 genera, 180 species. New World tropics.

# LILIANAE

## 6. Order Liliales

**6.A   Corsiaceae**—2 genera, 9 species. Chile and New Guinea.

**6.B   Smilacaceae**—The Green-brier Family. 10–12 genera, 300–320 species. Widespread in the tropics and subtropics, especially in the southern hemisphere.

Herbs, shrubs or woody vines, often prickly, often tendril-bearing. Leaves alternate (opposite), simple and entire, palmately veined, petiolate, stipulate, weakly sheathing. Plants synoecious or dioecious. Inflorescence of umbels or panicles, or flowers solitary and axillary. Flowers perfect or imperfect, regular. Floral tube sometimes present. Sepals 3, distinct, free or adnate to petals. Petals 3, distinct or adnate to sepals. Stamens 6, distinct, free or adnate to floral tube; pistillate flowers with 6 staminodes. Carpels 3 (rarely 1), connate; ovary superior (inferior) with 3 locules and 1–many axile ovules or locule 1 and ovules parietal; styles 3, distinct or 1 and entire to ± 3-lobed. Fruit a berry.

**6.C   Liliaceae** (sensu lato)—The Lily Family (Figure 16-9). Ca. 270 genera, 4000 species. Cosmopolitan.

§See page 436 for description and illustrations§

## 7. Order Asparagales

**7.A   Orchidaceae**—The Orchid Family (Figure 16-10). 700–1000 genera, 20,000 to 25,000 species. Cosmopolitan with the greatest diversity in the tropics.

§See page 442 for description and illustrations§

**7.B   Tecophiliaceae**

Herbs from corms, erect. Leaves basal and cauline, sheathing, linear, parallel-veined. Inflorescence of racemes and panicles with scattered bracts. Perianth parts 3 + 3, all petaloid, free or fused, radial. Stamens 6, free from perianth, some often reduced to staminodes, sometimes alternate with up to 6 extra staminodes, connivent and bent to one side of flower; anthers dehiscent by apical slits or pores. Ovary half-inferior. Fruit a capsule.

**7.C    Iridaceae**—The Iris Family (Figure 16-11). 60–88 genera, 1500 species. Cosmopolitan.

§See page 445 for description and illustrations§

**7.D    Xanthorrhoeaceae**—9–10 genera, 55 species. Australian region.

**7.E    Asphodelaceae**

Herbs from rhizomes, erect. Leaves mostly basal and subbasal, sheathing, linear, parallel-veined, sometimes succulent, reduced to bracts on flowering stems. Inflorescence of racemes, spikes and panicles with scattered bracts. Perianth parts 3 + 3, all petaloid, free or fused, radial. Stamens 6, free from perianth. Ovary superior. Fruit a capsule.

**7.F    Agavaceae**—The Agave Family (Figure 16-9F). 12–26 genera, 600 species. Widespread in dry subtropical areas.

Herbs, shrubs or trees. Leaves alternate or all basal, often forming dense basal or terminal rosettes, simple, basally sheathing, usually fibrous, often succulent, sometimes very stiff and hard. Plants synoecious, dioecious or polygamous. Inflorescence of racemes, spikes or panicles. Flowers perfect, regular or weakly irregular. Hypanthium sometimes well developed. Sepals 3, often petaloid, distinct, free or adnate to petals. Petals 3, distinct, free or adnate to sepals. Stamens 6, distinct, free or adnate to hypanthium. Carpels 3, connate; ovary superior to inferior with 3 locules and 1–many ovules per locule (rarely with 1 locule and 3 parietal ovules). Style 1 or stigma sessile. Fruit a capsule, berry or samara.

# COMMELINANAE

## 8. Order Arecales

**8.A    Arecaceae (Palmae)**—The Palm Family (Figure 16-12). 200–202 genera, 2700–3000 species. Pantropical with a few in subtropical or warm-temperate latitudes.

§See page 447 for description and illustrations§

## 9. Order Poales

**9.A    Rapateaceae**—16–17 genera, 80–100 species. Mostly tropical South America, disjunct in western Africa.

**9.B    Sparganiaceae**—The Bur-Reed Family (Figure 16-3). 1 genus, 13–20 species. North temperate regions and in Australia and New Zealand.

Submersed or emersed aquatic herbs. Leaves alternate and basal, two ranked, simple and entire, linear, basally sheathing, eligulate. Plants monoecious. Inflorescence of staminate and pistillate heads. Flowers imperfect, regular. Tepals 3–6, distinct, scale-like. Stamens 3–6, distinct. Carpels 1–2, connate; ovary superior with 1 or 2 locules and 1 apical or basal ovule per locule; styles 1 or 2. Fruit an achene.

**9.C    Typhaceae**—The Cattail Family (Figure 16-4). 1 genus, 10–15 species. Cosmopolitan.

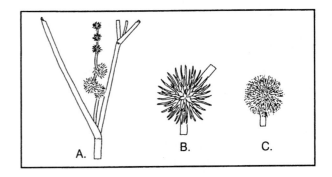

**Figure 16-3**    Examples of Sparganiaceae. A–C. *Sparganium eurycarpum*. A. Stem with cauline leaves and inflorescence. B. Pistillate head. C. Staminate head.

Semiaquatic herbs. Leaves alternate and basal, simple and entire, linear, basally sheathing, eligulate. Plants monoecious. Inflorescene a long-peduncled double spadix, the terminal spadix staminate, the lower spadix pistillate, each spadix with an early-deciduous, membranous spathe. Flowers imperfect, subtended by numerous slender scales or bristles. Sepals represented by the scales or bristles. Petals absent. Stamens 2–5, distinct or filaments connate. Capel 1; ovary superior with 1 locule and 1 apical ovule, ovary borne on a slender, long-pubescent stipe; style 1, undivided. Fruit a wind-dispersed, achene-like, one-seeded follicle, dehiscent after dispersal.

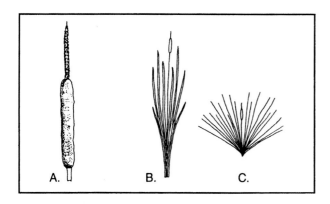

**Figure 16-4** Examples of Typhaceae. A–C. *Typha latifolia.* A. Double spadix with staminate flowers above and pistillate flowers below. B. Habit. C. Fruit with perianth bristles.

**9.D Bromeliaceae**—The Pineapple Family (Figure 16-13). 45–54 genera, 1500–2000 species. Mostly tropical and subtropical family endemic (with the exception of one species) to the New World.

§See page 450 for description and illustrations§

**9.E Thurniaceae**—1 genus, 2–3 species. Tropical South America.

**9.F Juncaceae**—The Rush Family (Figure 16-14). 7–9 genera, 300 species. Worldwide but most common in temperate latitudes; in the tropics mostly restricted to montane regions.

§See page 452 for description and illustrations§

**9.G Cyperaceae**—The Sedge Family (Figure 16-15). 70–102 genera, 4000–9300 species. Cosmopolitan.

§See page 454 for description and illustrations§

**9.H Hydatellaceae**—2 genera, 7 species. Australia, New Zealand, Tasmania.

**9.I Mayacaceae**—1 genus, 4–10 species. Warm temperate and tropical portions of North and South America and tropical Africa.

**9.J Eriocaulaceae**—The Pipewort Family. 9–13 genera, 1150–1200 species. Widespread in tropics and subtropics, few temperate.

Herbs. Leaves all basal or sub-basal, simple and entire, linear, not or scarcely sheathing. Plants monoecious (rarely dioecious). Inflorescence of scapose, involucrate heads. Flowers imperfect, regular or irregular, very small. Sepals 2–3, distinct or connate. Petals 2–3 (0). distinct or connate. Stamens 2–6 (1), distinct or filaments connate, adnate to the petals. Carpels 2–3, connate with 2–3 locules and 1 basal ovule per locule; style 1, 2–3 branched or style-branches again branched. Fruit a capsule.

**9.K Xyridaceae**—The Yellow-eyed Grass Family. 4–5 genera, 200–300 species. Pantropical and subtropical, with a few temperate representatives.

Herbs. Leaves alternate or all basal, two-ranked, simple and entire, basally sheathing, the blade flat, cylindrical or equitant. Plants synoecious. Inflorescence of long-peduncled spikes or heads. Flowers perfect, slightly irregular. Sepals 3, distinct, one hood-shaped, ± enclosing the rest of the flower parts, the two lateral sepals often smaller. Petals 3, distinct or connate. Stamens 3, opposite the petals; 3 staminodes sometimes present. Carpels 3, connate; ovary superior with 1 or 3 locules and numerous parietal, axile or free-central ovules; style 1, entire or 3-branched. Fruit a capsule.

**9.L Centrolepidaceae**—4 genera, 30–35 species. Southeast Asia to Australia, the East Indies and New Zealand; disjunct in southernmost South America.

**9.M Restionaceae**—The Restio Family. 30–42 genera, 400 species. Southern hemisphere, especially South Africa and Australia, and southeast Asia.

**9.N Flagellariaceae**—1 genus, 3 species. Old World tropics.

**9.O Joinvilleaceae**—1 genus, 2 species. Pacific islands.

**9.P Poaceae [Gramineae]**—The Grass Family (Figure 16-16 to 16-18). 651–657 genera, 10,000 species. Cosmopolitan.

§See page 456 for description and illustrations§

## 10. Order Commelinales

**10.A Commelinaceae**—The Spiderwort Family. 38–50 genera, 500–700 species (Figure 16-19). Cosmopolitan.

§See page 464 for description and illustrations§

**10.B Hanguanaceae**—1 genus, 1–2 species. Malaysia and Ceylon.

**10.C Philydraceae**—3–4 genera, 5 species. Australia and nearby islands.

**10.D Haemodoraceae**—14–16 genera, 100 species. Mostly southern hemisphere, in the New World extending northward to the southeastern United States.

**10.E Pontederiaceae**—The Water-hyacinth Family. 9 genera, 30–35 species. Widespread in tropical, subtropical and some temperate areas.

Floating or rooted aquatic or semiaquatic herbs. Leaves alternate, opposite or whorled, or all basal, basally sheathing, sometimes with stipule-like ligule; petioles sometimes inflated. Plants synoecious. Inflorescence a spike, panicle or solitary flower subtended by a spathe. Flowers perfect, regular or irregular. Hypanthium sometimes present. Sepals 3, petaloid, distinct or basally adnate to petals. Stamens 6 (1 or 3), distinct, free or adnate to hypanthium. Carpels 3, connate; ovary superior with 3 locules and 1–many ovules per locule or with 1 locule and many parietal ovules; style 1, entire or slightly lobed at apex. Fruit a capsule or an achene.

# 11. Order Zingiberales (APG)

**11.A Musaceae**—The Banana Family (Figure 16-5). 2 genera, 42 species. Old World tropics.

Herbs, sometimes very large and treelike with massive overlapping sheathing petioles. Leaves alternate, simple and entire but often torn into pinnate strips, pinnately veined, with parallel lateral veins; petioles sheathing, eligulate. Plants monoecious. Inflorescence a paniculiform cyme with 1–many spathes. Flowers imperfect, irregular. Hypanthium sometimes well developed. Sepals 3, petaloid but differing from petals. Petals 3, very unequal, two of them adnate to the sepals, the third free. Stamens 5 or 6, distinct, adnate to corolla. Carpels 3, connate; ovary inferior with 3 locules and numerous axile ovules; style 1, undivided. Fruit an elongated berry with a leathery exocarp.

**11.B Heliconiaceae**—1 genus, 100 species. New World tropics.

**11.C Strelitziaceae**—The Bird-of-Paradise Family. 3 genera, 7 species. Tropics of South America, Africa and Madagascar.

Herbs or trees without secondary growth. Leaves alternate, simple, entire but often torn into pinnate strips, long-petiolate, sometimes basally sheathing. Plants synoecious. Inflorescence of 1–several cymes, each enclosed by a boat-shaped spathe.

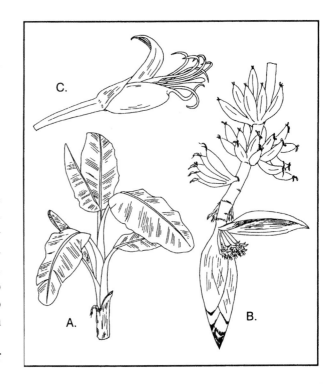

**Figure 16-5** Examples of Musaceae. A–B. *Musa xparadisiaca.* A. Pinnately veined leaves, B. Inflorescence with developing fruits. C. Staminate flower.

Flowers perfect, irregular. Hypanthium well developed. Sepals 3, petaloid but differing from petals. Petals 3, the 2 lateral petals connate, the odd petal distinct or nearly so. Stamens 5 or 6, the anthers linear, sometimes enclosed by the 2 lateral petals. Carpels 3, connate; ovary inferior with 3 locules and several to many axile ovules. Fruit a capsule; seeds with a brightly colored aril.

**11.D Lowiaceae**—1 genus, 6 species. South China, Malay peninsula and western Pacific.

**11.E Cannaceae**—The Canna Family. 1 genus, 50 species. Tropics and subtropics of New World.

Herbs. Leaves alternate, simple and entire, pinnately veined, sheathing, eligulate. Plants synoecious. Inflorescence a raceme or panicle. Flowers perfect, very irregular. Sepals 3, distinct, usually green. Petals 3, connate. Stamens 6 or 3, 5 or 2 of them sterile and petaloid in 2 or 1 series, the 2 inner staminodes connate nearly to the tip, all of them basally connate and adnate to the corolla; fertile stamen 1, petaloid with 1 asymmetrically attached anther sac, basally connate to other stamens and adnate to corolla. Carpels 3, connate; ovary inferior with 3 locules and many axile ovules; style 1, petaloid. Fruit a warty capsule.

**11.F   Marantaceae**—The Arrowroot Family. 30–32 genera, 170–400 species. Pantropical, especially well represented in the New World.

Herbs. Leaves alternate, 2-ranked, simple and entire, pinnately veined, sheathing with a pair of stipule-like ligules and a swollen joint [pulvinus] at the junction of petiole and blade. Plants synoecious. Inflorescence a spike or paniculiform cluster with 2-flowered cymules. Flowers perfect, very irregular. Sepals 3, distinct, green. Petals 3, distinct or connate. Stamens 4 or 5, 3 or 4 of them sterile and petaloid, variously connate, adnate to petals, appearing bilabiate and corolla-like; fertile stamen 1, petaloid, with a single asymmetrically attached anther sac, basally connate with staminodes and adnate to corolla. Carpels 3, connate; ovary inferior with 3 locules and 1 ovule per locule or 2 of the locules abortive, without ovules; style 1, undivided, often apically thickened or contorted. Fruit a capsule.

**11.G   Zingiberaceae**—The Ginger Family (Figure 16-6). 4550 genera, 1000 species. Pantropical, especially well represented in southeast Asia.

Herbs. Leaves alternate or all basal, 2-ranked, simple and entire, pinnately veined with a stipule-like ligule. Plants synoecious. Inflorescence of solitary flowers, spikes or racemes. Flowers perfect, very irregular, basally enwrapped by a calyx-like bract. Sepals 3, connate, usually green. Petals 3, connate. Stamens 5 or 3, 4 or 2 of them sterile and petaloid, 2 connate and 2 distinct or absent, adnate to corolla; fertile stamen 1, opposite the fused staminodes, basally adnate to corolla; filament often longitudinally grooved, enclosing the style. Carpels 3, connate; ovary inferior with 3 (2) locules and many axile ovules or with 1 locule and many parietal ovules; style 1, undivided. Fruit a berry or a capsule.

**11.H   Costaceae.** 4 genera, 150–200 species. Pantropical, especially well represented in the New World.

# References

Angiosperm Phylogeny Group II. 2003. An update of the Angiosperm Phylogeny Group classification for the orders and families of flowering plants: APG II. *Botanical Journal of the Linnean Society* 141:399–436.

Brummitt, R. K. 1992. *Vascular Plant Families and Genera.* Royal Botanic Gardens, Kew.

Chapman, G. P. 1993. *Grass Evolution and Domestication.* Cambridge University Press, New York and Cambridge.

Clayton, W. D. 1981. Evolution and distribution of grasses. *Annals of the Missouri Botanical Garden* 68:5–14.

———, and S. A. Renvoise. 1986. Genera graminum: grasses of the world. *Kew Bulletin, Additional Series* 13:1–389.

Connor, H. E. 1981. Evolution of reproductive systems in the Gramineae. *Annals of the Missouri Botanical Garden* 68:48–74.

Corner, E. J. H. 1966. *The Natural History of Palms.* University of California Press, Berkeley.

Crampton, B. 1974. Grasses in California. University of California Press, Berkeley. 200 pp.

Cronquist, A. 1981. *An Integrated System of Classification of Flowering Plants.* Columbia University Press. New York, pp. 1029–1242.

**Figure 16-6**   Examples of Zingiberaceae. A–B. *Hedychium flavescens.* A. Branch with pinnately veined leaves and inflorescence. B. Flower.

————, 1988. *The Evolution and Classification of Flowering Plants,* 2nd ed. New York Botanical Garden, Bronx, N.Y., pp. 450–500.

Cutler, D. F. 1966. Anatomy and taxonomy of the Restionaceae. *Jodrell Laboratory Notes* 4:1–25.

Dahlgren, R. M. T., and H. T. Clifford. 1982. *The Monocotyledons: A Comparative Study.* Academic Press, New York.

————, ————, and P. F. Yeo. 1985. *The Families of the Monocotyledons. Structure, Evolution and Taxonomy.* Springer-Verlag. New York, pp. 520

————, and F. N. Rasmussen. 1983. Monocotyledon evolution. Characters and phylogenetic estimation, pp. 255–395 *in* M. K. Hecht, B. Wallace and G. T. Prance (eds.), *Evolutionary Biology,* Vol. 16. Plenum Publ. Co.

Daubs, E. 1965. A monograph of the Lemnaceae. *Illinois Biological Monograph* 34:1–118.

de Wet, J. M. J. 1981. Grasses and the culture history of Man. *Annals of the Missouri Botanical Garden* 68:87–104.

den Hartog, C. 1970. *Sea-grasses of the World.* Verhandelingen der Koninklijke Nederlandsche Akadamie 59:1–275.

————, and F. von der Plas. 1970. A synopsis of the Lemnaceae. *Blumea* 18:355–368.

Dressler, R. L. 1981. *The Orchids. Natural History and Classification.* Harvard University Press, Cambridge, Mass.

————, 1993. *Phylogeny and Classification of the Orchid Family.* Timber Press, Portland, Oregon.

————, and C. H. Dodson. 1960. Classification and phylogeny in the Orchidaceae. *Annals of the Missouri Botanical Garden* 47:25–68.

Eiten, L. T. 1976. Inflorescence units in the Cyperaceae. *Annals of the Missouri Botanical Garden* 63:81–112.

Goldblatt, P. 1993. *The Woody Iridaceae. Nivenia, Klattia & Witsenia. Systematics Biology and Evolution.* Timber Press, Portland, Oregon.

Gould, F. W., and R. B. Shaw. 1983. *Grass Systematics.* Texas A&M University Press. College Station, Texas. 397 pp.

Haynes, R. R. 1977. The Najadaceae in the southeastern United States. *Journal of the Arnold Arboretum* 58:161–170.

————, 1978. The Potamogetonaceae in the Southeastern United States. *Journal of the Arnold Arboretum* 59:170–191.

Heywood, V. H. (ed.). 1993. *Flowering Plants of the World,* updated edition. Oxford University Press, New York. Pp. 269–325.

Hitchcock, A. S. 1951. *Manual of the Grasses of the United States,* 2nd ed. revised by A. Chase. U. S. Department of Agriculture Miscellaneous Publication 200.

Judd W. S., C. S. Campbell, E. A. Kellogg, P. F. Stevens, and M. J. Donoghue. 2002. *Plant Systematics: A Phylogenetic Approach.* 2nd ed. Sunderland, MA: Sinauer Associates.

Kaul, R. B. 1968. Floral morphology and phylogeny in the Hydrocharitaceae. *Phytomorphology* 18:13–35.

————, 1970. Evolution and adaptation of inflorescences in the Hydrocharitaceae. *American Journal of Botany* 57:708–715.

Kral, R. 1966. Eriocaulaceae of continental North America north of Mexico. *Sida* 2:285–332.

————, 1983. The Xyridaceae in the southeastern United States. *Journal of the Arnold Arboretum* 64:425–433.

Lawrence, G. H. M. 1951. *Taxonomy of Vascular Plants.* Macmillan, New York, N.Y., pp. 371–438.

Luer, C. A. 1972. *The Native Orchids of Florida.* New York Botanical Garden, New York.

————, 1975. *The Native Orchids of the United States and Canada excluding Florida.* New York Botanical Garden, New York.

Moore, H. E. 1973. The major groups of palms and their distribution. *Gentes Herbarium* 11:27–141.

Pohl, R. W. 1978. *How to Know the Grasses,* 3rd Edition. Wm. C. Brown Co., Dubuque, Iowa.

Rogers, G. K. 1983. The genera of Alismataceae in the southeastern United States. *Journal of the Arnold Arboretum* 64:387–424.

————, 1984. The Zingiberales (Cannaceae, Marantaceae, and Zingiberaceae) in the southeastern United States. *Journal of the Arnold Arboretum* 65:5–55.

Smith, L. B., and R. J. Downs. 1974. Bromeliaceae: Pitcairnoideae. *Flora Neotropica.* Monograph 14, part 1. Hafner Publ. Co., New York.

————. 1977. Bromeliaceae: Tillandsioideae. *Flora Neotropica.* Monographs 14, part 2. Hafner Publ. Co., New York.

————. 1979. Bromeliaceae: Bromelioideae. *Flora Neotropica.* Monograph 14, part 3. New York Botanical Garden, New York.

————, and C. E. Wood. 1975. The genera of Bromeliaceae in the southeastern United States. *Journal of the Arnold Arboretum* 56:375–397.

Soderstrom, T. R. 1981. Some evolutionary trends in the Bambusoideae (Poaceae). *Annals of the Missouri Botanical Garden* 68:15–47.

———, K. W. Hilu, C. S. Campbell, and M. E. Barkworth. 1987. *Grass Systematics and Evolution.* Smithsonian Institution, Washington, D.C.

Stevens, P. F. 1978. Generic limits in the Xeroteae (Liliaceae sensu lato). *Journal of the Arnold Arboretum* 59:129–155.

Stevens, P. F. (2001 onwards). Angiosperm Phylogeny Website, Version 5, May 2004 [and more or less continuously updated since]. http://www.mobot.org/MOBOT/research/APweb/.

Thieret, J. W. 1982. The Sparganiaceae in the southeastern United States. *Journal of the Arnold Arboretum* 63:341–356.

Tomlinson, P. B. 1979. Systematics and ecology of the Palmae. *Annual Review of Ecology and Systematics* 10:85–107.

———, and U. Posluszny. 1976. Genetic limits in the Zannichelliaceae (sensu Dumortier). *Taxon* 25:273–279.

Uhl, N. W., and J. Dransfield. 1987. *Genera Palmarum. A Classification of Palms based on the work of Harold E. Moore, Jr.* Allen Press, Lawrence, Kansas.

Wilder, G. J. 1975. Phylogenetic trends in the Alismatidae (Monocotyledonae). *Botanical Gazette* 136:159–170.

Wilson, K. A. 1960. The genera of Arales in the southeastern United States. *Journal of the Arnold Arboretum* 41:47–72.

Withner, C. L. (ed.). 1959. *The Orchids: a Scientific Survey.* The Ronald Press, New York.

# Araceae

## The Arum Family
### Alismatales ■

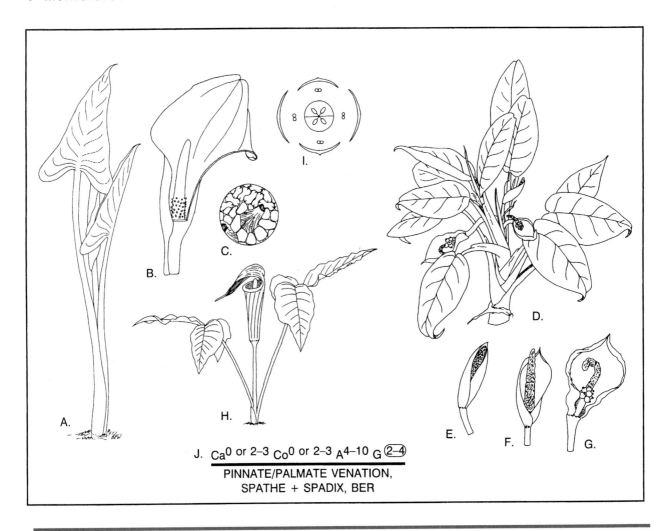

**Figure 16-7** Features of Araceae. A–C. *Zantedeschia aethiopica*. A. Basal leaves. B. Spathe and spadix with staminate flowers in upper portion and pistillate flowers below. C. Details of pistillate portion of spadix. D–G. *Agalonema* sp. D. Habit. E–G. Stages in the development of a flowering and fruiting spathe and spadix. H. Leaves and inflorescence of *Arisaema triphyllum*. I. Floral diagram of *Anthurium*. J. Generalized floral formula.

## FAMILY DESCRIPTION

Herbs, shrubs or woody vines, sometimes epiphytic. Leaves alternate, simple to pinnately or palmately lobed or compound, usually sheathing, sometimes with a well-developed ligule, sometimes a deciduous bladeless sheathing leaf [prophyll] basally enclosing each foliage leaf. Plants synoecious or monoecious (rarely dioecious). Inflorescence a spadix subtended or surrounded by a spathe. Flowers perfect or imperfect, regular, irregular or naked, very reduced. Tepals 0, 4 or 6, distinct or connate.

Stamens 1, 2, 4 or 8, distinct or connate; filaments short or anthers sessile. Carpels 1–many, connate; ovary superior or inferior and sunken into axis of spadix, with locules 1–many and 1–many axile, parietal, marginal, basal or apical ovules; style 1, short, or stigma sessile. Fruit a berry.

# FAMILY NOTES

The Araceae is a cosmopolitan family of about 110 genera and 1800–2450 species. By far the greatest diversity of species is in tropical and subtropical areas. The Araceae also includes several conspicuous temperate zone plants.

Members of the Araceae are mostly perennial herbs with pinnately or palmately veined leaves (Figures 16-7 A, D, H). The leaves are generally long-petiolate and may bear stipule-like structures. Many of the tropical species are vines or epiphytes; their adventitious roots anchor the plant to the trunk or branches of the host plant. The temperate species are erect herbs with basal leaves from a corm-like base.

Aroids have a large, often petaloid spathe [an expanded, often basally sheathing bract] that subtends a slender to very massive spadix (fleshy spike with reduced flowers; Figures 16-7 B, E–H). The small flowers may be perfect or imperfect. The perianth may be absent or composed of fleshy 4–6 lobes that are usually not differentiated into sepals and petals. Many members of the family are monoecious with staminate flowers on the upper half of the spadix and pistillate flowers on the lower half (Figures 16-7 B, G). The number of stamens ranges from one to six. The stamens are sometimes fused together into a disc-shaped structure. The ovary may be superior or sunken into the axis of the spathe. It usually matures as a berry (Figure 16-7 G).

Flowers in most aroids are insect pollinated. The spathe often serves as a visual flag to pollinators. Many aroids also have strong odors that provide olifactory cues to the pollinating insects. Some species, such as the calla-lily, are very sweetly scented. Several genera, however, have dark red-purple spathes and the odor of rotting meat. These are pollinated by carrion flies. Nectar is produced by the stigmas of some species.

Various members of the Araceae are grown as ornamentals or as food crops. Some of the tropical species grow in the dense shade of rain forests and are very shade-tolerant. Species of *Philodendron, Monstera, Caladium, Dieffenbachia* and various other genera are commonly grown in homes, offices and hotel lobbies for the tropical appearance of their foliage and because they require little attention. Some aroids are grown outdoors as ornamentals in areas with mild winters. The inflorescences of some species are extremely showy. The starch-filled corms of species in several genera are edible, e.g., *Colocasia esculenta* (taro) from which poi is made.

*Arisaema triphyllum* (Jack-in-the-pulpit; Figure 16-7 F) is one of the most familiar spring wildflowers of eastern North America. Both *Symplocarpus foetidus* and *Lysichiton americanum* have earned the common name, skunk cabbage, because of their unpleasant odor.

# Alismataceae

## The Water-Plantain Family
### Alismatales ■

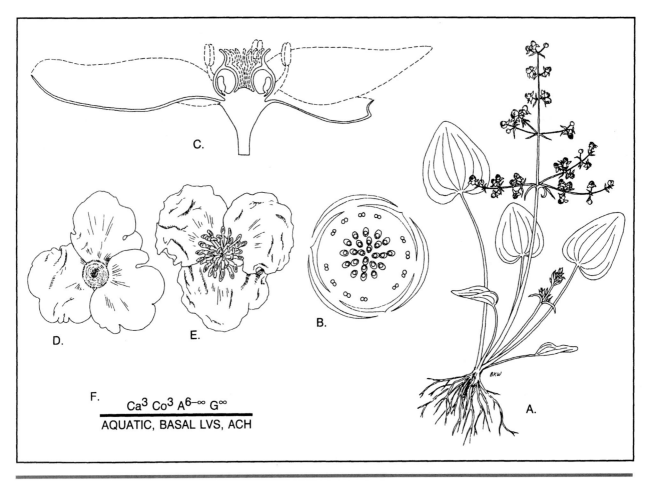

F. $Ca^3 \ Co^3 \ A^{6-\infty} \ G^\infty$
___
AQUATIC, BASAL LVS, ACH

**Figure 16-8** Features of Alismataceae. A–B. *Echinodorus berteroi*. A. Plant with basal leaves and scapose inflorescence. B. Floral diagram. C. Flower of *Alisma* sp. (l.s.). D–E. *Sagittaria* sp. D. Pistillate flower. E. Staminate flower. F. Generalized floral formula.

# FAMILY DESCRIPTION

Aquatic or semiaquatic herbs. Leaves all basal, simple and bladeless or with expanded blades, estipulate, sheathing. Plants synoecious or monoecious (rarely dioecious), inflorescence racemiform or paniculiform. Flowers perfect or imperfect, regular. Sepals 3, distinct. Petals 3, distinct. Stamens 6–many (3), distinct. Carpels 6–many, distinct; ovaries superior with 1 locule and 1 (2) basal ovule per carpel; style 1 per carpel, undivided. Fruit an achene (or rarely a follicle).

# FAMILY NOTES

The Alismataceae is a small family (12 genera, 75 species) of aquatic and semiaquatic plants widely distributed in temperate and tropical waters. The greatest diversity occurs in the New World.

The leaves of members of the Alismataceae are dimorphic. They are generally all basal, arising either from a corm or a stolon. In fully aquatic situations the leaves are generally elongate, linear and bladeless. In shallow water or on drying banks, the leaves are long-petiolate with generally expanded blades (Figure 16-8 A). The blades may be entire or sagittate and generally have several main veins from the base. Smaller lateral veins extend transversely across the leaf.

The flowers are 3-merous with an apocarpous gynoecium (Figures 16-8 B—D). The perianth is biseriate, with green sepals and white to pink, usually ephemeral, distinct petals. The number of stamens ranges from 6 in *Alisma* (Figure 16-8 C) to many in *Sagittaria* (Figure 16-8 E). The gynoecium consists of several to many small distinct carpels that may be arranged in a ring (Figure 16-8 C) or arrayed over the surface of a dome-shaped receptacle (Figures 16-8 B, D). In most genera the carpels ripen as achenes. The floral structure in this family is considered to be evolutionarily primitive.

The family has very limited economic importance. A few species are grown as ornamentals in water-gardens. The starchy corms of *Sagittaria* have been used for food by some tribes of American Indians. The fruits and corms are important foods for various kinds of wildlife. Common genera in North America are *Alisma* (water-plantain), *Sagittaria* (arrow-leaf), and *Echinodorus* (bur-head).

# Liliaceae

## The Lily Family
### Liliales ◼

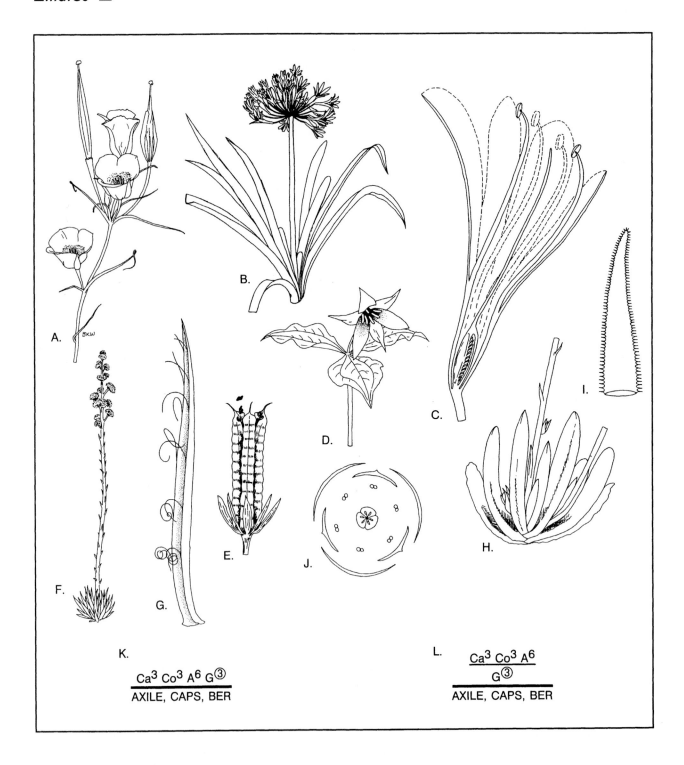

A.

B.

C.

D.

E.

F.

G.

H.

I.

J.

K.

$$\frac{Ca^3\ Co^3\ A^6\ G\text{③}}{\text{AXILE, CAPS, BER}}$$

L.

$$\frac{Ca^3\ Co^3\ A^6}{G\text{③}}$$

AXILE, CAPS, BER

# FAMILY DESCRIPTION

Herbs, shrubs, woody vines. Leaves alternate, whorled or all basal (opposite), simple, usually with sheathing bases, sometimes succulent or reduced to scales. Plants synoecious (rarely polygamous, or dioecious). Inflorescence of many kinds, often paniculate or racemose or umbellate or flowers solitary. Flowers perfect (or rarely imperfect), regular or somewhat irregular. Hypanthium sometimes present. Sepals 3 (2–5), distinct or basally adnate to petals, often petaloid. Petals 3 (2–5 or more), distinct or basally adnate to sepals, mostly petaloid. Stamens 6 (3–12), distinct or filaments ± connate, free or adnate to hypanthium. Carpels 3 (2–5), connate or nearly distinct and connate only at base; ovary superior (rarely inferior of half inferior) with 3 (2–5) locules and 1–many axile ovules per locule (rarely with 1 locule and 1–many parietal ovules); styles 3 (4–5), or 1 and entire or ± 3-lobed, or stigma sessile. Fruit a capsule, berry or samara.

# FAMILY NOTES

The Liliaceae as described above is a polymorphic family of worldwide distribution with about 280 genera and 4000 species. They are present from driest deserts to wettest tropics. Most of the Liliaceae are herbaceous plants, usually with bulbs, rhizomes, or corms and usually with parallel-veined leaves. The sepals and petals are usually similar in color and texture (Figure 16-9 C). There are generally 6 stamens, and the gynoecium is 3-carpellate. The degree of fusion of the carpels is highly variable. In some the carpels are nearly distinct; in others the carpels are wholly connate. The ovary matures into a capsule (Figure 16-9 E) or a berry.

There is much controversy regarding taxonomic limits within the Liliales. Many taxonomists have suggested that the Liliaceae should be broken into smaller, more cohesive groups, but there has been much disagreement regarding the circumscription of the families to be recognized. Cronquist (1981, 1988) accepted a very inclusive Liliaceae. Dahlgren et al. (1985) and Brummitt (1992) on the other hand, divided the same group of plants much more finely, recognizing more than twenty families where Cronquist recognized only one. In a similar fashion, Cronquist took a broader view of one of the segregate families that he did recognize, the Agavaceae, than did Dahlgren et al. and Brummitt.

Table 16-1 presents a summary of differences in interpretation of the Liliaceae *(sensu lato)*. We have included many of the families and genera that grow wild or in cultivation in the United States in this table. We have excluded some of the small, extralimital families that have been segregated from the Liliaceae and Agavaceae.

The lily family *(sensu lato)* is extraordinarily rich in ornamental species, and some are grown as food-crops. Genera of ornamentals include *Lilium, Tulipa, Hemerocallis, Convallaria, Narcissus, Hyacinthus, Hosta, Amaryllis, Agapanthus* and many others. *Allium* (onions, leeks, chives, garlic) and *Asparagus* are cultivated as food-crops. Indian tribes of the Pacific Northwest used the bulbs of *Camassia quamash* (camas lily) as a food source. In addition the family contains numerous colorful wildflowers such as species of *Lilium, Erythronium, Fritillaria, Trillium, Calochortus* and *Allium*.

**Figure 16-9**  Features of Liliaceae *(sensu lato)* (opposing page). A. Inflorescence of *Calochortus* (Liliaceae, *sensu stricto*) with radial flowers and immature capsules. B–C. *Agapanthus africanus* (Alliaceae). B. Habit with scapose umbel. C. Weakly bilateral flower (l.s.). D. Stem of *Trillium* sp. (Trilliaceae) with whorled, net-veined leaves and terminal flower. E. Capsule of *Zygadenus* (Melanthiaceae). F. Habit and inflorescence of *Agave* sp. (Agavaceae). G. Fibrous leaf of *Yucca* sp. (Agavaceae). H. Rosette of succulent leaves of *Aloe* sp. (Aloaceae). I. Leaf of *Aloe.* J. Floral diagram of *Lilium* (Liliaceae, *sensu stricto*). K. Generalized formula for flower with superior ovary. L. Generalized formula for flower with inferior ovary.

| Segregate family | Features | Representative genera |
|---|---|---|
| **Agavaceae (sensu stricto) | Rosette herbs, rosette shrubs, and rosette trees; rhizomes sometimes present. Leaves all basal or in terminal rosettes on short to tall, simple or branched trunks, basally sheathing, parallel-veined, fibrous and succulent, entire to coarsely toothed or fibrous-margined. Inflorescences a large, stout, bracteate raceme or panicle. Perianth parts 3 + 3, similar, free or fused, usually radial. Stamens 6, free or adnate to perianth. Ovary superior or inferior. Fruit a capsule or berry. | *Agave* (Fig. 16-9 F), *Cordaline*[1]. *Hesperaloe, Yucca* (Fig. 16-9 G). |
| *Alliaceae | Herbs, mostly from bulbs or bulb-like corms, erect. Leaves all basal or subbasal, sheathing, usually linear, parallel-veined. Inflorescence a scapose, cymose umbel subtended by 1 or more membranous bracts. Perianth parts 3 + 3, all petaloid, similar or dissimilar, free or fused, radial or bilateral, sometimes appendaged within. Stamens usually 6, free or adnate to perianth tube. Ovary superior. Fruit a capsule. | *Agapanthus, Allium, Bloomeria, Brodiaea, Muilla, Nothoscordum, Tritelei, Tulbaghia* |
| Aloaceae | Rosette herbs, rosette shrubs, and rosette trees; rhizomes sometimes present. Leaves all basal or in terminal rosettes on short to tall, simple or branched trunks, basally sheathing, parallel-veined, fibrous and succulent, entire to coarsely toothed or fibrous-margined. Inflorescences a large, stout, bracteate raceme or panicle. Perianth parts 3 + 3, similar, fused, radial or bilateral. Stamens 6, adnate to perianth. Ovary superior. Fruit a capsule. | *Aloe*[2] (Fig. 16-9 H–I) |
| *Alstroemeriaceae | Herbs from rhizomes, erect or twining. Leaves cauline, not sheathing, linear to lanceolate, parallel-veined, twisted with the lower side up. Inflorescence of umbel-like cymes, leafy-bracted. Perianth parts 3 + 3, all petaloid, more or less dissimilar, generally free, radial or bilateral. Stamens 6, free from perianth. Ovary inferior. Fruit usually a capsule. | *Alstroemeria, Bomerea* |
| *Amaryllidaceae | Herbs, usually from bulbs, erect. Leaves all basal or subbasal, sheathing, linear to sub-orbicular, parallel-veined. Inflorescence a scapose cymose umbel subtended by 1 or more membranous bracts. Perianth parts 3 + 3, all petaloid, free or fused, radial or weakly bilateral, sometimes appendaged within. Stamens 6, free or adnate to perianth tube. Ovary inferior. Fruit a capsule or a berry. | *Amaryllis, Crinum, Galanthus, Hymenocallis, Leucojum, Narcissus, Zephranthes* |

[1]Included in Asteliaceae by Dahlgren et al. (1985).
[2]Included in Asphodelaceae by Dahlgren et al. (1985).

**Table 16-1** Families sometimes included in the Liliaceae (sensu lato). Families included by Cronquist (1981, 1988) in Liliaceae (sensu lato) are noted with an asterisk (*). Families included in Agavaceae by Cronquist are noted with a double asterisk (**). Family circumscriptions follow Brummitt (1992) and Dahlgren et al. (1985), except as noted. Descriptions modified from Dahlgren et al. (1985).

| Segregate family | Features | Representative genera |
|---|---|---|
| *Anthericaceae | Herbs from rhizomes, erect. Leaves mostly basal and subbasal, sheathing, mostly linear, parallel-veined. Inflorescence of racemes, spikes and panicles with scattered bracts. Perianth parts 3 + 3, all petaloid, free or basally fused, radial. Stamens 6 (3), free or adnate to perianth tube. Ovary superior. Fruit a capsule. | *Anthericum, Johnsonia* |
| *Asparagaceae | Herbs, subshrubs, shrubs, and vines, erect or scrambling, sometimes spiny; branchlets photosynthetic, more or less leaflike. Leaves reduced to bladeless scales. Inflorescence of solitary flowers, umbels, or racemes. Perianth parts 3 + 3, all petaloid, free or basally fused. Stamens 6, free or fused to perianth. Ovary superior. Fruit a berry. | *Asparagus* |
| *Asphodelaceae | Herbs from rhizomes, erect. Leaves mostly basal and subbasal, sheathing, linear, parallel-veined, sometimes succulent, reduced to bracts on flowering stems. Inflorescence of racemes, spikes and panicles with scattered bracts. Perianth parts 3 + 3, all petaloid, free or fused, radial. Stamens 6, free from perianth. Ovary superior. Fruit a capsule. | *Asphodelus, Bulbine, Kniphofia* |
| *Colchicaceae | Herbs from corms, erect or twining. Leaves basal or cauline, sheathing, linear to lanceolate, parallel-veined, rarely tipped with tendrils. Inflorescence usually of racemes leafy-bracted. Perianth parts 3 + 3, all petaloid, similar, free or basally fused, radial or bilateral. Stamens 6, free or adnate to perianth. Ovary superior. Fruit a capsule. | *Colchicum, Gloriosa* |
| *Convallariaceae | Herbs from rhizomes, erect or spreading. Leaves basal or cauline, mostly non-sheathing, linear to ovate, parallel-veined. Inflorescence axillary or terminal, of solitary flowers, racemes, and panicles. Perianth parts 3 + 3 (2 + 2), all petaloid, free or fused, radial. Stamens 6 (4), free or adnate to perianth tube. Ovary superior. Fruit a berry (rarely a capsule). | *Clintonia,*[3] *Convallaria, Disporum*[3]*, Maianthemum, Medeola, Ophiopogon, Polygonatum, Smilacina, Streptopus*[3]*, Uvularia*[3]*.* |
| **Dracenaceae | Herbs, rosette shrubs, and rosette trees, sometimes climbing; rhizomes sometimes present. Leaves all basal or in terminal rosettes on short to tall, simple to much-branched trunks, sessile or with a sheathing base, parallel-veined, linear to ovate, fibrous and often succulent, entire to coarsely toothed or fibrous-margined. Inflorescences a large, stout, bracteate raceme or panicle. Plants sometimes dioecious or polygamodioecious. Perianth parts 3 + 3, similar, free or fused, usually radial. Stamens 6, free or adnate to perianth. Ovary superior or inferior. Fruit an indehiscent nutlet or a berry. | *Dasylirion,*[4] *Dracaena, Nolina,*[4] *Sansevieria* |

[3]Placed into Uvulariaceae by Dahlgren et al. (1985).
[4]Segregated as Nolinaceae by Dahlgren et al. (1985).

**Table** Continued
**16-1**

| Segregate family | Features | Representative genera |
|---|---|---|
| *Hemerocallidaceae | Herbs from rhizomes, erect. Leaves basal and cauline, sheathing, linear, parallel-veined. Inflorescence of terminal cymose panicles. Perianth parts 3 + 3, all petaloid, basally fused, radial or weakly bilateral. Stamens 6, adnate to perianth tube. Ovary superior. Fruit a capsule. | *Hemerocallis* |
| *Hostaceae | Herbs from rhizomes, erect. Leaves basal, sheathing, petioled, elliptic to ovate, parallel-veined. Inflorescence of terminal racemes. Perianth parts 3 + 3, all petaloid, basally fused, radial or weakly bilateral. Stamens 6, adnate to perianth tube. Ovary superior. Fruit a capsule. | *Hosta* |
| *Hyacinthaceae | Herbs from bulbs, erect. Leaves basal, sheathing, mostly linear to narrowly lanceolate, parallel-veined. Inflorescence of racemes or spikes. Perianth parts 3 + 3, all petaloid, free or fused, radial. Stamens 6, free or adnate to perianth. Ovary superior. Fruit a capsule. | *Camassia, Chlorogalum, Hastingsia, Hesperocallis, Hyacinthus, Muscari, Ornithogalum, Schoenolirion, Scilla* |
| *Hypoxidaceae | Herbs from corms or rhizomes, erect. Leaves basal, sheathing, linear, parallel-veined, hairy. Inflorescence scapose, of solitary flowers or umbels, spikes, or racemes borne on a hairy peduncle. Perianth parts 3 + 3, all petaloid, free or fused, radial. Stamens 6, free from perianth. Ovary inferior. Fruit a capsule or indehiscent. | *Hypoxis* |
| *Liliaceae (sensu stricto) | Herbs from bulbs, erect. Leaves basal and cauline, sheathing or non-sheathing, linear to ovate, parallel-veined, sometimes whorled. Inflorescence of solitary flowers, umbels, racemes, and panicles with scattered bracts. Perianth parts 3 + 3, all petaloid or the sepals more or less sepaloid, free, radial or weakly bilateral. Stamens 6, free from perianth. Ovary superior. Fruit a capsule. | *Calochortus*[5] (Fig. 16-9 A), *Erythronium, Fritillaria, Lilium, Tulipa* |
| *Melanthiaceae | Herbs from rhizomes, corms, or bulbs, erect. Leaves basal and cauline, sheathing or non-sheathing, linear to ovate, parallel-veined. Inflorescence of spikes, racemes, and panicles with scattered bracts. Perianth parts 3 + 3, all similar, petaloid, free, radial, sometimes very small. Stamens 6, free from perianth. Ovary superior or half-inferior, the carpels sometimes distinct in the upper half. Fruit a capsule. | *Aletris, Amianthium, Helonias, Melanthium, Narthecium, Schoenocaulon, Stenanthium, Tofieldia, Veratrum, Xerophyllum, Zigadenus* |

[5]Placed into Calochortaceae by Dahlgren et al. (1985).

**Table** Continued
**16-1**

| Segregate family | Features | Representative genera |
|---|---|---|
| **Phormiaceae | Rosette herbs, and subshrubs from stout rhizomes. Leaves basal and cauline, mostly restricted to the lower parts of the stem, basally sheathing, compressed and folded at base, parallel-veined, fibrous and somewhat succulent, entire or serrulate. Inflorescences a large, stout, bracteate panicle. Perianth parts 3 + 3, similar, free, radial or somewhat bilateral. Stamens 6, free Ovary superior. Fruit a capsule or berry. | *Phormium* |
| *Ruscaceae | Subshrubs, erect or spreading; branchlets flattened, photosynthetic, very leaflike. Leaves reduced to bladeless scales. Inflorescence of solitary flowers or umbels borne on the flattened stems, or racemes. Perianth parts 3 + 3, all small, free or fused. Stamens 6, free or fused to perianth, the filaments united into a tube. Ovary superior. Fruit a berry. | *Ruscus* |
| *Tecophiliaceae | Herbs from corms, erect. Leaves basal and cauline, sheathing, linear, parallel-veined. Inflorescence of racemes and panicles with scattered bracts. Perianth parts 3 + 3, all petaloid, free or fused, radial. Stamens 6, free from perianth, some often reduced to staminodes, sometimes alternate with up to 6 extra staminodes, connivent and bent to one side of flower; anthers dehiscent by apical slits or pores. Ovary half-inferior. Fruit a capsule. | *Odontospermum* |
| *Trilliaceae | Herbs from rhizomes, erect. Leaves cauline, non-sheathing, elliptic to cordate-ovate, parallel-veined with pinnate secondary veins, borne in a single whorl of 3–7 or more. Flowers solitary, terminal. Perianth parts 3 + 3 to 10 + 10, generally the number of parts in the perianth whorls equal in number to the leaves, free, radial or weakly bilateral, the sepals sepaloid, the petals petaloid or rudimentary. Stamens 6–20, free from perianth. Ovary superior. Fruit a berry or capsule. | *Trillium*[6] (Fig. 16-9 D) |

[6]Dahlgren et al. (1985) and Brummit (1992) include *Scoliopus* in the *Trilliaceae,* but this genus seems out of place there. Its family assignment remains in doubt.

**Table 16-1** Continued

# Orchidaceae

## The Orchid Family
### Asparagales ■

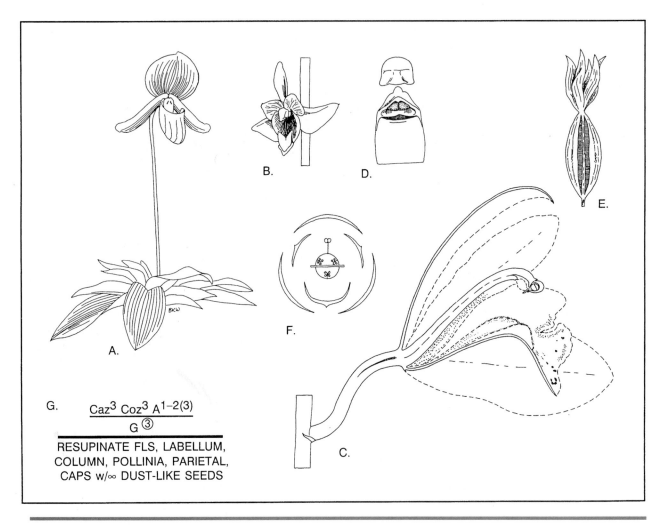

**Figure 16-10** Features of Orchidaceae. A. Habit of *Paphiopedilum;* note the pouch-like labellum. B. Flower of *Epipactis.* C. Flower of *Cymbidium* (l.s.). D. Column of *Cymbidium* with rostellum (top) removed. Note darkly shaded stigmatic chamber and pair of pollinia. E. Capsule of *Corallorhiza* releasing minute, dust-like seeds. F. Floral diagram of *Cymbidium*. G. Generalized floral formula.

## FAMILY DESCRIPTION

Herbs, terrestrial or epiphytic (rarely mycotrophic saprophytes). Leaves alternate or all basal (rarely opposite or whorled), simple, sheathing, sometimes ± succulent and attached to a swollen fleshy stem [pseudobulb]. Plants synoecious (very rarely polyg-amous or monoecious). Inflorescence a spike, raceme or panicle or flowers solitary. Flowers perfect (or rarely imperfect), very irregular. Sepals 3, distinct or connate, petaloid or green. Petals 3, distinct or connate, 2 of them similar, the third [labellum] dissimilar and often appendaged, saccate, spurred or otherwise elaborated. Stamens 1 or 2 (rarely 3) wholly adnate to stigma and style forming the column, anthers with pollen grains embedded in waxy masses [pollinia] or with pollen grains distinct or in small groups; pollinia sometimes covered by cap-like rostellum. Carpels 3, connate; ovary inferior with 1 locule and very many extremely tiny parietal ovules (less commonly with 3 locules and axile ovules); style 1, massive, together with adnate stamen tissue forming the column with the stigmatic surfaces variously positioned. Fruit a capsule with extremely many, very tiny seeds.

# FAMILY NOTES

The Orchidaceae is the largest family of flowering plants with approximately 835 genera and 20,000 to 25,000 species. Orchids reach a most bewildering diversity in the tropics and subtropics. A large percentage of the tropical species are epiphytic. Orchids are much less common and are represented by far fewer species in temperate latitudes. The orchids of temperate regions are almost always terrestrial.

Orchidaceae are among the most highly evolved of the monocots. The flowers are strongly zygomorphic (Figures 16-10 A–C). Both perianth whorls are petaloid, but the sepals generally differ from the petals in shape, and often in size and color as well. The two lateral petals are similar, but the third petal is generally very different in shape. The odd petal begins its development as the upper petal of the flower. However, the flowers of most orchids are **resupinate;** they have an inferior ovary with a half twist that turns the flower upside down as it develops. This is very important because the odd petal plays a major role in pollination. This petal, called the **lip** or **labellum** may be modified in many ways. It often serves as a landing platform for pollinating insects. Except in one genus, the number of stamens in an orchid flower is reduced to one or two and the androecium is wholly adnate to the stigma and style, forming a compound structure called the **column** (Figure 16-10 D). The pollen grains of most orchids are united into masses called **pollinia** (Figure 16-10 D). The pollinia are often concealed behind a cap-like sterile portion of the central stigma lobe called the **rostellum.** Pollinia are usually connected to a sticky glandular structure called a **viscidulum.** The functional stigma or stigmas are located on the undersurface of the column. The inferior ovary often contains thousands to millions of minute ovules.

Pollination in orchids involves many elaborate mechanisms by which pollinators, usually insects, transfer pollinia from flower to flower. Many orchids have specific pollinators. Pollination involves transfer of a pollinium to the body of the insect and thence to the stigma of another flower. Often the structure of the orchid flower corresponds precisely to the body of the pollinator. The pollinator's behavior while visiting the flower and gathering nectar (or other substances) brings a particular part of its body into contact with the pollinia or the sticky viscidulum. Adhesive substances on the stigma of the next flower visited remove the pollinium from the pollinator's body.

Orchid ovaries mature as peculiar capsules in which lateral slits form but the pericarp remains intact at both ends (Figure 16-10 E). The seeds are extremely small, sometimes no larger than fungal spores. They have very underdeveloped embryos and no endosperm. There are almost no stored food reserves. Under natural conditions, an orchid seed must establish a symbiotic relationship with a compatible mycorrhizal fungus immediately upon germination if it is to survive. During the early stages of its development the fungus supplies critical nutrients to the orchid seedling. Later the orchid may become fully independent or it may retain its mycorrhizal association throughout its life. Orchid seedlings may be grown under artificial conditions in the absence of the fungus by supplying the required nutrients. Often orchid seeds are germinated on nutrient agar in much the same way that fungi and bacteria are cultured.

There are three subfamilies of the Orchidaceae (Table 16-2). Subfamily Apostasioideae comprises only two genera and about 15 species of southeast Asia, the East Indies and northern Australia. The Cyprepedioideae is widespread in temperate and tropical areas with 4 genera and about 100 species. Cypripedium with several temperate species in North America and *Paphiopedilum* with several species in cultivation (Figure 16-10 A) are the largest genera. Both are called lady-slipper orchids.

| subf. Apostasioideae | subf. Cyprepedioideae | subf. Orchidoideae |
|---|---|---|
| Terrestrial; labellum only slightly different from lateral petals; stamens 2 or 3, filaments connate but anthers distinct; pollen grains separate, never forming pollinia; style tip free from stamens; rostellum absent; stigma 2–3 lobed; ovary 3-locular with axile placentation. | Mostly terrestrial; labellum pouch-shaped, very different from lateral petals; stamens 2, fully adnate to sides of column, a central staminode present; pollen grains separate or in groups, rarely forming pollinia; style fully adnate to column; rostellum absent; stigma unlobed or obscurely 3-lobed; ovary 3-locular with axile placentation or 1-locular with parietal placentation. | Terrestrial or epiphytic; labellum of many forms, usually very different from lateral petals; stamen 1, terminal on column, usually covered by rostellum; pollinia usually formed; style fully adnate to column; stigma lobes obscure, forming stigmatic cavity; ovary 1-locular with parietal placentation. |

Table 16-2   Comparison of the subfamilies of the Orchidaceae.

Of the three subfamilies the Orchidoideae is by far the largest with hundreds of genera and thousands of species.

The orchid family is prized for the remarkable diversity of its flowers. In addition to the numerous orchids that have been introduced from the wild into cultivation, many beautiful hybrids have been produced under artificial conditions. Many orchid species are interfertile, and intergeneric hybrids have been produced as well. The keeping qualities of the flowers make them very useful for cut flowers. The only economically significant product of the orchid family is vanilla, extracted from the fermented capsules of *Vanilla planifolia.* Some common genera of temperate terrestrial orchids are *Epipactis* (helleborine), *Platanthera* (rein orchis), *Spiranthes* (ladies' tresses), *Cypripedium* (lady-slipper) and *Corallorhiza* (coral-root). Some of the most common tropical orchids in cultivation are species of *Cattleya, Dendrobium, Epidendrum, Cymbidium,* and *Vanda.*

# Iridaceae

## The Iris Family
### Asparagales ▪

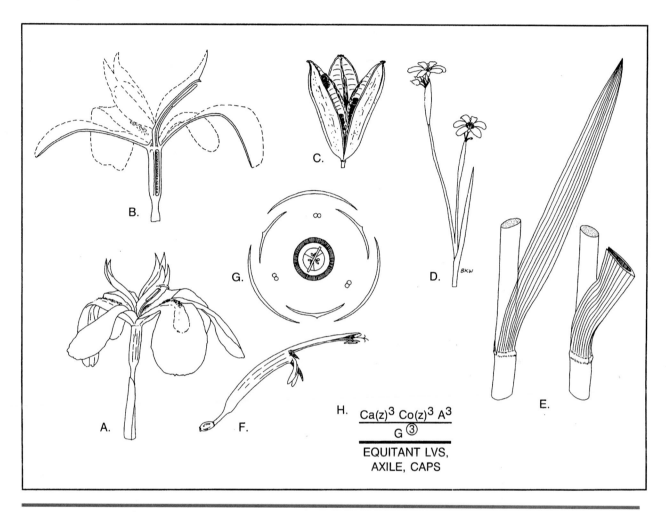

**Figure 16-11** Features of Iridaceae. A–C. *Moraea* sp. A. Intact flower with subtending bract and petaloid stigmas. B. Flower (l.s.). C. Capsule. D. Inflorescence of *Sisyrinchium* with radial flowers. E. Stem of *Iris* with equitant leaves (intact and x.s.). F. Bilateral flower of *Chasmanthe*. G. Floral diagram of *Sisyrinchium*. H. Generalized floral formula.

## FAMILY DESCRIPTION

Herbs or shrubs. Leaves alternate or basal, simple and entire, usually equitant, eligulate. Plants synoecious. Inflorescence of cymes, umbels, spikes or panicles, or flowers solitary, sometimes enclosed or subtended by 1 or more spathes. Flowers perfect, regular or irregular. Hypanthium usually well developed. Sepals 3, petaloid, distinct or connate. Petals 3, distinct or connate. Stamens 3, distinct or filaments connate, ± adnate to hypanthium. Carpels 3, connate; ovary inferior (or very rarely superior) with 3 locules and many axile ovules; style 1, undivided or 3-lobed, free or adnate to hypanthium; stigmas sometimes petaloid. Fruit a capsule.

# FAMILY NOTES

The Iridaceae, with 60–80 genera and 1500 species, are found on every continent, with a center of diversity in Africa. They are a family of showy-flowered monocots which have a very large diversity of flower shapes and pollination systems.

The iris family can easily be recognized by both vegetative and floral features. The leaves are characteristically **equitant** (Figure 16-11 E). They are basally sheathing and two-ranked. The blade of the leaf is folded lengthwise in such a way that only the undersurface is visible; the leaf appears to be attached edge-on to the stem. The upper surface is folded together and fused above the sheathing leaf-base. The flowers (Figures 16-11 A, B, D, F) are always showy with both sepals and petals petaloid. In some the sepals and petals closely resemble each other. In others, although both whorls are petaloid, they have very different shapes. The flower symmetry ranges from radial to strongly bilateral. There are three stamens and an inferior 3-carpellate ovary that matures as a capsule (Figure 16-11 C).

*Iris* and some other genera have petaloid stigmas which often are as large as or larger than the petals (Figures 16-11 A, B). Each stigma is situated directly above a sepal. The petals stand upright in the alternate position. The stigma's pollen-receptive surface is a small flap of tissue located on the side of the stigma that is pressed against the adjacent sepal. Between each sepal-stigma combination is a stamen. A nectar-gathering or pollen-gathering insect must push the stigma and sepal apart to get to a nectary or a stamen. In visiting more than one stigma-sepal-stamen complex, the insect is likely to transfer pollen from the stamen of one flower to the stigmatic flap of another. The combination of stigma and sepal resembles a bilabiate flower and is pollinated by large-bodied insects that visit such flowers. To a pollinator an *Iris* flower is three separate pollination units.

The importance of the Iridaceae lies chiefly in their showy flowers. Members of genera cultivated as ornamentals include *Crocus, Tigridia, Freesia, Gladiolus, Iris, Chasmanthe, Crocosmia, Sisyrinchium* and various others. The expensive spice, saffron is produced from the stigmas of *Crocus sativus.*

The family is currently circumscribed to include the Geosiridaceae, with a single species, from Madagascar.

# Arecaceae (Palmae)

## The Palm Family
### Arecales ■

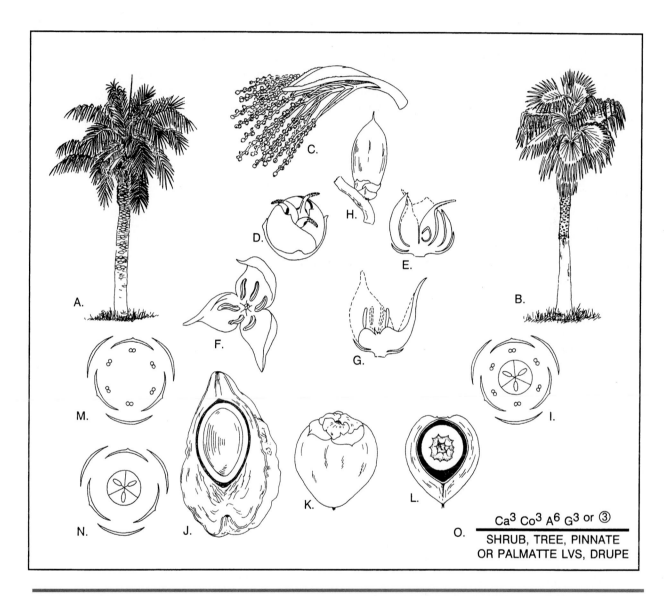

**Figure 16-12** Features of Arecaceae, A. Habit of *Phoenix dactylifera*, a feather palm. B. Habit of *Washingtonia filifera*, a fan palm. C–H. *Phoenix roebelinii*. C. Panicle of pistillate flowers and woody spathe. D–E. Pistillate flowers with 3 distinct carpels (intact and l.s.). F–G. Staminate flowers (intact and l.s.). H. Drupe. I. Floral diagram of *Washingtonia* (perfect flower). J. Drupe of *Cocos nucifera* (l.s.). K–L. Drupes of *Jubaea chilensis* (intact and l.s.). M–N. Generalized floral diagrams of staminate and pistillate flowers. O. Generalized floral formula.

# FAMILY DESCRIPTION

Shrubs, woody vines or trees, usually unbranched, sometimes with an undergroud rhizome. Leaves alternate, often forming dense terminal rosettes, simple to palmately or pinnately lobed or pinnately compound (rarely bipinnate), usually long-petiolate, estipulate, sheathing, often very large. Plants synoecious, monoecious, dioecious or polygamous. Inflorescence of spikes, racemes, heads, cymes or panicles, basally subtended by one or more spathes. Flowers perfect or imperfect, regular. Hypanthium sometimes present. Sepals 3 (2 or 4), distinct or connate. Petals 3 (2 or 4), distinct or connate, often sepaloid in color and texture (especially in pistillate flowers), stamens 6 (3 or 7–many), distinct or with connate filaments, free or adnate to corolla or hypanthium. Carpels 3 (1–10), distinct or connate; ovary superior with 1–3 locules and 1 basal or axile ovule per locule, or ovaries superior with 1 locule and 1 basal or marginal ovule per carpel; styles 3 (1–10), sometimes basally connate, or stigmas sessile. Fruit a 1-seeded (–3-seeded) drupe or berry, or ovaries coalescent into fleshy syncarp.

# FAMILY NOTES

The Arecaceae (Palmae) is a rather large family (200 genera and 2700–3000 species) with a pantropical distribution and some genera that range into subtropical or warm-temperate latitudes. Only a few are native to the United States. Arizona and California have one native species of palm *(Washingtonia filifera)* and there are several palms indigenous to Florida and adjacent regions of the southeastern United States.

Palms are among the most ancient monocots, with a fossil record extending back to at least the upper part of the Cretaceous period (about 65,000,000 years ago). Their fossilized pollen, leaves and trunks are found in many areas now far too cold or dry to support tropical vegetation.

Palms are unusual monocots in several respects. Their trunks (sometimes slender, sometimes very massive) are formed entirely by primary growth. In most species there is no increase in stem diameter as the tree grows, resulting in a columnar trunk. The leaf blades are plicate, folded like a fan when young. In some, the fan palms (Figure 16-12 A), the major veins arise in a palmate or subpalmate fashion from a shortened central rib **(costa)** and the blades are palmately lobed. In the feather palms (Figure 16-12 B), on the other hand, the blades are pinnately compound. These pinnately compound leaves do not develop in the same way as pinnately compound dicot leaves, however. The blade is simple before it fully expands. Weak zones of cells along some of the folds of the leaf tear apart as the leaf expands, ripping the leaf blade into leaflets.

The flowers are small and arranged in large, often paniculate clusters subtended by one or more large, often woody spathes (Figure 16-12 C). [A spathe is a sheathing bract that subtends or enwraps a flower or inflorescence.] Many are monoecious or dioecious. The flowers generally have a biseriate perianth of 3 sepals and 3 petals, but both are usually small and inconspicuous and often are sepaloid (Figures 16-12 D–F). The sepals and petals are often different in appearance from each other, however. In some monoecious species the staminate flowers are borne along the slender terminal portions of the panicle branches and the pistillate flowers are attached close to the main axis of the panicle. In this way the drupes (Figure 16-12 I), which are sometimes very massive, are supported as they mature. In others the staminate and pistillate flowers are variously mixed within the inflorescence.

This family is probably the third most important family in terms of human survival behind the grasses and the legumes. Many of the crops with which we are familiar grow poorly or not at all in the tropics. In many areas of the tropics palms provide food, fiber and shelter. Starch stored in the trunks of some palms is an important food source in some areas. Palm thatch is widely used throughout the tropics as a roofing material. Most palm species are used locally and do not have a major effect on the world trade. However, there are several major exceptions including *Phoenix dactylifera* (date palm), *Cocos nucifera* (coconut) and *Elaeis guineensis* (oil palm).

Numerous palm species are used as ornamentals in areas with warm climates. Commonly cultivated species include *Washingtonia* spp. (California and Mexican fan palms), *Roystonea regia* (royal palm), *Phoenix* spp. (ornamental date palms), *Chamaerops humilus* (European fan palm), *Cocos nucifera* (coconut palm), and many others. Some of the shrubby palms are grown indoors as house plants.

# Bromeliaceae

## The Pineapple Family
Poales ■

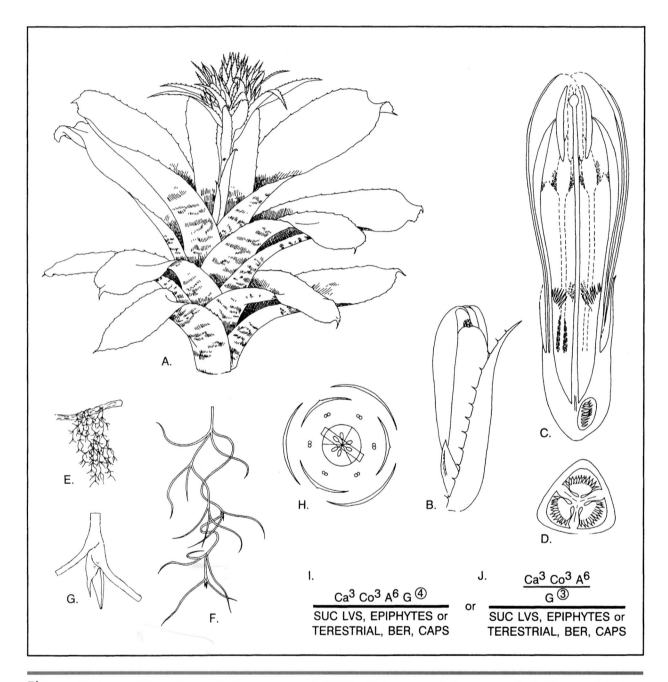

**Figure 16-13** Features of Bromeliaceae. A–D. *Achmea* sp. A. Habit with leaves and inflorescence. B–C. Flowers (intact and l.s.). D. Ovary (x.s.). E–G. *Tillandsia usneoides.* E. Epiphytic habit. F. Dangling stems and linear leaves. G. Capsule in leaf axil. H. Floral diagram of *Achmea.* I. Generalized floral formula for subfamilies Pitcairnioideae and Tillandsioideae. J. Generalized floral formula for subf. Bromelioideae.

# FAMILY DESCRIPTION

Herbs or shrubs (trees), often epiphytic. Leaves alternate, often forming dense basal rosettes, simple, basally sheathing, mostly somewhat succulent. Plants synoecious (dioecious). Inflorescence a spike, raceme or panicle. Flowers perfect (rarely imperfect), regular or irregular. Sepals 3, distinct or connate, green or petaloid. Petals 3, distinct or connate. Stamens 6, distinct or filaments connote at base, free or filaments ± adnate to petals. Carpels 3, connate; ovary superior to inferior with 3 locules and many axile ovules; style 1, undivided. Fruit a capsule or berry (sometimes coalescent into a dense multiple fruit).

# FAMILY NOTES

The Bromeliaceae is a mostly tropical and subtropical family of about 45–50 genera and 1500–2000 species endemic (with the exception of one species) to the New World. Some of the bromeliads are ground-dwelling plants of open habitats, but the family contains some of the most common epiphytic plants of tropical forests. They range as far northward as the southeastern United States and the southernmost mountains of Arizona. The Spanish-moss *(Tillandsia usneoides)* that festoons trees in the coastal regions of the Southeast is a bromeliad (Figures 16-13 E–G). One species of a South American terrestrial genus, *Puya* has an erect woody trunk and stands several meters tall.

Most bromeliads are acaulous herbs with leaves in tight rosettes (Figure 16-13 A). The leaves are usually leathery or fleshy and are covered with a thick cuticle. The leaf-margins of some members of the family bear sharp, spiny teeth. In some of the epiphytic species water collects in a well-like cavity formed by the overlapping leaf bases. Others absorb water directly from the air through hygroscopic hairs on the leaves.

The flowers are borne in spikes, racemes, or panicles and are often subtended by brightly colored petaloid bracts. The sepals are often green and herbaceous and are sharply differentiated from the petals (Figure 16-13 B). The petals are often appendaged within, sometimes with grooves or folds that enclose the filaments of the stamens (Figure 16-13 C). The ovary ranges from superior to wholly inferior.

The Bromeliaceae comprises three well-marked subfamilies that vary in several easily observable features (Table 16-3).

With the exception of *Ananas comosus* (pineapple), the family has limited economic importance. A pineapple is a dense multiple fruit in which the flowers of a spike coalesce and ripen together with the stem axis and subtending bracts. The cultivated pineapple is seedless because it is a sterile triploid. Various of the epiphytic species are cultivated indoors as ornamentals or as greenhouse curiosities. A few of the terrestrial species can be grown outdoors in warm-climate gardens. Fibers for cordage are processed from the leaves of several genera. The dried plants of Spanish-moss are sometimes used as packing material or as stuffing in upholstery.

| subf. Pitcairnioideae | subf. Tillandsioideae (Fig. 16-16 E–G, I) | subf. Bromelioideae (Fig. 16-16 A–D, H, J) |
|---|---|---|
| Mostly terrestrial plants with spiny-serrate leaves; never forming a trunk; ovary superior; fruit a capsule with winged seeds. | Mostly epiphytic plants with entire leaves; never forming a trunk; ovary superior; fruit a capsule with comose (hair-tufted) seeds. | Both epiphytic and terrestrial plants mostly with serrate leaves; some with a trunk; ovary inferior; fruit a berry. |

**Table 16-3** Comparison of the subfamilies of the Bromeliaceae.

# Juncaceae

## The Rush Family
## Poales ■

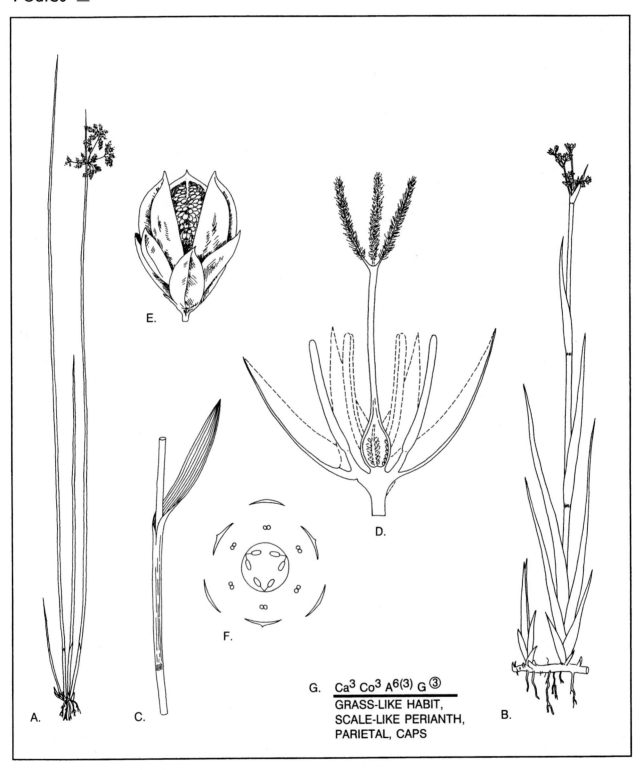

E.

D.

C.

F.

A.

B.

G.  $Ca^3 Co^3 A^{6(3)} G \underline{③}$
GRASS-LIKE HABIT,
SCALE-LIKE PERIANTH,
PARIETAL, CAPS

# FAMILY DESCRIPTION

Herbs (rarely shrubs), often semiaquatic. Leaves alternate or basal, usually three-ranked, simple and entire (rarely serrate), linear, basally sheathing, ligulate or eligulate; blades flat, terete, equitant or absent. Plants synoecious or rarely dioecious. Inflorescence variously paniculiform, sometimes condensed and headlike. Flowers perfect (rarely imperfect) regular. Sepals 3, distinct, scale-like. Petals 3, distinct, similar to sepals. Stamens 3 or 6, distinct or filaments basally connate. Carpels 3, connate; ovary superior with 1 locule and several—many parietal ovules or 3 basal ovules or with 3 locules and several–many axile ovules; style 1, 3-branched or stigmas sessile. Fruit a capsule.

# FAMILY NOTES

The Juncaceae, a family of 8–9 genera and about 300 species, is worldwide in distribution, but most common in temperate latitudes. In the tropics the family is mostly restricted to montane regions. The greatest generic diversity is in South America. Only two genera, *Juncus* (rush) and *Luzula* (wood-rush), occur in North America. Of these, *Juncus* is much more common, often occurring in marshes or other wetlands.

Leaves of the Juncaceae are quite variable. They have a well-developed basal sheath that may be open (e.g., *Juncus*) or closed (e.g., *Luzula;* Figure 16-14 C). In some species of *Juncus* most of the leaves are reduced to bladeless sheaths and photosynthesis takes place mainly in the sterns (Figure 16-14 B). In other species the leaves are flat and grass-like. In still others the blades are cylindrical or vertically folded and equitant (Figure 16-14 C) like those in the Iridaceae. The South African shrubby genus, *Prionium,* has stiff, flattened, serrate leaves.

Although the rushes often bear a superficial resemblance to grasses (Poaceae) and sedges (Cyperaceae), the flowers are distinctly different. Both grasses and sedges bear tiny flowers in spikelets. In the rushes the flowers are solitary or arranged in heads or various kinds of panicles, but never in spikelets. The perianth of the rushes is biseriate, with the 3 sepals and 3 petals all scalelike and similar in size, shape and color (Figure 16-14 A, F). In the other two families the perianth is very reduced: in the grasses to a pair of minute fleshy scales and in the sedges to 0–6 bristles. The ovaries of both grasses and sedges have only one ovule, and their fruits are indehiscent. In the rushes, on the other hand, the ovary has 3 to many ovules and matures as a capsule (Figure 16-14 E). In *Juncus* the ovary generally has several seeds per carpel whereas in *Luzula* there is only one ovule per carpel.

The rushes have little economic importance. They are rarely used as ornamentals. The stems of some rushes are occasionally woven together to make baskets or mats. Rushes generally are not weedy, but they are very common in wet areas. They provide important habitat for waterfowl and other wildlife. They are sometimes grazed but are not as palatable as grasses. In heavily grazed areas with wet soils, tough-stemmed rushes often displace more palatable species.

**Figure 16-14** Features of Juncaceae. A. Plant of *Juncus patens* with slender, cylindrical, leafless stems and leaves reduced to bladeless basal sheaths. B. Plant of *Juncus phaeocephalus* with flattened stems and 2-ranked equitant leaves. C. Stem of *Luzula* with flattened, grass-like leaves with closed basal sheaths. D. Flower of *Juncus* (l.s.). E. Loculicidal capsule of *Juncus acutus*. F. Floral diagram of *Juncus*. G. Generalized floral formula.

# Cyperaceae

## The Sedge Family
## Poales ◼

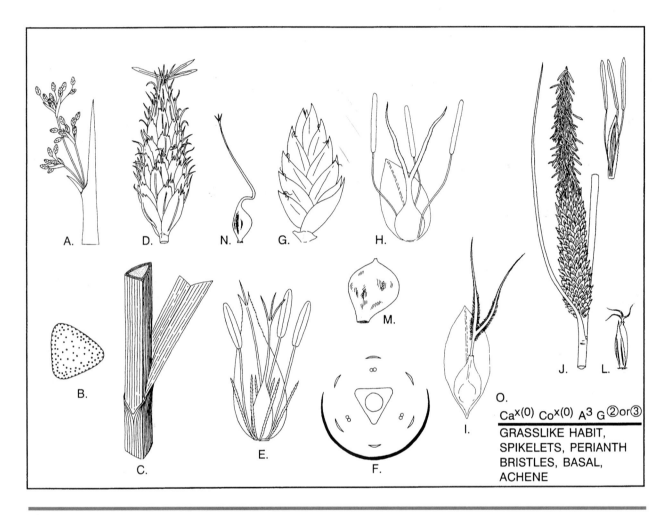

**Figure 16-15** Features of Cyperaceae. A–B. *Scirpus californicus.* A. Panicle of spikelets. B. Cross section of stem. C–D. *Scirpus robustus.* C. Stem and leaf illustrating closed leaf-sheath and channeled leaf-blade. D. Spikelet with spirally attached bractlets. E. Flower of *Scirpus* with subtending bractlet; note perianth of barbed bristles. F. Floral diagram of *Scirpus.* G. Spikelet of *Cyperus* with two-ranked bractlets. H. Flower of *Cyperus;* note absence of perianth bristles. I. Pistillate flower of *Carex* with subtending bractlets. Note that the ovary is enclosed in a hollow bractlet, the perigynium. J–L. *Carex spissa.* J. Androgynous compound spikelet. K. Staminate flower and subtending bractlet. L. Pistillate flower with subtending bractlets. M. Biconvex achene with deciduous style. N. Triangular achene with persistent style. O. Generalized floral formula.

## FAMILY DESCRIPTION

Herbs (rarely shrubs or woody vines), often semi-aquatic; stems often triangular. Leaves alternate or basal, 3-ranked, simple and entire (or minutely serrate), linear, basally sheathing with connate sheath margins, ligulate or eligulate; blades flat, terete, triangular or absent. Plants synoecious, monoecious or dioecious. Inflorescence of 1–many spikelets that are usually secondarily aggregated into racemes, spikes, heads or paniculiform to umbelliform clusters. Flowers perfect or imperfect, regular or naked, individually subtended by bractlets. Perianth parts reduced to bristles (rarely scales) or absent. Stamens 1–3 (4–many), distinct. Carpels 2 or 3, connate; ovary superior with 1 locule and 1 basal ovule; style 1 with 2 or 3 branches. Fruit an achene.

# FAMILY NOTES

The Cyperaceae is a large family with about 70–100 genera and 4000–9300 species. It has a cosmopolitan distribution and is especially well represented in wetlands of temperate and colder regions.

The Cyperaceae are grass-like plants with stems that are usually triangular in cross-section (Figure 16-15 B, C). [In *Eleocharis* and some other sedges the stem is round, flattened or several-angled.] The internodes are usually filled to the center with pithy tissue. The three-sided nature of the stem correlates with leaves arising in a 3-ranked pattern. The leaf-sheaths are closed; the margins of the sheath are fused together on the side of the stem opposite the leaf blade (Figure 16-15 C). There is considerable variation in the form of the leaf-blade. In some taxa the leaves are reduced to bladeless sheaths. In most members of the family the leaves are flattened but have a central V-shaped fold (Figure 16-15 C), especially near the stem. In some cases the leaves are triangular or cylindrical.

The inflorescence in the Cyperaceae consists of one to many **spikelets** (Figures 16-15 A, D, G, J). The spikelet consists of a central axis, the rachilla, to which are attached an overlapping series of scale-like bractlets. In many sedges the spikelets are simple with each bractlet subtending a tiny axillary flower. The perianth is very reduced; it is often absent or consists of a few bristles. These may be short and flattened or elongated and hairlike. The flower usually has three stamens that protrude from the spikelet when pollen is being released. The ovary is 2- or 3-carpellate and has one locule and a single basal ovule. The fruit is an achene. If the ovary is 2-carpellate, the achene is biconvex; if it is 3-carpellate the achene is ± triangular (Figures 16-15 M, N).

*Carex,* the largest genus in the family (1200–2000 species) has imperfect flowers. Staminate and pistillate flowers may be borne in the same or different spikelets. Mixed spikelets may be **gynecandrous** (pistillate flowers above and staminate flowers below) or **androgynous** (staminate flowers above and pistillate below; Figure 16-15 J). Both kinds of flowers lack a perianth. The staminate flowers are subtended by a single flattened, scale-like bract (Figure 16-15 K). Each pistillate flower is subtended by two bracts (Figures 16-15 I, L). The first is a flattened scale like those subtending the staminate flowers. The second, called a **perigynium,** is a hollow, sac-like structure that completely surrounds the ovary (Figure 16-15 I). The stigma is exserted through a hole in the end of the perigynium. The pistillate spikelet of *Carex* is considered to be a compound structure with each perigynium representing the end of a small lateral branch. In some species there is a short rachilla segment that extends past the ovary within the perigynium and in *Uncinia* the rachilla is exserted well beyond the ovary. The perigynium is a distinguishing feature for *Carex* and its close relatives.

The flowers of most sedges are wind-pollinated, producing large amounts of pollen and often bearing prominent stigmas. Members of a few genera have insect-pollinated flowers. In *Dichromena* an involucre of white bracts subtending a tight cluster of spikelets attracts pollinators. Nectar is not produced by these or any other sedges. Pollen is the "reward" for the pollinating insects.

The Cyperaceae includes species used in a variety of ways. The flattened pith of *Cyperus papyrus* was used in ancient Egypt as an early version of paper. The stems of various sedges including *C. papyrus,* can be bundled together to make rafts. The stems and leaves of various sedges are used in basketry, to make hats and mats, for thatch and in various other fashions. Edible tubers are produced by *Cyperus esculentus* (chufa), *Eleocharis tuberosa* (Chinese water-chestnut) and *Scirpus tuberosus.* Sedges are very important components of wetlands vegetation, providing food and other critical habitat requirements for a wide variety of wildlife.

# Poaceae (Gramineae)

## The Grass Family

Poales ■

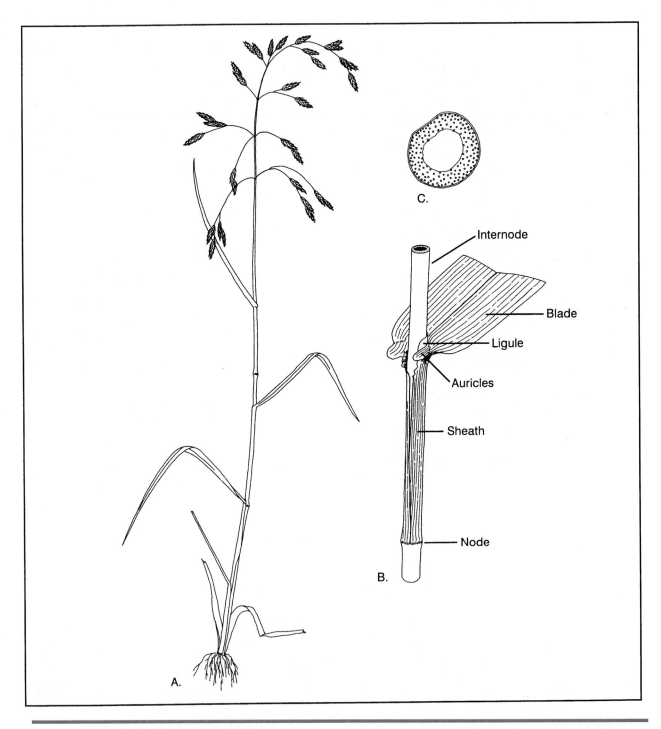

C.

Internode

Blade

Ligule

Auricles

Sheath

Node

B.

A.

**Figure 16-16** Vegetative features of Poaceae. A. Plant of *Bromus* illustrating the fibrous root system, enlarged nodes, two-ranked linear basally sheathing leaves and terminal panicle of spikelets. B. Details of a grass stem and leaf-base illustrating blade, open sheath, ligule and auricles. C. Cross section of stem illustrating scattered vascular bundles and hollow internode.

# FAMILY DESCRIPTION

Herbs, shrubs or trees (rarely climbing vines); stems terete or flattened on one side with nodes swollen. Leaves alternate or basal, 2 ranked, simple, entire, basally sheathing with overlapping (rarely connate) sheath margins, ligulate; blades flat (sometimes absent). Plants synoecious or polygamous (monoecious or dioecious). Inflorescence of spikelets that are secondarily clustered into spikes, racemes, or panicles (heads) (Figure 16-17 A); spikelets with 2 (0–1 or 3–7) basal bractlets (glumes) (Figure 16-17 B, E) and 1 or more flowers, each enclosed by a pair of bractlets (lemma and palea) (Figure 16-17 D, E), attached to a jointed rachilla (some or all bractlets rarely absent). Flowers perfect or imperfect, irregular (rarely regular). Perianth parts (lodicules) 2 (3), distinct, fleshy and becoming turgid at anthesis. Stamens 3 (1–2 or 4–6) distinct (rarely filaments connate in groups). Carpels 3 but usually apparently 2, connate; ovary superior with 1 locule and 1 basal ovule usually adnate to ovary wall; styles 2 (1 or 3), distinct or ± united. Fruit a caryopsis (achene, utricle, nut, or drupe), often shed from the plant together with the enclosing bracts.

# FAMILY NOTES

The Poaceae is a large and very diverse family with 650–660 genera, and about 10,000 species. It is cosmopolitan in distribution. Members of the grass family occur in an extraordinary diversity of environments from sea level to the tops of high mountains, from the equator to the polar regions, and from wetlands to very dry deserts. The plants range in size from minute annuals to giant bamboos. Grasses are the dominant wild plants over extensive areas of the earth's surface, and much of the arable land of the world is devoted to the cultivation of grasses.

Members of the grass family have a characteristic vegetative structure (Figure 16-16 A). Except in young plants the majority of the roots are adventitious and fibrous. The culms (erect stems) are cylindrical with enlarged solid nodes and narrower, often hollow internodes. Many grasses have rhizomes or stolons as well, enabling the plant to spread vegetatively. The culms are often stiff and fibrous, and sometimes are woody. Although some bamboos are treelike, with stems growing to 40 meters or more in height and 30 centimeters or more in diameter, all the growth is primary, and the basic stem structure is the same as in much smaller herbaceous grasses.

Grasses have several features that make them tolerant of grazing and mowing. Grass stems have intercolary meristems, located in the lower regions of the stems and leaves that enable these structures to increase in length from the base. This important feature allows grasses to grow back rapidly from the effects of being grazed or mowed. Additionally, grass stems commonly branch from the lowest nodes forming tillers (basal branches) that typically arise below the level of grazing or mowing. The tillers soon develop their own adventitious roots and become more or less independent of the parent culm. All of these features contribute the remarkable tolerance of the grasses to grazing and their ability to serve as the food base of many ecosystems.

The leaf sheaths of most grasses are open, with overlapping margins (Figure 16-16 B), though in some grasses (e.g., *Bromus*) the sheath margins are fused at least part of their length. Most grass leaves have a well-developed ligule at the junction of the blade and sheath. The ligules in many grasses are entire or toothed membranous scales. Ligules may be variously dissected or modified into bristles. Features of the ligules are often of taxonomic value. The upper margins of the sheath often bear a pair of small ear-like apical lobes called **auricles.** The blades of grass leaves are flat, often without a prominent midrib. In most herbaceous grasses the leaf-blades are attached directly to the sheath. In most bamboos the leaves have a short petiole-like structure (pseudopetiole) separating the blade from the sheath.

The epidermal cells of grass stems and leaves contain microscopic crystals of silicon dioxide called phytoliths. These minute sand grains have been important in the evolution of grazing animals. Over the lifetime of an animal that consumes large amounts of grass in its diet, the phytoliths gradually wear down the animal's teeth. Animals that use grasses as a main food source have evolved high-crowned teeth that in many cases continue to grow from the base as the crowns are worn away.

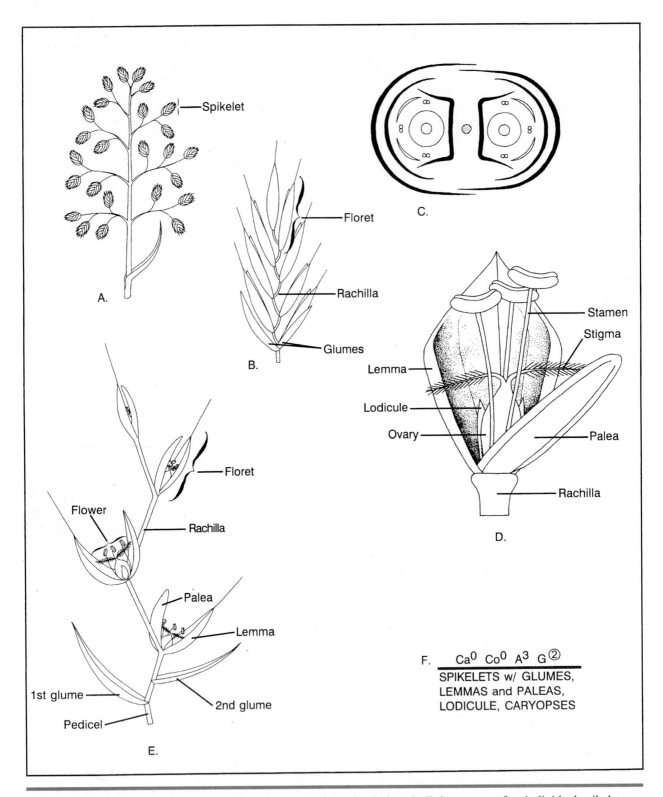

**Figure 16-17** Flower and spikelet features of Poaceae. A. Panicle of spikelets. B. Enlargement of an individual spikelet. C. Diagrammatically expanded spikelet. D. Enlarged floret. E. Diagram of two-flowered spikelet. F. Floral formula.

The floral formula in F reads:

$$\mathrm{Ca^0 \quad Co^0 \quad A^3 \quad G\textcircled{2}}$$

SPIKELETS w/ GLUMES,
LEMMAS and PALEAS,
LODICULE, CARYOPSES

Flowers in the Poaceae are borne in **spikelets.** Each spikelet has a jointed **rachilla** (central axis). Attached to opposite sides of the rachilla at the base of the spikelet are two bractlets called **glumes.** The **first glume** is attached to the rachilla slightly below the **second glume.** Above the glumes one or more **florets** are attached to the rachilla. Each floret typically consists of two bractlets, called the **lemma** and **palea,** and an enclosed flower. The flower parts are located between the lemma and the palea. The lemma is the larger of the two bracts and its margins overlap those of the palea. The palea is situated close to the rachilla of the spikelet.

The flower consists of a perianth of two (or three) scales called **lodicules,** usually three stamens and a superior ovary with usually two feathery styles. Grasses are wind-pollinated and in many grasses the lodicules play a very important role in pollination. At anthesis the lodicules become water-filled and expand, pushing the lemma and palea apart. This allows the anthers to dangle from the spikelet and the stigmas to protrude to the sides. After anthesis the lodicules contract and the lemma and palea close up around the developing fruit. The grass ovary contains a single ovule that in most genera is firmly attached throughout its periphery to the ovary wall. The ovary usually ripens as a **caryopsis** (grain). In a small minority of grasses the ovary ripens as an achene or utricle. A few bamboos produce drupes or nuts.

Because the flower is ephemeral and the sterile and fertile bracts usually tightly enclose or adhere to the developing fruit, grass identification depends heavily on characters of the bracts and spikelets (Figure 16-18). A very important feature is the number of florets per spikelet; this can be determined by counting the number of lemmas. The number of prominent veins (**nerves**) on the glumes and lemmas is often important in keys. The glumes and lemmas often bear bristle-like awns of various shapes, sizes and places of attachment (Figures 16-18 H–M).

Some grasses have one or more **sterile florets** that lack a functional gynoecium (Figures 16-18 O–R). A **fertile floret** is one that contains an ovary or a developing grain. Note that a fertile floret can either be perfect or pistillate by this definition and a sterile one can be either neuter or staminate. The bracts associated with these florets are often described by the term that describes the sexual condition of the floret; e.g., sterile lemma, fertile palea. Sometimes a sterile floret may consist merely of a sterile lemma with the palea and all of the flower parts suppressed.

Sterile florets may be inserted either below or above the fertile floret(s) on the rachilla. Some grasses even have sterile spikelets (Figure 16-18 N).

In most grasses the fruits fall together with some or all of the associated bracts. In some genera the rachilla of the spikelet disarticulates (breaks apart, shatters) above the glumes (Figure 16-18 A). Each floret falls as a unit, usually together with the piece of the rachilla to which it is attached. In other genera the entire spikelet falls as a unit, glumes and all (Figure 16-18 B). Most spikelets with two or more fertile florets disarticulate above the glumes. One-flowered spikelets may disarticulate above the glumes or fall away intact. If you are not sure which way the spikelet disarticulates, gently pull on the parts to see if the spikelet will prematurely fall off or fall apart. If a grass is reasonably mature this technique often works, but if the spikelets are immature they probably will merely tear apart when you pull on them.

## Grass Classification

Until recently the most widely followed system of classification for the grasses was that of Hitchcock (1935, 1951). In this system, which was based mostly on the study of temperate grasses, two subfamilies (Festucoideae and Panicoideae) were recognized. The working structure of this traditional system was tribes. Hitchcock recognized 14 tribes for grasses of the United States. This classification was based almost exclusively on characters of the spikelets and florets. All but the most recent floras in the United States follow Hitchcock's classification. However, although it is a workable system for identification, portions are now considered to be artificial and not representative of evolutionary relationships among the grass genera.

Recently much research on comparative physiology, anatomy, and development of grasses has led to major revisions in the systematics of the family. Unfortunately some of the characters that form the basis of the new grass systematics are not readily visible without much time and specialized laboratory equipment and techniques. Important characters in grass taxonomy include chromosome number and morphology, types of specialized cells of the leaf epidermis, type of photosynthetic pathway, arrangement of tissues in leaf cross-section, embryo anatomy, seedling morphology and germination requirements.

Although the modern grass classification is more representative of phylogenetic relationships than is

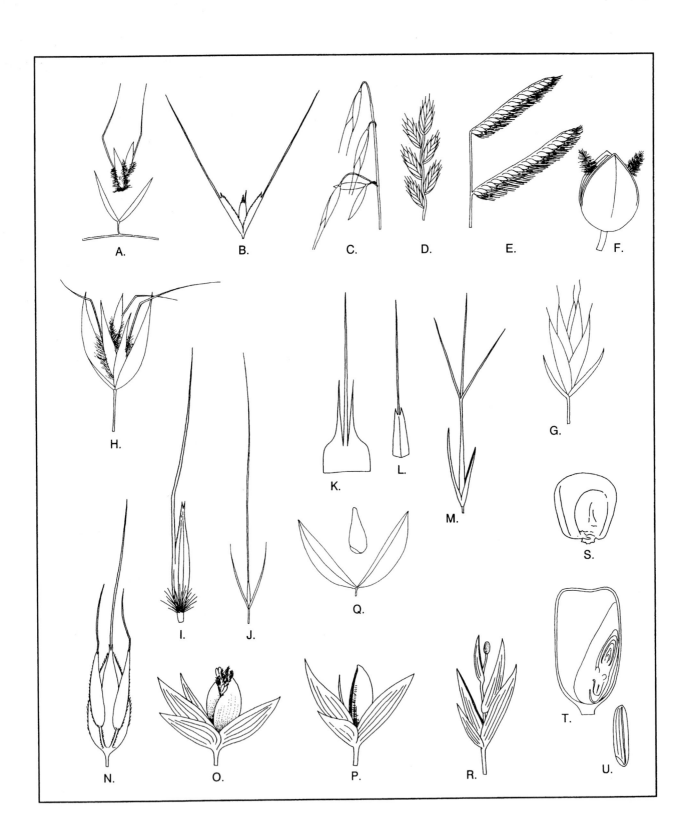

A.

B.

C.

D.

E.

F.

G.

H.

I.

J.

K.

L.

M.

N.

O.

P.

Q.

R.

S.

T.

U.

the system of Hitchcock, it is not as easy to use for identification. Most modern works on the grasses do not attempt to use the phylogenetic classification for identification. Instead the keys break the grasses into artificial groups similar in composition to the tribes used of Hitchcock.

# Survey of Subfamilies and Tribes of the Poaceae

The following is a summary of the tribes of grasses recognized in modern classifications of the family. There is some variation from one reference to another regarding the number of tribes and their inclusiveness. There is still controversy among agrostologists (grass specialists) regarding the exact placement of certain genera. This summary is restricted to grasses occurring in North America.

## Subfamily Bambusoideae

**Tribe Bambuseae**—Tropical herbs, or more often woody grasses or tall timber bamboos that are considered by agrostologists to be the most primitive. The leaves have a petiole-like base above the sheath. Some have the 6 stamens and 3 styles and 3 lodicules expected of a plant derived from a typical monocot flower. These grasses all have C-3 photosynthesis. They are most common in the tropics with only a few reaching the U.S. Examples include *Arundinaria, Bambusa* and *Phyllostachys.*

**Tribe Phareae**—A tropical tribe of herbaceous perennials. The leaves often are wide and some have pinnate looking venation. The leaves have a petiole-like base above the sheath. The spikelets are unisexual, one-flowered and borne in pairs with one staminate and the other pistillate. These grasses all have

C-3 photosynthesis. Only one genus, *Pharus,* occurs in the U.S.

## Subfamily Oryzoideae[1]

**Tribe Oryzeae**—As typically defined this tribe contains aquatic grasses *Oryza* and *Zizania* (cultivated rice and wild rice). They have one fertile floret per spikelet and reduced glumes. This tribe may be split into the Oryzeae (perfect flowers and 6 stamens) and the Zizaneae (imperfect flowers and 3 stamens). These grasses all have C-3 photosynthesis.

**Tribe Ehrharteae**—At least one source (Crampton, 1974) places the genus *Ehrharta* or veldt grass into subfamily Oryzoideae. The inclusion of Elrharteae in the Oryzoideae makes this subfamily quite heterogeneous. Veldt grass is a south African grass that has become weedy in central coastal California. It has a single fertile floret surrounded by two large sterile lemmas and glumes. These grasses all have C-3 photosynthesis.

## Subfamily Arundinoideae

**Tribe Arundineae**—A tribe of robust grasses with large panicles of dense plumose spikelets. The ligule is either a dense ring of hairs or a fringed membrane. They are widely distributed in subtropical areas. *Phragmites australis* (common reed) is the most widely distributed grass in the world, ranging from cool temperate latitudes to the tropics in both Old and New Worlds. Other genera include *Arundo, Cortaderia,* and *Molinia.* These grasses, all have C-3 photosynthesis.

**Tribe Danthonieae**—A tribe of subtropical and warm temperate areas that have several-flowered spikelets completely enclosed by large glumes and a

---

[1]Included by Clayton and Renvois (1986) in subf. Bambusoideae.

---

**Figure 16-18** Features of the Poaceae often used in keys (opposing page). A. Mature spikelet of *Avena* disarticulating above the glumes. B. Spikelet of *Polypogon* disarticulating below the glumes and falling intact. C. Portion of a panicle of pedicelled spikelets of *Avena.* D. Spike of sessile spikelets of *Lolium.* E. Secund (one-sided) spikes of sessile spikelets of *Bouteloua.* F. Dorsally compressed spikelet of *Paspalum.* G. Laterally compressed spikelet of *Festuca.* H–I. Spikelet and floret of *Avena;* lemma with long dorsal awn. J. One-flowered spikelet of *Nassella;* lemma with long terminal awn. K. Lemma of *Danthonia,* bifid at apex, awned between the lobes. L Lemma of *Bromus,* awned from a minutely bifid apex. M. Lemma of *Aristida,* with 3 terminal awns. N. Spikelets of *Sorghum,* one sessile and fertile, two stalked and staminate. O. Spikelet of *Setaria* with a fertile terminal floret and a sterile second floret represented by a lemma and a palea. P. Spikelet of *Panicum,* similar to that of *Setaria* except that the first glume is reduced in size and the sterile palea is absent. Q. Spikelet of *Phalaris* with tiny sterile floret below the solitary fertile floret. R. Spikelet of *Melica* with a rudimentary terminal floret. S–T. Caryopses (intact and l.s.) of *Zea.* U. Caryopsis of *Triticum.*

ligule represented by a line of hairs. Florets fall from the plant leaving the glumes behind. Genera include *Danthonia* and *Schismus.* Clayton and Renvoise (1986) included the Danthonieae in tribe Arundineae. Most genera have C-3 photosynthesis; three have C-4 photosynthesis.

**Tribe Centotheceae**—This tribe is similar to the Danthonieae except for a ligule represented by a fringed membrane. Only the genus *Chasmanthium* ranges into the United States. The Centotheceae was placed into its own subfamily by Clayton and Renvoise (1986). These grasses all have C-3 photosynthesis.

# Subfamily Chloridoideae (Eragrostoideae)

**Tribe Chlorideae**—A large tribe of tropical to temperate grasses with sessile, secund spikelets with one to several florets per spikelet. The lemmas are 3-nerved. *Spartina* is an important grass of salt marshes and Bermuda grass *(Cynodon)* is an important lawn grass and weed in the warmer parts of the world. Species of *Bouteloua* (grama grass) are dominant in much of the Great Plains and the southwestern U.S. These grasses all have C-4 photosynthesis.

**Tribe Eragrosteae** (Eragrostideae)—This tropical and warm-temperate tribe is characterized by spikelets of one to several fertile florets. Spikelets are usually borne in sessile and secund racemes on a single elongate rachis. In *Sporobolus* and *Muhlenbergia* the spikelets may be borne in dense spikelike panicles. The tribe produces some minor millets of Africa such as tef and finger millet *(Eleusine).* Other representative genera include *Eragrostis* and *Leptochloa.* These grasses all have C-4 photosynthesis.

**Tribe Aristideae**—*Aristida* (three-awn grasses) is the only genus in this tribe of questionable placement. Crampton (1974) and Clayton and Renvoise (1986) place this genus in the subfamily Arundinoideae whereas others place it here in the Chloridoideae. These grasses are characterized by their lemmas bearing three long awns and each spikelet consisting of a single fertile floret. These grasses all have C-4 photosynthesis.

**Tribe Orcuttieae**—A tribe that contains two genera of rare to extremely rare grasses, *Orcuttia* and *Neostapfia,* that are essentially endemic to the Central Valley of California. These are short annual grasses with broad lemmas and a rachilla that usu-

ally extends beyond the florets. They occupy vernal pools or moist areas. These grasses all have C-4 photosynthesis.

**Tribe Aeluropodeae**—A tribe of generally short, very spreading grasses of saline habitats. The ligule consists of a soft fringed membrane. The leaves are often very distinctly in two ranks on opposite sides of the culm **(distichous).** The most common of these in the U.S. is *Distichlis* (salt grass) of western salt marshes and saline flats. The Aeluropodeae were included in tribe Eragrostideae by Clayton and Renvoise (1986). These grasses all have C-4 photosynthesis.

**Tribe Unioleae**—The Unioleae is represented in the U.S. by the single genus *Uniola* (sea oats). They are coarse grasses of the coast of the southeastern U.S. from Louisiana to the Carolinas. They have a ligule represented by a ring of fine hairs and a many-flowered spikelet that disarticulates below the glumes. Clayton and Renvoise (1986) included the Unioleae in the tribe Eragrostideae.

# Subfamily Pooideae (Festucoideae)

**Tribe Poeae** (Festuceae)—This tribe is common in the temperate zone. It includes many common lawn, pasture, ornamental, and weed grasses. Its spikelets are usually several-flowered and are usually borne on pedicels. The glumes are unequal and one or both are shorter than the first lemma. The lemmas have 5-several nerves. Crampton (1974) recognizes the **Bromeae** as a tribe separate from the Poeae characterized by longer spikelets and the tendency for the paleas to adhere to the grain. Common genera include *Bromus, Festuca, Lolium,* and *Briza.* These grasses all have C-3 photosynthesis.

**Tribe Triticeae**—The barley tribe contains the grains wheat, rye, and barley. The spikelets are sessile and are borne on opposite sides of an elongated rachis. There may be one to several spikelets per node. This tribe contains the core of the classically defined tribe Hordeae. Only a few genera (e.g., *Lolium*) have be moved to other tribes. All of the genera in this tribe are more or less interfertile. Common genera include *Hordeum, Triticum, Secale, Elymus, Agropyron,* and *Aegilops.* These grasses all have C-3 photosynthesis.

**Tribe Agrostideae**—A large tribe found in all floras but most prominent in areas of warmer tropical or subtropical climates. They possess stalked spikelets with a single fertile floret surrounded by small glumes that are neither nor or thinner than usual. Common genera include *Agrostis, Ammophilla, Millium,* and *Phleum.* The genus *Stipa is* sometimes place in a separate tribe, Stipeae, based on the presence of a long prominent awn and a long cylindrical grain. Gould and Shaw (1983) and Clayton and Renvoise (1986) combine this tribe with the Aveneae. These grasses all have C-3 photosynthesis.

**Tribe Aveneae**—The oat tribe is very similar to the Poeae except they have large prominent glumes that totally enclose the lowest lemmas and usually the whole spikelet. The genus *Phalaris* is sometimes segregated as the tribe, **Phalarideae** (Crampton, 1974) based on its single-flowered spikelet and the presence of two tiny sterile lemmas below the fertile one. These grasses all have C-3 photosynthesis.

## Subfamily Panicoideae

**Tribe Paniceae**—The panic grass tribe includes the same taxa in the modern systems as in Hitchcock's system. This means that the unique visible characters of the spikelet can be used alone to define the tribe. This is not true for any of the tribes discussed above. Members of the Paniceae produce spikelets with a single fertile floret per spikelet, reduced glumes and prominent lemmas and paleas that are often thickened. The spikelets are dorsi-ventrally flattened unlike the lateral flattening among the tribes of the Festucoideae. This primarily tropical tribe contains a few minor small-seeded grains (millets), forage grasses and weeds. Representative genera include *Digitaria, Panicum, Paspalum, Pennisetum, Setaria,* and *Melinis.* Both C-3 and C-4 photosynthesis are present among genera of this tribe.

**Tribe Andropogoneae**—The bluestem tribe, like the panic grass tribe usually can be recognized by the nature of its spikelets alone. The spikelets are in pairs or threes with a pistillate or perfect sessile spikelet and one or two pedicellate staminate or sterile spikelets. In monoecious members both the sessile and pedicellate spikelets may be staminate. This primarily tropical tribe includes many important crops such as maize, sugar cane, sorghum, and the spice, lemon grass. Genera include *Andropogon, Bothriochloa, Cymbopogon, Sorghum, Zea,* and *Saccharum.* Both C-3 and C-4 photosynthesis are present among genera of this tribe.

# Economic Importance of the Poaceae

The grass family is unquestionably the most important family in the plant kingdom. It is the primary source of carbohydrates and vegetable protein for most people on the earth's surface. A list of grain crops indicates the vast importance of this family. Barley *(Hordeum vulgare),* maize *(Zea mays),* oats *(Avena sativa),* rice *(Oryza sativa),* rye *(Secale cereale),* sorghum *(Sorghum bicolor)* and wheat *(Triticum* spp.) are among the most commonly cultivated grains. Many other grasses are grown as grain crops in various areas of the world. The world's principal source of sugar is sugar cane *(Saccharum officinarum).* Various alcoholic beverages are derived by fermentation of malted grasses. Grain alcohol is widely used as an industrial solvent and as an additive to gasoline.

Additionally, the grass family is the source of most of the animal protein for large segments of the earth's population. Rangeland grasses, hay, and grains are a principal source of food for domesticated animals. Wild animals from field-mice to giant pandas and elephants depend on grasses for a large part of their diet. Grasses are the dominant plants in various ecosystems, ranging from the short-grass prairies to tropical savannas and bamboo forests.

Bamboos and other grasses are used in tropical lands in a multitude of ways, ranging from construction materials to water pipes, furniture and food (bamboo shoots). Grass thatch is used in many tropical areas as a roofing material. The tough, interlaced roots of prairie grasses hold soil together so well that pioneers in the Great Plains were able to construct sod houses. Some grasses are important in basketry.

There are many ornamental uses for grasses. Because of the durability of turf grasses and their ability to tolerate mowing, they are widely grown for lawns, playing fields, golf courses, etc. Some grasses are used as accent plants in bedding.

On the negative side, windborne grass pollen is a major cause of hay fever, and some grasses are aggressive and difficult-to-control weeds. Grasses that originated in the areas around the Mediterranean Sea have displaced much of the original herbaceous vegetation in California and other areas with a Mediterranean climate.

# Commelinaceae

## The Spiderwort Family
## Commelinales ■

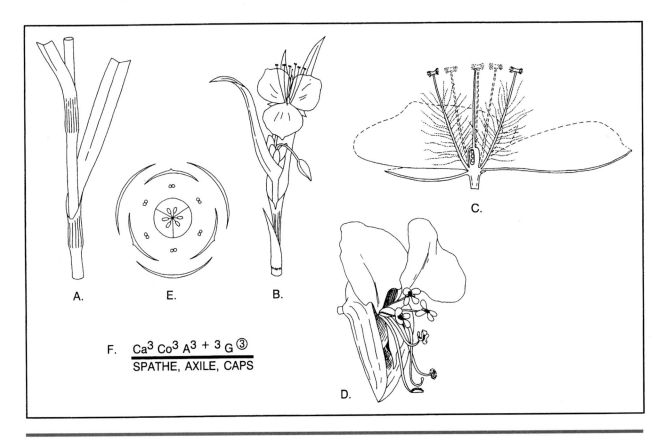

F. $\dfrac{Ca^3\ Co^3\ A^{3\ +\ 3}\ G\ \textcircled{3}}{\text{SPATHE, AXILE, CAPS}}$

**Figure 16-19** Features of Commelinaceae. A–C. *Tradescantia* sp. A. Stem and sheathing leaves. B. Inflorescence with sheathing bracts and radial flower. C. Flower (l.s.). D. Bilateral flower of *Commelina erecta* with subtending spathe and three stamens modified as staminodes. E. Floral diagram of *Tradescantia*. F. Generalized floral formula.

## FAMILY DESCRIPTION

Herbs with swollen nodes, sometimes ± succulent, sometimes twining. Leaves alternate, simple, entire, basally sheathing, the blade sometimes separated from the sheath by a pseudopetiole. Plants synoecious. Inflorescence of 1–many scorpioid or umbelliform cyme, each subtended by one or more spathes. Flowers perfect (rarely imperfect), regular or irregular. Sepals 3, distinct or connate, green. Petals 3, distinct (rarely connate). Stamens 6 (1–3), distinct (rarely filaments connate), often 1 or more reduced to staminodes. Carpels 3 (2), connate; ovary superior with 3 (2) locules and 1-few axile ovules per locule; style 1 undivided. Fruit a capsule (rarely a berry).

# FAMILY NOTES

The Spiderwort Family comprises about 50 genera and 50–700 species. It is widespread in tropical and subtropical latitudes with only a few temperate genera and species.

Members of the Commelinaceae are somewhat succulent plants, some of which have a more or less grass-like habit. The leaves have closed sheaths and linear to broadly ovate blades (Figure 16-19 A). The flowers are usually subtended by one or more boat-shaped spathes (Figure 16-19 B, D). The floral features of the family are relatively unmodified in comparison to those of most members of the Commelinidae. Unlike the petals of most members of the subclass, those of the Commelinaceae are color-ful. The flowers are insect-pollinated though they lack nectaries. Instead of nectar, pollinating insects gather pollen from the flowers. The flowers in most members of the family are radial (Figure 16-19 B), but in some species of *Commelina* they are bilateral (Figure 16-19 D).

Various members of the Commelinaceae are cultivated as ornamentals, some for their flowers and others primarily for their attractive foliage. Cultivated members of the family include species of *Tradescantia* (spiderwort), *Commelina* (dayflower), *Rhoeo* (boatflower), and *Zebrina* (wandering-Jew). Otherwise the family has very little economic importance.

# Exercises

## ■ Study Questions

### Araceae

1. What kinds of leaf venation do we find in the Araceae?
   How typical is this of monocots?

2. What is a spadix?
   What is a spathe?

3. Flowers in the Araceae are often _____-merous, but sometimes they lack a _____.

4. What kind of gynoecium does the Araceae have?
   How many carpels?
   Ovary position?

5. Ovaries in the Araceae generally mature as _____.

### Alismataceae

6. What is the habit of members of the Alismataceae?
   Leaves of these plants are all _____.

7. Flowers of the Alismataceae are _____-merous.

8. What kind of inflorescence characterizes the Alismataceae?

9. How many stamens might you find in a flower in the Alismataceae?

10. What kind of gynoecium characterizes the Alismataceae?
    Placentation?
    Ovary position?

11. What is the fruit type in the Alismataceae?

12. What features do members of the Alismataceae share with members of the Ranunculaceae? How do they differ?

## Liliaceae

13. Members of the Liliaceae are sometimes placed into several different families.

    Which families are recognized in the textbook?

    Which were recognized by your local key?

    Which in the family key in Appendix I?

    What features are used to separate these segregate families?

    What do they have in common?

    What is meant by Liliaceae *(sensu lato)?*

    What about Liliaceae *(sensu stricto)?*

14. Flowers in the Liliaceae *(sensu lato)* are _____-merous.

    In some the sepals and petals look almost exactly the same.

    How can we decide which are sepals and which are petals?

15. How many stamens do we find in Liliaceae flowers?

16. What kind of gynoecium characterizes the Liliaceae?

    How many carpels do we usually find per flower?

    Ovary positions?

17. Ovaries in the Liliaceae *(sensu lato)* generally ripen as _____.

    In a few genera they ripen as _____.

## Orchidaceae

18. Flowers in the Orchidaceae are __ -merous.

    What is the flower symmetry in the Orchidaceae?

    What is the "odd" petal called?

19. Flowers in most Orchidaceae are presented to the world upside down?

    What feature of the flower is responsible for this?

20. Where do we find the stamens of an orchid flower?

    Where is the stigma?

21. What pollinates orchid flowers?

22. What kind of gynoecium characterizes the Orchidaceae?

    How many carpels?

    What is the ovary position?

23. Ovaries in the Orchidaceae ripen as _____.

    What is unusual about the way these fruits dehisce?

    What features of the seeds of an orchid are unusual?

24. What role do fungi play in the reproduction of orchids?

    How do horticulturists get around the use of the fungi?

25. Many orchids hybridize freely in cultivation. What keeps these plants separate in nature?

26. List the families and species of monocots that you identified in the lab.

_____     _____

_____     _____

_____     _____

_____     _____

_____     _____

_____     _____

# Iridaceae

27. What term describes the leaves of the Iridaceae?

    If you were going to describe one of these leaves to a student in general botany, what would you say?

28. Flowers in the Iridaceae are _____-merous.

    In some the sepals and petals look almost exactly the same.

    How can we decide which are sepals and which are petals?

    What types of flower symmetry occur in the Iridaceae?

29. How many stamens do flowers in the Iridaceae have?

    How does this differ from the situation in the Liliaceae?

30. What kind of gynoecium characterizes the Iridaceae?

    How many carpels do we find per flower?

    What is the ovary position?

    What is the placentation?

31. What is the fruit type in the Iridaceae?

32. What is unusual about the stigmas of *Iris* and its closest relatives.

    What is the symmetry of an *Iris* flower?

    How many times must a pollinator visit an *Iris* flower to ensure pollination of all the ovules?

# Arecaceae

33. What is the habit of members of the Arecaceae?

    Contrast the stem anatomy of a palm tree with that of a dicot tree?

34. Palms have either leaves with _____ venation or leaves with _____ venation.

35. What is the sexual condition of most genera in the Arecaceae?

36. What kind of inflorescence characterizes the Arecaceae?

    What subtends the inflorescence?

37. What kinds of perianth do we find in flowers of the Arecaceae?

    How many stamens?

38. What kinds of gynoecium occur in the Arecaceae?

    How many carpels? Ovary position?

39. What are the fruits of the Arecaceae?

    Coconuts and dates are members of this family.

    How are they similar?

40. What is the traditional family name for the Arecaceae?

# Bromeliaceae

41. Many members of the Bromeliaceae are epiphytic. How do they live without soil?

42. Bromeliaceae flowers are _____-merous.

    How many stamens?

    How many carpels?

    Placentation?

    Ovary position?

43. What fruit types occur in the Bromeliaceae?

    What is a pineapple?

    Why don't pineapples produce seeds?

44. Spanish-moss and pineapples are in the same family. What do they have in common?

# Juncaceae

45. Describe the leaves in the Juncaceae.

    How do the leaf bases of *Juncus* differ from those of *Luzula?*

    Describe the variations in leaf cross section in *Juncus.*

46. Flowers of the Juncaceae are __-merous.

47. How would you describe the sepals and petals of the Juncaceae?

48. How many stamens would you expect to find in a flower in the Juncaceae?

49. What kind of gynoecium characterizes the Juncaceae?

    Number of carpels?

    Placentation?

    Ovary position?

50. What is the fruit type in the Juncaceae?

    How does the fruit of *Luzula* differ from that of *Juncus?*

## Cyperaceae

51. Stems in the Cyperaceae are typically _____ in cross section?

    They are generally _____ in the middle?

    The leaves are __-ranked.

52. What kind of leaf-sheaths are typical in the Cyperaceae?

53. What is the sexual condition of most genera in the Cyperaceae?

54. What is the primary inflorescence in the Cyperaceae?

    What subtends each flower?

55. What kind of the perianth do we find in flowers of the Cyperaceae?

    How many stamens?

56. What kind of gynoecium occurs in the Cyperaceae?

    How many carpels?

    How many ovules?

    Placentation?

    Ovary position?

57. Ovaries in the Cyperaceae mature as _____.

    How does fruit shape vary with carpel number?

58. What features would be most reliable to distinguish a member of the Cyperaceae from a member of the Juncaceae?

59. What is the usual sexual condition of members of the genus *Carex?*

    What feature is unique to the inflorescences of *Carex* and its closest relatives?

## Poaceae

60. Stems in the Poaceae are typically _____ in cross section?

    They are most commonly _____ in the middle?

    The leaves are _____-ranked.

61. What kind of leaf-sheaths are typical in the Poaceae?

    What structures are borne at the junction of leaf sheath and blade?

    How do the bases of the leaf blades of most bamboos differ from those in other grasses?

62. What is the primary inflorescence in the Poaceae called?

    Each primary inflorescence generally has a pair of _____ at the base.

    Each flower generally is subtended by a _____ and a _____.

    Which of these is most readily visible?

63. What is a floret?

    List its parts?

    What is the difference between a floret and a spikelet?

64. What are the perianth parts of a grass flower?

How are these important in pollination?

65. What kind of gynoecium occurs in the Poaceae?

How many carpels do we find per flower?

Ovary position?

What is unique about the ovule attachment?

66. The fruits of most grasses are _____.

How do these differ from achenes?

What are some examples of this fruit type that are important food crops?

67. What is the traditional family name for the Poaceae?

68. How would you determine if a plant is a grass or a sedge?

Based on vegetative features?

Based on reproductive features?

69. Grasses, sedges, and rushes are generally _____-pollinated.

70. List the families and species of monocots that you identified in the lab.

_____ _____

_____ _____

_____ _____

_____ _____

_____ _____

_____ _____

## Commelinaceae

71. Describe the leaves in the Commelinaceae.

What kind of bases do these leaves have?

What do the blades look like?

72. What subtends the inflorescence in the Commelinaceae?

73. Flowers in the Commelinaceae are __-merous.

Describe the sepals.

Describe the petals.

What is the floral symmetry?

In what ways does the perianth of Commelinaceae differ from the perianths of most other members of the Commelinidae?

74. What is the usual number of stamens in the Commelinaceae?

    In what ways are they sometimes modified?

75. What kind of gynoecium do members of the Commelinaceae have?

    Number of carpels?

    Placentation?

    Ovary position?

    Fruit type?

# Experimental Plant Systematics

chapter 17

Experimental systematics is the application of experimental techniques and specialized methods of gathering and interpreting data to the investigation of taxonomic problems. Taxonomists are often confronted with questions about the relationships among the plants they are studying. Although much can be learned from morphology, the need for additional forms of data has long been apparent. There is no clear-cut dividing line between traditional taxonomy and experimental systematics and even early taxonomists generally used data from as many sources as possible. During the last century, however, taxonomists relied to a much greater extent than in earlier times on experimental techniques that are designed to provide data applicable to systematic studies. This has been aided by the many technological advances that allow the gathering and interpretation of data undreamed of in earlier times.

The underlying bases of taxonomy have changed greatly in the centuries since Linnaeus and his contemporaries sought to organize their knowledge of plants into logical classifications. In the early history of taxonomy the primary sources of data for comparison and classification of plants were readily visible morphological features of the vegetative and reproductive parts. Species were generally assumed to be unchanging entities that could be described and cataloged in a fashion similar to inanimate objects. Variation was recognized but often was considered to be mere deviation from the "essence" of a species. These concepts began to change in the eighteenth century and were radically altered by the publication of Charles Darwin's *On the Origin of Species*. Taxonomy changed from the classification of a static set of objects to the study of an ever-changing assemblage of plants and animals with a history of many millions of years. Botanists came to realize that species, the basic units of classification, represent a diverse array of groupings of plants that vary both in time and over their geographical ranges. One of the major goals of taxonomy became the determination of the relationships among plants, particularly at the species and populational levels.

Experimental systematics developed as botanists sought methods of investigating these relationships. The experimental studies have several purposes:

1. to detect and explain patterns of variation in nature;

2. to define the limits of species;

3. to determine evolutionary relationships; and

4. to provide a classification in accordance with available data.

Taxonomists seek and use data from many different aspects of plants including structure, cytology, chemistry, molecule biology, physiology, ecology, and reproductive behavior. The gathering and interpreting of these many sources of data constitute experimental taxonomy.

In this chapter we survey the purposes and some of the methodologies employed in experimental taxonomy. The topic is very broad, and we cannot do more than provide an introduction. Detailed examples of experimental taxonomic studies can be found in the references listed at the end of this chapter and in numerous journal articles.

# Purposes of Experimental Taxonomy

The overall purpose of experimental taxonomy is to provide taxonomists with ways of answering questions about relationships among plants being studied. Listed below are some of the questions that taxonomists seek to answer:

1. What sets of plants constitute species within the taxonomic group being studied? How are these species similar? How do they differ? On what basis are the species recognized?

2. What are the evolutionary relationships among these species? What sorts of evidence can provide information about such relationships?

3. Do these species ever hybridize? If they don't, what prevents them from hybridizing? If they do, are the hybrids fertile, sterile, or somewhere in between? Has hybridization led to the formation of any new species within the group? Are there other consequences of hybridization in the group?

4. How do these plants reproduce? Do they undergo normal sexual reproduction or do they reproduce asexually? Are they self-fertile or do they require cross-pollination? If they are outcrossers, what pollinates them? Do species in the group differ in their reproductive patterns? How does their reproductive behavior relate to their evolutionary history?

5. What features vary within these plants? What is the extent of morphological variation? Are there microscopic features that vary? Are there internal, anatomical variables? Are there chemical components of the plants' cells that vary?

6. Have changes in the number or structure of the chromosomes accompanied evolution within the group? If so, how have these changes influenced the course of evolution of the plants? What kinds of changes have occurred?

7. What variation patterns can be detected within these plants? How much variation occurs in an individual plant? How much variation occurs within a population? What features vary in a geographical pattern? What features vary with ecological conditions? Are these variations genetically determined or are they environmentally induced? What do these patterns of variation reveal about the evolutionary history and relationships of these plants?

8. How can these patterns be interpreted? Which variables are useful in classifying and distinguishing taxa? At what taxonomic level should these variable features and other data be employed? How should these be used? Do the newly gathered and interpreted data support the taxonomic groups that have been recognized in the past? If not, what taxonomic and nomenclatural changes will have to be made?

# Biosystematics

A fundamental question in many taxonomic studies is, "What **species** are present among the group of plants with which I am working?" The species is the fundamental unit of classification and has been defined in several different ways. Much has been written and said about the basis for recognizing species. Two of the primary sets of criteria that have been used to recognize species are morphology (and any other measurable phenotypic characteristics) and the ability or lack of ability to interbreed. Most plant species were initially recognized on the basis of appearance; the individuals of one species looked different from those of other species. This is the **morphological species concept.** In the early part of the twentieth century many biologists began to use what is called the **biological species concept.** They proposed that individuals of species are bound together by the interchange of genes through crossbreeding. A biological species is composed of a set of interbreeding or potentially interbreeding individuals and populations and is separated from other biological species by breeding barriers.

A more recent view of species has focused on the lineage traced from parent to offspring. This **phylogenetic species concept** seeks to recognize species as monophyletic groups made up of an ancestor and all of its descendents. The Taxonomists would look for unique characteristics in this group that would make it recognizable from other groups of plants. These shared unique derived characteristics are called **synapomorphies.**

Many of the studies carried out today by taxonomists are designed to investigate the nature of species in particular groups of plants. **Biosystematics** is the area of experimental systematics that developed along with the biological species concept to investigate the nature of species and to use experimental methods to discern relationships. There are several aspects of biosystematic studies. One is to determine the nature of variation within the particular groups of plants being studied. To understand the variation, experiments are carried out to determine the basis and the reliability of characters in the group of plants being studied.

# Differentiating Environmental Influences from Inheritance

It is particularly important for a taxonomist to ascertain which characteristics of a plant are genetically determined and which ones are environmentally controlled (see discussion in Chapter 3). One approach that has been used is transplant studies. In one type of transplant study, plants that are evidently related but that grow in different localities and that differ in some of their features are grown together in a garden under uniform conditions. If the differences among the plants persist despite the uniform environment, the differences are evidently under genetic rather than environmental control. If, however, the differences disappear when the environmental conditions are uniform, the character variation is very likely induced by habitat differences. Another approach is to use reciprocal transplant gardens. Related plants from differing environments can be transplanted to the localities where their relatives grow. The appearance of the plants in these new environments can be compared with the appearance of their relatives that were already growing there. This too can help to distinguish inherited from environmentally induced variation. Experiments such as these may require a considerable investment of time and resources.

For those characters that are inherited, a taxonomist needs to understand at what level the observed variation occurs. Does it occur among different individuals within populations? Is a character fixed within individual populations but variable from one population to the next? Do populations in one geographical region have one character state whereas those in another have a different character state? Do the observed differences represent differences among species? Of particular interest are characters that vary on a geographical basis. The formation of geographical races is considered by many biologists to be the prelude to speciation (the formation of new species).

# Studies of Gene Exchange, Speciation, and Hybridization

The formation of barriers to exchange of genes between races is generally considered to be a necessary prelude to the differentiation of species. The barrier may be merely geographical. Distance or topographical obstacles may prevent plants of one population or race from breeding with those of another. However, various other kinds of barriers to gene exchanges can develop that are internal within the plants or in any hybrids that do form. The coexistence of related species in a particular locality can provide evidence about any barriers to hybridization. For example, if two related groups of plants are growing together and do not hybridize at all, they are effectively isolated from each other and can evolve independently. Such isolation is an indication that they constitute two different biological species.

Studies designed to investigate the nature of **isolating mechanisms** (barriers to gene exchange) involve many aspects of the biology of the plants. Prezygotic isolating mechanism prevent hybridization from happening. Chemical incompatibility between the pollen of one plant and the stigma of another may keep pollen tubes from forming. The style of one plant maybe too long for the pollen tube of another species to complete its growth. Internal incompatibility between the egg of one plant and the sperm of another may preclude fertilization. There may be differences in flower structure such that one plant is pollinated by a different insect (or other pollinating agent) than the other. Flowering of the plants may occur at different seasons. These barriers may be one hundred percent effective or they may be effective only part of the time. Occasional hybrids may form if a barrier breaks down.

Experimental hybridization under controlled greenhouse or garden conditions or involving plants in the wild may be used to test the effectiveness of barriers. A biosystematist may, for instance, hand-pollinate plants of differing floral structure to determine if factors other than geography or different pollinators are preventing hybrids from forming. Carefully designed hybridization experiments can be used to determine the extent to which hybrids can pass on their genes. However, plants that can hybridize under artificial conditions may not be able to do so in the wild.

The mere existence of hybrids is not an indication of ineffective barriers to gene exchange. As a result of postzygotic isolating mechanisms hybridization may occur between species without effective gene exchange. Hybrids that form in nature or under experimental conditions are often incapable of passing on their genes. Many plant hybrids are like mules—vigorous and healthy individuals but incapable of producing offspring. In other cases hybrids are weak and stunted and produce few or no flowers. First generation hybrids may be healthy and fertile but reproductive problems or stunting may develop among their offspring. A biosystematist can investigate these situations to determine what kinds of barriers to hybridization exist in natural populations and under experimental conditions and how effectively these barriers prevent gene exchange.

# Problems with the Biological Species Concept

A systematist sometimes must evaluate conflicting data in reaching conclusions about the nature of species. It would be convenient if all morphological species turned out to be biological species and vice versa. In many cases they are. Zoologists, to a large extent have adopted the biological species concept. However, botanists have long recognized that there are problems. Often species recognized on the basis of morphology have been found to interbreed with members of other morphological species. Some groups of plants are notoriously promiscuous (e.g., oaks). There is frequent hybridization, sometimes involving plants that are morphologically very different (semispecies). In other cases seemingly homogeneous morphological species contain two or more groups of individuals that are morphologically indistinguishable but incapable of interbreeding (sibling species).

The ability to interbreed is not necessarily a yes or no proposition. There is a continuum from plants that are fully infertile to those that are completely intersterile. Barriers to hybridization between two species may be stronger in some populations than in others. Although hybrids between two species may produce fewer or less healthy offspring than their parents, some interbreeding and gene flow may still occur. Although hybrids may form readily, the hybrids may not be fertile. Plants that are able to interbreed under greenhouse conditions may never do so in the wild because they grow in different regions or different habitats. Just because plants are capable of interbreeding does not mean that the plants actually do interbreed in nature. Still another problem involves plants that bypass sexual reproduction altogether. Such **apomictic** plants do not fit into the biological species concept because their reproduction does not involve breeding at all.

# Cytogenetic Studies

Chromosomal characters have had a key role in biosystematic investigations since the techniques of study were perfected in the 1920's and 30's. Cytogenetic studies (investigations of chromosomal structure, number, behavior, and evolution) can provide much information of use to a taxonomist. Three aspects of cytological studies are commonly used: chromosome number, karyotype analysis, and chromosome behavior in meiosis.

Chromosome numbers are usually obtained either from meristems (most commonly root-tips) or from developing microsporocytes (the cells that undergo meiosis in anthers). Cells that are undergoing mitosis or meiosis are chemically treated to stop the nuclear divisions. The chromosomes are strained and microscope slides are prepared. For most plants it is relatively easy to determine the chromosome number. Surveys of chromosome numbers within a genus or higher level taxon often reveal instances where numerical chromosomal changes have taken place. Chromosome number changes that are detected in such surveys can be very informative about the pathways along which evolution has proceeded within a particular group of plants.

Changes in chromosome number are of two different kinds. **Aneuploid changes** occur when there has been a stepwise change in chromosome number that has accompanied the differentiation of races or species. For example, most of the species of a genus may have twelve pairs of chromosomes per cell (a set of twelve inherited from each parent). Such a plant is a diploid (symbolized as $2n = 24$). The **base number** for such a genus is symbolized as $x = 12$. The presence of related species with eleven or thirteen pairs of chromosomes is an indication of chromosomal structural changes that have reduced or increased the number of chromosomes from the base number and modified the distribution of genes on the chromosomes. **Euploid changes** in chromosome number involve whole sets of chromosomes and result in **polyploidy.** If a plant has four sets of chromosomes it is a tetraploid; if it has six sets it is a hexaploid, if eight sets an octoploid, etc.

Chromosome counts have been reported for many species of plants. Indices are available in which chromosome counts reported by botanists around the world have been compiled. Supplementary indices are issued every couple of years to add newly reported counts. A partial list of available chromosome number indices is included below:

Cave, M. S. (ed.). 1958–65. Index to Plant *Chromosome Numbers for 1956–1964;* vol. 1 (nos. 1–4 + supplement)—1956–59 + Previous to 1956; vol. 2 (nos. 5–9)—1960–64. University of North Carolina Press, Chapel Hill. Each year's reports issued separately, unbound, mimeographed.

Darlington, C. D., and A. P. Wylie. 1955. *Chromosome Atlas of Flowering Plants.* George Allen & Unwin, London.

Federov, A. A., et al. (eds.). 1969. *Chromosome Numbers of Flowering Plants.* Academy of Sciences of the U.S.S.R., Komarov Botanical Institute, Leningrad.

Goldblatt, P. (ed.). 1981. Index to Plant Chromosome Numbers 1975–1978. *Missouri Botanical Garden Monographs in Systematic Botany* vol. 5.

———. (ed.). 1984. Index to Plant Chromosome Numbers 1979–1981. *Missouri Botanical Garden Monographs in Systematic Botany* vol. 8.

———. (ed.). 1985. Index to Plant Chromosome Numbers 1982–1983. *Missouri Botanical Garden Monographs in Systematic Botany* vol. 13.

———. (ed.). 1988. Index to Plant Chromosome Numbers 1984–1985. *Missouri Botanical Garden Monographs in Systematic Botany* vol. 23.

——— and D. E. Johnson. (eds.). 1990. Index to Plant Chromosome Numbers 1986–1987. *Missouri Botanical Garden Monographs in Systematic Botany* vol. 30.

———. (eds.). 1991. Index to Plant Chromosome Numbers 1988–1989. *Missouri Botanical Garden Monographs in Systematic Botany* vol. 40.

Moore, R. J. (ed.). 1973. Index to Plant Chromosome Numbers 1967–1971. *Regnum Vegetabile* vol. 90. International Bureau for Plant Taxonomy and Nomenclature, Utrecht, Netherlands.

Moore, R. J. (ed.). 1974. Index to Plant Chromosome Numbers for 1972. *Regnum Vegetabile* vol. 91. International Bureau for Plant Taxonomy and Nomenclature, Utrecht, Netherlands.

Ornduff, R. (ed.). 1967. Index to Plant Chromosome Numbers for 1965. *Regnum Vegetabile* vol. 50. International Bureau for Plant Taxonomy and Nomenclature, Utrecht, Netherlands.

Ornduff, R. (ed.). 1968. Index to Plant Chromosome Numbers for 1966. *Regnum Vegetabile* vol. 55. International Bureau for Plant Taxonomy and Nomenclature, Utrecht, Netherlands.

Karyotype studies involve examining each and every chromosome in the nucleus of a cell and comparing their morphologies. Chromosomes may differ in the location of the centromere, in size, and in the amount and location of staining and nonstaining portions. Mitotic cells are used for karyotype studies. The comparison of the karyotype of one species with that of another can be very helpful in determining whether structural chromosomal changes have accompanied evolution. The amount of similarity or dissimilarity in karyotype is a strong indicator of relationship and may help to explain the chromosome behavior observed during meiosis.

Examination of meiotic chromosomes can reveal information about both chromosome number and chromosome behavior. In a meiotically normal species, chromosomes synapse and subsequently separate as the cells go through the various steps that ultimately lead to pollen production. Chromosome number can be determined either before paired chromosomes have separated (late prophase I or metaphase I) or after separation (anaphase I, metaphase II or anaphase II).

The analysis of chromosome behavior is usually done along with chromosome counting. The behavior of the chromosomes during meiosis is studied and any differences from normal are noted. Meiosis in experimentally generated or natural hybrids can reveal the presence of changes such as inversions and translocations that result in pollen with genetic imbalances and sterility. One way in which one species become isolated from another is for the segments of chromosomes to be transferred to nonhomologous chromosomes (translocations). In other cases a chromosomal segment is repositioned within a chromosome, sometimes in an inverted position (inversions). When these changes have accompanied the separation of two species, exact pairing of chromosomes within the cells of each species is normal. However, in interspecific hybrids pairing may be hampered or prevented, and when it does occur abnormal meiosis may occur. In the microsporocytes of a hybrid one often sees unpaired chromosomes (univalents), clusters of chromosomes (multivalents) that form unusual configurations such as chains, X-shapes or rings, or chromosomal bridges between the poles in anaphase. Meiotic irregularities can explain the sterility of hybrids observed in pollen and garden studies. The degree to which chromosomes synapse and the presence of meiotic irregularities can reveal much about the chromosomal differentiation of races or species.

# Hybrid Speciation

Hybrids present a difficult problem for plant systematists. McDade (1990, 1992) has made crosses between species of *Aphelandra* in the greenhouse and was able to show how characteristics of the hybrids were generally intermediate between the two parents. A cladistic analysis is focused on speciation as a splitting event and is not designed to recognize hybrids. However, McDade's (1995) work has shown that hybrids tend to "jump" from parent to parent lineages in a cladistic analysis or appear sister to one of the two parents.

Polyploids often rise through hybridization. As described above, hybrids with structural differences in the chromosomes they have received from the parent plants are likely to have irregular meiosis. Synapsis may be incomplete or it may fail entirely. Such a hybrid is generally sterile. Occasionally, however, the entire chromosome complement in a hybrid individual or some or its cells may be doubled [through various means that are not discussed here]. Following the doubling, each parental chromosome then has an exactly matching chromosome with which it can synapse. In one generation a ster-

ile hybrid can give rise to a fully fertile offspring with twice as many chromosomes as its parents. Such a plant can become a new species in a single generation. A new species formed in this fashion is an amphidiploid (allopolyploid). Because of the difference in chromosome number between the polyploid and the parent species, the hybrid is generally well isolated from its parents. Polyploid speciation is very common among plants. In some biosystematic studies natural polyploids have been duplicated in experimental crosses between the plants suspected of being the parents.

Polyploidy involving the genes of only one plant species sometimes occurs. This is **autopolyploidy.** It is often accompanied by meiotic irregularities and reduced fertility. Undoubtedly autopolyploidy has been involved in the evolution of some kinds of plants but it is less common and less important as a speciation mechanism than is allopolyploidy.

# Biochemical Systematics

The biochemical interactions that take place within any living organism are exceedingly complex. They are under the combined control of the organism's DNA and its environment. The entire set of chemical reactions of even a single cell has never been wholly documented. However, enough is known of internal chemistry of plants for systematists to realize that physiological processes and chemical compounds can provide a wealth of characters for comparative studies. Recently many taxonomists have been looking to the chemistry of plants for characters that might help to resolve taxonomic problems. One advantage of chemical characters is that they are closer to gene action and presumably less influenced by environmental factors than are morphological features. A major disadvantage of chemical features is the difficulty in learning the techniques and the expensive laboratory equipment required. Chemical tests are often very time consuming, and as a result it is sometimes difficult to get large enough sample sizes to be statistically significant. Biochemical characters can be grouped for systematic studies as **physiological processes, micromolecular studies,** and **macromolecular studies.** Each has its own role and methods of study.

Physiological processes that vary among plants in patterns useful to a systematist include photosynthesis, various hormonal interactions, and numerous biosynthetic pathways. Several different photosynthetic pathways have evolved within various groups different kinds of plants living in similar environments. Most species use a pathway called C-3 photosynthesis. Alternate pathways called C-4 photosynthesis and CAM (Crassulacean Acid Metabolism) have evolved independently in several plant families. Biosynthetic pathways are the enzyme-mediated steps by which cells transform the structure of various biochemicals. Variations in these pathways determine what kinds of chemicals that can be synthesized by the cells of a particular kind of plant.

Organic compounds range from small, comparatively simple molecules to enormously large and complicated chemical structures. Micromolecules include numerous types of compounds such as simple sugars, amino acids, lipids, various pigments, scented compounds, and toxins. Many of these compounds can yield data of taxonomic significance. Macromolecules are the giants of the molecular world. These include nucleic acids (DNA and RNA), proteins, and polysaccharides (complex sugars).

# Micromolecular Studies

Several classes of chemicals, often grouped together as secondary plant products, are the subjects of micromolecular studies. These include flavonoids (a group of pigments found in almost all higher plants), terpenoids (including scented components of essential oils), alkaloids, and non-protein amino acids. In specific groups of plants, other types of chemicals may be important (e.g., mustard oils in the Brassicaceae, betalain pigments in the Caryophyllidae). Many of these chemicals have a role in defending plants from herbivores. Each group of compounds requires its own set of techniques for extraction, separation, purification, and identification. Plants often have several different representatives of a particular class of compounds that must be separated before they can be identified. To separate chemically related compounds, a taxonomist carrying out micromolecular studies often uses one or another type of **chromatography:** paper chromatography, thin-layer chromatography, high pressure liquid chromatography [HPLC], or gas chromatography. Identification of the compounds usually requires specialized training and equipment (e.g., a nuclear resonance magnetic spectrophotometer).

# Macromolecular Studies

The use of macromolecules in systematic studies is a result of several major scientific developments. The first are the many theoretical and practical advancements in molecular biology that have made biotechnology and recombinant DNA household words. The second is the development of cladistic theory (see chapter 4) that has removed much of the subjectivity of systematics and given phylogeny and classification a firm scientific basis. The third is the computer revolution. Innovative computer software and modern high-speed computers together enable scientists to analyze the large amounts of data generated in macromolecular studies. An in-depth discussion of the techniques for macromolecular studies is beyond the scope of this text. We recommend the excellent reference by Crawford (1990) and Hillis, Moretz and Mable (1996) for more detailed surveys of these techniques. The field of molecular systematics is advancing very rapidly, and new approaches in data acquisition and analysis are constantly being developed.

Studies of macromolecules vary in the amount of information that can be gathered and in the ease of acquisition of these data. Macromolecules are composed of numerous molecular subunits that may be arranged in many different combinations. DNA and RNA are built up of long chains of nucleotides that are attached to a backbone of sugars (deoxyribose or ribose) and phosphate groups. The linear arrangement of the nucleotides contains the genetic code. Evolutionary change is accompanied by mutations that affect the sequence of the nucleotides. The molecular composition of proteins, which consist of chains of amino acids, are determined by segments of a cell's DNA, and changes in the DNA *often* cause corresponding changes in protein structure. [Some DNA mutations are silent; different base pair combinations code for the same amino acids.] Although there have been major efforts toward the use of amino acid sequences of proteins in determining relationships, the principal thrust of most recent studies has been directed toward DNA.

DNA is located in several places within a plant cell. The largest amount, nuclear DNA comprises the chemical backbones of the chromosomes. It is composed of hundreds of millions of base pairs comprising thousands of genes plus a variable amount of non-coding DNA, transposable elements, and other units with distinctive nucleotide sequences. The nuclear DNA contains most of the protein-coding genes in a plant cell. Two of the smaller organelles, the mitochondria and chloroplasts, have their own DNA, which is independent of that in the nucleus. The DNA from each of these parts of the cell is a potential source of taxonomic data.

The nuclear DNA (nDNA) has the largest number of genes and the greatest amount of information, but it has several disadvantages as a taxonomic tool. Nuclear DNA is enormously large and complex. There is a great deal of variation in the amount of nDNA from one plant group to another and the distribution of genes on the chromosomes is highly variable as well. Much of the nDNA is present in only two copies per cell (on homologous chromosomes). Other parts may be duplicated two to many times, sometimes with variation from copy to copy. The nuclear ribosomal DNA (rDNA), for instance, which codes for ribosomal RNA, exists in hundreds to thousands of copies per cell. Additionally, once per generation the nDNA of one plant is rearranged and split apart in meiosis (differently in each cell) and recombined with that from another plant. As a result of recombination, mutations that arise in separate individuals may end up in a subsequent generation in the same gene on the same chromosome.

Mitochondrial DNA (mtDNA) and chloroplast DNA (cpDNA) are much smaller molecules. The genes are generally not duplicated, or only to a limited extent, on the organelles' DNA, but because there are several to many of each of these organelles per cell, each DNA molecule exists in multiple copies. Non-nuclear DNA does not undergo repeated rearrangements through meiosis and sexual reproduction. Mitochondria and chloroplasts are generally inherited from only one parent. In most cases the only mitochondria and chloroplasts present in a zygote come from the egg cell alone; the sperm cell contributes only its nucleus. Mutations that arise in separate individuals will not be recombined onto the same gene. As a result, the evolution of mtDNA and cpDNA follows a branching pattern whereas nDNA has reticulate evolution.

Chloroplast DNA has become the molecule of choice for many botanical studies. It is a comparatively small (as DNA's go), circular molecule. It is structurally conservative but has enough mutations to provide phylogenetically informative characters. Except for a pair of inverted repeated segments that are present in most plants, the genes exist in single copy form. See Olmstead and Palmer (1994) for an excellent review of the use of cpDNA in systematic studies.

Mitochondrial DNA, a similar molecule, is widely used in animal studies, but its genes evolve very slowly in many plant groups and provide few taxonomically useful characters. In addition the structural arrangement of the mtDNA in plants is labile and may be affected by the addition of fragments of DNA from the chloroplasts and nucleus. The mtDNA in plants has the added drawback of having various repeated segments that make analysis much more difficult.

## Techniques for Studying Macromolecules

Some techniques are designed to summarize the amount of change that has taken place in particular macromolecules. These studies do not determine the precise changes that have taken place, but instead determine the effects of additive changes. Related species of plants have homologous segments of DNA and homologous proteins. These are molecules that share a common ancestor in the plants' ancestors. Comparison of these molecules can suggest how much the species have diverged.

### Protein Electrophoresis

**Electrophoresis** is a method of comparing compounds that have differing electrical charges. Protein molecules vary in electrical charge depending on their amino acid composition. Protein extracts can be placed into an aqueous gel (most commonly starch or polyacrylamide gels) and are exposed to an electric current. The protein molecules will migrate through the gel in response to the electric current in proportion to their charge. Alternatively the proteins are placed into an ionic detergent (sodium dodecyl sulfate-SDS) which removes the charge differential and are exposed to the electric current. Proteins of different molecular weight migrate at different rates through the gel. Electrophoresis is an inexpensive and relatively easy way to separate protein samples. Proteins that have been separated electrophoretically can be stained to make the bands in the gel visible.

Electrophoresis will separate proteins that differ in charge or molecular weight, but it cannot prove that two samples that migrate the same way are homologous. Two different proteins may migrate at the same rate but have chemical differences. Use of more than one electrophoretic technique may reveal some of these hidden differences.

**Isozymes** (homologous proteins coded by different loci or proteins suspected of being homologous) and **allozymes** (allelic variants of a particular protein formed at the same locus) can both be analyzed in this way. Isozyme/allozyme analysis is often used by systematists to study the degree of genetic variation within and between populations or species. These analyses can also be used to support or refute hypotheses of hybrid origin of a species. For example, if two species suspected of being the ancestors of an allotetraploid hybrid differ in some of their enzymes, the presence of both sets of enzymes in the polyploid would support (not prove) the hypothesis of origin of the species (provided that other possible ancestral taxa do not also produce these chemicals).

### DNA Hybridization

DNA molecules have sometimes been studied by using molecular hybridization techniques. Double-stranded DNA is heated gently, forcing the strands to dissociate. If the dissociated DNA of one species is mixed with the dissociated DNA of another, and then cooled, some hybrid molecules are formed with homologous segments condensing to form double-stranded molecules. If the two species are very closely related and there have been few molecular changes in the DNA structure, the strands from one plant will show a high affinity for the strands of the other. If the DNA of the two species is quite different, though, the hybrid molecules tend to dissociate readily. The amount of heating needed to dissociate hybrid DNA molecules is a measure of their similarity, and presumably of the closeness of relationship of the plants. The higher the temperature that is required to dissociate the hybrid DNA, the more closely related the plants are presumed to be.

### Restriction Enzyme Studies

Restriction endonucleases are enzymes that selectively digest (cut apart) DNA molecules at specific sites. The sites are determined by the presence of a particular nucleotide sequence in the DNA molecule. Numerous restriction endonucleases are known and commercially available, each specific for a different nucleotide sequence. Restriction sites are the locations on a DNA molecule where the nucleotide sequence specific for a particular endonuclease occurs. Restriction fragments are the segments of DNA produced by digestion of a DNA molecule by a restriction endonuclease.

### Restriction Fragment Length Polymorphism (RFLP)

In a study of restriction sites, total DNA or cpDNA is extracted from a plant sample and digested singly by one or more endonucleases, resulting in numerous restriction fragments. These are then separated by electrophoresis (see below) and transferred to a nylon membrane. Genetic probes consisting of cloned fragments of $^{32}$P-labeled DNA are added one by one and allowed to recombine with fragments that have matching base pair sequences. Autoradiographs are exposed and developed, revealing a series of bands where labeled probes have hybridized with restriction fragments. Variation in the bands can be used as characters in phylogenetic analysis. Linkage maps can be constructed that show whether different plants have the same sequence of genes on their chromosomes or cpDNA molecules. In some cases this reveals the presence of inversions, deletions, or other structural changes.

### Restriction Site Analysis

The number of and location of restriction sites on a DNA molecule can be determined for a series of taxa or populations. Typically a battery of different restriction enzymes is employed to maximize the number of restriction sites located on the molecule. The restriction enzymes are employed singly and sequentially, and the resulting restriction fragments are separated electrophoretically. The DNA is stained with ethidium bromide, a chemical that in combination with DNA fluoresces under ultraviolet light. Photographs of the resulting band pattern can be compared for the presence or absence of particular restriction sites.

The location of the various restriction sites on the DNA molecule can be mapped relative to each other. Inversions, deletions, and insertions can be detected through this process. As an example, Jansen and Palmer (1987) determined through restriction site analysis that the cpDNA molecule in all members of the Asteraceae except the small subfamily Barnedesioideae have a large inverted segment. The Barnedesioideae and all other angiosperms lack this inversion. This inversion apparently occurred early in the evolution of the Asteraceae. The Barnedesioideae are the surviving members of the original Asteraceae stock.

The presence or absence of particular restriction sites can also be used as characters in phylogenetic analysis. These data can be cladistically analyzed to determine relationships among the taxa. As an example, Jansen and Palmer used these methods to analyze relationships among the tribes of the Asteraceae (Jansen and Palmer, 1988; Jansen, Michaels, and Palmer 1991).

### Gene Sequencing

The most detailed and complicated studies of macromolecules are those in which the exact chemical structures of protein molecules or nucleic acids are determined. The information derived from such studies can be of much use to a taxonomist. Until recently the expense and time involved in such analyses puts them beyond the realm of most taxonomic studies. In the late 1980's breakthroughs in biotechnology including development of the Polymerase Chain Reaction (PCR) and equipment for automated gene sequencing brought DNA analysis into systematist's laboratories and made available a set of new tools for the systematic toolbox.

Typically an individual gene or designated section of the DNA is chosen for the analysis. One of the most commonly used of these is *rbc*L (a slowly evolving chloroplast DNA gene of about 1500 base pairs that codes for the large subunit of the enzyme ribulose biphosphate carboxylase/oxygenase). Because this enzyme is essential in photosynthesis, it is found throughout the plant kingdom. Mutations occur in the *rbc*L gene, but the molecule is conservative in its structure. The methods for extraction, purification, amplification, and sequencing of the *rbc*L gene are beyond the scope of this text but are discussed in various references cited below (e.g., Chase et al., 1993; Doyle, 1993; Olmstead et al., 1992; Olmstead and Palmer, 1994; Soltis et al., 1993). The individual base substitutions are character state changes that can be analyzed cladistically.

Another set of molecules that have considerable utility in providing nucleotide sequence data for phylogenetic analysis are the nuclear ribosomal DNAs and the related ribosomal RNAs (Hamby and Zimmer, 1992). Like the rbcL gene, rDNA is widespread and has sufficient variation for phylogenetic analysis. Parts of the rDNA molecules are conservative and evolve slowly, whereas another part (a non-transcribed spacer region) changes much more rapidly. The slowly evolving regions can be used to investigate taxa at higher levels of the classification hierarchy, and the spacer region more closely related taxa at lower levels.

Nuclear genes with low copy number are being used more commonly in systematic studies. These studies essentially take isozyme analysis to the DNA level. Sometimes these single copy genes are duplicated into a small set of genes called a gene family. The phosphoglucoisomerase (PGI) gene (Gottlieb and Ford 1996), alcohol dehydrogenase (ADH) gene and the granule-bound starch synthase I (GBSSI) gene have all been used in systematic studies (Peralta and Spooner 2001, Mason-Gamer, Weil and Kellogg 1998).

The ability to obtain molecular data has to some extent surpassed the abilities of systematists to analyze those data. There are several different methods of attempting to estimate the one correct phylogeny of a group of plants. Several different parsimony methods, molecular distance analysis, and maximum likelihood methods all have been used. Controversies surround the application of each method. Cladistic analysis of macromolecular data requires the use of high speed computers and well-designed software. The large data sets that result from macromolecular studies generate enormous number of possible cladograms, only one of which can represent the true phylogeny of the group of plants being investigated. As an example, parsimony analysis of a large data set may require months of computer time to arrive at a solution that may not represent the shortest possible tree.

# Noncoding Nuclear Sequences

High copy non-coding sequences of nuclear DNA, such as microsatellites or variable number tandem repeats (VNTR) appear to evolve very quickly and are useful in determining relationships of populations. These regions can be examined using several Polymerase Chain Reaction (PCR) techniques such as Randomly Amplified Polymorphic DNA (RAPDs) or Inter-Simple Sequence Repeats (ISSR) to obtain a DNA "fingerprint" of an individual that can be used for comparison within and between populations. This techniques has become valuable in tracing within species lineages and examining past plant migrations, as well as in the identification of cultivars for use in the horticultural trade (Fernandez, Figueiras and Benito, 2002; Wolfe 1998).

# Plant Microcharacters

The oldest and still the most important sources of information used to study plant variation are the gross morphological features of the plants themselves. These are the **morphological characters.** Characters of this type can be found on any part of the plant. Morphological characters have been artificially divided into two types. **Macro characters** are those visible to the naked eye aided only with a 10× magnifier or dissecting scope, whereas **microcharacters** are either below that range of magnification or require special equipment and training to distinguish them. Macrocharacters include almost all those used in identification keys and are the subjects of Chapters 3 and 11. Microcharacters are playing an increasingly important role in explaining the variation found in plants. Microcharacters include fine details of leaf, stem and root anatomy, and inclusions in cells made visible through the use of specialized sectioning and staining techniques. Some of these features can be seen in the compound microscope, and others have been made visible for the first time through the use of electron microscopes. Both scanning and transmission electron microscopes provide data useful to taxonomists. The great depth of field and high resolution of the scanning electron microscope allows details of stomates, trichomes and other epidermal features, pollen grains, seed coats and many other structures to be observed and photographed. Transmission electron microscopes make visible such fine structures as the inner structure of pollen exines.

# Modern Technology and the Interpretation of New Characters

An important area of experimental systematics is the discovery and testing of previously unknown characters. As sophisticated technology has developed, many previously unknown or unmeasurable features have become potential taxonomic data. The scanning electron microscope, for instance, has enabled botanists to examine in great detail minute plant structures that were formerly difficult or impossible to see with a light microscope. Sophisticated equipment for extraction and purification of chemicals places chemical characters within the grasp of the systematist. The search for characters has become limited only by the amount of specialized training and resources the taxonomist possesses. More and more, taxonomic problems are being addressed by groups of specialists since no one person can efficiently master all the specialties required.

Not all newly discovered characters are taxonomically useful. They must be tested to determine the amount and nature of variation and the taxonomic levels at which that variation takes place. The desirability of a particular character must sometimes be weighed against the cost in time, effort and money of acquiring the data. With very expensive and time-consuming data-gathering techniques comes the problem of limited sample size. Until a character is tested its usefulness is unknown. A character may not be taxonomically important even though it was measured with expensive equipment. Taxonomists must keep in mind the purposes for which the data are being gathered and not become enamored by the data gathering techniques themselves.

With the increase in the types and amounts of data available has come a need to analyze information efficiently and accurately. Although the human mind is remarkably versatile in dealing with patterns of variation, particularly in morphological characters, there are limits in the number of features that can be evaluated simultaneously. The addition of chemical characters, microcharacters, and other products of experimental techniques often exceeds a taxonomist's ability to correlate the many sorts of information. Technological advances help in the processes of data analysis. The explosive growth of computer technology has made many forms of data analysis practical for taxonomic study. Sophisticated programs for data base management and statistical analysis have reduced the time required for sorting and comparing data.

Interpretation of the results still requires the careful judgment of a trained systematist. Ultimately a systematist must examine the data gathered in the contexts of the plants being studied and the questions being asked. Of critical importance are asking the right questions, designing the proper experiments, choosing the best data-gathering techniques and correctly analyzing the results.

# References

Buth, D. G. 1984. The application of electrophoretic data in systematic studies. *Annual Review of Ecology and Systematics* 15:501–522.

Camp, W. H., and C. L. Gilly. 1943. The structure and origin of species. *Brittonia* 4:323–385.

Chase, M. W., and 41 co-authors. 1993. Phylogenetics of seed plants: an analysis of nucleotide sequences from the plastid gene *rbc*L. *Annals of the Missouri Botanical Garden* 80:528–580.

Clausen, J., D. D. Keck, and W. M. Heisey. 1940. *Experimental Studies on the Nature of Species. I. Effect of Varied Environments on Western North American Plants.* Carnegie Inst. Washington Publ. 520.

Crawford, D. J. 1990. *Plant Molecular Systematics.* John Wiley and Sons, New York.

Cronquist, A. 1978. Once again, what is a species?, pp. 3–20 *in Biosystematics in Agriculture.* Beltsville Symposia in Agricultural Research no. 2, Allenheld, Osmun & Co., New York.

Davis, P. H., and V. H. Heywood. 1963. *Principles of Angiosperm Taxonomy.* D. Van Nostrand Co., Princeton, New Jersey.

Doyle, J. J. 1993. DNA, phylogeny, and the flowering of plant systematics. *Bioscience* 43:380–389.

Fairbrothers, D. E. 1977. Perspectives in plant serotaxonomy. *Annals of the Missouri Botanical Garden* 64:147–160.

Fernandez M. E., A. M. Figueiras, and C. Benito. 2002. The use of ISSR and RAPD markers for detecting DNA polymorphism, genotype identification and genetic diversity among barley cultivars with known origin. *Theoretical Applications in Genetics* 104:845–851.

Gibbs, R. D. 1974. *Chemotaxonomy of Flowering Plants.* 4 vols. McGill-Queen's Univ. Press, Montreal.

Goldman, N. 1990. Maximum likelihood inference of phylogenetic trees, with special reference to a poisson process model of DNA substitution and parsimony analysis. *Systematic Zoology* 39:345–361.

Gottlieb, L. D. 1977. Electrophoretic evidence and plant systematics. *Annals of the Missouri Botanical Garden* 64:161–180.

Gottlieb, L. D., and V. S. Ford. 1996. Phylogenetic relationships among the sections of Clarkia (Onagraceae) inferred from the nucleotide sequences of PgiC. *Systematic Botany* 21:45–62.

Grant, V. 1981. Plant Speciation. 2nd ed. Columbia Univ. Press, New York.

Grant, W. F., ed. 1984. *Plant Biosystematics.* Academic Press, Toronto.

Hagen, J. B. 1983. The development of experimental methods in plant taxonomy, 1920–1950. *Taxon* 32:406–416.

Hamby, R. K., and E. A. Zimmer. 1992. Ribosomal RNA as a phylogenetic tool in plant systematics, pp. 50–91 *in* P. S. Soltis, D. E. Soltis, and J. J. Doyle (eds.). *Molecular Systematics of Plants.* Chapman and Hall, New York.

Harborne, J. B., and B. L. Turner. 1984. *Plant Chemosystematics.* Academic Press, London.

Hawkes, J. G. (ed.). 1968. *Chemotaxonomy and Serotaxonomy.* Systematics Association Special Vol. No. 2. Academic Press, London.

Heywood, V. H. (ed). 1968. *Modern Methods in Plant Taxonomy.* Academic Press, London.

———— (ed.). 1973. *Taxonomy and Ecology.* Systematics Association Special Publication no. 5. Academic Press, London.

————, and D. M. Moore (eds.). 1984. *Current Concepts in Plant Taxonomy.* Systematic Association Special Volume 25. Academic Press, London.

Hillis, D. M. 1987. Molecular versus morphological approaches to systematics. *Annual Review of Ecology and Systematics* 18:23–42.

————, and C. Moritz. 1991. *Molecular Systematics.* Sinauer Associates, Sunderland, Massachusetts.

Hillis, D. M., C. Moritz, and B. K. Mable. 1996. *Molecular Systematics,* 2nd ed. Sinauer, Sunderland, MA.

Jackson, R. C. 1971. The karyotype in systematics. *Annual Review of Ecology and Systematics* 2:327–368.

————. 1976. Evolution and systematic significance of polyploidy. *Annual Review of Ecology and Systematics* 7:209–234.

Jansen, R. K., H. J. Michaels, and J. D. Palmer. 1991. Phylogeny and character evolution in the Asteraceae based on chloroplast DNA restriction site mapping. *Systematic Botany* 16:98–115.

————, and J. D. Palmer. 1987. A chloroplast DNA inversion marks an ancient evolutionary split in the sunflower family (Asteraceae). *Proceedings of the National Academy of Science* 84:5818–5822.

————. 1988. Phylogenetic implications of chloroplast DNA restriction site variation in the

Mutisieae (Asteraceae). *American Journal of Botany* 75:753–766.

Lewin, R. A. 1981. Three species concepts. *Taxon* 30:609–613.

Lewis, W. H. (ed.). 1980. *Polyploidy: Biological Relevance.* Basic Life Sciences Vol. 13. Plenum Press, New York.

Martin, P. G., and J. M. Dowd. 1991. Studies of angiosperm phylogeny using protein sequences. *Annals of the Missouri Botanical Garden* 78:296–337.

Mason-Gamer, R. J., C. F. Weil, and E. A. Kellogg. 1998. Granule-bound starch synthase: Structure, function, and phylogenetic utility. *Mol. Biol. Evol.* 15:1658–1673.

McDade, L. 1990. Hybrids and phylogenetic systematics. I. Patterns of character expression in hybrids and their implications for cladistic analysis. *Evolution* 44:1685–1700.

McDade, L. 1992. Hybrids and phylogenetic systematics. II. The impact of hybrids in cladistic analysis. *Evolution* 46:1329–1346.

McDade, L. 1995. Hybridization and phylogenetics. In *Experimental approaches to plant systematics* (Monographs in Systematic Botany from the Missouri Botanical Garden, Vol. 53.). P. C. Hoch and A. G. Stephenson (eds), 305–331. Missouri Botanical Garden, St. Louis.

Moore, D. M. 1978. The chromosomes and plant taxonomy, pp. 39–56 *in* H. E. Street (ed.), Essays in Plant Taxonomy. Academic Press, London.

Olmstead, R. G., H. J. Michaels, K. M. Scott, and J. D. Palmer. 1992. Monophyly of the Asteridae and identification of their major lineages inferred from DNA sequences of *rbc*L. *Annals of the Missouri Botanical Garden* 79:249–265.

———, and J. D. Palmer. 1994. Chloroplast DNA systematics: a review of methods and data analysis. *American Journal of Botany* 81:1205–1224.

Pankhurst, R. J. 1992. *Practical Taxonomic Computing.* Cambridge University Press, New York and Cambridge.

Peralta, I. E., and D. M. Spooner. 2001. Granule-bound starch synthase (GBSSI) gene phylogeny of wild tomatoes (*Solanum* L. section *Lycopersicon* (Mill.) Wettst. subsection *Lycopersicon*). *Am. J. Got.* 88:1888–1902.

Radford, A. E. 1986. *Fundamentals of Plant Systematics.* Harper & Row, New York.

Raven, P. H. 1974. Plant systematics: 1947–1972. *Annals of the Missouri Botanical Garden* 61:166–178.

Ross, H. H. 1974. *Biological Systematics.* Addison-Wesley Publ. Co., Reading, Mass.

Sivarajan, V. V. *Introduction to the Principles of Plant Taxonomy.* 2nd ed. edited by N. K. B. Robson. Cambridge University Press, Cambridge and New York.

Smith, P. M. 1976. *The Chemotaxonomy of Plants.* Elsevier, New York.

———. 1978. Chemical evidence in plant taxonomy, pp. 19–38 *in* H. E. Street (ed.), *Essays in Plant Taxonomy.* Academic Press, London.

Solbrig, O. T. 1971. *Principles and Methods of Plant Biosystematics.* Collier-Macmillan, Toronto.

Soltis, D. E., D. R. Morgan, A. Grable, P. S. Soltis, and R. Kuzoff. 1993. Molecular systematics of Saxifragaceae sensu stricto. *American Journal of Botany* 80:1056–1081.

Soltis, P. E., D. R. Soltis, and J. J. Doyle (eds.). 1992. *Molecular Systematics of Plants.* Chapman and Hall, New York.

Stace, C. A. 1978. Breeding systems, variation patterns and species delimitation, pp. 57–78 *in* H. E. Street (ed.), *Essays in Plant Taxonomy.* Academic Press, London.

Stebbins, G. L. 1950. *Variation and Evolution in Plants.* Columbia Univ. Press, New York.

———. 1971. *Chromosomal Evolution in Higher Plants.* Edward Arnold, London.

Stuessy, T. F. 1979. Ultrastructural data for the practicing plant systematist. *American Zoologist* 19:621–635.

———. 1990. *Plant Taxonomy. The Systematic Evaluation of Comparative Data.* Columbia University Press, New York.

Turner, B. L. 1977. Chemosystematics and its effect upon the traditionalist. *Annals of the Missouri Botanical Garden* 64:235–242.

Valentine, D. H. 1978. Ecological criteria in plant taxonomy, pp. 1–18 *in* H. E. Street (ed.), *Essays in Plant Taxonomy.* Academic Press, London.

Wet, J. M. J. de. 1971. Polyploidy and evolution in plants. *Taxon* 20:29–35.

Wolfe, A. 1998. Using ISSR markers in studies of natural populations: A workshop for the 1998 ASPT/BSA meeting in Baltimore, MD.

# Exercises

1. With the assistance of your instructor, choose a journal article in which the author(s) have used experimental data-gathering techniques.

   Give a complete literature citation for the article.

2. What are the questions that the author(s) were seeking answers for?

3. What sorts of data-gathering methods were employed to address the questions?

4. Were the questions satisfactorily addressed?

5. Were there unanswered questions?

6. What were the systematic applications of the experimental results?

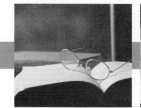

# Revisions and Monographs

## chapter 18

**M**uch taxonomic research is directed toward the study of particular taxonomic groups. These studies are designed to assemble information about the members of the group, to interpret the data, to provide an updated classification of the group and to modify or correct the nomenclature to reflect any reinterpretations that have resulted. Studies of this sort are published as **revisions** and **monographs.** They serve as a basic information source for many other studies.

A monographic study generally begins when a taxonomist encounters a group of plants with one or more taxonomic "problems." Often the beginning of a study involves unidentified or misidentified specimens that do not fit well into known taxa. A taxonomic problem may involve unexplained patterns of variation, a lack of correspondence between keys and descriptions in different floras, new data that call into question the existing interpretations of the plants, nomenclatural difficulties, or any other information that suggests to the taxonomist that the plants are in need of study. Many plants, particularly tropical taxa, have never been studied in a comprehensive fashion. Some species have not been examined in detail since they were first discovered and others have never even been described.

The size of the taxonomic group being studied ranges from a portion of a genus to a family or occasionally to an even larger taxon. Sometimes the group being studied is defined regionally (e.g., a revision of the North American species of a genus with a worldwide distribution). It may involve a section or subgenus or a group of closely related genera. Often a study that begins with a small group ultimately expands to encompass a larger assemblage of plants, particularly if additional taxonomic problems are encountered that lead the investigator outside the original limits of the group being studied. A taxonomist who intensively studies a particular taxonomic group may become the world's expert on that assemblage of plants.

There is no clear-cut boundary between a revision and a monograph. A monograph is generally considered to be a more comprehensive study involving a greater scope of data than a revision, but titles of studies may not accurately reflect the depth of the research. A regional study or a treatment of a subgenus or section is usually designated as a revision rather than a monograph.

The various aspects of a monographic study correspond to areas of taxonomy considered in several of the chapters in this book. You may wish to refer to these chapters for more detailed coverage of the subject matter. Some of the portions of a monographic study take place sequentially and others are simultaneous. Often the investigations feed back on one another with discoveries in one area of the study stimulating further research in other areas.

## Nomenclature and Literature Search

An early aspect of any monographic study involves searching published literature to find out what has already been discovered regarding the taxa being investigated. This involves tracking down published names and descriptions pertaining to the group under study and other references pertaining to the taxonomy of the plants. These are not just the names currently being used but any synonyms as well. There are several major sources for this information that a taxonomist should consult. These are listed below with brief summaries.

1. *Index Kewensis Plantarum Phanerogamarurn.* 2 volumes, Oxford, 1893–95 with 19 supplements, 1900–1991. An alphabetical index to generic and species names of angiosperms and gymnosperms published anywhere in the world from 1753 onward. It does not list infraspecific taxa. It includes very abbreviated references to the place of publication of each name. The two original volumes and the earliest supplements listed certain names as synonyms, but these should not be accepted uncritically. *Index Kewensis* is an essential starting point for anyone beginning a nomenclatural search.

   An electronic version of *Index Kewensis* on Compact Disc was published in 1993 as an interactive database for use with MS-DOS formatted microcomputers. It brings together information from the original index and all the supplements and is current through 1992. It has an easy to use format that allows more than just the alphabetic searches by taxon name that are possible with the hard cover versions of the index. To anyone who has laboriously pored over the *Index* and all the supplements looking for information, the electronic version is a welcome addition.

2. *Index Filicum.* Copenhagen, 1906, with supplements to 1933. An index to the names of ferns, comparable to Index Kewensis.

3. *Gray Herbarium Card Index.* Cambridge, Massachusetts, continuously updated. Only plants on the western hemisphere are included in the index. Genera, species and infraspecific taxa are listed along with place and date of publication. The earlier cards have been printed in book form.

   An electronic database version entitled *Harvard Gray Herbarium Index of New World Plants* became available in 1993 on the INTERNET. The Gray Index *(www.huh.harvard.edu/databases/)* is now incorporated into the International Plant Names Index.

4. *Genera siphonogamarum.* Berlin, 1900–1907. An accounting of the families and genera of plants named prior to this date and arranged according to the Engler and Prantl classification system. See Taxon 22(1):19–50, 1973, for a comparison of families and higher taxa of some of the more recent systems.

5. *Index Nominum Genericorum.* 3 volumes. The Hague, 1979. A modern index to generic names that gives important information about their places and dates of publication and about their type species.

6. *Index Londenensis to the Illustrations of Flowering Plants, Ferns, and Fern Allies.* Oxford. 1921–1931, with one two-volume supplement in 1941. An index to all the earlier illustrations of vascular plants. It is especially valuable for locating names of plants used in horticulture.

7. *A Dictionary of Flowering Plants and Ferns,* ed. 8. J. C. Willis: revised by H. K. Airy Shaw. Cambridge, England, 1973. A listing of all genera and families since 1753. Names considered to be synonyms are indicated with reference to the preferred name. Date and place of publication are not included.

8. The International Plant Names Index *(www.ipni.org)* is designed to provide an online database of the names and bibliographic details of all vascular plants. This project is jointly conducted by the Royal Botanic Gardens, Kew and the Australian National Herbarium.

9. *The Plant Book.* D. J. Mabberley, Cambridge, England, 1987. Similar to *A Dictionary of Flowering Plants and Ferns,* but family circumscriptions and classification largely follow Cronquist (1981). Most older synonyms are omitted.

10. *Vascular Plant Families and Genera,* R. K. Brummitt, Kew, England, 1992. Similar to *A Dictionary of Flowering Plants and Ferns,* but family and generic circumscriptions and classification are based on the arrangement of families accepted at the Royal Botanic Gardens at Kew. Most older synonyms are omitted. Included is a family by family listing of accepted genera and an indication of how the each is classified in eight different systems from Bentham and Hooker to Cronquist. An outline of the eight systems is included as well.

11. *Taxonomic Literature.* F. A. Stafleu. Utrecht, The Netherlands, 1967. A selective guide to botanical publications arranged alphabetically by author. Each entry has information about the

botanist, his or her nationality, a list of major publications authored by the botanist and the institution where the botanist's herbarium and type specimens are now located. (Not as complete as the following reference.)

12. *Taxonomic Literature,* 2nd ed. F. A. Stafleu. 7 volumes, 1976–1988. This edition is much expanded with additional botanists listed and many additional citations. This is an extremely valuable reference for tracking down obscure early literature.

*Taxonomic Literature.* Supplements I (A–Ba) and II (Be–Bo). Stafleu, F. A., and E. A. Mennega, 1992–3. Koeltz Scientific Books, Konigstein, Germany. (Supplementary to Vol. 1 of Taxonomic Literature, 2nd ed., A–G.)

In addition to the problem of locating binomials and types, botanists often must locate sources of information on the biology and economic values of plants. These may be scattered throughout a library. Good library technique is essential in finding what you need to know about a given taxon. The indices and other literature available in your library and your own skill in using them can be limiting factors. Fortunately there are many sources of help and new ones are being developed. Many libraries are part of interlibrary loan arrangements, and literature not available at one institution can often be found at another. Computerized data searches can be very helpful in this aspect of plant study.

Sometimes the literature search reveals that a taxonomist is already actively studying the plants or may have recently completed a revision or a monograph. Depending upon the circumstances this may bring about a termination of the study by the botanist who was just beginning to research the project, a collaborative effort involving both workers, or a partitioning of the study. In any case the discovery that someone is already working on the group should involve communication between all concerned so as to avoid conflicts. Occasionally botanists with differing taxonomic philosophies choose to pursue independent research on the same plants, but this is usually avoided.

If there are no conflicts, the next step is to gather together copies of the pertinent references. Of particular importance are the publications in which new taxa were described or nomenclatural changes proposed. Some references may be in the form of well-known and easily obtained journals and books, but others may be obscure publications that are rare and difficult to locate. The availability of many early taxonomic references in microfiche format has made the location of some early literature much easier. Some references may be obtained by interlibrary loan, but occasionally a trip to a major botanical reference library may be necessary. This aspect of the literature search, though time-consuming, is very important in determining the nomenclatural history of the group. Important references may be written in various languages. Many early workers published their studies in Latin. References written in foreign languages cannot be omitted from the literature search, even if the researcher is not familiar with the language.

In addition to the location of nomenclatural references, the literature search also involves the accumulation of published information about the structure, biology and geographical distribution of the plants. These data are often of considerable importance. They can help an investigator to avoid duplicating research that has already been done. On the other hand, published information can suggest avenues of investigation that may be of value in a monographic study.

# Herbarium Studies

One of the initial steps in a monographic study is to survey specimens of the taxa to be investigated. Dried plant specimens serve as a basic data source for several kinds of information. Individually, specimens serve as a permanent record of the morphology of particular plants. Collectively, specimens document patterns of variability, flowering and fruiting times, ecological relationships and geographical range. To a taxonomist beginning a monographic study, specimens provide valuable data for planning the investigation. Variation patterns that do not correspond to accepted taxonomic interpretations of the group being investigated help the taxonomist to focus the study. Unidentified or misidentified specimens may lead to problem taxa. The initial herbarium survey may be carried out at the investigator's home institution or during a visit to a major herbarium.

A major aspect of the initial herbarium and literature surveys is to help the researcher to plan field studies. Data from specimens can help the investigator to determine when and where to travel to find live plants of the taxa being studied. The specimens provide documentation of localities where the plants have been collected previously and when the plants were in reproductive condition. Some taxa may be so common that they can easily be relocated. Others may occur in very limited geographical ranges, and without data from previously collected specimens they may be difficult or impossible to find.

Although the initial herbarium survey may involve specimens deposited in only one herbarium, sooner or later in a study though the investigator will have to consult specimens from other herbaria. Many herbaria loan specimens to other institutions for taxonomic research. Such loans are generally made to institutions rather than to individuals. The curator or director of the herbarium at the investigator's home institution requests from the director of the lending institution that specimens of particular taxa be sent for study. Most herbaria are willing to loan specimens for a limited period of time (often six months or a year). The lending institution prescribes the terms under which the specimens are provided.

Specimens received on loan should be carefully handled and protected from damage. Valuable references for anyone carrying out herbarium studies are *A Guide to The Practice of Herbarium Taxonomy.* (Leenhouts, 1968) and *The Herbarium Handbook* (Bridson and Forman, 1992). The investigator should make a record of the data from all of the specimens examined. If the information is stored in a computer database, it can be accessed in various forms. The investigator is expected to annotate the specimens before returning them. This generally involves affixing small labels bearing the plant's correct name, the investigator's name and the date. The lending institution profits from the loan by receiving a set of accurately identified specimens.

An important reference for taxonomists who plan to borrow specimens for their research is **Index Herbariorum** (Holmgren et al., 1990; www.nybg.org). There are two parts to this index. *Part I, The Herbaria of the World,* is arranged (in edition 8) by country and city. For each herbarium there is a brief description of the number and types of specimens in the collection, the name of the director and members of the professional staff and their specializations and the loan policies of the institution. If the herbarium contains historically important collections of particular early botanists, this information is usually listed. For each herbarium an acronym is listed; for instance, the acronyms of the Field Museum in Chicago, the New York Botanical Garden and the University of Arizona are F, NY and ARIZ, respectively. *Part II, Collectors,* is a six-volume index of collectors arranged in alphabetical order. The earlier volumes were published in the 1950's and are much out of date for modern collectors but extremely useful for earlier field workers. For each collector the herbaria where his or her collections are known to be deposited are listed by their acronyms. The index of collectors is very helpful in tracking down type specimens and other historically important collections.

Although some herbaria are willing to loan any specimens in their collections, some are unwilling to loan types. The investigator may have to travel to the institution where the type is deposited or request a photograph of the specimen. Microfiche photographs of the type specimens deposited at some of the major herbaria are available. Many images of types are now available at individual herbaria through the Internet. For some major herbaria a listing of all the type specimens known to be on deposit is available as electronic databases on the Internet. One cannot assume that such lists include all the types on deposit. Often a researcher will discover among the specimens received on loan from an herbarium that there are unmarked types. These are type specimens (often isotypes or syntypes, less commonly holotypes) that are filed among the ordinary specimens of the institution. Such specimens can be recognized as types only if the researcher has studied the literature in which the names were published.

In the latter part of a study, specimens serve as the basis for preparation of keys, descriptions of taxa, range maps and other portions of the final manuscript. These specimens serve as a reservoir of data that if properly organized can assist the taxonomist in presenting information to other workers. Specimens gathered as a part of a researcher's own field studies are often very important in interpreting the variations seen on borrowed specimens. The investigator may ultimately distribute duplicates of his or her own collections to herbaria from which specimens have been borrowed. These specimens can serve as documentation of the investigator's taxonomic concepts and of evidence acquired in the course of the study.

# Field Studies

Field investigations are a very important aspect of most monographic studies. Much information of use to a taxonomist can be gained only from examination of the plants growing their natural environments. With field work a botanist can determine the amount of variation that occurs within populations of plants. The researcher can observe and record interactions of the plants with other organisms such as visitation of the flowers by pollinators, the effects of herbivores and of other kinds of plants with which the members of the study group occur. Natural hybridization among members of the study group may suggest relationships or explain the existence of difficult-to-identify specimens. Observations made in the field can suggest subjects for future laboratory or greenhouse experiments.

Field studies also provide the opportunity to gather specimens for dissection or experimental investigation. Seeds or live specimens may be brought back from the field to be grown under controlled conditions to test for variability, for hybridization experiments, to test breeding systems or for other investigations. Samples may be chemically preserved for anatomical or cytological studies. Bulk specimens may be gathered for extraction and comparative analysis of various chemicals (e.g., alkaloids, pigments, oils, etc.). A researcher carrying out field work can ensure that the specimens gathered represent statistically valid samples. Often the investigator will discover variations not represented in available herbarium samples. For example, collectors may have overlooked or avoided gathering specimens too large to fit conveniently onto a standard herbarium sheet.

When an investigator goes into the field, he or she has the opportunity to examine gaps in the known distribution of the plants. Sometimes these turn out to be real gaps that represent habitats unsuitable for the plants or areas where, for reasons unknown, the plants do not occur. In other cases, though, the field studies will enable the researcher to fill in the gaps by documenting the presence of species in areas where they had not previously been collected. For species known from only one or a few specimens, the field work may enable the taxonomist to determine whether the plants are really rare or merely undercollected.

The types of data that can be gathered from field studies are often dependent on the amount of time that can be spent at a particular location. Some field expeditions involve much travel with comparatively little time spent at any one place. In such a trip the investigator may be attempting to gather specimens from as many taxa as possible over a broad geographical area. Field studies may, on the other hand, involve spending days or weeks studying the plants of a limited area. Often such investigations involve experimental studies of pollination mechanisms, hybridization, or population dynamics. The choice of studies to be carried out in the field is dependent on the kinds of plants being studied, where they grow, the amount of information already known about them, the types of information desired, the investigator's budget, and the time available for the research.

There are sometimes frustrations as well as rewards involved in field work. A collector may have only sketchy preliminary data for guidance regarding where to look for some plants. After much searching he may come up empty-handed. The plants may no longer occur where they have been collected previously. Urban expansion, agricultural development, deforestation, or other historical changes may have eliminated the habitats where the plants once occurred. The plants may actually be extinct or their range may have been much diminished. Sometimes, despite good planning the collector will arrive too early or too late for the plants or during an unusually dry season. Weather and travel conditions may leave much to be desired. The plants may grow in countries where governmental restrictions make collecting difficult or impossible. Civil wars or political instability may preclude travel to some areas. Budgetary restrictions or other time commitments may prevent a researcher from going to places he or she would like to visit or from staying as long as desired.

# Experimental Studies

Once the investigator has determined the nature of the taxonomic problems that need to be studied, he or she may undertake experimental studies. The researcher may attempt to hybridize members of different taxa, may carry out transplant studies, test breeding systems, study pollination mechanisms or perform other experiments that require living plants. Experiments on living plants may be carried out in the field or in experimental gardens, greenhouses or growth chambers. Other studies may involve dried or chemically preserved specimens. These investigations may involve chemical analyses, anatomical studies, chromosomal studies, the examination of microcharacters with an electron microscope, or various other data-gathering techniques. Data from studies of both living and preserved specimens may be subjected to statistical analysis. Relationships may be investigated by traditional inferential methods or by cladistic approaches. Increasingly monographic studies also involve molecular phylogenetic analyses of DNA.

# Preparation and Publication of the Revision or Monograph

As various aspects of the study progress, the investigator begins to accumulate data relevant to the taxonomic problems that were detected early in the investigation. Sometimes the data indicate the existence of previously unrecognized problems as well. This may lead to additional field studies or experimentation. Eventually, explanations for enough of the problems accumulate that the end of the study is in sight. Final determinations have to be made regarding the taxonomic limits within the study group, classification of the taxa recognized and any nomenclatural adjustments that must be made. An investigator may not be able to resolve all of the taxonomic problems and may occasionally be forced to make tentative decisions. The publication of the accumulated results in the form of a monograph or a revision is the final product of the study.

The format of the published monograph or revision varies somewhat depending upon the methods used by the investigator and whether portions of the study are published separately. The following features are generally included. There is an introduction that summarizes previous studies of the plants and taxonomic problems that were investigated. If experimental methods were employed in the study, the methods are described and the results presented and discussed. The implications of the various types of data gathered in the study are analyzed.

A taxonomic treatment follows. The generic name and any synonyms are listed together with references to the places and dates of publication of the names. A standard botanical description of the genus is presented and this is followed by a key to the included taxa. Each species and infraspecific taxon is presented with the correct name, publication data, information about the type specimen(s) and a list of synonyms together with their publication data and types. A standard botanical description is presented for each taxon along with a statement of the plant's geographical distribution, ecological range, and flowering period. Often there is a brief presentation of additional information about the taxon (e.g., economic uses). The publication may include a list of all the specimens examined during the study or a list of representative specimens. There may be photographs or line drawings of the taxa and maps showing the ranges of the plants. New taxa may be described in the publication, and nomenclatural changes made that were necessitated by the study. After the presentation of the taxa there may be a list of excluded and doubtful names. This is followed by a list of references cited in the publication.

# The Importance of Monographic Studies

A monographer is a specialist who provides valuable information to the remainder of the scientific community. Monographers serve the important role of gathering up the widely scattered information that is known about a group of plants and summarizing it in a well-organized format. A revision or monograph is generally the most important single source of information for someone seeking to know more about a group of plants. Stuessy (1975) pointed out the many contributions to the general advance of knowledge that monographers make.

The relationship between floras and monographic research is reciprocal. Just as a flora can serve to point out taxa in need of further study, revisionary and monographic studies provide a foundation for the treatment of many of the plants that occur in the area covered by the flora. The author of a flora cannot be a specialist in all groups of plants. Published research by various members of the taxonomic community assists the author(s) of a flora in the interpretation of many of the taxa in the study area.

# References

Benson, L. 1962. *Plant Taxonomy. Methods and Principles.* Ronald Press, New York.

Bridson, D., and L. Forman. (eds.) 1992. *The Herbarium Handbook,* revised edition. Royal Botanic Gardens, Kew.

Davis, P. H., and V. H. Heywood. 1965. *Principles of Angiosperm Taxonomy.* D. Van Nostrand Co., New York.

Holmgren, P. K., N. H. Holmgren, and L. C. Barnett. 1990. *Index Herbariorum,* Part 1, 8th Edition. International Association for Plant Taxonomy, New York. Regnum Vegetabile vol. 120.

Jones, S. B., and A. E. Luchsinger. 1979. *Plant Systematics.* McGraw-Hill Book Co., New York.

Leenhouts, P. W. 1968. A guide to the practice of herbarium taxonomy. *Regnum Vegetabile* 58:1–60.

Pankhurst, R. J. 1992. *Practical Taxonomic Computing.* Cambridge University Press, New York and Cambridge.

Radford, A. E. 1986. *Fundamentals of Plant Systematics.* Harper and Row, New York.

————, W. C. Dickison, J. R. Massey, and C. Ritchie Bell. *Vascular Plant Systematics.* Harper & Row, New York.

Sivarajan, V. V. *Introduction to the Principles of Plant Taxonomy.* 2nd ed. edited by N. K. B. Robson, Cambridge University Press, Cambridge and New York.

Stace, C. A. 1980. *Plant Taxonomy and Biosystematics.* University Park Press, Baltimore.

Stuessy, T. F. 1975. The importance of revisionary studies in plant systematics. *Sida* 6:104–113.

————. 1990. *Plant Taxonomy. The Systematic Evaluation of Comparative Data.* Columbia University Press, New York.

————. 1994. *Case Studies in Plant Taxonomy.* Columbia University Press, New York.

————, and K. S. Thomson (eds,). 1981. *Trends, Priorities and Needs in Systematic Biology,* 2nd ed. Association of Systematics Collections, Lawrence, Kansas.

Turner, B. L. 1971. Training of systematists for the seventies. *Taxon* 20:123–130.

# Exercises

1. Botanical journals in which monographs and revisions are often published include **Brittonia, Madroño, Rhodora, Sida,** and **Systematic Botany.** Examine a revision or monograph published in one of these journals or in another journal suggested by your instructor.

   a.  Give a complete literature citation for the publication you are examining. Include the title, author, journal issue and inclusive pages.

   b.  Is the paper a revision or a monograph? How did you decide?

   c.  Did the author indicate the nature of particular taxonomic problems that needed to be solved?

   d.  What experimental data-gathering methods were used by the author?

   e.  Did the author describe any new taxa in this paper?

       Were there any nomenclatural changes?

   f.  Does the paper include maps? illustrations?

   g.  Were there any aspects of the paper that you did not understand? You may wish to discuss these with your instructor.

2. With the help of your instructor, choose a small genus (one or two species) for study.

   a.  Carry out a literature search on this genus. Locate all of the published names pertaining to the genus.

   b.  Examine a group of herbarium specimens of this genus (from your school's herbarium). From the literature survey and your examination of the specimens, did you detect anything that might constitute a taxonomic problem?

   c.  On the basis of specimen data and literature, plan a field expedition to sample members of this genus over its entire geographical range. When would be the best time to take the trip? What sorts of samples and observations would you want to gather? (Your instructor may ask you to collect samples from your local area.)

   d.  Read Chapter 17 (Experimental Plant Systematics). Design experimental studies that you would want to carry out as a part of your revision. Justify your choice of experimental methods. What facilities would you need to carry out these studies? How long would these investigations take? Can you estimate how much they would cost?

   e.  On the basis of specimens available in your school's herbarium (and specimens that you may have collected), prepare the introduction and the taxonomic portion of a "revision" of your genus.

# Glossary

**Abaxial**—The side of an organ away from the axis (e.g., the undersurface of a leaf).

**Aberrant**—Differing from the form normal or usual for a taxon.

**Abortive**—Imperfectly developed; defective; barren.

**Acaulescent**—Stemless or apparently stemless; sometime the stem is subterranean or protrudes only slightly.

**Accessory bud**—Any bud in the axil of a leaf in addition to the axillary bud.

**Accessory fruit**—A fruit or group of fruits that ripen together with tissue not derived from the pericarp (e.g., fleshy hypanthium of an apple or a rose hip, fleshy receptacle of a strawberry, fleshy calyx of mulberry flowers, etc.).

**-aceae**—Standardized suffix indicating the scientific name of a family.

**Achene**—A single-seeded indehiscent dry fruit with the seed free from the pericarp except for the placenta (Fig. 10-16 A–D).

**Achlorophyllous**—Without chlorophyll.

**Acropetal**—Centripetal, with the most mature structures toward the outside (or base) and the least mature toward the center (or apex).

**Actinomorphic**—Radially symmetric, regular; with the parts in a perianth whorl all alike with 2 or more lines of symmetry (Fig. 10-6 A).

**Acuminate**—Tapering to a point with angle becoming progressively smaller (Figs. 3-8 B; 3-9 B)

**Acute**—Forming an angle less than 90 degrees, sharp-pointed (Figs. 3-8 A; 3-9 A).

**Acyclic**—Arranged in spirals, not in whorls.

**Adaxial**—The side of an organ toward the axis (e.g., the upper surface of a leaf).

**Adherent (adhesion)**—Two or more structures of different kinds clinging together or growing together during development, but not truly adnate.

**Adnate (adnation)**—The fusion of unlike parts (e.g., stamens adnate to petals); the opposite of free.

**Adventitious roots**—Roots that originate from other than the interior of other roots (e.g., roots that develop from a stem or leaf; Fig. 3-1 C, E).

**Aestivation**—The arrangement of the perianth or its parts in the bud.

**Aggregate fruit**—The ripened product of an apocarpous gynoecium (Fig. 10-18 A–D). A simple aggregate fruit develops from the ovaries of the component carpels without involvement of additional tissues. An aggregate accessory fruit ripens together with non-pericarp tissue.

**Akene**—Achene.

**-ales**—Standardized suffix indicating the scientific name of an order.

**Allopatric**—Of different lands; used to describe populations, races, species, etc., living in different places.

**Allopolyploidy**—The formation of a new species via doubling of chromosomes in a hybrid; polyploid speciation (q.v.); amphidiploidy.

**Allozymes (alloenzymes)**—Different forms of the same enzyme formed by different alleles at the same locus. Often used to study amount of variation within and between populations at or below the species level. Allozymes can be separated by electrophoresis.

**Alternate leaves**—Type of phyllotaxy characterized by one leaf per node, the commonest condition (Fig. 3-5 B).

**Alternate stamens**—Stamens positioned between the petals or in line with the sinuses of a lobed, sympetalous corolla (Fig. 10-7 D).

**Alternation of generations**—The alternation between a haploid phase (gametophyte) and diploid phase (sporophyte) in the sexual reproductive cycle of a plant.

**Ament**—A catkin.

**Amentiferae**—A polyphyletic group of angiosperm families characterized by the production of flowers in aments (catkins).

**Amphidiploidy**—The formation of a new species via doubling of chromosomes in a hybrid, resulting in plants that are functionally diploid; polyploid speciation (q.v.); allopolyploidy.

**Amphitropous**—A type of ovule that is adnate to the funiculus in an inverted position but with the micropyle not against the funiculus.

**-anae**—Suffix used by some authors to denote the scientific name of a superorder.

**Anastomosing**—Reticulate, forming a network as in a net-veined leaf (Fig. 3-6 A, B).

**Anatropous**—A type of ovule that is adnate to the funiculus in an inverted position with the micropyle against the funiculus.

**Androecium**—Collective term for all the stamens of a flower.

**Androgynous**—Used to describe spikes of *Carex* (Cyperaceae) that have staminate florets at the top and pistillate below.

**Anemophilus**—Flowers that are characteristically wind-pollinated, usually with perianth reduced.

**Aneuploidy**—A change in chromosome number amounting to a gain or loss of less than a full set. Often the result of uneven translocation of chromosome arms.

**Angiosperm**—A vascular seed-bearing plant that bears its ovules enclosed in an ovary derived from the bases of one or more carpels.

**Anisomeris**—Usually applied to determinate organs with few parts. The condition when different whorls the perianth have different numbers of parts (e.g., 4 petals and 2 sepals).

**Anisophylly**—Production of large and small leaves by the same plant (as in *Selaginella;* Fig. 8-11 A).

**Annual**—An herbaceous plant that grows from seed, reproduces and then dies in one growing season, usually in less than one calendar year.

**Annual ring**—The yearly increment of new wood (xylem) added to the stem of a woody plant by cambial activity.

**Annulus**—A thickened band or zone of cells in the sporangial wall of a true fern (Fig. 8-1 D).

**Anterior**—Front or abaxial; on the side away from the axis or toward a subtending bract.

**Anterior lip**—The abaxial (lower) lip of a bilabiate flower.

**Anther**—The pollen bearing portion of a stamen, composed of one or two pollen sacs joined to a connective (Fig. 10-7).

**Antheridium**—A spherical multicellular sperm-producing sex organ borne by gametophytes of ferns, fern allies and certain non-vascular plants.

**Anthesis**—At the time of expansion of the flowers; often used to mean the whole period of flowering.

**Anthocarp**—A fruiting structure in which the fruit proper is surrounded by tissue derived from surrounding perianth, bracts or receptacle.

**Anthocyanin**—A class of complex water-soluble pigments that are colored blue or red, usually dissolved in the cell sap within a vacuole.

**Anthoxanthin**—A class of water-soluble pigments that are colored yellow to red. They are chemically similar to the anthocyanins.

**Antrorse**—Bent forward of upward, often used to describe prickles or hairs.

**Aperturate**—Used in the description of pollen grains with one or more thin spots or gaps in the exine.

**Apetalous**—Without petals or any trace of petals (Fig. 10-3 A).

**Apex**—Tip.

**Apical**—At the tip.

**Apical placentation**—A placentation type found in both simple and compound ovaries in which one or more seeds or ovules are attached at the top of the ovary (Fig. 10-8 L).

**Apiculate**—Apex with a short, abrupt, flexible point.

**Apocarpous gynoecium**—A gynoecium composed of 2 or more wholly distinct carpels, usually considered to be a primitive condition (Fig. 10-8 E).

**Apomict**—A plant that generally produces seeds without fertilization of an egg by a sperm nucleus. Pollination may or may not be required for seed formation.

**Apomorphy (apomorphic character state)**—An evolutionarily derived character state, important for determining relationships in cladistic analyses. In a character transformation series with mare than two states, a particular character state may be apomorphic relative to some states and plesiomorphic (q.v.) relative to others.

**Apopetalous corolla**—Corolla of distinct petals; choripetalous; polypetalous (Fig. 10-3 B).

**Aposepalous calyx**—Calyx of distinct sepals.

**Apostemonous androecium**—Androecium of distinct stamens.

**Appendages**—Any attached subsidiary or secondary part. In many Asteraceae the style branches bear terminal appendages with sweeping hairs.

**Appressed**—Lying flat against a surface.

**Arborescent**—Treelike, large and woody with a trunk.

**Archegonium**—The flask-like multicellular egg-containing sex organ borne by the gametophytes of bryophytes, ferns, fern allies and most gymnosperms.

**Areole**—(1) The pad-like axillary buds of cacti from which spines are produced (Fig. 17-7 D, G). (2) The open spaces between the veins of a net-veined leaf.

**Aril**—A structure arising from the region of attachment of the hilum and funiculus, often forming an extra seed coat; it may be dry or fleshy.

**Aristate**—Bearing a stiff awn or bristle, often used to describe leaf apices.

**Articulated**—A structure with one or more joints or conspicuous breaks in its continuity; the structure often breaks (disarticulates) at these joints.

**Artificial classification system**—A system of classification based on a few characters chosen *a priori* without regard to the "naturalness" of the groups thus created; classifications designed for a single purpose.

**Ascending**—Rising at an oblique angle or curving upward.

**Asexual reproduction**—Reproduction not involving the fusion of sperm or egg to form a zygote.

**Attenuate**—Drawn out to a long slender point with nearly parallel sides; used to describe leaf apices and bases.

**Auriculate**—With rounded lobes (auricles) at base. In grasses the auricles are lobe-like projections of the sheath opposite the attachment of sheath and blade.

**Autopomorphy (autapomorphic character state)**—A derived character state unique to a single taxon in a cladistic analysis.

**Autogamy (autogamous)**—Breeding system expressing the tendency to set seed as a result of self-pollination. Varies from facultative autogamy, the ability to self-pollinate, to obligate autogamy in which self-pollination is unavoidable.

**Autopolyploidy**—The formation of a polyploid through doubling of the chromosomes in a non-hybrid plant, resulting in offspring with multiple sets of homologous chromosomes.

**Autotrophic**—Able to produce food and incorporate energy through photosynthesis.

**Awl-like leaves**—Subulate; leaves that are short and thick, tapering to a point; resembling the tip of a leather-punch.

**Awn**—A stiff bristle. Used in the Asteraceae for a pappus of stout bristles or hard needle-like structures. Used in the Poaceae for the bristly appendages often borne on glumes and lemmas.

**Axile placentation**—A placentation type found in compound ovaries in which the placental area of the ovary is attached to an axis derived from the connote margins of the component carpels. Such an ovary is divided into two or more locules by septa (Fig. 10-8 C–E).

**Axil**—Angle between a petiole (or other lateral structure) and the stem (or axis) to which it is attached.

**Axillary bud**—A bud borne in a leaf axil (Fig. 3-2 A, B); lateral bud. Branches arise from axillary buds.

**Axis**—The middle of a structure (e.g., the rachis of a compound leaf, the center of a compound ovary).

**Baccate**—Berrylike, a fruit that is fleshy or pulpy.

**Banner petal**—The uppermost (and outermost) petal of a papilionaceous corolla (Fig. 12-18 G); also known as the standard.

**Barbed**—Bearing short bristles or prickles pointing backward, often clinging readily to flesh or clothing.

**Bark**—The outer layer of a woody stem, usually with one or more suberized cork layers that prevent water loss and that protect the inner living tissues from mechanical damage; all tissues from the phloem outward. Bark can have several configurations or patterns: **smooth**—no break in the outer bark; **furrowed**—with anastomosing or rectangular ridges; **scaly**—peeling horizontally (ring-barked); **flaky**—broken into small, thin flakes; **plated**—broken into large, thick flakes.

**Basal**—At the base.

**Basal leaves**—Leaves attached directly to a short below-ground stem or rhizome; apparently from the ground (Fig. 3-5 A).

**Basal ovule**—An ovule borne at the base or bottom of an ovary chamber or locule; characteristic of basal placentation.

**Basal placentation**—A placentation type found in both simple and compound ovaries in which one or more seeds or ovules are attached at the bottom of the ovary (Fig. 10-8 K).

**Basifixed anther**—An anther joined at its base to the filament.

**Basipetal**—Centrifugal, with the most mature structures toward the center (or apex) and the least mature toward the outside (or base).

**Beak**—A long prominent, usually non seed-bearing, apical portion of an ovary as in Geraniaceae and some Brassicaceae. In some Asteraceae the pappus is borne at the end of a slender beak on the achene.

**Beard**—Bearing a tuft of long hairs as on some petals.

**Berry**—A pulpy or fleshy indehiscent fruit without a stony endocarp (Fig. 10-15 A). Berries may be derived from simple or compound ovaries and contain one to many seeds.

**Betacyanin**—A class of red to blue nitrogen-containing water-soluble pigments. They are nitrogenous counterparts of the anthocyanins. In the angiosperms known only from some families of the Caryophyllidae.

**Betalain**—A group of water-soluble nitrogenous pigments comprising the betacyanins and betaxanthins.

**Betaxanthin**—A class of yellow or red-orange water-soluble pigments similar to the betacyanins. They are nitrogenous counterparts of the anthoxanthins.

**Biaperturate**—A pollen grain with two apertures.

**Bicarpellate**—Composed of 2 carpels.

**Biconvex**—Lens-shaped; flattened but convex-curved on both surfaces.

**Biennial**—An herbaceous plant that requires two years to complete its life cycle. In the first season germination is followed by vegetative growth only with accumulation of stored food reserves. In the second season the plant flowers, sets seed and dies.

**Bifid**—Deeply divided at apex into two parts or lobes.

**Bifoliolate**—A compound leaf with two leaflets.

**Bilabiate**—Shape of sympetalous corolla or calyx with lobes oriented in two groups; two-lipped (Fig. 10-4 C).

**Bilateral symmetry**—Zygomorphic; irregular; with only one line of symmetry (Fig. 10-6 B).

**Bilobed**—Cut down the middle into two lobes.

**Binomial**—A species name consisting of a generic name and a specific epithet; popularized by Linnaeus.

**Binucleate pollen**—A pollen grain that is shed from the anther with 2 nuclei or cells: a tube nucleus and a generative nucleus.

**Biosystematics**—The experimental use of cytological, genetic, morphological, chemical and ecological information etc., to explain variation in nature.

**Bipinnately compound**—A compound leaf with leaflets attached to secondary rachises; pinnate, with the leaflets again pinnate (Fig. 3-3 C).

**Biradial**—Isobilateral (q.v.).

**Biseriate**—In two series; perianth with both calyx and corolla.

**Bisexual flower**—Perfect flower, with a functional androecium and gynoecium; bisporangiate.

**Bisporangiate**—Producing both megaspores and microspores; bisexual; perfect.

**Bitegmic**—An ovule with two integumentary coats.

**Biternately compound**—A leaf divided into three parts, each of which is again divided into three parts.

**Blade**—The flattened expanded part of a leaf (Figs. 3-2 B; 3-3 A), petal (Fig. 10-4 A), sepal, etc.

**Bract**—(1) A modified leaf associated with flowers. It differs from the foliage leaves in size, shape, color, texture, or some other features. (2) The dry, often quite small leaflike structure that subtends a cone scale in the Pinaceae and certain other conifer families.

**Bracteate, bracteolate**—Bearing bracts or bractlets.

**Bracteole**—Bractlet.

**Bractlet**—A small bract; often used for any secondary bracts.

**Branchlets**—The ultimate branches.

**Bud scales**—Scale-like modified leaves or leaf-bases that often cover axillary or terminal buds.

**Bulbel**—A small bulb attached to a larger bulb.

**Bulb**—A short vertical underground stem with fleshy storage leaves attached (Fig. 3-1 G)

**Bulbil**—A fleshy asexual reproductive structure borne on the aerial part of a plant; bulblet.

**Bundle scar**—A small scar within a leaf scar left by a vascular bundle that entered the petiole of the fallen leaf.

**Bur**—A fruiting structure covered with spines or prickles.

**Caducous**—Falling at the time a flower opens (e.g., the sepals of the Papaveraceae, the perianth cap of *Eucalyptus*).

**Callosity**—A thickened raised area differing in color and texture from the surrounding area (e.g., the thickened "grains" on the inner sepals of some *Rumex* species).

**Calyculum**—A series of small outer bracts in some Asteraceae that subtend the principal bracts of the involucre.

**Calyptra**—A cap or lid covering a structure (e.g., the perianth cap of *Eucalyptus*).

**Calyx**—The collective term for the sepals of a flower; the outermost whorl or spiral of flower parts; the perianth when only one whorl is present.

**Calyx lobes**—The separate sepal-tips of a synsepalous calyx.

**Calyx tube**—The connate portion of a synsepalous calyx between the spreading of the lobes or limb and the base, varying in shape from disc-like to cup-shaped or long-tubular; sometimes used in the sense of hypanthium.

**Cambium**—The lateral meristem responsible for secondary growth and increase in diameter in woody plants.

**Campanulate**—Used to describe a corolla (calyx, hypanthium, involucre) with the shape of a bell, the tube very short or absent and the throat greatly expanded (Fig. 10-5 C).

**Campylotropous**—A type of ovule that is attached in an inverted position but is not adnate to the funiculus.

**Canescent**—Bearing grayish or hoary pubescence.

**Capillary**—Very slender and threadlike.

**Capillary bristles**—A type of pappus in the Asteraceae composed of very slender bristles.

**Capitate flower clusters**—Flowers in tight heads or headlike clusters.

**Capitate stigma**—A headlike stigma, enlarged relative to the style.

**Capitulescence**—A cluster of heads; often used to describe the secondary inflorescence of the Asteraceae.

**Capitulum**—A short dense inflorescence of sessile flowers (Fig. 10-13 F); a head.

**Capsule**—A dry dehiscent fruit derived from a compound ovary, extremely variable in size, shape, number of seeds and nature of dehiscence (Fig. 10-17 A-F).

**Carinate**—Keeled.

**Carpel**—The basic units of the gynoecium, usually interpreted as leaflike megasporophylls with infolded or inrolled connate margins bearing one or more ovules on the inside (Fig. 10-8). Carpels are the innermost parts of a complete flower.

**Carpophore**—Remnants of the floral axis to which are attached the mericarps of fruits of Apiaceae (Fig. 13-7 G).

**Caruncle**—A swollen structure borne near the hilum of some seeds: resembles an aril.

**Caryopsis**—A single-seeded indehiscent dry fruit with the seed coat wholly adnate to the pericarp (Fig. 10-16 F, G).

**Category**—Any level in a classification hierarchy.

**Catkin**—A spike-like or racemelike inflorescence of apetalous staminate or pistillate flowers, often with scaly bracts and wind pollination. Catkins are often lax and drooping.

**Caudate**—Tailed; bearing a long slender protuberance.

**Caudex**—A persistent stem base of a perennial herb, located at or below the soil surface and producing a new set of leaves each season.

**Caulescent**—Possessing a stem visible above the ground.

**Cauline leaves**—Leaves attached to an above-ground stem, as opposed to basal leaves.

**Cell**—(1) A locule; a cavity within a hollow structure such as an anther or an ovary. (2) The fundamental unit of all living organisms, in plants usually bounded by a cell wall.

**Cellular endosperm**—Endosperm that develops as distinct cells from its origin from the fusion nucleus without going through a free-nuclear stage

**Cellulose**—A complex carbohydrate that is a major component of cell walls.

**Centrifugal sequence**—Development from the inside to the outside or from the apex to the base (literally, fleeing from the center); basipetal.

**Centripetal sequence**—Development from the outside toward the inside or from the base toward the apex (literally, seeking the center); acropetal.

**Chaff**—Dry, scale-like structures; used collectively for the receptacular bractlets of the Asteraceae (Fig. 15-2).

**Chaffy**—Dry and scale-like, not green and lifelike. In Asteraceae used to describe receptacle with receptacular bracts.

**Character**—Any attribute of an organism that can be consistently measured, counted or otherwise defined.

**Character state**—One of the expressions of a character; a character may have as few as two states (e.g., presence-absence) or it may have several to many states (e.g., petal color).

**Character transformation series**—A hypothesized sequence of changes of states in a character from the most plesiomorphic (primitive) to the most apomorphic (derived). The sequence may be linear or branched.

**Chasmogamy (Chasmogamous flowers)**—Production of flowers that open and expose stamens and styles to the environment and that are potentially cross-pollinated. The opposite of cleistogamy.

**Chemotaxonomy**—The use of comparative biochemistry in systematic studies.

**Chloroplast DNA**—A circular DNA molecule found in chloroplasts, cpDNA.

**Choripetalous**—Corolla of distinct petals; apopetalous; polypetalous (Fig. 10-3 B).

**Chromosomes**—Structures located in the nucleus of a plant or animal cell that carry the genetic information encoded in the molecular structure of DNA. Chromosomes are generally very diffuse except during meiosis and mitosis when they contract and divide. Each living cell contains 2 or more chromosomes.

**Ciliate**—Pubescent along margins with slender spreading hairs.

**Ciliolate**—Pubescent along margins with very short hairs.

**Circinate vernation**—Coiled in the bud with the apex in the center; the condition producing fern "fiddleheads."

**Circumboreal**—Occurring in northern latitudes of both the Old and New Worlds.

**Circumscissile**—A type of dehiscence in which the top of a fruit or other structure breaks in a ring and falls off.

**Cladistics**—A type of analysis of phylogenetic relationships in which monophyletic taxa are grouped on the basis of shared derived features.

**Cladode**—A flattened stem acting as if it were a leafblade; also called a cladophyll.

**Cladogram**—A graphic display of a hypothesis of relationship among a group of taxonomic entities in the form of a branching diagram in which the branching sequence is based on inferred historical connections between the entities as evidenced by shared, uniquely derived features (synapomorphies). A cladogram can be rotated at any node (q.v.) and still have the same information content.

**Clasping**—Base of a cordate or sagittate sessile leaf with basal lobes extending around the stem (Fig. 3-4 D).

**Class**—A rank in the classification hierarchy above the order and below the division, names characterized by the suffix, -opsida.

**Classification hierarchy**—A system of nested taxonomic ranks (Tables 1-1; 2-1). From largest (most inclusive) to smallest (least inclusive) the major ranks (and the suffixes used to designate them) are: Division (-ophyta), Class (-opsida), Order (-ales), Family (-aceae), Genus (no standardized suffix) and Species (no standardized suffix). Additional taxonomic ranks are often included as needed [e.g., Subclass (-idae), Subfamily (-oideae), Subspecies (no standardized suffix), etc.].

**Clavate**—Shaped like a "caveman's club," narrow at one end and expanded at the other.

**Claw**—The stalk or constricted basal portion of a petal or sepal (Fig. 10-4 A).

**Cleft**—Divided about halfway to the middle, as in a leaf or corolla.

**Cleistogamy (cleistogamous flower)**—Production of flowers that do not open and are self-pollinated. The opposite of chasmogamy.

**Clique**—A group of mutually compatible characters that support the same hypothesis of relationship in character compatibility cladistic analysis.

**Clone**—An individual derived by vegetative means from another individual. The parent and offspring are genetically identical.

**Coalescence**—Connation or cohesion.

**Coherent, cohesion**—Two or more similar structures clinging together or grown together during development but not truly connate.

**Collateral buds**—A condition in which two or more buds are positioned side by side in the same leaf axil.

**Collection number**—The number given by a field collector that references the plant collected to the information recorded in a field notebook.

**Column**—A structure in an orchid flower derived by adnation of stamens to the style and stigma (Fig. 19-11 D).

**Columnar**—A tree trunk that is unbranched with leaves at the top (e.g., many palms).

**Coma**—A tuft of hairs.

**Comose**—Bearing a tuft of hairs (e.g., seeds of milkweeds).

**Complete flower**—A flower with all four floral whorls (calyx, corolla, androecium and gynoecium).

**Compound cyme**—A compound inflorescence in which the component simple inflorescences are cymes (Fig. 10-13 K–P).

**Compound dichasium**—A compound cyme that repeatedly forms flowers in pairs arising from the axils of opposite bracts on the pedicels of the preceding flowers; dichasial cyme (Fig. 10-13 K).

**Compound inflorescence**—An inflorescence consisting of two or more simple inflorescences with a common axis or branching point (e.g., a panicle or compound umbel).

**Compound leaves**—Leaves with two or more blades (leaflets) attached to one petiole (Fig. 3-3 B–D).

**Compound ovary**—A structure derived from the connote bases of two or more carpels.

**Compound pistil**—A structure derived by partial or complete connation of two or more carpels.

**Compound umbel**—A compound inflorescence with small umbels (umbellets) at the ends of the branches (rays) of a larger umbel (characteristic inflorescence of the Apiaceae; Fig. 10-13 F).

**Compressed**—Flattened.

**Conduplicate**—Folded together lengthwise.

**Cone**—A strobilus; a reproductive structure with a central axis and a series of structures such as sporophylls and cone scales.

**Cone scale**—One of the individual units of a cone of a pine or other conifer, derived from a reduced branch; also called an ovuliferous scale.

**Connate (connation)**—The fusion of two or more structures of the same kind (e.g., a sympetalous corolla is made of connate petals); the opposite of distinct.

**Connate-perfoliate**—Opposite sessile leaves with fused bases completely encircling the stem.

**Connective**—The portion of an anther between the pollen sacs (Fig. 10-7 A).

**Connivent**—Clustered closely together but not fused (e.g., connivent stamens have their anthers pressed together).

**Contorted aestivation**—Floral parts spirally twisted in the bud.

**Convoluted aestivation**—Floral parts rolled inward from one side to the other.

**Cordate**—In the shape of a valentine heart with the notch at the base; also used for a leaf base with a notch between two rounded basal lobes (Fig. 3-9 F).

**Coriaceous**—With a thick and leathery texture.

**Corm**—A short erect enlarged underground stem with leaves absent or scalelike (Fig. 3-1 F).

**Corolla**—Collective term for the petals of a flower; the second whorl or spiral of parts of a complete flower; the inner sterile whorl or spiral.

**Corolla lobes**—The separate petal-tips of a sympetalous corolla (Fig. 10-4 B, C).

**Corolla throat**—The flaring portion of a sympetalous corolla (Fig. 10-4 B, C).

**Corolla tube**—The cylindrical portion of a sympetalous corolla (Fig. 10-4 B, C).

**Corona**—Apparent third sterile whorl of perianth derived from appendages of either petals or filaments.

**Corymbiform cyme**—A more or less flat-topped compound cyme resembling a corymb (Fig. 10-13 P).

**Corymb**—A short broad more or less flat-topped indeterminate inflorescence with pedicels of different lengths (Fig. 10-13 D); sometimes used for determinate inflorescences of similar appearance.

**Costa**—A rib or prominent nerve; especially the shortened central midrib of some palm leaves.

**Cotyledons**—The seed leaves.

**Crassinucellate**—A condition in which the nucellus or megasporangial tissue remains several layers thick when the ovules mature; considered to be relatively primitive.

**Crenate**—Margin of leaf blade with rounded teeth (Fig. 3-10 B).

**Crenulate**—Margin of leaf blade with very small rounded teeth (Fig. 3-10 C).

**Crisped**—Leaf margin strongly undulate, often wrinkled.

**Culm**—Jointed, often hollow stems of a grass.

**Cultigen**—A plant or variety of plant that is known only in cultivation, at least in the area covered by a particular flora or manual.

**Cultivar**—A taxonomic unit recognized by the International Code of Nomenclature of Cultivated Plants, but accorded no official recognition in formal botanical classification systems. Used to describe races or selected forms of cultivated plants; the name is enclosed within single quotes (e.g., *Tagetes patula* 'Summer Gold'). Informally used incorrectly as synonym of cultigen.

**Cuneate**—Wedge-shaped; a base or shape in which sides converge to a point and the sides are straight.

**Cupule**—A little cup; used to describe involucre that subtends or surrounds the ovary or ovaries in the Fagaceae.

**Cuspidate**—A leaf apex abruptly tapering to a stout sharp point.

**Cyathium**—A type of inflorescence characteristic of the genus *Euphorbia* in which a cuplike involucre bearing one or more nectaries encloses several to many tiny naked staminate flowers consisting of a single stamen each clustered around a solitary naked pistillate flower (Fig. 13-9 L–M).

**Cyclic**—Parts in whorls.

**Cyme**—Any of several types of determinate inflorescences; a simple cyme consists of a three-flowered cluster with the oldest flower in the center.

**Cymose**—Determinate.

**Cymule**—A very small, condensed cyme (e.g., clusters of flowers in catkin of *Alnus*).

**Cypsella**—The achene of a member of the Asteraceae, developed from an inferior ovary and usually bearing a pappus.

**Cytogenetics**—The study of chromosome numbers, morphology and behavior, particularly as applied in taxonomic investigations.

**Cytology**—The study of cells, as used in taxonomy the study of chromosomes.

**Deciduous**—Falling all at once, as the leaves of a non-evergreen plant.

**Decompound**—Several times compound.

**Decumbent**—Reclining or lying on the ground, but with the tip ascending.

**Decurrent**—Extending down along an axis (e.g., decurrent leaves have a flange of blade tissue that extends down the stem below the node to which the leaf is attached).

**Decussate**—Leaves opposite in four ranks with successive pairs at right angles.

**Dehiscence (dehiscent)**—Splitting opening of a fruit or anther in a regular pattern, releasing the seeds or pollen contained within.

**Deliquescent**—(1) Becoming liquid. (2) Very much branched (e.g., the stem axis of an elm or oak tree).

**Deltoid**—With the shape of an equilateral triangular.

**Dentate**—Bearing teeth that spread at right angles to the margin (Fig. 3-10 G).

**Denticidal capsule**—A capsule that opens at the tip by teeth (Fig. 10-17 E).

**Denticulate**—Bearing small teeth that spread at right angles to the margin (Fig. 3-10 H).

**Determinate**—An inflorescence in which a terminal (or central) flower forms first, using up the terminal bud, thereby arresting further growth of the axis; additional flowers arise from the axils of bracts formed before the differentiation of the terminal bud; cymose (Fig. 10-12 B).

**Determination**—The process of determining the correct name for a plant; identification.

**Diadelphous**—A stamen arrangement characteristic of most papilionoid Fabaceae in which nine stamens are connote by their filaments and the tenth is distinct; in two sets (Fig. 12-18 J).

**Diagnostic feature**—A feature or character of a plant that can be used to distinguish one taxon from another.

**Dichasial cyme**—A compound cyme which repeatedly forms flowers in pairs arising from the axils of opposite bracts on the pedicels of the preceding flowers; compound dichasium (Fig. 10-13 K).

**Dichotomous key**—A series of paired mutually exclusive statements that divides a set of objects into progressively smaller subsets. Used in plant identification to separate the name or description of one plant or group of plants from the names or descriptions of other plants.

**Dichotomous venation**—Veins that fork into equal pairs and that do not anastomose.

**Didynamous**—With four stamens in two pairs of unequal length.

**Diffuse**—Loosely branched or spreading; of open growth.

**Digitate**—Handlike; palmately lobed or compound.

**Dimorphic**—Occurring in two forms (e.g., ferns with sterile foliaceous fronds and reduced fertile fronds; Fig. 8-14 I, J).

**Dioecious**—A sexual condition in which staminate and pistillate flowers occur on separate individuals; a term properly applied to plants, not flowers.

**Diploid**—A tissue or plant with two sets of chromosomes in each cell (characteristic of sporophytes).

**Disarticulate**—To break apart into segments as in rachilla of grass spikelets.

**Disc florets (disc flowers)**—see Disk floret.

**Disciform head**—A type of head in the Asteraceae bearing both filiform florets and disk florets (Fig. 15-7 C).

**Discoid**—Having the form of a disc (e.g., a discoid stigma).

**Discoid head**—A type of head in the Asteraceae bearing only disk florets (Fig. 15-7 A).

**Disk**—(1) An elevated, more or less fleshy, often nectar-bearing outgrowth from the receptacle that surrounds the base of the pistil in some flowers. (2) The collective assemblage of disk florets in the heads of some Asteraceae.

**Disk florets (disk flowers)**—The tubular, usually actinomorphic and usually perfect flowers of some Asteraceae (Figs. 15-2; 15-6; 15-7A).

**Dissected**—Divided into many slender segments.

**Distichous**—In two vertical series or ranks (e.g., the leaves of grasses).

**Distinct**—Not fused to parts of the same type or whorl (e.g., petals that fall individually are distinct); the opposite of connate.

**Diurnal**—Opening or active only during daylight hours.

**Divaricate**—Spreading widely, often at right angles to an axis.

**Divided**—Margin indented nearly or all the way to the midrib or base.

**Division**—The most inclusive taxon recognized within the plant kingdom; the equivalent of phylum (used in animal kingdom), name characterized by the suffix, -ophyta.

**Dorsal**—Back side; relating to the back or outer surface of a part or organ.

**Dorsiventral**—With a top and bottom surface.

**Double**—A term used to describe flowers that have more than the usual number of petals or petaloid perianth elements; often selected for horticultural purposes.

**Double fertilization**—A process restricted to flowering plants in which one sperm cell unites with the egg cell to form a zygote and the second sperm cell unites with two polar nuclei to form a fusion nucleus. The latter gives rise to the endosperm.

**Doubly (crenate, serrate, or dentate)**—Teeth with smaller teeth on their margins (Fig. 3-10 F).

**Drupe**—A fleshy, usually 1-seeded indehiscent fruit with the pericarp differentiated into an exocarp, a fleshy mesocarp and a stony endocarp that encloses the seed(s); a stone-fruit (Fig. 10-15 E, F).

**Drupelet**—One of the component small fruits in an aggregate fruit such as a raspberry.

**Dwarf shoot (Short shoot)**—A lateral stem axis characterized by little or no elongation of internodes.

**-eae**—Standardized suffix indicating the scientific name of a tribe.

**Egg**—The female gamete; in angiosperms one of the cells of the embryo sac formed next to the micropyle; in bryophytes, fern allies, ferns and most gymnosperms formed in an archegonium.

**Elaboration**—Functional and structural specialization of a structure (e.g., insect-trapping leaves of a pitcher plant; long-plumose styles that function in dispersal of *Clematis* achenes).

**Elater**—A hygroscopic filamentous structure that aids in spore dispersal as in *Equisetum* (Fig. 8-10 E).

**Ellipsoid**—A three-dimensional structure with an elliptical outline.

**Elliptic**—A flat structure with the form of a closed curve that is widest in the middle and narrows equally toward both ends (Fig. 3-7 F, I).

**Emarginate**—With a shallow notch at the apex (Fig. 3-8 F).

**Embryo**—The plantlet in a seed, derived from a zygote (except in apomictic species); the young sporophyte in ferns or fern allies that is still attached to the gametophyte.

**Embryo sac**—The female gametophyte within the mature ovule of an angiosperm; the egg is produced within the embryo sac.

**Emersed**—Sticking out of water.

**Endemic**—Restricted in distribution to a particular geographical region.

**Endocarp**—An inner layer of pericarp tissue differentiated in some fruits as a stone-like or parchment-like layer (Fig. 10-14 B).

**Endosperm**—A starch- or oil-containing tissue present in seeds of most angiosperms; derived from the fusion nucleus; sometimes called the albumin.

**Entire**—A smooth continuous margin without teeth (Fig. 3-10 A).

**Entomophilous flowers**—Flowers that are visited and pollinated by insects.

**Epappose**—Without a pappus.

**Ephemeral**—Of short duration. (1) Herbaceous plants that grow for a short time and then either die entirely (annuals) or die back to the ground level (perennial ephemerals). (2) Flowers that persist for only one day (e.g., flowers of spiderwort).

**Epicalyx**—An involucel of bractlets that immediately subtend the calyx of an individual flower (e.g., strawberry, many Malvaceae).

**Epigynous (epigyny)**—A condition in which the perianth and stamens are adnate to the top or sides of the ovary; all flowers with an inferior ovary are epigynous (Fig. 10-10 C, D).

**Epipetalous stamens**—Stamens that are individually adnate to the corolla (Fig. 10-7 C).

**Epiphyte**—An air plant; a plant growing on another plant without deriving direct nutrition from the host plant.

**Epistemonous petals**—Petals that are individually adnate to a stamen column (e.g., flowers of Malvaceae).

**Equitant**—Leaves that are vertically folded and apparently inserted edge-on to the stem; such leaves are two-ranked in Iridaceae.

**Erose**—Margin with irregular teeth, appearing to have been gnawed.

**Estipulate (exstipulate)**—Without stipules.

**Euploid**—A change in chromosome number amounting to the gain or loss of one or more full sets.

**Eusporangiate**—The condition where a sporangium originates from several cells, the common condition in most spore-bearing plants.

**Even-pinnate leaves**—A pinnately compound leaf with a pair of terminal leaflets.

**Evergreen**—Remaining green throughout the year, even in the dormant season (if any); not losing all the leaves at one time.

**Excurrent**—(1) Extending beyond the margin or tip, as a mucro or awn. (2) A type of branching pattern, characterized by pines, in which the trunk is unbranched and branches arise perpendicular to the trunk.

**Exfoliate**—To peel off in shreds, thin layers or plates, as the bark from a tree trunk.

**Exine**—The outer wall of a pollen grain, often ornamented in ways characteristic for particular taxa.

**Exocarp**—The outer layer or "skin" of a fruit when two or more layers can be distinguished (Fig. 10-14 B).

**Exserted**—Sticking out; projecting from the corolla; the opposite of included.

**Exstipulate (estipulate)**—Without stipules.

**Extra-floral nectary**—Nectary glands not associated with flowers (e.g., some plants have nectaries on the leaves).

**Extrorse anther dehiscence**—The opening of the anthers on the side away from the floral axis.

**Falcate**—Sickle shaped.

**False septum**—An ovary septum not derived from infolded carpel margins, usually derived as an ingrowth from the placentae.

**Family**—A rank in the classification hierarchy above the genus and below the order; names characterized by the suffix, -aceae.

**Farinose**—Mealy; surface with small granular water-filled vescicles which may dry out to become powdery scales.

**Fascicle**—A condensed cluster of structures (e.g., a fascicle of pine needles).

**Female gametophyte**—Megagametophyte.

**Fern**—An ancient group of vascular plants with megaphyllous leaves and a life cycle characterized by free-living gametophyte and sporophyte phases. The spores are borne in sporangia on modified or unmodified leaves.

**Fern allies**—A polyphyletic or paraphyletic assemblage of ancient vascular plants with microphyllous leaves or no leaves at all. The life cycle is similar to that of the ferns with an alternation free-living gametophyte and sporophyte phases. The sporangia are borne in various arrangements on the plants.

**Fernlike**—With leaves large and pinnate; usually several times compound.

**-ferous**—Suffix meaning to carry or to bear.

**Fertile**—Bearing reproductive structures (stamens, ovules, fruits, etc.); possessing pollen or developable ovules. In grasses descriptive of floret that can produce a caryopsis.

**Fertile lemma**—In grasses a lemma that subtends a floret that can produce a caryopsis.

**Fertilization**—The union of two gametes resulting in a zygote.

**Fibrillose**—With fine fibers.

**Fibrous roots**—A root system of several to many relatively slender roots of about the same diameter (Fig. 3-1 B).

**Fiddlehead**—A coiled immature fern frond that resembles the narrow end of a violin; the result of circinate vernation.

**Field notebook**—A small bound notebook in which field observations and collection information are recorded.

**Filament**—The stalk of a stamen which bears an anther; any slender thread-like structure.

**Filiform**—Threadlike, long and very slender.

**Filiform floret (filiform flower)**—A type of pistillate flower in the Asteraceae with a very slender corolla tube and little or no corolla limb (Figs. 15-6; 15-7 C).

**Fimbriate**—Fringed.

**Fistulose**—Enlarged and hollow cylinder.

**Flabellate**—Fan-shaped, flattened or branching in one plane.

**Flaccid**—Limp, floppy, lacking turgor.

**Flavonoids**—A family of pigments found in almost all higher plants; often used comparatively in chemotaxonomic studies.

**Flora**—The plants growing in a region; a list of the plants growing in an area; such a list with descriptions and keys for identification (more properly called a manual).

**Floret**—(1) In grasses the unit composed of a lemma and palea and the small flower they enclose. (2) In Asteraceae the individual small flowers of the head. (3) Any very small flower of a dense inflorescence.

**Flower**—An axis bearing one or more carpels or one or more stamens or both and usually one or two series of perianth parts.

**Flower bud**—A bud that develops into a flower.

**Foliaceous**—Leaflike in color and texture, often used to describe sepals or bracts.

**Follicle**—A dry dehiscent fruit derived from a single carpel that dehisces either along the suture (e.g., *Delphinium*) or along the midrib (*Magnolia*), but not both (Fig. 10-17 I, J).

**Forma**—The lowest-ranking taxon in the taxonomic hierarchy, sometimes designated to note an unusual mutant or "sport" within a species; a little-used taxonomic rank in most modern studies.

**Free**—Not fused to other kinds of structures (e.g., stamens free from petals); opposite of adnate. Sometimes used loosely in the sense of distinct.

**Free veins**—Veins that diverge and do not rejoin.

**Free-central placentation**—A placentation type found in compound ovaries with only one locule in which ovules are attached to a free-standing axis in the middle of the ovary and septa are absent; derived from axile placentation (Fig. 10-8 I, J).

**Free-nuclear endosperm**—Multiplication of endosperm nuclei by mitosis without cell-wall formation.

**Fronds**—Leaves of ferns; sometimes used for any large leaf (e.g., palm fronds).

**Fruit**—A ripened ovary along with any adnate parts.

**Fruticose**—Shrubby or shrublike in the sense of being woody.

**Fugacious**—Falling or withering very early.

**Funiculus**—The stalk by which an ovule is attached to an ovary wall or placenta.

**Funnelform**—Used to describe a corolla (calyx, hypanthium, involucre) with the shape of a funnel, the tube slender, the throat gradually widening upward (Fig. 10-5 B).

**Furrow**—A long, narrow groove.

**Fusion**—The evolutionary or developmental union of originally separate structures. Of two patterns: connation (cohesion), the union of similar parts and adnation (adhesion), the union of dissimilar structures.

**Fusion nucleus**—The nucleus or cell within an embryo sac formed from the union of the two polar nuclei with a sperm nucleus in the process of double fertilization; the first endosperm nucleus.

**Galea**—A hooded or helmet-shaped projecting upper lip of a zygomorphic sympetalous corolla; a structure of similar appearance derived from other perianth structures (e.g., upper sepal of *Aconitum*).

**Gametangia**—The reproductive structures of plants in which sperm or eggs are produced by mitosis.

**Gamete**—A sex cell; an egg or a sperm.

**Gametophyte**—The multicellular haploid phase of a plant life cycle which produces sperm or eggs or both by mitosis.

**Gametophyte-dominant alternation of generations**—A type of life cycle in which the haploid, gamete-forming phase is free-living and dominant, and the diploid, spore-producing phase is dependent; characteristic of bryophytes and some other non-vascular plants.

**Gamopetalous**—Sympetalous; a corolla of connate petals (Fig. 10-3 C).

**Genotype**—The summation of genetic characters of an individual.

**Genus (pl. genera)**—A rank in the taxonomic hierarchy below the family and above the species. Its name has no standardized suffix, but is always capitalized and underlined or italicized.

**Glabrate (glabrescent)**—Pubescent when young but losing the hairs in maturity.

**Glabrous**—Smooth, without pubescence (Fig. 3-11 A).

**Gland**—A structure that secretes a liquid substance (e.g., volatile oils, resins, nectar, etc.). The gland may retain the substance internally or exude it as a thin film or droplets.

**Glandular**—Pubescent with stalked or sessile gland-tipped trichomes (Fig. 3-11 K, L).

**Glandular punctate**—Dotted with sessile or embedded glands.

**Glaucous**—Covered with a bluish or whitish "bloom" that can sometimes be rubbed off; possessing the bluish cast characteristic of organs with such a coating.

**Globose**—Spherical.

**Glochid**—A minute barbed spine characteristic of some cacti.

**Glomerule**—A dense rounded cluster of flowers.

**Glume**—Either of the two basal bracts of a grass spikelet that do not directly subtend flowers; a small chaff-like bract.

**Glutinous**—Sticky.

**Gymnosperm**—A group, probably representing several evolutionary lineages, characterized by the production of seeds that are not borne in carpels (e.g., pines, ginkgos, cycads, gnetums).

**Gynandrium**—Column of an orchid derived by adnation of stamen, stigma and style tissue.

**Gynecandrous**—Used to describe spikes of *Carex* (Cyperaceae) that have pistillate flowers at the top and staminate flowers below.

**Gynobasic style**—A style that appears to arise directly from the receptacle or ovary base rather than from the apex of the ovary (e.g., style of most Boraginaceae and Lamiaceae).

**Gynoecium**—Collective term for the carpels of a flower: may be monocarpous, composed of one carpel; apocarpous, composed of 2 or more distinct carpels; or syncarpous, composed of 2 or more connate carpels.

**Gynostegium**—A structure found in Asclepiadaceae and some Aristolochiaceae formed from adnation of stamens to the stigma (Fig. 14-12 A, B).

**Habit**—The shape a plant takes (i.e., herb, shrub, tree, vine).

**Half inferior ovary**—Calyx and other whorls adnate to the lower half of the ovary; considered to be inferior in keys.

**Halophyte**—A plant tolerant of or adapted to excessive mineral soils in the soil solution.

**Haploid**—A tissue or plant in which each cell has one set of chromosomes (the nuclear condition of spores, gametophytes and gametes).

**Hastate**—A type of leaf base with a pair of basal lobes diverging sharply from the remainder of the blade (Fig. 3-9 H).

**Head**—A short dense simple inflorescence of sessile flowers attached to a common receptacle; may be indeterminate or determinate; a capitulum.

**Helicoid**—Coiled in a spiral pattern.

**Helicoid cyme**—A one-sided determinate inflorescence that uncoils as it matures; often used interchangeably with scorpioid cyme.

**Hemiparasite**—A parasitic plant that is green and photosynthetic, manufacturing its own carbohydrates but drawing moisture or other substances from its host.

**Herb (herbaceous)**—A plant with little or no aboveground perennial woody tissue, usually dying or dying back to the ground level at the end of the growing season; some herbs have much secondary growth but stems die at end of growing season; others have subterranean woody rhizomes.

**Herbarium**—A collection of dried plant specimens ranging in size from a bound notebook to a museum with millions of specimens.

**Hesperidium**—A type of berry derived from a superior ovary with a thick rind and locules filled with swollen juice-filled hairs; generally restricted to *Citrus* and its close relatives (Fig. 10-15 D).

**Heterospory (heterosporous)**—Production of two distinctly different types of spores, the larger ones called megaspores which give rise to megagametophytes, and the smaller ones called microspores which give rise to microgametophytes (e.g., as in *Selaginella* and all seed bearing plants).

**Hexaploid**—A polyploid cell, tissue, plant or taxon characterized by having six sets of chromosomes.

**Hilum**—The scar on a seed that marks the point of attachment of the funiculus.

**Hip**—The fleshy ripened hypanthium and enclosed achenes of *Rosa* (Fig. 10-18 D, E).

**Hirsute**—Pubescent with coarse, stiff or shaggy trichomes (Fig. 3-11 F).

**Hispid**—Pubescent with stiff, bristly, usually stout-based trichomes (Fig. 3-11 E).

**Hoary**—Covered with a close white or whitish pubescence; appearing frosty.

**Holotype**—The specimen deposited in a specific herbarium designated by the author of a taxon to serve as the permanent reference point for the name of the taxon.

**Homoplasy**—The result of parallel or convergent evolution in which two or more kinds of organisms independently acquire the same (or apparently the same) character state. Homoplasy sometimes also results from reversals when in a particular lineage a character reverts from an apomorphic to a plesiomorphic state. In phylogenetic analyses homoplasy can cause character conflicts or misinterpretation of relationships.

**Homospory (homosporous)**—Production of spores which are all morphologically similar (e.g., as in *Lycopodium* and most ferns).

**Horizontal classification**—A grouping created when individuals or taxa of the same rank are grouped on the basis of overall similarity or some other criterion. A vertical classification is the hierarchy created when such taxa are grouped and ranked into successively more inclusive taxa.

**Hyaline**—Thin and transparent or translucent.

**Hybrid**—The offspring of two parent plants that are genetically more or less different from each other.

**Hydrophyte**—A plant adapted to life in water; an aquatic.

**Hygroscopic**—The tendency to change form or size with changes in moisture (e.g., annulus of a fern sporangium).

**Hypanthium**—A structure derived by adnation of the bases of the perianth elements and stamens; the shape varies from disc-like to cup-shaped, flask-like or long-tubular (Fig. 10-10 B, D).

**Hypogynous disc**—A fleshy nectar-bearing disc subtending the ovary in some plants.

**Hypogyny (hypogynous)**—A condition in which the perianth elements and stamens are inserted directly on the receptacle and are wholly free from the gynoecium; no hypanthium is present, but stamens may be adnate to the corolla (Fig. 10-10 A).

**-idae**—Standardized suffix indicating the scientific name of a subclass.

**Identification**—The process of determining the name associated with a particular plant; determination.

**-iflorae**—Suffix used by some botanists to denote the scientific name of a superorder.

**Imbricate**—Overlapping, as shingles on a roof.

**Imperfect**—A flower lacking either stamens or carpels; unisexual; monosporangiate.

**-inae**—Standardized suffix indicating the scientific name of a subtribe.

**Inaperturate**—A pollen grain without an aperture.

**Incised**—With a margin deeply and sharply cut.

**Included**—Not protruding, as stamens not projecting from the corolla; the opposite of exserted.

**Indehiscent**—Not splitting open with a regular pattern or not splitting open at all.

**Indeterminate**—An inflorescence in which the terminal bud remains active during the development and all flowers originate from lateral buds (Fig. 10-12 A); the most mature flowers occur at the base or outside of the flower cluster; racemose (s.l.).

**Indurate**—Hardened.

**Indusium**—A flap of sterile tissue that covers the sorus in some ferns (Fig. 8-1 C).

**Inequilateral**—A leaf base characterized by sides that do not match; oblique (Fig. 3-9 G).

**Inferior ovary**—Condition that results from adnation of the outer whorls of the flower to the ovary wall (Fig. 10-10 C, D).

**Inflated**—Puffed up; bladdery.

**Inflorescence**—Any kind of flower cluster.

**Infraspecific taxon**—A taxonomic group smaller than a species (e.g., subspecies, variety, form).

**Ingroup**—In cladistics a monophyletic group under investigation for character polarity by comparison with an outgroup (q.v.).

**Insectivorous**—A plant that traps and digests insects.

**Insertion (Inserted)**—The method of attachment of one structure to another.

**Integument**—The covering of an ovule.

**Internal phloem**—Phloem occurring on the pith side of the xylem as well as to the outside; possessing bicollateral vascular bundles.

**Internode**—The region of a stem between two nodes (Fig. 3-2 A, B).

**Interrupted**—Not continuous.

**Introduced**—Not native to the region in which it is growing.

**Introrse anther dehiscence**—The opening of the pollen sacs on the surface toward the floral axis (e.g., the anthers of Asteraceae).

**Involucel**—A secondary involucre; a small involucre around part of a compound cluster; an epicalyx of bractlets immediately subtending a flower.

**Involucrate**—Subtended or surrounded by an involucre.

**Involucre**—One or more whorls of bracts immediately subtending a flower or inflorescence, often forming a cup-like structure (Fig. 10-11 A).

**Involute**—Descriptive of a flat structure such as a leaf that is rolled inward toward the upper surface (e.g., the leaves of many grasses; Fig. 3-10 L).

**Irregular**—Bilaterally symmetric; zygomorphic (Fig. 10-6 B).

**Isobilateral**—A type of symmetry in which a flower can be divided into two equal parts along two planes only (e.g., flower of *Dicentra* with two unequal pairs of petals at right angles; Fig. 11-11 H).

**Isomerous**—All whorls of a flower except the gynoecium with the same definite number of parts.

**Isotype**—A duplicate specimen of a holotype, collected at the same time and place.

**Isozyme (isoenzyme)**—Different forms of the same enzyme formed at different locations on the chromosomes of a plant or on different kinds of plants. Often used to study amount of variation within and between populations at or below the species level. Isozymes can be separated by electrophoresis.

**Jointed**—With nodes or breaks in the outline.

**Karyotype**—The form and number of chromosomes.

**Keel**—The pair of apically connate and basally distinct lower petals of a papilionoid flower (Fabaceae) that form a unit resembling the prow of a boat (Fig. 12-18 G).

**Keeled**—Bearing a longitudinal ridge like the bottom of a boat.

**Key**—A series of paired mutually exclusive statements that divides a set of objects into progressively smaller subsets. Used in plant identification to separate the name or description of one plant or group of plants from the names or descriptions of other plants.

**Kingdom**—The highest rank in the classification hierarchy.

**Labellum**—The odd petal of an orchid flower (Fig. 19-11 A–C), developmentally the upper petal but appearing as the lower because the inferior ovary has a half twist, usually much elaborated; the lip.

**Laciniate**—Cut into narrow pointed lobes.

**Laminar placentation**—Arrangement of ovules along the surfaces of the septa of an ovary rather than restricted to definite placentae.

**Laminar stamen**—Broad petal-like stamen with anthers embedded in flattened tissue (Fig. 10-7).

**Lamina**—Blade; flat surface.

**Lanate**—Pubescent with long tangled curly trichomes; wooly.

**Lanceolate**—A shape much longer than broad, widest near the base and tapering to the apex (Fig. 3-7 E).

**Lateral**—On or at the side.

**Lateral bud**—A bud borne in a leaf axil (Fig. 3-2 A, B); axillary bud.

**Lateral veins**—Secondary veins of a pinnately veined leaves.

**Later homonym**—A scientific name that exactly duplicates an earlier name; later homonyms are illegitimate and cannot be used.

**Latex**—Milky or sometimes brightly pigmented juice produced in specialized cells or tubules.

**Leaf**—The primary lateral appendages of a stem axis, usually flattened and in most plants serving as the primary organs of photosynthesis; of two types.

**Leaf base**—In taxonomy the base of the leaf blade; in morphology the attachment of the leaf to the blade, the base of the petiole.

**Leaf bud**—A vegetative bud.

**Leaf scar**—The scar on a twig left when a leaf falls (Fig. 3-2 A).

**Leaflet**—One of the individual blades of a compound leaf (Fig. 3-3 B–D).

**Lectotype**—A specimen chosen by a later researcher to serve as if it were the holotype from among the specimens available to the original publishing author of a scientific name when (a) the holotype has been lost or destroyed or (b) no holotype was designated.

**Legume**—A dry dehiscent fruit derived from a single carpel that dehisces along both the midvein and the suture (Fig. 10-17 K, L).

**Lemma**—In grass spikelets the lower of two bracts that together enclose the flower (Fig. 20-7 C).

**Lenticel**—The specialized openings in the bark of twigs that provide a passage for gas exchange.

**Lenticular**—Lens-shaped; biconvex.

**Leptosporangiate**—A type of sporangial development in which sporangia and stalk arise from a single cell; restricted to the more advanced groups of ferns.

**Ligneous**—Possessing woody tissues.

**Ligulate floret**—A type of flower characteristic of the tribe Lactuceae of the Asteraceae; it is perfect and has a strap-shaped corolla limb with 5 apical lobes (Figs. 15-6; 15-7 D).

**Ligulate head**—A type of head in the Asteraceae bearing only ligulate florets (Fig. 15-7 D).

**Ligule**—A strap-shaped structure. (1) In Asteraceae the strap-shaped limb of a ray corolla or ligulate corolla. (2) In Poaceae a projection from the top of the sheath in grasses and similar plants. (3) In heterosporous Lycophyta a fleshy protuberance on the leaf base.

**Limb**—The expanded flat part of an organ, particularly the spreading lobes of a sympetalous corolla (Fig. 10-4 B).

**Line of dehiscence**—A mark or groove along which a fruit splits open.

**Linear**—Long and narrow with sides parallel or nearly so (e.g., leaves of most grasses; Fig. 3-7 C).

**Lip**—(1) One of the two halves of the limb of a bilabiate corolla or calyx, called the upper (posterior) lip and the lower (anterior) lip (Fig. 10-4 C). (2) The odd petal of an orchid flower (Fig. 19-11 A–C), developmentally the upper petal but appearing as the lower because the inferior ovary has a half twist, usually much elaborated; labellum.

**Lip cells**—Thin cells in the jacket of many fern sporangia that separate or break apart as the annulus contracts, causing the spores to be ejected.

**Lobe (lobed)**—(1) A large projecting segment of the margin of a leaf or other flattened structure; an especially prominent tooth. (2) Separate tips of a synsepalous calyx, sympetalous corolla, involucre, etc. (Fig. 10-4 B, C).

**Lobule**—A small lobe.

**Locule**—Chamber on the inside of an ovary, sometimes filled with air and sometimes filled with various kinds of juicy or fleshy tissues.

**Loculicidal**—Dehiscence of a fruit along the back between the septa.

**Lodicules**—The minute remnants of the perianth of a grass flower which push apart the lemma and palea when turgid.

**Loment**—A type of indehiscent fruit derived from a single carpel that breaks transversely into one-seeded segments (Fig. 10-17 M); considered to be a modified legume (e.g., *Desmodium*).

**Long shoot**—A stem or twig with well-developed internodes.

**Lower lip**—The abaxial portion of a bilabiate corolla or calyx (Fig. 10-4 C).

**Lumper**—A taxonomist who tends to view similarities as much more important than differences, and consequently recognizes larger, more broadly defined taxonomic groups than do most other taxonomists.

**Male gametophyte**—Microgametophyte.

**Manual**—A book containing keys for identification of the flora of an area, often along with descriptions.

**Marcescent**—Withering, but the remnants persistent.

**Marginal placentation**—A type of placentation found only in simple pistils in which the ovules are attached to the folded margins of the carpel (Figs. 10-7 C, D; 10-8 A, B).

**Maritime**—Occurring in or along the sea.

**Markedly**—A structure that is developed to such a degree that it is easily visible with the proper aid.

**Megagametophyte**—The haploid derivative of a megaspore that produces the egg cell(s) by mitosis; in angiosperms, the embryo sac.

**Megaphyll**—A type of leaf usually characterized by branching veins and a gap in the stele of the axis where the leaf originated.

**Megasporangium**—The sporangium of a heterosporous plant that contains the larger of the two types of spores; in angiosperms, an ovule.

**Megaspore**—The larger of two spores produced by a heterosporous plant.

**Megasporocyte (megaspore mother cell)**—A cell located in a megasporangium that undergoes meiosis forming megaspores.

**Megasporophyll**—A leaf that bears megasporangia; in angiosperms, a carpel.

**Meiosis**—The process in which chromosomes are duplicated, exchange segments through crossing-over and divide in two successive nuclear divisions with each of the resulting cells having half as many chromosomes as the one cell had before the division began. In plants the cells that undergo meiosis are sporocytes (spore mother cells), and the products of meiosis are spores. Each of the four spores is genetically unique.

**Membranaceous**—Thin, rather soft and pliable, often more or less translucent.

**Mericarp**—One of the one-carpellate units of a schizocarp that break away from the axis of the ovary.

**Meristem**—A region of cell division, and thus growth of a plant. Terminal buds occur at the tips of stems and roots. Lateral meristems (cambia) result in increase in diameter.

**-merous**—Suffix indicating number of parts, usually referring to the corolla (e.g., flowers 5-merous).

**Mesocarp**—The middle layer of the pericarp of a fruit in which three layers can be distinguished (Fig. 10-14 B).

**Mesophyte**—A plant adapted neither to excessive amounts of moisture nor to excessive drought. Intermediate between hydrophytes and xerophytes.

**Microgametophyte**—The haploid derivative of a microspore that produces the sperm cells by mitosis; in angiosperms the pollen grain.

**Microphyll**—A leaf characterized by having a single unbranched vein and leaving no gap in the xylem of the stem axis.

**Micropyle**—A pore in the integument of an ovule; in angiosperms the entryway for the pollen tube into the ovule; in gymnosperms the receptive area in pollination.

**Microsporangium**—The sporangium of a heterosporous plant that produces the smaller of the two types of spores; in angiosperms the anther sacs.

**Microspore**—The smaller of the two spores produced by a heterosporous plant.

**Microsporocyte (microspore mother cell)**—A cell located in a microsporangium that undergoes meiosis forming microspores. Often studied in meiosis to determine chromosome number.

**Microsporophyll**—A leaf that bears microsporangia; in angiosperms, a stamen.

**Midrib**—The main vein or rib of a leaf or leaflike part; a continuation of the petiole.

**Mitosis**—The process in which the chromosomes of a nucleus are duplicated and then divided equally between two daughter nuclei.

**Mixed bud**—A bud with both leaf and flower primordia.

**Monadelphous**—Stamens with filaments connate into a tube surrounding the ovary and style (e.g., stamens in Malvaceae; Fig. 10-7 B).

**Monocarpous**—A gynoecium composed of 1 carpel (Fig. 10-8 D).

**Monoecious**—A sexual condition in which staminate and pistillate flowers occur on the same individual; a term properly applied to plants, not flowers.

**Monograph**—An in-depth taxonomic study of a particular taxonomic group using various traditional and experimental methods.

**Monophyletic**—In cladistic analyses a natural taxonomic group composed of an ancestral species and *all* of its descendants; holophyletic. Many taxonomists also use the word monophyletic in the sense of paraphyletic (q.v.) to include taxa derived from a common ancestor but not necessarily including all of the descendants.

**Monosporangiate**—Producing either megaspores or microspores; imperfect; unisexual.

**Monotypic**—Any taxon of a classification hierarchy that includes only a single subordinate taxon (e.g., a family with only one genus).

**Mucronate**—A type of leaf apex with a short abrupt point (mucro; Fig. 3-8 E).

**Multifid**—Repeatedly lobed into many small segments.

**Multilocular**—Having several to many locules or inner chambers.

**Multiple fruit**—A structure derived from the coalescence of the ovaries of several to many flowers, often with accessory tissues; a ripened inflorescence (Fig. 10-18 F, G).

**Multiseriate**—In several to many series.

**Mycorrhizal association**—A symbiotic association of a fungus and the root of a vascular plant. It is essential for the survival of many plants (e.g., Ericaceae).

**Mycotrophic**—Dependent upon a mycorrhizal association.

**Naked bud**—A bud without bud scales.

**Naked flower**—A flower without any perianth.

**Native**—A plant present or presumed to have been present in an area prior to the beginning of the recorded history of the area. Assumed not to have been introduced to the area by man.

**Natural classification system**—A classification designed to reflect a "plan of nature" or the "real" or "natural" order of living organisms. Such classifications were largely phenetic in design though much character weighting generally took place in their construction.

**Naturalized**—A plant that has been introduced into an area since the beginning of the recorded history of the area and that has become established as a component of the vegetation of the area.

**Nectariferous disc**—A fleshy nectar-secreting disc beneath, around or on top of the ovary of a plant.

**Nectary**—A glandular structure that secretes a fluid containing sugars or amino acids (nectar). Nectaries in flowers often provide a "reward" to pollinating insects that visit the flowers. Extrafloral nectaries occur on non-reproductive structures.

**Needle leaves**—Leaves that are long, narrow and thickish, resembling a needle (Fig. 3-7 A).

**Neotropical**—Growing in the New World tropics.

**Neotype**—A specimen chosen by a later researcher to serve in place of a holotype when all specimens available to the original publishing author of a scientific name have been lost or destroyed. A neotype loses status if original material is rediscovered.

**Nerve**—A longitudinal vein of a leaf, bract or other structure.

**Net-veined leaves**—Leaves that have a network of veinlets that anastomose; reticulate-veined (Fig. 3-6 A, B).

**Neutral flower**—A flower that does not produce functional stamens or pistils.

**Nocturnal**—Active or open only at night.

**Node**—(1) A joint where a leaf is or was attached to a stem (Fig. 3-2 A, B); (2) A branching point in a cladogram, representing a speciation event.

**Nomenclatural synonym**—Names based upon the same type specimen. Generally all have the same epithet.

**Nucellus**—The megasporangium tissue in an ovule; the megasporocyte differentiates from nucellar tissue. In some seeds the nucellus matures as a storage tissue called perisperm.

**Nuclear endosperm**—Endosperm that develops first by mitosis without cell wall formation with subsequent cell wall formation.

**Numerous stamens**—Generally applied to a number of stamens more than twice as many as the petals, usually more than 10.

**Nut**—An indehiscent, usually one-seeded fruit with a hard, bony pericarp (Fig. 10-16 K, L).

**Nutlet**—A small or diminutive nut; one of the one-seeded segments of the ovary of a member of the Lamiaceae or Boraginaceae (Fig. 10-16 Q).

**Ob**—A Latin prefix, usually signifying inversion (e.g., oblanceolate, a spear-head shaped structure attached at its narrow end).

**Obcompressed**—Flattened perpendicular to a radius (e.g., achenes of some Asteraceae that are flattened at right angles to a radius of the head).

**Obcordate**—A shape resembling a valentine heart attached by the narrow end; an apex characterized by a deep notch.

**Oblanceolate**—A shape much longer than broad, narrow at the base and widening toward the apex (Fig. 3-7 G).

**Oblique**—A leaf base with sides that do not match; inequilateral (Fig. 3-9 G).

**Oblong**—A shape longer than broad with sides nearly or quite parallel most of their length; rectangular (Fig. 3-7 D).

**Obovate**—A shape 2–3 times longer than broad, narrow at the base and broadening toward the apex, resembling an egg attached by its narrow end (Fig. 3-7 J).

**Obovoid**—A solid that is shaped like an egg attached at its narrow end, widest in its upper half.

**Obscure (obscurely)**—Not readily apparent, visible with difficulty.

**Obsolete**—Not evident or apparent but present at least in embryonic stages.

**Obtuse**—A broad category of apices and bases referring to angles greater than 90°; any blunt apex or base, including rounded, truncate, etc. (Figs. 3-8 C; 3-9 C).

**Ocrea**—A nodal sheath formed by fusion of two stipules, characteristic of some Polygonaceae (Fig. 17-11 B).

**Octoploid**—A polyploid cell, tissue, plant or taxon characterized by having eight sets of chromosomes.

**Odd pinnate**—Pinnately compound with one terminal leaflet.

**-oideae**—Standardized suffix indicating the scientific name of a subfamily.

**Opposite leaves**—Type of phyllotaxy characterized by two leaves per node, diametrically across the stem from each other (Fig. 3-5 C).

**Opposite stamens**—Stamens positioned directly in front of the petals or lobes of a sympetalous corolla (Fig. 10-7 E).

**-opsida**—Standardized suffix indicating the scientific name of a class.

**Orbicular**—Circular in outline (Fig. 3-7 K).

**Order**—A rank in the classification hierarchy above the family and below the class; names characterized by the suffix, -ales.

**Ornithophilous**—Flowers pollinated by birds.

**Orthotropous**—A type of ovule that is attached to the funiculus in an erect position, not bent over.

**Ostiole**—A pore (e.g., the hole at the end of a fig).

**Our, ours**—Taxa occurring in the region covered by the flora or manual in use.

**Outgroup**—A monophyletic taxon that is used in a phylogenetic analysis to determine which of two or more states of a character in the group being investigated (the ingroup) is apomorphic. Ideally the sister group of the ingroup is used in this analysis; sometimes more than one outgroup must be used.

**Oval**—Broadly elliptic, with broadly rounded ends.

**Ovary**—The ovule-bearing part of a pistil.

**Ovate**—A shape 2–3 times longer than broad, broadest at the base and narrowing toward the apex, resembling an egg attached by its broad end (Fig. 3-7 H).

**Ovoid**—A solid that is oval or ovate in outline; egg-shaped.

**Ovulate**—Bearing or referring to ovules (e.g., an ovulate cone).

**Ovule**—An integumented megasporangium which after fertilization matures as a seed.

**Ovuliferous scale**—The structure in a gymnosperm cone derived from a branch that bears ovules on its surface; a structure homologous to a simple microsporangiate cone.

**Paleaceous**—Chaffy; covered with or composed of dry membranaceous scales (e.g., a pappus of scales; a paleaceous receptacle in the Asteraceae).

**Palea**—In a grass spikelet the upper of the two bractlets that enclose a grass flower, the lower being the lemma.

**Paleobotany**—The study of plant fossils.

**Paleotropical**—Growing in the Old World tropics (Africa, India, SE Asia).

**Pale**—A receptacular bractlet in the Asteraceae.

**Palm-like**—Resembling a palm tree with a single trunk and a terminal cluster of leaves; columnar.

**Palmate**—Radiating from a common point of origin; used for major veins of a leaf (Fig. 3-6 B) or leaflets of a compound leaf.

**Palmately compound**—A compound leaf with all of the leaflets arising from a common point of attachment (Fig. 3-3 D).

**Palmately lobed**—With lobes radiating from a common point of origin (Fig. 3-10 J); usually associated with palmate venation.

**Palmatifid**—Palmately lobed about halfway to the base.

**Panicle**—An indeterminate branching raceme; often used loosely for any pyramidal compound inflorescence whether determinate or indeterminate (Fig. 10-13 H).

**Paniculiform cyme**—A compound cyme that in overall outline resembles a panicle; an inflorescence with an indeterminate main axis and determinate secondary branches; a thyrse (Fig. 10-13 N).

**Pantropical**—Growing in tropical regions throughout the world.

**Papilionaceous**—A flower type characteristic of Fabaceae subf. Papilionoideae characterized by a banner petal (standard), two wing petals, and two keel petals.

**Papillose**—Bearing microscopic rounded projections.

**Pappus**—A peculiar modified calyx consisting of dry scales, bristles or awns characteristic of the Asteraceae (Figs. 15-1 B; 15-4). (Similar structures of some Valerianaceae and some Dipsacaceae).

**Parallel venation**—With the principal veins usually parallel and close together, and secondary veinlets more or less perpendicular to the primary veins (Fig. 3-6 C).

**Paraphyletic**—A taxonomic group that includes an ancestral species and some, but not all, of its descendants. In cladistic analyses such a group is rejected in favor of monophyletic taxa. Many currently recognized plant taxa are probably paraphyletic.

**Parasite**—A plant that draws water and nutrients from another plant (the host). Parasitic flowering plants include internal stem parasites with vegetative tissues within the host's stems, external stem and leaf parasites that have most of their vegetative structures on the outside, and root parasites that invade the host's root system to varying degrees.

**Paratype**—A specimen not formally designated as a type but cited along with the type collection in the original description of a species or lower level taxon. In the event of loss or destruction of all the other original specimens, an investigator can designate a paratype as lectotype.

**Parietal placentation**—A placentation type found in compound ovaries in which the placental areas are attached to the side walls of the ovary (Fig. 10-8 F–H). Such an ovary usually has one locule. Sometimes broadly used to include marginal placentation.

**Parsimony criterion**—A way of choosing between conflicting hypotheses, often used in cladistic analyses. Given a choice of two or more cladograms (hypotheses of relationship) the one involving the fewest assumptions is most likely to be correct. Also called Ockham's Razor (it was first proposed by William of Ockham in 1347) or the principle of simplicity.

**Parted**—Margin dissected nearly all the way to the midrib or base.

**Pea-like**—Papilionaceous.

**Pectinate**—Arranged like teeth of a comb.

**Pedicel**—The stalk of an individual flower in an inflorescence (Fig. 10-11); the stalk of a spikelet in the Poaceae or Cyperaceae (Fig. 20-7).

**Peduncle**—The stalk of a flower cluster (Fig. 10-11) or of a solitary flower.

**Pellucid dots**—Sessile or embedded glands that are clear and transparent or nearly so.

**Peltate**—Attached to the stalk away from the margin (e.g., a mushroom; Fig. 3-9 J).

**Pentadelphous**—Stamens connote by their filaments in five bundles.

**Pepo**—A large berry derived from an inferior ovary, characterized by a thick rind and soft, often watery or fibrous contents and many seeds (restricted to the Cucurbitaceae; Fig. 10-15 C).

**Perennial**—A plant without a fixed life span, potentially capable of indefinite growth; some reach reproductive maturity in the first season of growth and others live for several to many years before reproducing.

**Perfect**—A flower with both a functional androecium and a functional gynoecium.

**Perfoliate**—A stem passing through the blade of a leaf (Fig. 3-4 C).

**Perianth**—The collective term for all of the outer sterile parts of a flower, comprising the calyx and the corolla when both whorls are present.

**Pericarp**—The wall of a ripened ovary; the fruit coat (Fig. 10-14).

**Perigynium**—A sac-like hollow bract that encloses a pistillate flower in *Carex* (Cyperaceae) and its close relatives.

**Perigynous (perigyny)**—A condition in which the perianth and stamens are adnate, forming a hypanthium that is free from the ovary; all perigynous flowers have a superior ovary.

**Perisperm**—Stored food in a seed derived from the nucellus or integuments.

**Petaloid appendages**—Flattened petal-like structures attached to the involucral glands of the cyathium of some species of *Euphorbia*.

**Petaloid**—With the color and texture commonly associated with petals.

**Petal**—One unit of the inner whorl of the perianth, usually colored and more or less showy.

**Petiolate**—A leaf with a petiole (Fig. 3-4 A).

**Petiole**—The stalk of a leaf (Figs. 3-2 B–D; 3-3 A–D).

**Petiolule**—The stalk of a leaflet.

**Phenetic**—The grouping of organisms on the basis of overall similarities or differences, without regard to genetic or phylogenetic characteristics.

**Phenogram**—A dendrogram portraying the relative similarities among a group of organisms or taxa as calculated by some clustering algorithm.

**Phenotype**—The visible or otherwise measurable features of an organism which represent the interaction between the organism's genotype and its past and present environment.

**Phloem**—The food-conducting tissue of a vascular plant.

**Phyllary**—One of the involucral bracts of the head of a member of the Asteraceae (Fig. 15-2).

**Phyllode**—A flattened blade-like petiole or petiol-erachis that serves in place of a normal leaf blade as a primary photosynthetic surface (e.g., some species of *Acacia*).

**Phyllotaxy**—The arrangement of leaves on the stem.

**Phylogenetic classification system**—A classification system in which attempts have been made to group organisms according to their evolutionary relationships; most have been constructed in part on the basis of subjective criteria. Cladistics (also called phylogenetic systematics) involves objective criteria for evaluating relationships but has not yet generated a system of classification for all the flowering plants.

**Phylogeny**—The evolutionary history of a group of organisms.

**Phylum**—A taxon in the classification hierarchy of animals that is equivalent in rank to the **division** of the plant classification hierarchy.

**Pilose**—Pubescent with soft straight hairs (Fig. 3-11 B).

**Pinna**—A primary division of a pinnately compound leaf; a pinna, itself may be further divided or compound.

**Pinnately compound**—A compound leaf with the leaflets attached to a central rachis (Fig. 3-3 B).

**Pinnately divided**—Deeply lobed in a pinnate fashion.

**Pinnately lobed**—With lobes arrayed along a central midrib like the vanes of a feather (Fig. 3-10 I).

**Pinnate**—Constructed in a feather-like pattern. Used for the venation pattern of a leaf in which lateral veins diverge from a single midvein and for compound leaves in which leaflets are attached to a central rachis.

**Pinnatifid**—Lobed in a pinnate fashion.

**Pinnatisect**—Cut all the way to the midvein in a pinnate fashion.

**Pinnule**—An ultimate leaflet of a compound leaf that is 2× or more pinnate.

**Pistillate flower**—Having one or more pistils but no functional stamens.

**Pistil**—The visual unit of the gynoecium, composed of stigma, style and ovary; it consists of one or more carpels; if it is monocarpellate with a single placenta it is a simple pistil; if it represents 2 or more connate carpels it is a compound pistil.

**Pith**—The tissue derived from the ground meristem and located in the center of dicot stems. Several kinds of pith can be recognized: solid pith is parenchymatous throughout; diaphragmed pith is solid throughout but has cross partitions differently colored or textured from the remainder of the pith; chambered pith is pith with cross partitions dividing empty cavities; spongy pith has irregular small cavities; hollow pith represents a breakdown of the cells resulting in a cavity.

**Placentation**—The arrangement of ovules within an ovary.

**Placenta**—The tissue within an ovary to which ovules are attached.

**Plaited**—Plicate.

**Plane**—Flat (Fig. 3-10 K).

**Plesiomorphy (plesiomorphic character state)**—An evolutionarily primitive state of a character. In a character transformation series the original state is plesiomorphic to all others. In a transformation series with more than two states, a particular character state may be apomorphic (q.v.) relative to some states and plesiomorphic relative to others.

**Plicate**—Folded like a fan.

**Plumose**—Branched like a feather, with a central shaft and fine spreading bristles (e.g., pappus type in Asteraceae with each bristle resembling a tiny feather).

**Polar nuclei**—The two nuclei of an embryo sac that combine with a sperm nucleus to form the fusion nucleus which gives rise to the endosperm.

**Pollen**—The structures produced in anthers or microsporangiate cones that result from the maturation of a microspore; 2- or 3-nucleate microgametophyte of a gymnosperm or angiosperm enclosed within a spheroidal to ellipsoid cell wall.

**Pollen tube**—Tubular structure that grows from a pollen grain and transports the sperm nuclei to the egg in seed plants.

**Pollination**—The process by which pollen grains are transported from a pollen sac (in an anther or pollen cone) to the receptive structure of the female (stigma or micropyle).

**Pollinium**—A coherent mass of pollen shed as a unit as in Asclepiadaceae (Fig. 14-12 C) and Orchidaceae (Fig. 19-11 D).

**Polyadelphous**—Stamens connate by their filaments in three or more bundles.

**Polygamodioecious**—A sexual condition in which a species produces staminate and pistillate flowers as well as perfect flowers, usually on separate plants.

**Polygamous**—A sexual condition in which an individual plant produces both perfect and imperfect flowers (e.g., Asteraceae with radiate or disciform heads).

**Polynomial**—A species name consisting of a generic name and two or more adjectives or descriptive phrases. Use of polynomials was abandoned by most botanists after Linnaeus popularized the use of binomial names.

**Polypetalous**—Apopetalous; choripetalous; a corolla of distinct petals (Fig. 10-3 B).

**Polyphyletic**—A non-natural taxonomic group hypothesized to have an origin in two or more different evolutionary lines; the most recent common ancestor of the two lineages is excluded from the group. Such a taxonomic grouping is generally rejected when it is demonstrated to be polyphyletic and the component taxa are reorganized into monophyletic groups. Some plant taxa currently recognized may actually prove to be polyphyletic when subjected to thorough analysis.

**Polyploid**—A tissue or plant in which each cell has three or more sets of chromosomes.

**Polyploid speciation**—The origin of a new species through hybridization of two parent taxa followed by a doubling of the chromosomes in the offspring. The original hybrid is often highly sterile, but the doubling of the chromosomes restores fertility and genetically isolates the hybrid from both parents. Amphidiploidy; allopolyploidy.

**Pome**—The fleshy accessory fruit of inferior-ovaried members of the Rosaceae derived from the adnation of the hypanthium to the ovary wall (Fig. 10-15 B). The flesh of a pome is mostly hypanthium tissue.

**Population**—A group of potentially interbreeding individuals, usually occupying a definitive area.

**Poricidal**—A structure that opens by means of pores (e.g., stamens of Lauraceae and Ericaceae; capsule of *Papaver*).

**Posterior lip**—The adaxial (upper) lip in a bilabiate flower.

**Posterior**—At or toward the back; opposite the front.

**Prickle**—A small, often weak, spinelike body originating from the epidermal or cortical layers and borne irregularly on the bark or epidermis (Fig. 3-12 E).

**Primordium**—A group of cells differentiated in a meristematic region that are the first recognizable stages of the organ they are destined to form (e.g., a leaf-primordium).

**Procumbent**—Trailing or lying flat but not rooting.

**Prostrate**—A general term for lying flat on the ground.

**Protandry (Protandrous)**—A condition in which the anthers of a flower mature and shed pollen before the stigmas of the same flower are receptive.

**Prothallus**—The gametophyte stage of ferns, usually small and flat, which bears the sex organs, the antheridia and archegonia.

**Protogyny (Protogynous)**—A condition in which the stigmas of a flower are receptive to pollen before pollen from that flower is released.

**Pseudanthium**—A flower cluster that seemingly appears as a single flower (e.g., a head in the Asteraceae or a cyathium of *Euphorbia*).

**Puberulent**—Pubescent with very short hairs (Fig. 3-11 H).

**Pubescence**—The broad term for any type of plant hairiness (Fig. 3-11 B–L).

**Pubescent**—Bearing plant hairs (trichomes; Fig. 3-11 B–L); sometimes used specifically for a covering of soft or downy hairs.

**Punctate stigma**—Undifferentiated or dot-like stigma.

**Punctate**—Bearing sessile or embedded glands (pellucid dots).

**Pungent**—(1) Ending in a stiff sharp point or tip. (2) Acrid to the taste or smell.

**Pyxis**—A capsule with circumscissile dehiscence, the top falling away as a lid.

**Quadrifoliolate**—A compound leaf with four leaflets.

**Raceme**—An unbranched indeterminate inflorescence with a rachis and pedicellate flowers (Fig. 10-13 B).

**Racemose**—With flowers in racemes; used broadly to mean indeterminate.

**Rachilla**—The diminutive rachis of a spikelet in the Poaceae and Cyperaceae.

**Rachis**—The axis of a pinnately compound leaf (Fig. 3-3 B, C) or of an elongated inflorescence (Fig. 10-11 B).

**Radial symmetry**—Actinomorphic; regular; with the parts in a perianth whorl all alike with 2 or more lines of symmetry (Fig. 10-6 A).

**Radiate head**—A type of head in the Asteraceae bearing disk florets in the center and ray florets around the periphery (Fig. 15-7 B).

**Radical**—Belonging or pertaining to the roots; sometimes used for basal leaves.

**Radicle**—The embryonic root of a seed or seedling.

**Ray**—(1) One of the branches of the main umbel of a compound umbel (e.g., Apiaceae; Fig. 10-13 G). (2) A ray floret (Asteraceae).

**Ray floret**—A type of pistillate or neuter flower in the Asteraceae with a spreading, often 3-lobed ligule (anterior lip) composed of 3 connate petals; usually without any posterior lip (Figs. 15-2; 15-6; 15-7 B).

**Receptacle**—(1) The more or less enlarged end of a pedicel to which the parts of a flower are attached: (2) The enlarged peduncle apex in the Asteraceae (and some other families) to which bracts and flowers are attached.

**Recurved**—Bent or curved downward or backward.

**Reflexed**—Abruptly bent or curved downward or backward.

**Regular**—Actinomorphic; radially symmetric; with the parts in a perianth whorl all alike with 2 or more lines of symmetry (Fig. 10-6 A).

**Reniform**—Kidney shaped, broader than long, broadly rounded and notched at the base (Fig. 3-7 L).

**Replum**—The septum in a silique or silicle.

**Resupinate**—Turned upside down during development (e.g., flowers of most orchids).

**Reticulated**—Forming a network; net-veined.

**Retrorse**—Bent or turned backward or downward, often used to describe prickles or hairs.

**Retuse**—Bearing a shallow notch at the apex.

**Revision**—A taxonomic study of a particular taxonomic group that is updates the classification and nomenclature of the group.

**Revolute**—With the margins curled under (Fig. 3-10 M).

**Rhizome**—An underground horizontal stem involved in perennation or vegetative reproduction (Fig. 3-1 C).

**Rhombic (rhomboid)**—Diamond-shaped.

**Rib**—A prominent vein.

**Root**—A cylindrical plant organ without nodes or internodes and with branches originating from the inside; usually serving as the principal organ of attachment and absorbtion.

**Root-parasite**—A flowering plant that parasitizes the root system of another plant. Some are hemi-parasites that have green photosynthetic leaves and others are complete parasites that draw all their nutrients and water from the host.

**Rosette**—A tight cluster of leaves radiating from a central area of attachment; rosettes are usually basal (e.g., dandelion), but may be borne at the end of a stem (e.g., a palm tree).

**Rostellum**—The hollow cap borne at the end of the column and covering the pollinia in the Orchidaceae, derived from sterile anthers.

**Rotate**—Wheel shaped; used to describe a corolla (calyx, hypanthium, involucre) with a very short or obsolete tube and a flat, circular limb (Fig. 10-5 F).

**Rounded**—An apex or base with a smooth continuous curve (Figs. 3-8 D; 3-9 D).

**Rugose**—Roughened; often used to describe leaves with the veinlets prominently raised on the undersurface.

**Runner**—A slender stolon or rhizome that takes root at the nodes.

**Saccate**—Bag-shaped; with a pouch.

**Sagittate**—Like an arrowhead; triangular with basal lobes pointing downward or concavely curved toward the stalk; used to describe bases and shapes (Fig. 3-9 I).

**Salverform**—Used to describe a corolla (calyx, hypanthium, involucre) with a slender tube and an abruptly expanded flat limb (e.g., corolla of *Phlox*; Fig. 10-5 E).

**Samara**—An indehiscent winged fruit (Fig. 10-16 H–J).

**Samaroid schizocarp**—A winged fruit that breaks apart into one-carpellate segments (e.g., *Acer*; Fig. 10-16 M); schizocarpic samara.

**Scabrous**—Pubescent with short triangular hairs that are rough and sand-papery to the touch (Fig. 3-11 G).

**Scale**—(1) Small dry flattened usually scarious structures. (2) Small triangular leaves that are appressed to the branchlets (e.g., *Juniperus*; Fig. 3-7 B). (3) The ovule-bearing structures of a gymnosperm cone.

**Scandent**—Climbing without the aid of tendrils.

**Scape (scapose)**—Leafless peduncle arising from the ground level; it may bear bracts or be completely naked.

**Scar**—The roughened spot left behind when a structure breaks away from another structure (e.g., a leaf scar on a twig).

**Scarious**—With a dry membranous texture, often translucent; the texture of a dry onion peel.

**Schizocarp**—A type of fruit derived from a compound ovary that breaks apart into indehiscent one-carpellate units (mericarps), each of which contains one or more seeds (e.g., fruits of Apiaceae, *Malva;* Fig. 10-10 M–P).

**Schizocarpic capsule**—A type of fruit derived from a compound ovary which breaks apart into one-carpellate segments, each of which dehisces, releasing the enclosed seeds (Fig. 13-9 F).

**Schizocarpic samara**—A winged fruit that breaks apart into one-carpellate segments (e.g., *Acer;* Fig. 10-16 M); samaroid schizocarp.

**Sclerophyll**—A firm leaf with relatively large amounts of internal strengthening tissue. Such a leaf is stiff even when physiologically wilted.

**Scorpioid cyme**—A circinately coiled determinate inflorescence in which the apparent rachis of the inflorescence is a sympodial axis derived from pedicel bases (Fig. 10-13 L, M); helicoid cyme.

**Scurfy**—Covered with scales.

**Secondary veins**—The veins that arise from the largest veins.

**Section**—A rank in the taxonomic hierarchy between genus and species; a less inclusive taxon than a subgenus. The name has no standardized suffix and is always underlined or italicized.

**Secund**—One-sided; used for inflorescences that have all or most of the flowers on one side.

**Seed**—A ripened ovule containing an embryo within a seed coat, often with additional storage tissues (e.g., endosperm).

**Seed coat**—Testa, the covering of a seed.

**Segment**—One of the parts of a leaf, calyx, corolla, perianth, etc. that is deeply divided but not truly compound.

**Semiaquatic**—A plant that grows rooted in water or wet soil but has stems and leaves that extend well beyond the water.

**Sensu lato (s.l.)**—In the broad sense; as broadly defined.

**Sensu stricto (s.s.)**—In the restricted sense; as narrowly defined.

**Sepal**—One unit of the outer whorl of the perianth, usually greenish but sometimes very showy.

**Sepaloid**—Used to describe a petal or bract that resembles a "typical" sepal in color, size, shape, etc.

**Septate**—Having one or more internal septa or cross partitions.

**Septicidal capsule**—A dry fruit that dehisces along the septa and not directly in the locules.

**Septum**—A partition; used particularly for the partitions derived from carpel wall tissue that divide a compound ovary into 2 or more locules.

**Series (-seriate)**—A set or whorl of similar structures (e.g., sepals, stamens, etc.).

**Serrate**—Saw-toothed; bearing upwardly curving teeth (Fig. 3-10 D).

**Serrulate**—Serrate with very small teeth (Fig. 3-10 E).

**Sessile**—Without a stalk; positioned directly against another structure (Fig. 3-4 B).

**Seta**—A bristle.

**Setose**—Covered with bristles.

**Sheath**—A long, more or less tubular structure surrounding another structure (e.g., in *Pinus* the cluster of bud scales around the base of a fascicle of needles; in *Equisetum* the connate bases of a whorl of leaves; in Poaceae and elsewhere the base of a leaf that enwraps the stem).

**Sheathing**—A leaf base that enwraps the stem to which it is attached (Fig. 3-4 E).

**Shoot**—The aerial part of a plant adapted for support and photosynthesis; composed of stem and leaves.

**Short shoot**—A short lateral branch with very short internodes, often sharply differentiated from long shoots (e.g., *Pinus, Cedrus, Ginkgo*); dwarf shoot.

**Shrub**—A low woody plant with one to many relatively slender trunks.

**Silicle**—A silique twice as long as broad or less (Fig. 10-17 G).

**Silique**—A capsular fruit type characteristic of the Brassicaceae that has two locules separated by a membranous septum (replum), parietal placentation, and sides (valves) that completely fall away from the septum at maturity (Fig. 10-17 H).

**Simple cone**—A cone with the spore-bearing or ovule-bearing structures borne on a primary axis.

**Simple cyme**—A two- to three-flowered unit with the oldest flower in the center; a simple dichasium (Fig. 10-13 J).

**Simple inflorescence**—Flowers attached directly (either sessile or pedicellate) to the sides or apex of the primary axis; an unbranched inflorescence (e.g., raceme, spike, head, simple cyme, simple umbel).

**Simple leaves**—Leaves with only one blade (Fig. 3-3 A); simple leaves vary from entire to deeply lobed but there is always a flange of blade tissue connecting adjacent lobes. (The leaflets of compound leaves are wholly separate).

**Simple ovary**—The ovary of a simple pistil.

**Simple pistil**—A pistil derived from only one carpel, bearing a single, usually marginal placental region.

**Simple umbel**—An inflorescence in which all the pedicels radiate from a central point of attachment (Fig. 10-13 E).

**Sinuate**—Wavy-margined in 2 dimensions.

**Sinus**—A notch between two structures (e.g., the area between two lobes of a leaf; the space between two petals).

**Sister group**—A monophyletic taxon that is hypothesized to be the closest relative of a given taxon. Sister taxa are hypothesized to share a common ancestor not shared with any other taxa.

**Sorus (pl., sori)**—A cluster of sporangia on a fern leaf, usually located on the undersurface of the frond (Fig. 8-1 B–C).

**sp.**—Abbreviation for species (singular).

**Spadix**—A thick fleshy spike, often associated with a spadix, the characteristic inflorescence of the Araceae (Fig. 19-7 B–C, E–G).

**Spathe**—A basally sheathing bract, common in some monocot families, sometimes petaloid (Figs. 19-6 C, 19-7 B, E–H).

**Spatulate**—Shaped like a spatula with an expanded apex and a tapering base.

**Speciation**—The origin of new species by various evolutionary processes.

**Species (pl., species)**—A unit in the taxonomic hierarchy smaller than the genus. Defined variously depending on the nature of the research being carried out. As treated in floras it is a group of plants comprising one or more populations that is recognizable by a set of discrete or continuously varying characters and not overlapping in most of these characters with other similar groups. It may be defined in other works by the breeding barriers that separate one group of populations from another. A species defined in a flora may or may not correspond with one defined by breeding behavior. In many cases no information on breeding is known.

**Species name**—A two-word name consisting of a genus name and a specific epithet; it should be underlined or italicized.

**Specific epithet**—The second part of a species name; a Latin word, usually either an adjective or a possessive noun; usually written in lowercase and underlined or italicized.

**Sperm**—A male gamete, produced in plants by mitosis within a gametophyte.

**Spike**—An unbranched elongated simple inflorescence with sessile flowers attached directly to a rachis (Fig. 10-13 C).

**Spikelet**—The primary inflorescence in the Cyperaceae and Poaceae with a series of scalelike bractlets attached to a rachilla and subtending tiny apetalous flowers.

**Spine**—A woody modified leaf or part of a leaf; sometimes used loosely to mean any sharp-hard structure including thorns and prickles (Fig. 3-12 C).

**Splitter**—A taxonomist who tends to view differences as much more important than similarities and consequently recognizes smaller, more narrowly defined taxonomic groups than do most other taxonomists.

**Sporangiophore**—In *Equisetum* the peltate structure that bears reflexed sporangia.

**Sporangium**—A structure borne on a diploid sporophyte plant within which spores are produced by meiosis; a spore-bearing sac.

**Spore**—A haploid reproductive body produced by meiosis in a sporangium and giving rise by mitosis to a gametophyte. It may be released directly into the environment (as in ferns) or retained on the parent plant (as in the megaspores of seed plants).

**Sporocarp**—A receptacle containing sporangia, produced by certain aquatic ferns.

**Sporophyll**—A leaf that bears one or more sporangia.

**Sporophyte-dominant alternation of generations**—A type of life cycle in which the diploid, spore-producing phase is free-living and dominant, and the haploid, gamete-producing phase is dependent; characteristic of seed plants.

**Sporophyte**—The multicellular diploid phase of a plant life cycle that produces spores by meiosis.

**spp.**—Abbreviation for species (plural).

**Spur**—A tubular or sac-like projection from a petal or sepal; it usually contains a nectar-secreting gland.

**ssp.**—Abbreviation for subspecies.

**Stamen**—The unit of the androecium, typically composed of an anther and a filament (or anther sometimes sessile); the pollen-producing part of a flowering plant (Fig. 10-7).

**Staminate flower**—Having one or more stamens but no functional pistil.

**Staminode**—A sterile stamen that does not produce pollen. It may be vestigial or variously modified (e.g., petaloid staminodes of *Canna*).

**Standard**—Banner petal; the uppermost (and outermost) petal of a papilionaceous corolla; the petals of an *Iris* flower.

**Stele**—The primary vascular structure of a stem or root together with any tissues (such as pith) that may be enclosed.

**Stellate**—Pubescent with hairs that branch from the base and resemble tiny stars (Fig. 3-11 J); often used for any types of branched hairs.

**Stem**—The main axis of a plant, bearing leaves and flowers, characterized by nodes and internodes and by branches that originate from externally produced buds.

**Sterile**—Devoid of functional reproductive structures.

**Stigma (stigmatic)**—The pollen-receptive portion of a pistil.

**Stipe**—The stalk of an organ; used in the ferns for the petiole; used in Capparaceae and elsewhere to describe a stalk borne above the receptacle on which the ovary is inserted, in some cases derived from the receptacle and in others from the base of the ovary.

**Stipitate glandular**—Pubescent with stalked glandular hairs (Fig. 3-11 K).

**Stipitate**—Borne on a stalk or stipe.

**Stipular spines**—Spines derived from modified stipules, borne on a stem in pairs at the nodes (Fig. 3-12 D).

**Stipulate**—A leaf with stipules (Fig. 3-2 C).

**Stipule scar**—A mark left on a stem when a stipule breaks off (Fig. 3-2 B).

**Stipules**—The basal appendages borne in pairs by leaves of some plants (Fig. 3-2 B, C).

**Stolon**—A horizontal stem that roots at the nodes and gives rise to a new plant at its tip (Fig. 3-1 E).

**Stoma (pl., stomata)**—A pore in the epidermis of a leaf or stem, surrounded by a pair of guard cells, functioning in gas exchange.

**Strigose**—Pubescent with straight appressed hairs (Fig. 3-11 D).

**Strobilus**—A conelike aggregate of sporophylls, often with overlapping, scale-like units.

**Stylar appendages**—Non-stigmatic structures borne on the style branches of most Asteraceae, often pubescent with brush-like sweeping hairs.

**Style**—More or less elongated part of a pistil between the stigma and the ovary.

**Style branches**—The result of incomplete fusion of the component carpels of a compound pistil in which the tips of the styles are distinct.

**Stylopodium**—A disc-like to long-tapering enlargement borne atop the ovary at the base of the styles (e.g., some Apiaceae).

**Sub**—Prefix meaning almost (e.g., a subsessile leaf is almost sessile but has a very short petiole).

**Subclass**—A rank in the taxonomic hierarchy between the class and order; names characterized by the suffix -idae.

**Subfamily**—A rank in the taxonomic hierarchy between the family and genus; a more inclusive taxon than the tribe; names characterized by the suffix -oideae.

**Subgenus**—A rank in the taxonomic hierarchy between genus and species; a more inclusive taxon than a section. The name has no standardized suffix and is always underlined or italicized.

**Subglabrous**—Almost glabrous; very sparsely and inconspicuously pubescent.

**Subopposite**—Leaves that are paired but offset such that they are not truly opposite.

**Subsessile**—Bearing a very short petiole.

**Subshrub**—A weakly woody shrub, or a plant that is mostly herbaceous but woody at the base.

**Subspecies**—A rank in the taxonomic hierarchy below the species; used to designate geographical races; often used interchangeably with varieties but when both are used subspecies is the more inclusive taxon; written together with the species name and designated by the abbreviations ssp. or subsp. (e.g., *Yucca whipplei* ssp. *parishii*).

**Subtend**—To stand at the base of, as a bract at the base of a flower; the flower is in the axil of the bract or the bract is attached to the pedicel of the flower.

**Subtribe**—A rank in the taxonomic hierarchy between the tribe and genus; name characterized by the suffix -inae.

**Subulate**—Awl-shaped, thickish and tapering from base to apex.

**Succulent**—Thickened and juicy with stored water (e.g., leaves of *Aloe,* stems of cacti).

**Suffrutescent, suffruticose**—A plant that is low and somewhat woody, at least at the base; a sub-shrub; woody at base with herbaceous stems produced annually.

**Superorder**—A rank in the taxonomic hierarchy between the subclass and order; names characterized by the suffixes -anae or -iflorae.

**Superior ovary**—An ovary that is not adnate to the bases of the other flower parts (Fig. 10-10 A, B).

**Superposed buds**—Condition when one bud forms above another at a node.

**Suture**—A line of fusion or dehiscence.

**Sweeping hairs**—Hairs present on the style branches of some Asteraceae that gather pollen from the anther tube as the style elongates.

**Syconium**—A multiple fruit/inflorescence characteristic of *Ficus* (figs) with numerous small imperfect flowers enclosed in an infolded hollow inflorescence axis that becomes fleshy at maturity (Fig. 10-18 F).

**Sympatric**—Together in the same land; used to describe populations, races, species, etc., living in the same place.

**Sympetalous**—A corolla of petals that are connate, at least at the base (Fig. 10-3 C).

**Symplesiomorphy (symplesiomorphic character state)**—A plesiomorphic character state shared by two or more taxa.

**Sympodial**—A type of branching pattern where the principal stem at each node is determinate and a branch simulates a continuation of that main stem; the resulting axis is actually a series of lateral branches (e.g., stem axis of a grape vine with the tendrils as determinate stems; axis of a scorpioid cyme; Fig. 10-13 L, M).

**Synangia**—A group of connate sporangia.

**Synapomorphy (synapomorphic character state)**—An apomorphic character state shared by two or more taxa, considered in cladistic studies to be evidence of relationship.

**Syncarp**—An aggregate of more or less coherent pistils, each of which retains its integrity (e.g., some Annonaceae).

**Syncarpous**—A gynoecium composed of 2 or more carpels that are connate at least at the base (Fig. 10-8 F–I).

**Synoecious**—A sexual condition in which the flowers on a plant are perfect; a term properly applied to plants, not flowers.

**Synonym**—A name proposed for a plant but not used because (a) it is predated by another validly published name or (b) it was applied in a different circumscription of the taxon (e.g., a name published in a genus or rank different from that accepted in the treatment being used).

**Synsepalous**—A calyx of connate sepals.

**Syntype**—One of several specimens collectively designated as types by the publishing author of a taxon; these may represent several duplicates from the same collection or several different collections. When syntypes have been designated, a modern researcher must choose one of them as a lectotype. Designation of syntypes is no longer permissible under the International Code of Botanical Nomenclature.

**Systematics**—The study of relationships among organisms and the classification and naming of those organisms.

**Tailed**—Caudate; bearing a long slender appendage (e.g., in some Asteraceae the anther sacs have long basal appendages).

**Tap-root**—An enlarged vertical main root that is noticeably larger in diameter than the lateral roots (Fig. 3-1 A).

**Tautonym**—A binomial in which the specific epithet exactly repeats the generic name; tautonyms are considered to be illegitimate names.

**Taxon (pl., taxa)**—A group of organisms of any rank given formal recognition in a classification hierarchy (e.g., class, order, family, genus, species, variety).

**Taxonomic synonym**—Names based upon different type specimens that in the opinion of a taxonomist are applicable to the same taxon.

**Taxonomy**—The description, identification, naming and classifying of objects or organisms.

**Tendril**—A slender cylindrical structure by which some climbing plants cling to supporting structures (Fig. 3-12 A); morphologically a tendril may be derived either from stem or leaf tissue. Most commonly tendrils twist around the structures to which they are attached but they may be claw-like (e.g., *Doxantha*) or bear adhesive disks (e.g., *Parthenocissus*).

**Tenuinucellate**—A condition in which the nucellus or megasporangial tissue is only one cell layer thick when the ovules mature; considered to be relatively advanced.

**Tepal**—A segment of a perianth that is not clearly differentiated into petals and sepals (e.g., a cactus flower) or in which the sepals and petals, though distinguishable by position are very similar in size, color and texture (e.g., a tulip flower).

**Terete**—Round in cross section; cylindrical or nearly so, but sometimes tapering to one or both ends.

**Terminal**—At the apex.

**Terminal bud**—A bud borne at the end of a more or less elongated stem (Fig. 3-2 A).

**Terminal bud scale scar**—Scars left on a twig when the scales of a terminal bud fall off (Fig. 3-2 A).

**Ternate**—Divided in threes.

**Ternately decompound, ternately divided**—Repeatedly divided in threes.

**Terpenoids**—A family of biochemicals found in many plant groups including volatile oils and various other substances. Often used in chemotaxonomic studies.

**Terrestrial**—Adapted to and living on land.

**Testa**—Seed coat.

**Tetradynamous**—A condition of the stamens in most Brassicaceae in which 4 stamens have long filaments and 2 have short filaments.

**Tetrad**—A cluster of four spores, the products of meiosis. Pollen grains of some plants are shed as tetrads.

**Tetraploid**—A plant or tissue whose cells have 4 sets of chromosomes.

**Thallus**—A flattened or rounded structure not differentiated into leaves or stems (e.g., plant body of Lemnaceae).

**Thorn**—Woody sharp-pointed structure modified stem (Fig. 3-12 B); may be terminal or lateral (developed from axillary buds); sometimes used loosely to mean spine or prickle.

**Throat**—The expanded part of a sympetalous corolla between the tube and the limb (Fig. 10-4 B).

**Thyrse**—A paniculiform cyme; an inflorescence with an indeterminate main axis and determinate secondary branches (Fig. 10-13 N).

**Tomentose**—Pubescent with tangled woolly hairs (Fig. 3-11 I); sometimes used for any type of dense, felt-like pubescence (e.g., stellate-tomentose).

**Topotype**—A specimen of a plant collected from the same locality as the holotype but on a different date. A topotype has no formal standing.

**Torus**—Receptacle.

**Translator arms**—Non-cellular structures that connect the pollinia of adjacent anthers to a gland on the side of the gynostegium of Asclepiadaceae.

**Translucent oil glands**—Pellucid dots; glandular punctations.

**Tree**—A woody plant with one to few relatively massive trunks and a more or less elevated crown.

**Tribe**—A rank in the taxonomic hierarchy between family and genus; when used together with subfamilies, tribe is the less inclusive taxon; used primarily in large complex families such as the Asteraceae and Poaceae; name characterized by the suffix -eae.

**Trichome**—Any type of plant hair (except for root hairs).

**Trifoliolate**—A compound leaf with three leaflets.

**Trigonous**—Triangular in cross section.

**Trinucleate pollen**—A pollen grain that is shed from the anther with 3 nuclei or cells; a tube nucleus and two sperm nuclei.

**Triploid**—A polyploid cell, tissue, plant, species, etc. with three sets of chromosomes.

**Truncate**—Squared off; appearing as if cut off at apex or base (Figs. 3-8 G; 3-9 E).

**Tube**—The cylindrical part of a sympetalous corolla below the throat; the connate part of a synsepalous calyx (Fig. 10-4 B).

**Tuber**—A short swollen subterranean stem (e.g., a potato; Fig. 3-1 D).

**Tubercle**—A protruding bump on a surface.

**Tubular**—A corolla, calyx, involucre, etc. with little or no change in diameter from base to apex (Fig. 10-5 A).

**Turbinate**—Shaped like an inverted cone; top-shaped.

**Turgid**—Swollen.

**Twig**—A young woody stem.

**Twining**—A pattern of spiral stem growth in vines that climb by wrapping around twigs or other supporting structures.

**Type collection**—Type specimen.

**Type locality**—The geographical location where a type specimen was collected.

**Type specimen**—A specimen designated to serve as a reference point for a scientific name.

**Type variety, type subspecies**—The variety or subspecies containing the type specimen of the species of which it is a part. The varietal or subspecific epithet is the same as the specific epithet and is written without an authority (e.g., *Eriogonum baileyi* S. Wats. var. *baileyi*). Sometimes described as typical variety or subspecies but not necessarily "typical" as in commonly encountered.

**Umbel**—An indeterminate inflorescence with pedicels arising from a common central point of attachment (Fig. 10-13 E).

**Umbellet**—One of the small umbels of a compound umbel (Fig. 10-13 G).

**Umbelliform cyme**—A determinate inflorescence with the appearance of an umbel; a determinate inflorescence with the primary branches arising as in an umbel; a cymose umbel (Fig. 10-13 O).

**Umbo**—A conical projection arising from the surface of a cone scale in *Pinus*.

**Uncinate**—Bearing a hook at the tip.

**Undulate**—With margin wavy in three dimensions (up and down; Fig. 3-10 N).

**Uniaperturate**—A pollen grain with a single aperture.

**Uniseriate**—In one ring (e.g., a uniseriate perianth with sepals only).

**Unitegmic**—An ovule with only one integument.

**Upper lip**—The adaxial portion of the limb of a bilabiate corolla (Fig. 10-4 C).

**Urceolate**—Urn-shaped, used for corollas, involucres, etc. that are short and broad but narrow both at the apex and the base (Fig. 10-5 D)

**Utricle**—A bladdery achene with the pericarp loose and fragile (Fig. 10-16 E).

**Valvate**—(1) A type of capsule or legume dehiscence in which segments (valves) separate. (2) Arrangement of sepals, petals, bracts, cone scales, etc., with margins that meet edge to edge but do not overlap.

**Valve**—(1) A segment of the external part of a capsule or legume that separates from other such units when the fruit dehisces, especially applied to the sides of a silique or silicle that break free from the replum (Fig. 10-17 G, H). (2) The three enlarged inner sepals covering the fruit of *Rumex* (Polygonaceae; Fig. 17-11 I).

**var.**—Abbreviation for variety.

**Variety**—One of the lowest ranking taxa in the taxonomic hierarchy, representing a geographical race within a species; often used more or less interchangeably with subspecies, but the less-inclusive taxon when both are recognized. Formerly used by some taxonomists for sporadic mutants or "sports" and sometimes confused with the horticultural cultigen and cultivar.

**Vein**—A strand of vascular tissue (xylem and phloem) in a leaf, petal, etc.

**Velutinous**—Pubescent with soft straight hairs that give the surface a velvety texture.

**Venation**—The pattern of veins on a leaf, petal, etc.

**Ventral**—Adaxial; the side toward the axis (e.g., the upper surface of a leaf); the opposite of dorsal.

**Vernation**—The arrangement of leaves in the bud.

**Versatile anthers**—Hung or attached in the middle and capable of turning like the blades of a propeller.

**Vertical classification**—The ranking of taxa in a hierarchical arrangement of progressively more inclusive taxa.

**Verticillate**—Arranged in whorls or verticils.

**Verticil**—A whorl (e.g., inflorescence in many Lamiaceae).

**Vescicle**—A small bladdery sac or cavity filled with air or fluid.

**Vestiture (vesture)**—Any kind of pubescence.

**Villous**—Pubescent with slender curved or wavy but not matted hairs (Fig. 3-11 C).

**Vine**—A plant with long slender weak stems that are unable to support their own weight and that are supported by other plants or structures or that trail across the ground.

**Viscidulum**—Glandular structure attached to the pollinia of an orchid; adheres to body of pollinating insect.

**Viscid**—Sticky.

**Viviparous**—Used to describe seeds that germinate while still on the parent plant, as in some mangroves.

**Wanting**—Absent.

**Whorl**—A ring of similar structures.

**Whorled leaves**—Type of phyllotaxy characterized by three or more leaves per node, symmetrically distributed around the stem (Fig. 3-5 D).

**Wing**—(1) Either of the two lateral petals in a papilionaceous flower (Fabaceae; Fig. 12-18 G). (2) A thin dry membranous appendage of a structure (e.g., the airfoil of a samara).

**Woody**—Possessing a considerable percentage of cells with thick secondary walls; usually with an active vascular cambium.

**Xerophytic (xerophyte)**—Plants adapted to growing in dry situations.

**Xylem**—The water- and mineral-transporting tissues of a plant, usually also important in structural support of the plant. Wood is secondary xylem.

**Zygomorphic**—Bilaterally symmetric; irregular; divisible into equal halves along only one plane only (Fig. 10-6 B).

**Zygote**—A fertilized egg resulting from fusion of a sperm and egg, in plants giving rise by mitosis to the sporophyte.

# Index

Note: Italicized page numbers indicate illustrations.

Antherdia, 140
Anthericaceae, *439*
*Anthericum, 439*
Anthers, 190, 194
Anthocarp, 258
Anthocyanins, 247
Anthophyta, 126
*Anthriscus cerefolium,* 407
*Anthurium, 432*
*Antigonon,* 269
*Antirrhinum,* 401
*Antirrhinum majus, 400*
Apetalous, 192
*Aphelandra,* 478
Apiaceae, *13,* 67–68, *204,* 205, 371,
    406–07
Apiales, 371–72, 406–07
Apical placentation, 197
*Apium graveolens,* 407
Apocarpous gynoecium, 196
Apocynaceae, 68, 365, 395
*Apocynum androsaemifolium,* 365
Apomictic plants, 476
Apomorphies, 61, 64
Aponogetonaceae, 424
Apopetalous, 192
Apostasioideae, 443, *444*
Apple, *203,* 322–23
Apricot, 322
Aquifoliaeae, 371
Aquifoliales, 371
*Aquilegia,* 239
*Aquilegia eximia, 238*
Araceae, 423, 432–33
*Arachis hypogaea,* 319
Araliaceae, 67–68, 371
Araucaria Family, 171
Araucariaceae, 171
Arbor vitae, 182
Arbutoideae, 386, *387*
*Arbutus, 385*
*Arceuthobium,* 252
Archaefructaceae, 221
*Archaefructus,* 221
Archegonia, 140, 167
*Arctostaphylos, 385,* 386, 387
*Arctostaphylos luciana, 385*
Arctoteae, *376,* 380
*Arctotheca,* 378
Arctotideae, 380
*Arctotis, 376,* 378
Arecaceae, *13,* 187–88, 426, 447–49
Arecales, 423, 426, 447–49
Arecidae, 421–29
Areoles, 266
*Argemone,* 242
*Arisaema triphyllum, 432,* 433
*Aristida, 461,* 462
Aristideae tribe, 462
*Aristolochia, 225*
*Aristolochia californica, 225*
Aristolochiaceae, 221, 225–26

*Armeria maritima, 251*
*Armoracea rusticana,* 340
Arrow-grass Family, 424
Arrow-leaf, 435
Arrowroot Family, 429
*Artemisia dracunculus,* 378
Artichoke, 378
Artificial classification, 89–90
*Artocarpus altilis, 328*
Aruhm Family, 423
Arum Family, 432–33
*Arundinaria,* 461
Arundineae, 462
Arundineae tribe, 461
Arundinoideae, 461–62
*Arundo,* 461
*Asarum caudatum, 225*
Asclepiadaceae, 68, 266, 310, 366,
    394–95
*Asclepias, 205, 394,* 395
*Asclepias syriaca, 394*
Ash, *204*
*Asimina triloba, 224*
Asparagaceae, *439*
Asparagales, 423, 425–26, 442–46
*Asparagus,* 437, *439*
Aspen, 316
Asphodelaceae, 426, *439*
*Asphodelus, 439*
*Aspidotis,* 154
Aspleniaceae, 146, 160
*Asplenium platyneuron, 146*
*Asplenium rhizophyllum, 146*
*Aster, 13,* 376
Aster tribe, 381–82
Asteraceae, *13,* 13, 204, 372, 374–83, 481
Asterales, 13, 372, 374–83, 408–09
Astereae, *376,* 381–82
Asteridae, 13, 129
Asterids, *223*
    asterid II, 373–83
    basal to lineages, 360–64
    Euasterids I, 364–70
    Euasterids II, 370–73
    overview, 359
Asteroideae, 380–83
*Astragalus,* 320
*Athyrium,* 158
*Atriplex, 259,* 260
*Atropa belladonna,* 403
Atropine, 403
Auricles, 457
Australian-fuschia, 344
Austrobaileyaceae, 223
Austrobaileyales, 221, 223–24
Author(s), 15
Autopolyploidy, 478
Autopomorphy, 64
*Avena, 461*
*Avena sativa,* 463
Aveneae tribe, 463
Avocado, 235

Axil, 28
Axile placentation, 197
Axillary bud, 29
*Azolla filiculoides, 147*
Azollaceae, 147

# B

*B. juncea,* 340
*B. rapa,* 340
Baby blue-eyes, 405
Baby's breath, 255
Baby's tears, 331
Bachelor's buttons, 378
Balanopaceae, 282
Balanophoraceae, 252
Bald-cypress, 182
Balsaminaceae, 361
Bamboo, 463
*Bambusa,* 461
Bambuseae tribe, 461
Bambusoideae, 461
Banana Family, 428
Banner, 318
Barberries, 237
Barberry Family, 226, 236–37
Barbeyaceae, 284
Barclayaceae, 223
Bar-graphs, 41
Barley, 463
Barley tribe, 462
Barnedesia tribe, 379
Barnedesieae, 379
Barnedesioideae, 379, 481
Basal, 31
    angiosperms, 223–24
    asterids, 360–64
    eudicots, 222
    monocots, 423–25
    placentation, 197
    rosids, 276–78
    tricolpates, 226–28
Base number, 477
Basellaceae, 249
Basil, 399
Bataceae, 286
*Batis maritima, 286*
Bay leaf, 235
Bayberry Family, 285
Bayesian analysis, 70
Beak, 340
Bean, 319
Bean Family, 317–20
Beardtongue, 401
Bedstraw, 393
Beech, 335
Beech Family, 284–85, 334–35
Beet, 260
Begonia Family, 284
Begoniaceae, 284
Belchnaceae, 145–46
Bell peppers, 403

*Ecballium elaterium,* 333
*Echeveria,* 297
*Echinocystis,* 333
*Echinodorus,* 435
*Echinodorus berteroi, 434*
*Echinodorus rostratus, 199*
*Echinops,* 380
Edible fig, 328
*Edrianthus,* 409
Eel-grass Family, 424
Effective publication, 19
Egg nucleus, 188, 190
Eggplants, 403
*Ehretia,* 397
Ehretiaceae, 397
Ehretioideae, 397
*Ehrharta,* 461
Ehrharteae tribe, 461
Eichler, W., 95
Eidicots, 222
Elaeagnaceae, 283
*Elaeis guineensis,* 449
Elaeocarpaceae, 283
*Elaphoglossum,* 158
Elatinaceae, 282
*Elatostema,* 331
Elderberry, 411
Electrophoresis, 480–81
*Eleocharis,* 455
*Eleocharis tuberosa,* 455
*Eleusine,* 462
Elliptic leaves, 33
Elm Family, 283–84
*Elymus,* 462
Emarginate leaf, 33, *34*
Embryo, 190
Embryo sac, 188
*Emmenanthe,* 405
Empetraceae, 364
Enations, 137
Endive, 378
Endocarp, 202
Endosperm, 190
Engler, Adolf, 95
Entire margin, 35
*Ephedra californica,* 172
*Ephedra viridis,* 172
Ephedraceae, 172
*Epidendrum,* 444
Epigynous, 198
*Epilobium, 306,* 307
*Epilobium angustifolium,* 307
*Epipactis, 442,* 444
Epipetalous, 194, *195*
*Episceae, 366*
Epistemonous, 342
Equisetaceae, 142, 151–52
Equisetales, 151–52
Equisetophyta, 128, 137, 141, 142, 151–52
Equisetopsida, 151–52
*Equisetum,* 152

*Equisetum arvense, 151*
*Equisetum laevigatum, 151*
Equitant, 446
Eragrosteae tribe, 462
Eragrostideae, 462
*Eragrostis,* 462
Eragrostoideae, 462
Eremolepidaceae, 252
*Erica,* 387
Ericaceae, 361, 385–87
Ericales, 361–64, 385–91
Ericoideae, *387*
*Eriobotrya,* 322
Eriocaulaceae, 427
*Eriodictyon,* 405
*Eriogonum,* 269
*Eriogonum fasciculatum, 199, 268*
*Erodium, 204, 302,* 303
*Erodium cicutarium,* 303
*Erymgium,* 407
*Erysimum,* 340
*Erythronium, 437, 440*
Erythroxylaceae, 281
Escallonia Family, 372
Escalloniaceae, 294, 295, 299, 372
Escallonioideae, 294
*Eschscholzia,* 241
*Eschscholzia caespitosa,* 84–85
*Eschscholzia californica, 240,* 242
Estipulate, 30
Euasterids I, 364–70
Euasterids II, 371–73
*Eucalyptus, 205,* 206, *304,* 305
Eucommiaceae, 364
*Eucrypta chrysanthemifolia, 404*
*Eugenia,* 305
Eupatorieae, *376,* 383
*Eupatorium,* 383
Eupatorium tribe, 383
*Euphorbia, 308,* 309–10
*Euphorbia lathyris, 308*
*Euphorbia marginata, 308*
Euphorbiaceae, 266, 281, 308–10
Euphorbiales, 275
Euploid changes, 477
Eupomatiaceae, 224
Eupteleaceae, 226
European fan palm, 449
Eurosid I, 279–86
Eurosid II, 286–90
*Euryale,* 231
Eusporangiate ferns, 141
Evening-primrose, 307
Evening primrose Family, 278, 306–07
Everlasting tribe, 381
Evidence
    character variation, 38–46
    characters and character states, 25–26
    vegetative terminology, 26–37
*The Evolution and Classification of*
    *Flowering Plants* (Cronquist), 98–99
Evolutionary systematics, 94–108

Evolutionary trends, in flowering
    plants, *104*
*Exbucklandia,* 301
Exine, 190
Exocarp, 202
Experimental studies, 491–92
Experimental systematics
    biochemical systematics, 478–79
    biosystematics, 474–75
    cytogenetic studies, 476–78
    environmental influences vs.
        inheritance, 475
    hybrid speciation, 478
    modern technology, new characters,
        and, 483
    molecular studies, 479–82
    noncoding nuclear sequences, 482
    overview, 473
    plant microcharacters, 483
    problems with, 476
    purposes of, 474
    studies, 475–76
Experts, 77

## F

Fabaceae, *13,* 187–88, *205,* 206, 283,
    317–20
Fabales, 275, 283, 317–20
Fagaceae, 284–85, 334–35
Fagales, 284–86, 334–38
*Fagopyrum esculentum,* 269
*Fagus,* 335
*Fagus sylvatica, 334*
*Fallugia,* 322
*Fatsia japonica, 371*
*Feijoa sellowiana,* 305
*Felicia,* 378
Fennel, *204,* 407
Fern leaves, 140
Ferns and fern allies
    life cycles, 137–40
    taxonomy, 140–47
Ferns, sporangia of, 141
*Ferocactus, 265*
Fertile florets, 459
Fertile fronds, 140
Fertile stamens, 194
*Festuca, 461,* 462
Festuceae, 462
Festucoideae, 462–63
Fibrous root, 27, *28*
*Ficus,* 206, *207, 327,* 328, 329
*Ficus carica,* 328
Fiddlehead, 140
Field notebook, 115
Field studies, 491
Fiesta flower, 405
Fig, 328
Fig Family, 327–29
Figwort Family, 369, 400–401
Filament, 194

Poison-sumac, 346
Poisonwood, 346
Pokeberry Family, 248
Polar nuclei, 188
Polarization, 66
Polemoniaceae, 362, 388–89
*Polemonium,* 389
Pollen, 166, 167, 194
Pollen grain, 190
Pollen sacs, 166, 167
Pollen tube, 190
Pollination, in angiosperms, 190
Pollinia, 395, 443
Polyadelphous stamens, 194
*Polygala dalmaisiana, 283*
Polygalaceae, 283
Polygamous flowers, 192
Polygonaceae, *199,* 250, 268–69
Polygonales, 247, 250–51, 268–69
*Polygonatum, 439*
*Polygonum,* 269
*Polygonum coccineum, 268*
Polymerase Chain Reaction (PCR), 482
Polypetalous, 192
Polyphyletic, 66–67
Polyploids, 478
Polyploidy, 477
Polypodiaceae, 141, 146, 159–60
Polypodiales, 137, 144, 153–60
Polypodiidae, 142, 144
Polypodiophyta, 128, 141, 143–47, 153–60
Polypodiopsida, 143, 153–60
*Polypodium,* 160
*Polypodium californicum, 159*
Polypody, 160
Polypody Family, 159–60
*Polypogon, 461*
*Polystichum,* 158
*Polystichum munitum, 157*
Pome, 203
Pomegranate Family, 278
*Poncirus,* 344
Pondweed Family, 424
Pontederiaceae, 428
Pooideae, 462–63
Poor man's orchid, 403
Poplar, 316
Poppy Family, 226, 240–42
*Populus,* 316
*Populus alba, 315,* 316
*Populus fremontii, 315*
*Populus tremuloides,* 39
Poricidal capsule, *205,* 206
*Portulaca, 205*
*Portulaca grandiflora, 264*
*Portulaca oleracea, 263,* 264
Portulacaceae, 247, 249, 263–64
Posidoniaceae, 424
*Potamogeton, 424*
Potamogetonaceae, 424

Potato, 403
Potato Family, 370, 402–03
*Potentilla,* 322, *323*
*Potentilla anserina, 323*
Pot-marigold, 378
Prairie grass, 463
Prantl, Karl, 95
Prickle, 36
Prickly poppy, 242
Prickly-pear, *265*
Pride-of-Madeira, 397
Primrose, 391
Primrose Family, 362, 390–91
*Primula,* 391
Primulaceae, 362, 390–91
Principal of simplicity, 65
*Prionium,* 453
Priority, of publication, 18
*Prodromus Systematis Naturalis Regni Vegetabilis,* 91
Pronunciations, of scientific names, 16–17
*Prosopis, 317*
Protea Family, 227
Proteaceae, 227
Proteales, 226–27
Protein electrophoresis, 480–81
Prothallus, 140
*Prunella vulgaris, 399*
Prunoideae, 322, *324*
*Prunus, 203,* 322, *324*
*Prunus amygdalus, 324*
*Prunus avium, 324*
*Prunus caroliniana, 324*
*Pseudotsuga,* 180
*Pseudotsuga menziesii, 179*
*Psidium cattleianum, 304*
*Psidium guavaja, 305*
Psilophyta, 137, 141, 142–43
Psilotaceae, 142–43
Psilotophyta, 128, 141
*Psilotum nudum, 142*
*Psittacanthus, 251*
*Ptelea, 204, 343,* 344
*Ptelea trifoliata, 344*
Pteridaceae, 145, 153–54, 160
*Pteridium aquilinum, 155,* 156
Pteridophyta, 137, 141
*Pteris,* 154
*Pterocarya, 285*
Puberulent, 36, *37*
Pubescent, 36
Publication, 18, 19
Pumpkin, 333
*Punicaceae,* 278
Purslane Family, 249, 263–64
*Puya, 451*
Pygmyweed, 297
*Pyracantha,* 322, 323, *324*
Pyrethrum, 378
*Pyrola secunda, 385*

Pyrolaceae, 386, *387*
Pyroloideae, *387*
*Pyrus,* 322

## Q

Quaking aspen, 39, 316
Qualitative characters, 25
Quantitative characters, 25, 40–41
*Quercus, 204, 334,* 335
*Quercus agrifolia, 334*
*Quercus suber, 335*
Quiinaceae, 282
Quilwort Family, 143
Quince, 322
Quinoa, 260

## R

*R. purshiana, 326*
Raceme, 200
Racemose, 199, *200*
Rachilla, 459
Rachis, 30, 198, 199–200
Radial flower, 193–94
Radiate heads, 377
Radish, 340
Rafflesia Family, 290
Rafflesiaceae, 290
Rafflesiales, 290
Ragweeds, 378
Ramie, 331
Randomly Amplified Polymorphic DNA (RAPDs), 482
Range, 44
Rangeland grasses, 463
Ranking, 5
Ranunculaceae, 226, 238–39
Ranunculales, 226, 236–42
*Ranunculus,* 239
*Ranunculus aquatilis, 38*
*Ranunculus californicus, 238*
Rapateaceae, 426–27
*Raphanus, 339,* 340
*Raphanus sativus, 340*
Raspberry, 322
Ratany Family, 280
Ray flowers, 377
*rbcL,* 482
Receptacle, 191
Recorded information, 4
Red Mangrove Family, 281
Redstem filaree, 303
Rein orchids, 444
Rejected names, 18
Rendel, Alfred, 96
Reniform leaves, 33
Replum, 340
Reproduction, 39
Resedaceae, 286–87
Restionaceae, 427

Strict consensus tree, 64
*Striga,* 401
Strigose, 36, *37*
Strobili, 165, 167
Stromatopteridaceae, 144
Styles, 188, 196
Stylidiaceae, 372
Stylopodia, 407
Styracaceae, 363
Subsessile, 31
Subshrub, 27
Subspecies, 16
Subtends, 198
Succulent, 27
Suffrutescent, 27
Sugar beets, 260
Sugar cane, 463
Sumac, 346
Sumac Family, 289, 345–46
Sundew Family, 250, *251*
Sunflower, 378
Sunflower Family, 372, 374–83
Sunflower tribe, 382–83
Superior ovaries, 197–98
Surface feature terms, 36–37
Surianaceae, 283
Swamp bay, 235
Sweet alyssum, 340
Sweet violet, 314
Sweet-flag Family, 423
*Swertia,* 365
Sword fern, 158
Sycamore Family, 226–27
Syconium, 206, *207,* 328
*Symingtonia,* 301
Symmetry, of corolla, 193
Sympetalous, 192, 193
*Symphoricarpos,* 411
*Symphyandra,* 409
Symplesiomorphies, 61, 66
Symplocaceae, 364
*Symplocarpus foetidus,* 433
Sympodial, 200
Synamorphies, 474
Synapomorphies, 61, 66
Syncarpous gynoecium, 196
Synoecious flowers, 191–92
Synonyms, 19
Synoptic keys, 127–29
Synsepalous, 192
Syntypes, 18
*Systema* (de Candolle), 91
Systematics. *See also* Experimental
   systematics
   biochemical, 478–79
   biosystematics, 474–75
   cladistic or phylogenetic, 92
   numerical phenetic, 92–94
   overview, 91–92
   traditional or evolutionary, 94–108
*Syzygium, 304,* 305
*Syzygium aromaticum,* 305

# T

Tageteae, 382
*Tagetes, 204,* 378
Takhtajan, Armen, 98, 101, 103, *105–07*
Takhtajan's system, 98
*Talauma,* 233
*Talinum,* 264
Tamaricaceae, 250
Tamarisk Family, 250
*Tamarix, 250*
*Tanacetum cineriifolium,* 378
Tanbark oak, 335
Tansy-ragwort, 378
Tap root, 27, *28*
Tapioca, 310
*Taraxacum officinale,* 378
Taro, 433
Tarragon, 378
Tautonyms, 19
Taxaceae, 170, *171*
Taxodiaceae, 171, 181–82
*Taxodium,* 182
Taxon (taxa), 5
Taxonomic key, 4–5
*Taxonomic Literature,* 488–89
Taxonomic synonyms, 19
Taxonomy, goal of, 473
*Taxus bacata, 171*
Tea Family, 362
Teasel Family, 372
Tea-tree, 305
*Tecomaria capensis, 368*
Tecophiliaceae, 425, *441*
*Tectaria,* 158
Tendril, 36
Terminal bud, 29
Terms, definitions, 27–37
Terpenoids, 479
Tetradynamous, 340
*Tetragonia tetragonioides, 261,* 262
Tetrameristaceae, 364
*Thalictrum,* 239
*Thalictrum fendleri, 238*
*Thamnosma,* 344
*Thamnosma montana, 343*
Theaceae, 362
Theligonaceae, 366
Thelypteridaceae, 145, 160
Theophrastaceae, 362
Theophrastus, 90
Thistle Tribe, 380
Thorn, 36
Thorne, Robert, 67, *98,* 101, 103, *107*
Thorne's system, 101, *102*
Three-awn grasses, 462
3-nucleate condition, 190
Throat, 193
*Thuja,* 182
Thurniaceae, 427
Thyme, 399

Thymelaeaceae, 287
*Thymus officinalis,* 399
*Tiarella, 293,* 294
*Tibouchina, 279*
Tickseed, 319
*Tigridia,* 446
*Tilia americana, 288*
Tiliaceae, 287–88
*Tillandsia usneoides, 450,* 451
Tillandsioideae, 451
Tillers, 457
Tipu tree, 319
*Tipuana tipu,* 319
Tissue layers, 188
*Tmesipteris, 142*
Toadflax, 401
Tobacco, 403
*Tofieldia,* 440
Tomatillo, 403
Tomatoes, *203,* 403
Tomentose, 36, *37*
*Torreya californica, 171*
Touch-me-not Family, 361
Tovariaceae, 287
*Toxicodendron,* 346
*Toxicodendron diversilobum, 345*
*Toxicodendron radicans, 345*
*Toxicodendron vernicifluum, 346*
*Toxicodendron vernix, 345*
Tracheophytes, 137, *166*
*Trachymene,* 407
*Tradesacntia, 464,* 465
Traditional classification, vs.
   phylogenetic, 66–68
Traditional systematics, 94–108
Translator arms, 395
Trapaceae, 278
*Trautvetteria,* 239
Tree, 27
Tree Fern Family, 145
Tree length, 66
Tree mallow, *204*
*Tribulus terrestris, 280*
*Trichostemma lanatum,* 399
*Tridax,* 378
Trifoliate-orange, 344
*Trifolium,* 319
Trilliaceae, *436, 441*
*Trillium, 436,* 437, *441*
*Trillium grandiflorum,* 14
Trimeniaceae, 224
*Triodanis, 408,* 409
*Triosteum,* 411
Triploid fusion nucleus, 190
*Tritelei, 438*
Triticeae tribe, 462
*Triticum, 204, 461,* 462, 463
Triuridaceae, 425
Trochodendraceae, 228
Trochodendrales, 228
*Trollius,* 239
Tropaeilaceae, 286